# Geomorphology and Climate

*Edited by*
**Edward Derbyshire**
*Reader in Physical Geography*
*University of Keele*

*A Wiley–Interscience Publication*

**JOHN WILEY & SONS**
London · New York · Sydney · Toronto

Copyright © 1976, by John Wiley & Sons Ltd.

All rights reserved.

No part of this book may be reproduced by any means, nor transmitted, nor translated into a machine language without the written permission of the publisher.

*Library of Congress Cataloging in Publication Data:*

Derbyshire, Edward
 Geomorphology and climate.

 'A Wiley–Interscience publication.'
 1. Climatic geomorphology. I. Title.
GB406.D37      551.4      75–4523

ISBN 0 471 20954 6

Photosetting by Thomson Press (India) Limited, New Delhi and printed in Great Britain by J. W. Arrowsmith Ltd., Bristol

CHAPTER ONE

# Geomorphology and Climate: Background

EDWARD DERBYSHIRE

## 1.1 Introduction

The central concern of climatic geomorphology is the extent to which variations in the elements of climate, notably solar energy and moisture, are reflected in geomorphic processes to produce distinctive suites of landforms. The narrow view of the subject matter of climatic geomorphology, which arises from undue emphasis on the continental European tradition stressing global regionalization and broad qualitative associations of landforms and climate, is likely to minimize its methodological significance (Morgan, 1970). On the other hand, the broadest views of the subject, equating it with the greater part of dynamic and historical geomorphology (glacial, periglacial, arid and fluvial geomorphology of both current and Pleistocene landscapes) may leave the subject so loosely defined that its identity becomes lost, again leading to an underestimation of its potential significance (e.g. Clayton, 1971).

The mainstream of literature in climatic geomorphology (Stoddart, 1969; Rohdenburg, 1971; Rathjens, 1971; Derbyshire, 1973) has been dominated by two great methodological traditions: first, the application of the concept of zonality based essentially on climatological and ecological principles and, second, the inductive definition of climate–process provinces based on assumed general relationships between the efficacy of selected geomorphic processes and standard climatic means.

## 1.2 The zonal concept

On a global scale, variations in characteristics of the atmosphere–lithosphere interface may be described as zonal, in that they conform to the broad zonation of solar-derived energy and water availability from equator to poles. A notably early attempt to quantify relationships of this kind was Humboldt's study of vertical zonation of temperature in mountain areas in 1817. While work at this time was preoccupied with temperature zonation, some workers (including Dove in 1846 and Linsser in the late 1860s) recognized the moisture factor as paramount in the definition of some climates.

The stimulus to these developments in climatological study was the need to understand the nature of plant growth and plant distribution. With the exception of xerophiles, the physiological classification of vegetation by de Candolle, dating from 1874, is based on temperature tolerance limits which can be directly related to the major thermal zones of the earth set out by Supan in 1879. Wladimir Köppen's map of 1884, which included considerations of seasonality, was a direct stimulus to the mapping of zonal vegetation. Both Drude and Schimper in the last decade of the nineteenth century recognized the great zonal vegetation associations of the world on a physiognomic basis, zoned according to both moisture and temperature gradients. Thus, the best known classification of world climates (Köppen, 1923) is essentially phytogeographical and based on de Candolle's thermal zones.

The phytogeographically-based climatic classifications can be compared with the concept of zonal soils, made up of great soil groups (Dokuchaev; see Glinka (1915) and Marbut (1928, 1935)), the terminology of which is partly phytophysiognomic and partly pedological. The relationship of soil bodies to mean climate, acting through natural vegetation, is a strong one at the continental and global scales. The zonal soil groups, with their intrazonal variations, are conceptually similar to Clements's (1936) climatic climax of the zonal vegetation formations and their edaphic or physiographic sub-climaxes (Tansley, 1929 and 1935) respectively. However, just as the development of phytosociological methods (Whittaker, 1953; Poore, 1955–56; Becking, 1957; Odum, 1964) has underlined the inadequacy of Clements's view of the association for dynamic study of plant communities at the field and regional scales (Pears, 1968), so the concept of zonal soils is not a practical tool in field soil classification. The dependence of a zonal classification, however refined, on the relative importance of processes assumed to have produced the soils inhibits its practical application. The view of many field pedologists on this question is expressed by Leeper (1964, pp. 21–22) in the following words:

The textbook, orthodox classification of Northern Hemisphere writers is along different lines. They do not consciously list the properties of profiles ... in order to find a self-consistent answer. On the contrary, they already know the answer, whether by intuition or by copying an earlier worker who must have been similarly inspired. A product of such a grouping is the 'chernozem', which is a soil of the following properties ... But the chernozem is also *defined* as belonging to perennial grassland under a continental climate with very cold winters, warm and moist summers, and about 18 inches of precipitation annually. If it is found in a different climate we are not allowed to call it a chernozem, however many features of the profile there may be in common. Clearly, the naming of a soil has here been confused with naming a geographical region; and the geographer's attempt to draw boundaries around what he regards as a natural region is notoriously a matter for individual decision.

There is abundant evidence that many soils of the world contain relic elements in their profiles (Morrison and Wright, 1967; Yaalon, 1971; cf. Ollier, Chapter 5) arising from the time lag in the response of pedogenic processes to environmental changes, notably those affecting climatic–edaphic–vegetation relationships. Accordingly, soil classification is being based increasingly on the properties of the soils in the field (Leeper, 1956; Northcote, 1960), the basic unit varying from the modal soil (Kubiëna, 1953) to the soil series or polypedon

(US Department of Agriculture 1960), although some fundamental deficiencies of the USDA system have been set out by Webster (1968).

Conceptually, the present condition of climatic geomorphology appears in many ways comparable to the state of climatic classification in the early years of this century, the Anglo-American school of phytogeography in the 1930s and soil science before 1950. The great syntheses of climatic geomorphology represented by the work of Julius Büdel (notably his 1948 and 1963 papers) and Tricart and Cailleux (1965, 1972) are monoclimactic in concept and global in scale. Criteria, as with the early climatic and vegetation classifications, are mixed. Büdel maps landform assemblages partly on climatic criteria. In its emphasis on the importance of the role of vegetation and in the extended treatment accorded to vertical zonation, or *étagement* (cf. Büdel (1968) and Derruau (1968); for developments of this idea at di erent scales, see, for example, Bik (1967), Grishankov (1973), Kotarba and Starkel (1972), Hastenrath and Wilkinson (1973) and Morariu and Mac (1974)), the classification of Tricart and Cailleux is rather more sophisticated. Inevitably, however, it is dependent on several non-morphological criteria. Like the obsolete soil classifications of Marbut (1928, 1935) and Robinson (1949), the bases of the mapped entities are processes, and factors assumed to control them, rather than the nature of the entities themselves. Moreover, just as the classifications designed to represent vegetation associations and soil groups at the continental scale break down at the regional and field scales, so too do those of climatically-grouped landform suites. Thus, in the present state of knowledge of geomorphological dynamics, especially in the humid and wet–dry tropics (Tricart, 1972; Thomas, 1974), the scale factor is a major constraint on the application of the principle of zonality. In cryonival regimes, despite the wealth of distinctive microforms and associated sedimentary structures which appear to show a notable degree of sensitivity to, and a relatively simple dependence upon, the temperature climate (Poser, 1948; Tricart, 1963 and 1969; Tricart and Cailleux 1965 and 1972), the wide range of process environments and the variability of the process balance is evident from several major reviews published in recent years (Troll, 1958; Embleton and King, 1975; Péwé, 1969; Washburn, 1973; Ives and Barry, 1974). Even in glaciated regions, the determination of climate as a factor in variation in the most common and typical landforms such as cirques is a problem of some considerable complexity (Chapter 15).

## 1.3 Spatial and temporal scales of variation in processes and landforms

While climatic geomorphology may have reached a stage in its development when the 'moribund nature of classification' (Hare, 1973) should be admitted, it is justifiable to view as useful existing landform classifications based on the intrinsic properties of landforms (Passarge, 1926) together with the dominant processes responsible for their formation (Rathjens, 1971). As the most satisfactory maps of climate, vegetation and soils have such a basis, landform maps used to test the strength of the climatic factor in morphogenesis should use landform criteria. To employ climatic–geomorphic or pedological–biogeographical criteria is, at least in part, to beg the question.

In a review of geomorphological mapping, Tricart (1965) described maps on a scale of 1:500,000 or greater as essentially morphostructural, climatically specific landforms being lost because of generalization (cf. Büdel (1948) and Murphy (1968, 1971). At scales between 1:500,000 and 1:5,000, structure can be regarded as given. These maps display one or more of the following data types: morphometric, morphographic (e.g. degree-of-slope maps), morphogenetic (incorporating laboratory results etc.) and chronological. While regional morphogenetic maps of phenomena produced under a single, distinctive morphoclimatic regime (e.g. fluvial (Pels, 1964); glacial (Derbyshire and coworkers, 1965); periglacial (Kaiser, 1960); arid (Grove, 1958; Grove and Warren, 1968)) or maps attempting a more comprehensive coverage of forms (Tricart, 1965; Tricart and Vogt, 1967; Brown and Crofts, 1973) may be of considerable interest in terms of climatogenetic geomorphology, morphometry is potentially of greater interest to geomorphologists seeking to define the influence of climate upon current landform change. Despite the suggestive results of Peltier's (1962) objectively derived data on mean relief, mean slope and mean number of drainageways per unit distance to differentiate glacial and tropical landscapes from others in terms of number/slope relationships, applications of the method remain rare and the deficiency is recognized by Smith and Atkinson (Chapter 13) as a major one in the study of limestone landscapes. At the regional scale, application of this method to slope gradients holds promise as a means of determining the role of climate in landform, as Melton (1960) and Kennedy (Kennedy and Melton, 1972) have shown, although the process links remain inferential (Kennedy, Chapter 6).

Over the past quarter of a century, the treatment of spatial geomorphic data has evolved from the representation of elements such as slope facets (Waters, 1958; Savigear, 1965) as 'a specialised form of topographical mapping' (Tricart, 1965), to a coherent form of background data for analysis of specific problems, both pure (e.g. Brunsden and Jones (1972) and Derbyshire and coworkers (1975)) and applied (e.g. Dearman and Fookes (1974) and Brunsden and coworkers (1975)). The range of spatial analytical techniques now appears to be widening rapidly (Chorley, 1972) and may well redress an imbalance particularly evident in British research work between process geomorphology and areal geomorphology (Mather, 1972). While this group of techniques has obvious relevance to climatic–geomorphic problems (Evans, 1972; see also Chapter 15), the extent to which the results are intellectually satisfying will depend on the confidence with which the climate–process–form links underlying them can be elucidated. For this purpose, climatic geomorphology can be regarded as a specialized branch of process geomorphology as well as a generalized extension of it (Büdel, 1963).

Recent work on slopes and slope processes exemplifies the general trend towards systematic measurement of component variables in geomorphic research. In reviewing the literature on rates of slope retreat, Young (1974) presents figures on rates of soil creep, solifluction, surface wash, solution and landsliding under different climates. Surface wash appears to be the dominant process in savanna and semi-arid climates, but it is exceeded in importance by creep in humid temperate lands while under rainforest both are rapid. Solution is important in all humid climates, but the precise quantitative significance of solifluction, both in present and former periglacial climates, has not been determined. While slopes which suffer catastrophic denudation appear to be affected most by events with a recurrence interval of 10–50 years, there is little information on the relative importance of catastrophic and continuous processes in slope retreat. This, together with a general dearth of studies designed to establish the relationship between form and process in the light of soil mechanics principles, constitutes a major obstacle to the establishment of the role of climate in slope evolution (Carson, Chapter 4). Change of slope form and gradient with time constitutes a particularly challenging problem to the climatic geomorphologist because climate acts largely through ground hydrology which itself varies widely with bedrock type and the nature of the regolith (Kennedy, Chapter 6), especially in its degree of weathering as it affects porosity (Carson and Kirkby, 1971; Kirkby, 1973). Hydrological slope models underline the inadequacies of the present state of knowledge of the relationship of climate (evapotranspiration) to soil water storage acting through the vegetation cover (Kirkby, Chapter 8).

Detailed work on weathered mantles provides ample demonstration of the inadequacy of the evidence of form alone as a basis for testing the central relationship of climatic geomorphology (Ollier, Chapter 5; Thomas, Chapter 14). Following a study of periodic morphogenetic features of postglacial age in a savanna landscape in Papua, Mabbutt and Scott (1966), for example, concluded that a uniformitarian, monogenetic explanation of landforms and their associated soils was inadequate, even in low latitudes, and that morphogenetic systems erected on broad landscape traits are suspect because changes in stability, process and pedogenesis demand detailed field study of slopes and correlative deposits (cf. Louis's (1973) review).

It was suggested long ago (de Martonne, 1913) that the forms of fluvial erosion are sensitive indicators of variations in certain climatic elements, although four decades elapsed before the first attempts were made to test this contention quantitatively. Chorley (1957) found a direct relationship between drainage density and amount and intensity of precipitation and, using a comparative regional approach (Chorley and Morgan, 1962) explained differences in terms of variations in runoff intensities produced by varying rainfall amounts and differences in relief. At the same time, Melton (1957) explained most of the variation in drainage density in the southwestern United States in terms of one direct climatic variable (rainfall intensity) and two indirect climatic variables (percentage of bare surface area and infiltration), although it was recognized that the influence of particular rock types on the latter variable might be considerable. There is some evidence to suggest that permeable and impermeable rocks behave as two distinct sub-populations within a single climatic region and are responsible for the detailed variation of the spatial pattern produced by the climate, notably the precipitation/evaporation balance (Gregory, Chapter 10).

Traditional climatic and climatogenetic geomorphology, in its emphasis on landform classification in global zones defined climatically, biotically or in terms of types and rates of geomorphic processes (Strahkov, 1967), has produced a result of considerable geographical and palaeoclimatic interest: landforms and their correlative deposits constitute the essential underpinning of much Pleistocene climatic reconstruction, both static and dynamic (e.g. Poser (1948), Büdel (1959), Butzer (1957, 1958, 1963, 1971), Barry (1960), Pels (1966), Butler and coworkers (1973), Lamb and Woodroffe (1970), Derbyshire (1971, 1972); and Andrews and coworkers (1972)). However, morphological regionalization has contributed little of the kind of information required to test the strength of the relationship between macroclimatic and mesoclimatic parameters and landform assemblages. Much recent work in this field has been directed toward resource appraisal so that regions have been defined on phytological and pedological as well as geomorphological criteria (e.g. Kondracki and Richling (1972); for critique, see Speight (1974)). Nevertheless, there are signs of a movement towards areal mapping based on long-term field measurement and using a framework of systems theory as a means of establishing geosystems (Gvozdetskiy and coworkers, 1971) as a basis for the synthetic establishment of areal climatogenetic units.

Hare (1973) has recently shown that climatology has emerged from a stage of development similar to that of traditional climatic geomorphology to enter one based on the understanding of climate and climatic variation in terms of energy and moisture balances. The reluctance of geomorphologists to move in the same direction is explained in terms of the length of the geomorphic time-scale and the magnitude–frequency relationship (Wolman and Miller, 1960) which encourages the employment of stochastic methods rather than those based on energy considerations. The energy–moisture balance approach appears to be particularly appropriate to climate–weathering–soil–plant relationships. While the factors and processes of rock weathering are now known in some detail (e.g. Keller, 1957), fundamental principles, such as those affecting the thermodynamic and kinetic stability of rock and regolith components (Curtis, Chapter 2), have not been applied in the investigation of specific surface and soils studies (cf. Ollier, Chapter 5). It is important to recognize that the study of biological factors in rock weathering is also in a rather retarded state (Ivashov, 1973).

An outline of the potential of general systems theory in geomorphological research has been given by Chorley and Kennedy (1971) and its potential in model building indicated in some detail by Chorley (1967, 1971, 1972). The development of this approach has been very uneven and it is particularly poorly developed in climatic geomorphology. Attempts to use hardware models with a specific climatic–geomorphic framework (e.g. Gavrilović (1972)) are quite rare. This is true of process–response models (Kirkby, 1971 and 1973), despite early initiatives in the use of mathematical models in the study of slope degradation (Scheidegger, 1961) and the evolution of drainage networks (Leopold and Langbein, 1962) and the general increase in the use of numerical methods in geomorphology (Chorley, 1966; Doornkamp and King, 1971), although the potential of the approach in examining climate–landform relationships is evident enough from the contributions of Trudgill (Chapter 3) and Kirkby (Chapter 8).

A particular problem is provided by the range in relaxation times within and between geomorphic systems (Allen, 1974) in comparison with those, say, in soil systems. Response to the climatically-stimulated extreme event, such as prolonged intense rainfall on hillslopes, the once in a century flood, a glacier surge or a succession of extended but milder winters in middle latitude mountains, may be immediate and the period between such events dominated by slow processes of modification (Starkel, Chapter 7). Variations in magnitude and frequency of specific meteorological or climatic events may result in a composite of forms expressing a spatial and temporal hierarchy (Douglas, Chapter 12). This was noted by Rapp (1960) in his classic study of current mass movements in northern Sweden. As a quantitative, long-term study of a dynamic landsurface in relation to meteorological variations, this work remains unique, although the approach has been applied in a more limited and localized way from time to time. For example, in the case of basally-eroded shale cliffs in east Devon, England, a long-term programme of mapping and analysis of related surface variables (slopes, soils, vegetation) has established that a condition of approximate balance exists between wave removal and replacement by shallow sliding, creep and small-scale rotational shear slide. Age and species composition of the vegetation on the cliff-face provide critical evidence of periodicity and rates of movement (Derbyshire and coworkers, 1975). Shallow sliding and small-scale rotational shears occur on a time scale of 1–3 years and correlate with high tides and severe easterly winds. Extreme storm events (high tides with easterly winds and heavy rainfall lasting several days) may produce excessive removal and oversteepening of the lower cliff such that large-scale shear slides are triggered. In this case, the cliff profile becomes composite: the rough balance between removal and supply of debris is reestablished on the lower cliff while on the middle cliff occasional sliding and creep produce accumulation and partial stabilization by vegetation ensues. The time-scale in this case appears to be of the order of 30 years (Figure 1.1). In addition, sections of the cliff show evidence of having moved by shear sliding on an even larger scale. While there is no reliable evidence with which to date this order of movement, there is no known reference to them in the written record so that the time-scale involved may be of the order of some hundreds of years. Rates of movement for both bare and vegetated parts of the cliff, compounded of movements of the first and second order, suggest an average rate for retreat of the cliff top of 0·63 m/yr with a maximum of 1·03 m/yr, most of the surficial cliff debris and its vegetation mat being *in transit* under prevailing extreme conditions of climate and tide. Thus, the cliff illustrates

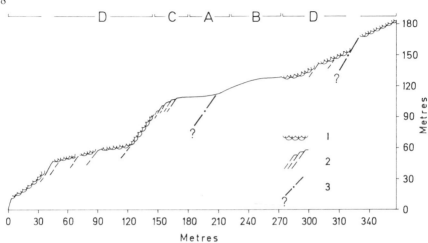

Figure 1.1. Natural scale profile of the cliff at St. Mary's Bay, Brixham, Devon. Key to symbols: 1—Movements on time-scale 1–3 years; 2—Movements on a time-scale of 30 years; 3—Possible movements of third temporal order and magnitude. Suggested queue state of debris in period 1968–1974 is shown by letters at top of figure, following Thornes (1971).

juxtaposed low-magnitude–high-frequency and high-magnitude–low-frequency morphological components which can be ascribed to meteorological events of particular scale and incidence.

The status of climatic elements as variables in the denudation system will vary from dependent to independent according to geographical and time scales (Kalesnik, 1961; Schumm and Lichty, 1965). Specific studies designed to demonstrate the scale relationship are few but include Douglas's (1966) work in eastern Australia and Morgan's in west Malaysia (1973 and Chapter 11). Both Morgan and Douglas find that climate is dominant at the macroscopic scale (G-scale value $< 3\cdot 5$: Haggett and coworkers (1965)), although this may not apply universally (see Ahnert, 1970), while Morgan suggests that climate and lithology may be of equal value at small scales ($G = 10$), a view supported by Douglas (Chapter 12). This is the scale of Tricart's VIth landform order (1965) at which morphodynamics are dominated by interaction of climate and lithology as influenced by vegetation. The process–landform relationships at this scale remain something of a black box (Gregory and Walling, 1974), a situation which is now capable of improvement in view of recent technological advances in automatic, continuous recording instrumentation and the application of systems theory. The inherent difficulties, notably that of scale, in relating energy inputs to sediment outputs above the small plot scale are outlined by Douglas in Chapter 9, although the general feasibility of using this approach at a regional scale on the basis of frequency–intensity–duration patterns of meteorological/climatic events within the framework of the denudation system is demonstrated by Morgan (Chapter 11). Trudgill (Chapter 3) recommends the study of microscale relationships as these are most likely to be in equilibrium with the

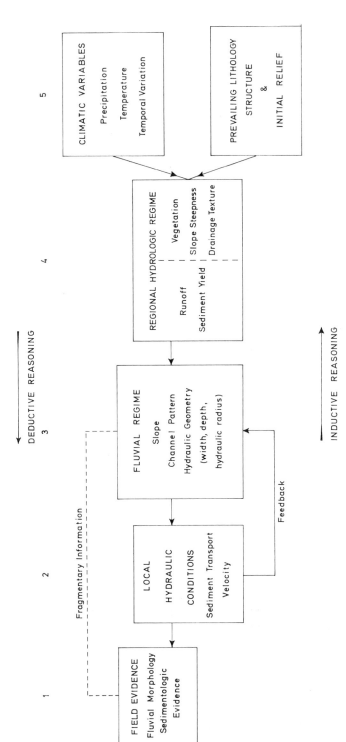

Figure 1.2. Chain of natural processes and human reasoning connecting fluvial geomorphic field evidence to climatic conditions. Redrawn after Baker (1974).

present climate: the higher the scale order, the higher the proportion of the landscape resulting from polygenetic interaction. In the study of fluvial terraces, Baker (1974) concludes that, in relating field evidence to formative climatic variables, information loss occurs through time and by negative feedback. Using a five-point scale, with field evidence as level 1 (Figure 1.2) deductive reasoning establishes level 2 information, but the implications for levels 3–5 are much more speculative. Given the accumulating research literature on interaction at the 4th and 5th levels, (climate and hydrology), the particularly cryptic links appear to lie on either side of level 3 (fluvial regime).

Given that the time-scale problem may be met in part by long-term institution-based experiments (see below), and recognizing the methodological dificiencies of the landform classification approach outlined above, future work would be most fruitful if concentrated on establishing climate–process–form links. While the great syntheses of world climatic geomorphology have emphasized 'the recognition of spatial land form problems and an analysis of their origin' (Zakrzewska, 1971) and so lie within the geographical geomorphology tradition (Russell, 1949), their sounder parts rest on an impressive foundation of research into landforms produced under a variety of distinctive climatic regimes (Dury, 1972), much of it process-based and so in the geological geomorphic tradition as defined by Hammond (1965).

## 1.4 Climatic data

To the process geomorphologist concerned with the determination of the climatic and meteorological factors in landform evolution, a fundamental question concerns the nature of the climatic/meteorological and landform data available or selected. There is a great wealth of data available on climatic elements and on hydrological variables, to which has been added more recently data on specific processes such as slope erosion (e.g. Leopold (1962) and Leopold and Emmett (1965)).

### 1.4.1 Standardized data

Factual data on the climatic elements vary enormously in quality, in temporal and spatial frequency, in number of parameters measured and in length of the instrumental record. The bulk of the data is available in the publications of the state meteorological offices, services or institutes. These range in style and content from the *Daily Weather Report*, published by the British Meteorological Office and containing records from 55 first-order stations for the six-hourly observations of the previous day, to the *World Weather Records*, which contain monthly values of temperature, pressure and precipitation for a decade (published by the Smithsonian Institution in Washington DC and by the US Weather Bureau).

The data collated and published by state meteorological services is derived

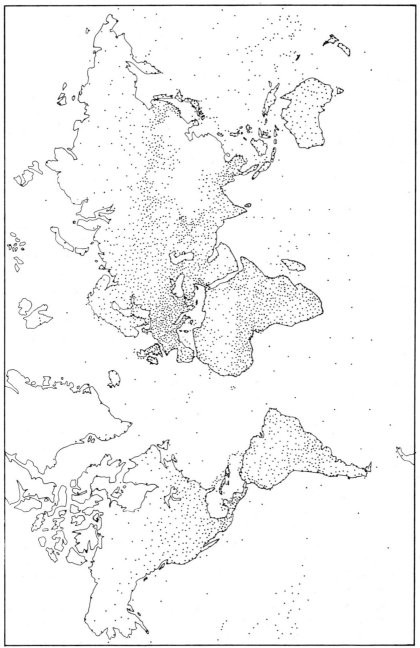

Figure 1.3. The global network of surface, synoptic observing stations. Modified from Atkinson by Barry and Perry (1973).

from a variety of sources, the principal being the meteorological stations, ordered according to the range and frequency of their observations. First order meteorological stations make manual observations every hour, self-recording (autographic) observations, or both. Because of the need for a 24-hour weather watch and for three-dimensional (upper air) observations, such stations are run by a variety of central or local government bodies such as departments of transport (air traffic control). The distribution of such stations is extremely uneven, being particularly sparse in high alpine environments and over vast tracts of the arctic and subarctic (Figure 1.3). Second order weather stations observe the standard surface elements (air temperature, dew point, pressure, duration of sunshine, air and ground minimum temperatures, wind direction and force, precipitation, cloud type and amount and visibility) on a regular basis but aerological work is not undertaken. Manual observations may be limited to two in twenty four hours, although most such stations supply autographic records of selected elements and may record additional information, some of particular interest to geomorphologists, such as ground and earth temperatures at selected depths, and evaporation. Stations of this type are often run by institutions such as river authorities, local government offices, power generation establishments, forestry services and commissions, coastguard stations, museums and universities, some of which will supply data at modest cost. Third order meteorological stations observe only a selection of the standard elements, notably atmospheric pressure, air temperature, precipitation and duration of sunshine. Stations of this type are frequently run by civic authorities, sometimes as adjuncts to air pollution measurement stations.

The hierarchy of weather stations numbers over 500 in the United Kingdom, the monthly means of their data being readily available in condensed form as the *Monthly Weather Report*. A further source of data is the network of rain gauges, now numbering more than 5000 in the United Kingdom, some of which are read manually each day while others record for a seven-day period. Many gauges are tended by amateur observers as well as by river authorities. This network is an important complement to the meteorological stations, greatly intensifying the recording coverage of a notoriously variable weather element (Rodda, 1967 and 1969). Rainfall data, usually available for 24-hour periods, and sometimes for particular meteorological events where autographic records are kept, is readily available and is first order data in the study of drainage basin hydrographs. Rainfall data from all sources is usually published on an annual basis in a form such as *British Rainfall* (HMSO, 1861). Selected climatological data of potential value to geomorphologists (notably rainfall and evaporation) is also available, often as ancillary information, in a variety of publications such as the *Surface Water Yearbook* of Great Britain (e.g. Water Resources Board, 1974) and annual reports of research institutes (e.g. Natural Environment Research Council, 1973), and selected synoptic indices are summarized in CRUMB (Climatic Research Unit, 1972; for a list of international sources of climatic data see Stringer (1972) and Climatic Research Unit (1973)).

*1.4.2 Utilization of Standardized data*

Despite the impressive volume of climatological data for much of the earth's surface, and particularly for the middle latitudes, relatively little of it has been utilized by climatologists (Barry and Perry, 1973) and even less has been applied to geomorphological analysis.

Selected gross parameters have been employed in several well known studies of global geomorphic patterns. Peltier (1950) selected mean annual temperature and precipitation, graphing zones to show relative intensities of chemical and mechanical weathering processes ('weathering regions') as a function of these two parameters. The same technique was employed to infer the relative intensity of mass movement, wind action and pluvial erosion and the graphs combined to derive nine morphogenetic regions. Leopold and coworkers (1964) used a similar graphic technique to zone dominant geomorphic processes in terms of mean annual temperature and precipitation, as did Wilson (1969, 1973a) in establishing his climate–process systems. Both of these works draw attention to the difficulty of extending the relationship to distinctive landform suites. Tanner (1961) has classified climates on the basis of water availability by plotting mean precipitation against mean potential evaporation but makes no claim to rigorous correlation of the resultant classes with landform assemblages.

Another approach using standard climatic data is the mapping of geomorphologically critical values of selected elements (Visher, 1937 and 1945). The world maps of Common (1966) show the incidence of limits such as 0 °C and 221 °C temperatures and 50 mm and 127 mm precipitation for annual and seasonal periods. While the climatically-controlled distribution of perennially frozen ground and the limit of pack ice coincides quite closely with distinctive cryonival landscapes (Troll, 1948; Williams, 1961) and coastlines (Davies, 1964 and 1972) respectively, the morphogenetic regionalization suggested by Common's sixteen indices is very complex and no climatogenetic relationships are suggested.

Explanations of regional variations in geomorphic processes in terms of standard climatic means or data derived from them have been attempted at several scales and with varying degrees of success. Corbel (1964) distinguished erosional provinces on the basis of temperature range, moisture availability and relief, while both Fournier (1960) and Strahkov (1967) used sediment yield as a basis for world maps of erosion. Strahkov found significant shifts in denudation rates at selected critical mean annual values of temperature and precipitation. Fournier found a high correlation between world sediment yield patterns and selected indices derived from precipitation means and seasonal incidence and frequency of precipitation.

Magnitude and frequency of occurrence of elements of the 'effective climate', especially rainfall events and erosivity (intensity and energy) of rain (Wischmeier and Smith, 1958), have been recognized as being of critical importance in respect of specific geomorphic processes such as those associated with river channels (Leopold and Maddock, 1953; Leopold and coworkers, 1964), hillslopes (Carson and Kirkby, 1971) and soil erosion by rainsplash and overland flow (Wischmeier and Smith, 1958; Hudson, 1971; Roose, 1967).

Mean annual sediment yield has been shown to vary systematically with mean annual precipitation. For example, Langbein and Schumm (1958) were able to show that peak sediment yield is associated with semi-arid catchments, yield falling off at lower precipitation levels and also at higher levels as the associated increase in vegetation cover acts as a limitation on yield. In reviewing this topic, Wilson (1973b) has tested the relationship using data from some 1500 drainage basins. This work suggests that the relationship between mean annual precipitation and mean annual sediment yield cannot be graphed for the earth as a whole owing largely to the variable status of the main controlling factors (of which climate is but one) from one basin to another.

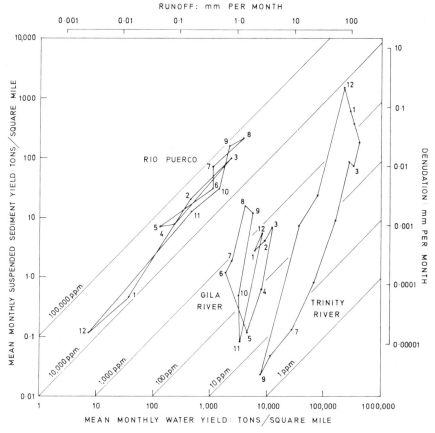

Figure 1.4. Sedihydrogram for rivers in three climatic regimes: Trinity River, California (mediterranean); Gila River, New Mexico (semi-arid continental); and Rio Puerco, New Mexico (arid continental). Redrawn from Wilson (1972).

Using a dynamic classification of climate which stresses air mass sources and their relative dominance (Strahler, 1965; Oliver, 1970), Wilson (1973b) examined the relationship for the world's seasonal climates (mediterranean, tropical wet–dry and continental). He tentatively concluded that fluvial erosion rates are much higher in seasonal than in non-seasonal climates (Fournier, 1949), continental climates showing a peak with a mean annual precipitation of 250–500 mm (cf. Langbein and Schumm, 1958), the figure for mediterranean climates being 1270–1650 mm and for tropical wet–dry 1780–1905 mm. The seasonal rhythm of erosion and runoff in drainage basins under different climates has been expressed in a log/log plot of mean monthly sediment yield and mean monthly water yield, the sedihydrogram (Wilson, 1972). In Figure 1.4 the sedihydrogram for part of the Trinity River catchment, California (mediterranean climate), shows notable runoff in all months but low to very low sediment yields during the rainless summer. Dessicated soils and reduced plant cover after the dry season result in the higher erosion and turbidity of the streams for the same water yield in late summer compared to early summer. The two sedihydrograms from New Mexico are from a region of continental climate with a relatively wet summer. The curve for the Gila River (centre), a semi-arid region with a mean annual precipitation of 508 mm, shows sediment yield peaks during winter–spring melting and the summer storm season. Sediment concentration is greatest at the summer peak. The curve for the Rio Puerco, a very dry region with precipitation of less than 304 mm/yr, displays a definite but much suppressed spring peak (3). With the sparse plant cover, the highest sediment yields coincide with the summer storm peaks.

Based as it is on standard records, this approach has considerable climatic–geomorphic potential, particularly in relation to comparative catchment erosion studies at a variety of geographical and time scales (Gregory and Walling, 1974), especially where instrumentation can be designed to take account of the interaction of landform and contemporary process at the field scale.

*1.4.3 Data specific to experimental requirements*

With the growth of specific process studies at the field scale, it has become evident that a good deal of the available standard meteorological data, both in their form and in their temporal and spatial distribution, fails to meet the needs of systematic research in geomorphology (Table 1 in Chorley (1966)). Indeed, the lack of appropriate and precise climatic data is admitted in a good deal of published work of climatic–geomorphic type. It is true that, for students of fluvial catchments, a considerable volume of hydrological data is available in addition to standard meteorological data and that there has recently been a general move toward the adoption of standardized observational practices. Even in this relatively advanced field of process geomorphology, however, specific additional instrumentation is usually essential (Gregory and Walling, 1973). Location of recording sites on the basis of stratified areal

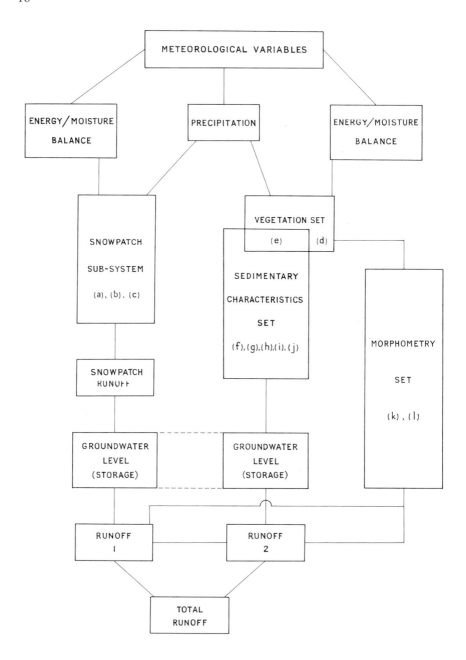

Figure 1.5. Major components of the nivation cirque/snowpatch system. Important set components include snowpatch morphology, albedo and density (a, b, c), percentage vegetation cover (d), bulk density of L/A horizons (e), texture, thickness, infiltration capacity and plasticity index of regolith (f, g, h, i), sediment in transport (j), basin shape (slope population) and slope erosional change (k, l).

sampling and the long-term monitoring of such sites is essential to a proper understanding of surface process rates (Young, 1974) and this is particularly so in assessing the precise role of individual meteorological events over a broad magnitude scale.

Continuous or semi-continuous electronic systems offer the best means of overcoming the problems of temporal sampling (cf. Schumm, 1964), with sensors designed to monitor essential variables within the regolith, on the landsurface and in the atmosphere immediately above the surface. While considerable developments in hydrometeorological measuring methods have occurred in the past ten years (for a review, see Gregory and Walling, 1973), a wide range of temperature, radiation, heat flow and wind sensors are also available for field application (Monteith, 1972). Generalized atmospheric and soil sensing systems provide a valuable adjunct to process studies in hydrology and geomorphology by providing data on the local incident climate. Systems of this type (Strangeways, 1972) have been used in the study of the effect of morphological factors (slope and altitude) on rainfall (Natural Environment Research Council, 1974) as part of a long-term catchment study on Plynlimon, central Wales, in water and energy balance studies of a cirque glacier in South Georgia (Natural Environment Research Council, 1972) and in the study of the growth of needle ice (Soons and Greenland, 1970). In catchments dominated by snowbank-induced processes, the energy balance approach is particularly appropriate in investigating the links between landform process rates, snowbank budget, microclimate and (bridging the gap between macroclimate and microclimate (Holmes and Dingle, 1965; Barry, 1970)) synoptic (air mass) climate (Figure 1.5 and Chapter 15), using dynamic indices such as the weather types of Lamb (1950) and the *Grosswetterlagen* of Hess and Brezowsky (1969). Adoption of this approach in the study of the geomorphology of snowy and frost-susceptible catchments is only in its initial stages. That it has not been attempted in other types of terrain is not surprising as the required instrumentation, wealth of data produced and the volume of computation involved in relating the data to erosion and deposition on the one hand and synoptic indices on the other as a basis for predictive models is formidable. A team approach and provision for long-term experimentation on a time-scale of several decades is demanded, which would require appropriate commitment at the level of the institution or research council.

## 1.5 Conclusion

In a recent exhaustive review of the literature of climatic geomorphology, Stoddart (1969) has shown that, despite the great volume of recent work on the effects of climatic shifts (Dury, 1972) (some of which is dependent on the recognition of contentious climate-diagnostic landforms (Wilhelmy, 1958), a concept which raises the problem of equifinality which, in turn, is bound up with questions of scale (Douglas, Chapter 12; Thomas, Chapter 14)), the fundamental stumbling block of the vast majority of work in climatic geomorphology is ignorance of the nature of the form–change sequence during meteorological events and climatological sequences of varying magnitude and frequency. It is interesting and relevant to note that the realization of a similar fundamental deficiency has been recognized by soil scientists (Bridges, 1974) and that it has become apparent largely in the context of scale (Courtney, 1974).

Ahnert (1973) has suggested that the recent concern of geomorphologists with the relationships between parameters and systems, using a 'functional' method and leading to increased confidence in prediction, is an essential stage in the growth of the discipline. It is generally true of a developing science that theoretical and experimental work go hand in hand, maintaining a rough balance, each acting as check and stimulus on the other. The evidence contained in the reviews in this volume suggests that research in climatic geomorphology is deficient both in its factual (process) base and in its failure to apply experimental, functional methods to climatogenetic problems. Significant further advance will be dependent on correcting this imbalance.

**References**

Ahnert, F., 1970, Functional relationships between denudation, relief, and uplift in large mid-latitude drainage basins, *Am. J. Sci.*, **268**, 243–263.

Ahnert, F., 1973, Inhalt und Stellung der funktionalen Methode in der Geomorphologie, *George. Zeits. Beiheft*, **33**, 105–113.

Allen, J. R. L., 1974, Reaction, relaxation and lag in natural sedimentary systems: general principles, examples and lessons, *Earth Science Rev.*, **10**, 263–342.

Andrews, J. T., R. G. Barry, R. S. Bradley, G. H. Miller and L. D. Williams, 1972, Past and pesent glaciological responses to climate in Eastern Baffin Island, *Quat. Res.*, **2**, 303–314.

Baker, V. R., 1974, Paleohydraulic interpretation of Quaternary alluvium near Golden Colorado, *Quat. Res.*, **4**, 94–112.

Barry, R. G., 1960, The application of synoptic studies in palaeoclimatology: a case study of Labrador–Ungava, *Geogr. Annlt.*, **42A**, 36–44.

Barry, R. G. 1970, A framework for climatological research with particular reference to scale concepts, *Trans. Inst. Brit. Geogr.*, **49**, 61–70.

Barry, R. G. and A. H. Perry, 1973, *Synoptic Climatology: methods and Applications*, Methuen, London, 555 pp.

Becking, R. W., 1957, The Zurich-Montpellier School of phytosociology, *Bot. Rev.*, **23**, 411–488.

Bik, M. J. J., 1967, Structural geomorphology and morphoclimatic zonation in the Central Highlands, Australian New Guinea, in *Landform Studies from Australia and New Guinea*, J. N. Jennings and J. A. Mabbutt (Eds.) Australian National University Press, Canberra, 26–47.

Bridges, E. M., 1974, A new soil classification for England and Wales, *Area*, **6**, 29–31.

Brown, E. H. and R. S. Crofts, 1973, Land shape of Britain on map, *Geogr. Mag.*, **44**, 137–140.

Brunsden, D. and D. K. C. Jones, 1972, the morphology of degraded landslide slopes in southwest Dorset, *Quart. J. Eng. Geol.*, **5**, 205–222.

Brunsden, D., J. C. Doornkamp, P. G. Fookes, D. K. C. Jones and J. M. H. Kelly, 1975, Geomorphological mapping techniques in highway engineering, *J. Inst. Hway. Engnrs.*, in press.

Budel, J., 1948, Das System der Klimatischen Geomorphologie, *Verhandl. Deutscher Geographie*, **27**, 65–100.

Büdel, J., 1959, Climatic zones of the Pleistocene, *Intnl. Geol. Rev.*, **1**, 72–79.
Büdel, J., 1963, Klimagenetische Geomorphologie, *Geogr. Rundschau*, **15**, 269–285.
Büdel, J., 1968, Geomorphology—principles, in *Encyclopedia of Geomorphology*, R. W. Fairbridge (Ed.), Reinhold, New York, 416–422.
Butler, B. E., G. Blackburn, J. M. Bowler, C. R. Lawrence, J. W. Newell and S. Pels, 1973, *A geomorphic map of the Riverine Plain of South-eastern Australia*, Australian National University Press, 39 pp.
Butzer, K. W., 1957, Mediterranean pluvials and the general circulation of the atmosphere, *Geogr. Annlr.*, **37**, 48–53.
Butzer, K. W., 1958, Quaternary stratigraphy and climate in the Near East, *Bonner Geogr. Abhl.*, **24**, 1–157.
Butzer, K. W., 1963, Climatic–geomorphologic interpretation of Pleistocene sediments in the Eurafrican subtropics, *Viking Fund Publ. Anthropology*, **36**, 1–27.
Butzer, K. W., 1971, *Environment and Archeology*, 2nd ed., Methuen, London, 703 p.
Carson, M. A. and M. J. Kirkby, 1971, *Hillslope Form and Process*, Cambridge Univ. Press, 475 pp.
Chorley, R. J., 1957, Climate and morphometry, *J. Geol.*, **65**, 628–638.
Chorley, R. J., 1966, The application of statistical methods to geomorphology, in *Essays in Geomorphology*, G. H. Dury (Ed.), Heinemann, London, 275–387.
Chorley, R. J., 1967, Models in geomorphology, in *Models in Geography*, R. J. Chorley and P. Haggett (Eds.), Methuen, London, 59–96.
Chorley, R. J., 1971, The role and relations of physical geography, *Progress in Geography*, **3**, 87–109.
Chorley, R. J. (Ed.), 1972, *Spatial Analysis in Geomorphology*, Methuen, London, 393 pp.
Chorley, R. J. and B. A. Kennedy, 1971, *Physical Geography: A Systems Approach*, Prentice-Hall, London, 370 pp.
Chorley, R. J. and M. A. Morgan, 1962, Comparison of morphometric features, Unaka Mountains, Tennessee and North Carolina, and Dartmoor, England, *Bull. Geol. Soc. Am.*, **73**, 17–34.
Clayton, K. M., 1971, Geomorphology—a study which spans the geology/geography interface, *J. Geol. Soc. London.*, **127**, 471–476.
Clements, F. E., 1936, Nature and structure of the climax, *J. Ecol.*, **24**, 252–284.
Climatic Research Unit, 1972 continuing, *Monthly Bulletin (CRUMB)*, **1**, University of East Anglia, Norwich, England.
Climatic Research Unit, 1973, *Monthly Bulletin (CRUMB)*, **2**, S 7, University of East Anglia, Norwich, England.
Common, R., 1966, Slope failure and morphogenetic regions, in *Essays in Geomorphology*, G. H. Dury (Ed.), Heinemann, London, 53–81.
Corbel, J., 1964, Érosion terrestre, étude quantitative (méthodes—techniques—résultats), *Ann. Geogr.*, **73**, 385–412.
Courtney, F., 1974, New soil classification; was it worth It?, *Area*, **6**, 205–206.
Davies, J. L., 1964, A morphogenetic approach to world shorelines, *Zeits. f. Geom.*, **8**, 127–142.
Davies, J. L., 1972, *Geographical Variation in Coastal Development*, Oliver and Boyd, Edinburgh, 204 pp.
Dearman, W. R. and P. G. Fookes, 1974, Engineering geological mapping for civil engineering practice in the United Kingdom, *Quart. J. Eng. Geol.*, **7**, 223–256.
Derbyshire, E. 1971, A synoptic approach to the atmospheric circulation of the last glacial maximum in south-eastern Australia, *Palaeogeogr., Palaeoclim., Palaeoecol.*, **10**, 103–124.
Derbyshire, E., 1972, Pleistocene glaciation of Tasmania, review and speculations *Austr. geogr. Studies*, **10**, 79–94.
Derbyshire, E. (Ed.), 1973, *Climatic Geomorphology*, Macmillan, London, 296 pp.

Derbyshire, E., M. R. Banks, J. L. Davies and J. N. Jennings, 1965, Glacial map of Tasmania, *R. Soc. Tasm., Spec. Pub. 2*, 11 pp.
Derbyshire, E., L. W. F. Page and R. J. C. Burton, 1975, Integrated field mapping of a dynamic landsurface: St. Mary's Bay, Brixham, in *Environment, Man and Economic Change*, A. D. M. Phillips and B. J. Turton (Eds.), Longmans, London, 48–77.
Derruau, M., 1968, Mountains, in *Encyclopedia of Geomorphology*, R. W. Fairbridge (Ed.), Reinhold, New York, 737–739.
Doornkamp, J. C. and C. A. M. King, 1971, *Numerical Analysis in Geomorphology: An Introduction*, Edward Arnold, London, 372 pp.
Douglas, I., 1966, Denudation rates and water chemistry of selected catchments in eastern Australia and their significance for tropical geography, *Unpublished Ph.D. dissertation, Australian National University*, Canberra, 472 pp.
Dury, G. H., 1972, Some current trends in geomorphology, *Earth Sci. Rev.*, **8**, 45–72.
Embleton, C. and C. A. M. King, 1975, *Periglacial Geomorphology*, Edward Arnold, London, 203 pp.
Evans, I. S., 1972, General geomorphometry, derivatives of altitude and descriptive statistics, in *Spatial Analysis in Geomorphology*, R. J. Chorley (Ed.), Methuen, London, 17–90.
Fournier, F., 1949, Les facteurs climatiques de l'érosion du sol, *Bull. Ass. Géogr. Français*, **203**, 97–103.
Fournier, F., 1960, *Climat et érosion: la relation entre l'érosion du sol par l'eau et les précipitations atmosphériques*, Presses Univ. France, Paris, 201 pp.
Gavrilović, D., 1972, Experimente zur Klimageomorphologie, *Zeits. f. Geomorph.*, **16**, 315–331.
Glinka, K. D., 1915, *Pedology*, St. Petersburg.
Gregory, K. J. and D. E. Walling, 1973, *Drainage Basin Form and Process*, Edward Arnold, London, 456 pp.
Gregory, K. J. and D. E. Walling (Eds.), 1974, Fluvial processes in instrumented watersheds, *Inst. Brit. Geogr., Sp. Pub. 6*, 1–6.
Grishankov, G. Ye., 1973, The landscape levels of the continents and geographical zonality, *Soviet Geography*, **14**, 61–78.
Grove, A. T., 1958, the ancient erg of Hausaland and similar formations on the south side of the Sahara, *Geogr. J.*, **124**, 58–522.
Grove, A. T. and A. Warren, 1968, Quaternary landforms and climate on the south side of the Sahara, *Geogr. J.*, **134**, 194–208.
Gvozdetskiy, N. A., K. I. Gerenchuk, A. G. Isachenko and V. S. Prebrazhenskiy, 1971, The present state and future tasks of physical geography, *Soviet Geogr.*, **12**, 257–266.
Haggett, P., R. J. Chorley and D. R. Stoddart, 1965, Scale-standards in geographical research: a new measure of areal magnitude, *Nature*, **205**, 844–847.
Hammond, E., 1965, What is a landform? Some further comments, *Prof. Geogr.*, **17**, 12–13.
Hare, F. K., 1973, Energy-based climatology and its frontier with ecology, in *Directions in Geography*, R. J. Chorley (Ed.), Methuen, London, 171–192.
Hastenrath, S. and J. Wilkinson, 1973, A contribution to the periglacial morphology of Lesotho, southern Africa, *Biul. Peryglac.*, **22**, 157–167.
Hess, P. and H. Brezowsky, 1969, Katalog der Grosswetterlagen Europas, *Ber. dtsch. Wetterd.* (Offenbach), **15** (113), 56 pp.
HMSO, 1861 continuing, *British Rainfall*, Her Majesty's Stationery Office, London.
Holmes, R. M. and A. N. Dingle, 1965, The relationship between the macro and the micro climate, *Agric. Met.*, **2**, 127–133.
Hudson, N., 1971, *Soil Conservation*, Batsford London, 320 pp.
Ivashov, P. V., 1973, The significance of biological factors in the weathering of rocks and minerals, *National Aeronautics and Space Administration, Technical Translation*, 29 pp.

Ives, J. D. and R. G. Barry, 1974, *Arctic and Alpine Environments*, Methuen, London, 999 pp.

Kaiser, K., 1960, Klimazeugen des periglazialen Dauerfrostbodens in Mittel-und Westeuropa, *Eiszeitalter und Gegenwart*, **11**, 121–141.

Kalesnik, S. V., 1961, The present state of landscape studies, *Soviet Geogr.*, **2**, 24–34.

Keller, W. D., 1957, *The Principles of Chemical Weathering*, Lucas Bros., Columbia, Missouri, 111 pp.

Kennedy, B. A. and M. A. Melton, 1972, Valley asymmetry and slope forms of a permafrost area in the Northwest Territories, Canada, in *Polar Geomorphology*, R. J. Price and D. E. Sugden (Eds.), *Inst. Brit. Geogr. Sp. Pub. 4*, 107–121.

Kirkby, M. J., 1971, Hillslope process–response models based on the continuity equation, in *Slopes: forms and process*, D. Brunsden (Ed.), *Inst. Brit. Geogr., Sp.Pub. 3*, 15–30.

Kirkby, M. J., 1973, Landslides and weathering rates, *Dept. Geogr., Univ.Leeds, Working Paper 34*.

Kondracki, J. and A. Richling, 1972, Synthetic physico-geographical research, *Geographica Polonica*, **22**, 13–25.

Köppen, W. 1923, *Die Klimate der Erde, Grundriss der Klimakunde*, de Gruyter, Berlin, 369 pp.

Kotarba, A. and L. Starkel, 1972, Holocene morphogenetic altitudinal zones in the Carpathians, *Studia Geom. Carpatho-Balcanica*, **6**, 21–35.

Kubiëna, W. L., 1953, *The Soils of Europe*, Murby, London, 317 pp.

Lamb, H. H., 1950, Types and spells of weather around the year in the British Isles: annual trends, seasonal structure of the year, singularities, *Quart. J. R. Met. Soc.*, **76**, 393–429.

Lamb, H. H. and A. Woodroffe, 1970, Atmospheric circulation during the last Ice Age, *Quat. Res.*, **1**, 29–58.

Langbein, W. B. and S. A. Schumm, 1958, Yield of sediment in relation to mean annual precipitation, *Trans. Am. Geophys. Union*, **39**, 1076–1084.

Leeper, G. W., 1956, The classification of soils, *J. Soil Sci.*, **7**, 59–64.

Leeper, G. W., 1964, Introduction to soil science, 4th ed., *Melbourne Univ. Press*, Melbourne, 253 pp.

Leopold, L. B., 1962, The vigil network, *Intnl. Ass. Sci. Hydr. Bull.*, **7**, 5–9.

Leopold, L. B. and W. W. Emmett, 1965, Vigil network sites: a sample of data for permanent filing, *Intnl. Ass. Sci. Hydr. Bull.*, **10**, 12–21.

Leopold, L. B. and W. Langbein, 1962, The concept of entropy in landscape evolution, *U.S.Geol.Surv. Prof.Pap. 500-A*, 20 pp.

Leopold, L. B. and T. Maddock, 1953, The hydraulic geometry of stream channels and some physiographic implications, *U.S.Geol.Surv. Prof.Pap. 252*, 56 pp.

Leopold, L. B., M. G. Wolman and J. P. Miller, 1964, *Fluvial processes in geomorphology*, Freeman, San Francisco, 522 pp.

Louis, H., 1973, Fortschritte und Fragwürdigkeiten in neueren Arbeiten zur Analyse fluvialer Landformung besonders in den Tropen, *Zeits. f. Geomorph.*, **17**, 1–42.

Mabbutt, J. A., and R. M. Scott, 1966, Periodicity of morphogenesis and soil formation in a savannah landscape near Port Moresby, Papua, *Zeits. f. Geomorph.*, **10**, 69–89.

Marbut, C. F., 1928, A scheme for soil classification, *1st. Intnl. Congr. Soil. Sci.*, **4**, 1–31.

Marbut, C. F., 1935, Soils of the United States, in US Dept. Agric., *Atlas of American Agriculture*, part III, 98 pp.

Martonne, E. de, 1913, Le climat—facteur du relief, *Scientia*, **1913**, 339–355.

Mather, P. M., 1972, Areal classification in geomorphology, in *Spatial Analysis in Geomorphology*, R. J. Chorley (Ed.), Methuen, London, 305–322.

Melton, M. A., 1957, An analysis of the relations among elements of climate, surface properties, and geomorphology, *Geol. Dept., Columbia Univ., New York, Tech. Rep. 11*, 1–102.

Melton, M. A., 1960, Intravalley variation in slope angles related to microclimate and erosional development, *Bull. Geol. Soc. Am.*, **71**, 133–144.
Monteith, J. L., 1972, *Survey of Instruments for Micrometeorology*, (Intnl. Biol. Prog. Handbook 22), Blackwell, Oxford, 263 pp.
Morariu, T. and J. Mac, 1974, On the dominant and secondary present-day modelling of Roumania's relief, *Studia Geom. Carpatho-Balcanica*, **8**, 85–94.
Morgan, R. P. C., 1970, Climatic geomorphology: its scope and future, *Geographica*, **6**, 26–35.
Morgan, R. P. C., 1973, The influence of scale in climatic geomorphology: a case study of drainage density in West Malaysia, *Geogr. Annlr.*, **55A**, 107–115.
Morrison, R. B. and H. E. Wright (Eds.), 1967, *Quaternary Soils*, Intnl. Assoc. Quat. Res., VII Congr., Proc. 9, 338 pp.
Murphy, R. E., 1968, Landforms of the world, (Annals Map Supplement 9), *Ann. Ass. Am. Geogr.*, **58**, 198–200.
Murphy, R. E., 1971, Regions of erosional and depositional landforms, *Sci. Rep. Tohoku Univ.*, 7th. ser., Geog., **20**, 213–220.
Natural Environment Research Council, 1972, Overseas catchments, South Georgia—Antarctica, *Inst. Hydrol. Res. Rep.*, 1971–1972, 14–15.
Natural Environment Research Council, 1973, Instrumentation, *Inst. Hydr. Res. Rep.*, 1972–1973, 41–45.
Natural Environment Research Council, 1974, Instrumentation: automatic weather station, *Inst. Hydr. Res. Rep.* 1973–1974, 31–32.
Northcote, K. H., 1960, Factual key for the recognition of Australian soils, *C.S.I.R.O., Aust. Soils Div. Rep.*, 60 pp.
Odum, E. P., 1964, The new ecology, *Bioscience*, **14**, 14–16.
Oliver, J., 1970, A genetic approach to climate classification, *Ann. Ass. Am. Geogr.*, **60**, 615–637.
Passarge, S., 1926, Morphologie der Klimazonen oder Morphologie der Landschaftsgurtel?, *Petermanns Geogr. Mitt.*, **72**, 173–175, (English transl. in Derbyshire, 1973).
Pears, N. V., 1968, Some recent trends in classification and description of vegetation, *Geogr. Annlr.*, **50A**, 162–172.
Pels, S., 1964, The present and ancestral Murray River System, *Austr. geogr. Studies*, **2**, 111–119.
Pels, S., 1966, Late Quatenary chronology of the Riverine Plain of southeastern Australia, *J. geol. Soc. Aust.*, **13**, 27–40.
Peltier, L. C., 1950, The geographic cycle in periglacial regions as it is related to climatic geomorphology, *Ann. Ass. Am. Geogr.*, **40**, 214–236.
Peltier, L. C., 1962, Area sampling for terrain analysis, *Prof. Geogr.*, **14**, 24–28.
Péwé, T. L. (Ed.), 1969, *The Periglacial Environment*, McGill–Queen's Univ. Press, Montreal, 487 pp.
Poore, M. E. D., 1955–56, The use of phytosociological methods in ecological investigations, *J. Ecol.*, **43**, 226–244, 245–269, 606–651; **44**, 28–50.
Poser, H., 1948, Boden- und Klimaverhaltnisse in Mittel- und Westeuropa wahrend der Würmeiszeit, *Erdkunde*, **2**, 53–68.
Rapp, A., 1960, Recent development of mountain slopes in the Karkevagge and surroundings, northern Scandinavia, *Geogr. Annlr.*, **42**, 65–200.
Rathjens, S. (Ed.), 1971, *Klimatische Geomorphologie*, Wissenschaftliche Buchgesellschaft, Darmstadt, 485 pp.
Robinson, G. W., 1949, *Soils; their origin, constitution and classification*, Murby, London.
Rodda, J., 1967, The rainfall measurement problem, *Intnl. Ass. Sci. Hydr. Pub.*, **78**, 215–231.
Rodda, J., 1969, The assessment of precipitation, in *Water, Earth and Man*, R. J. Chorley (Ed.), Methuen, London, 130–134.

Rohdenburg, H., 1971, Einführung in die Klimagenetische Geomorphologie, *Lenz-Verlag, Giessen*, 350 pp.
Roose, E. J., 1967, Dix années de mésure de l'érosion et du ruissellement au Sénégal, *L'Agronomie Tropicale*, **22**, 123–152.
Rudberg, S., 1972, Periglacial zonation—a discussion, *Gottinger Geogr. Abh.*, **60**, 221–233.
Russell, R. J., 1949, Geographical geomorphology, *Ann. Ass. Am. Geogr.*, **39**, 1–11.
Savigear, R. A.G., 1965, A technique of morphological mapping, *Ann. Ass. Am. Geogr.*, **55**, 514–538.
Scheidegger, A. E., 1961, *Theoretical geomorphology*, Springer–Verlag, Berlin, 333 pp.
Schumm, S. A., 1964, Seasonal variations of erosion rates and processes on hillslopes in western Colorado, *Zeits. f. Geomorph., Supplementband 5*, 215–238.
Schumm, S. A. and R. W. Lichty, 1965, Time, space and causality in geomorphology, *Am. J. Sci.*, **263**, 110–119.
Soons, J. M. and D. E. Greenland, 1970, Observations on the growth of needle ice, *Water Resources Res.*, **6**, 579–593.
Speight, J. G., 1974, A parametric approach to landform regions, in Brown, E. H. and R. S. Waters, *Progress in Geomorphology, Inst. Brit. Geogr. Spec. Pub. 7*. 213–230.
Stoddart, D. R., 1969, Climatic geomorphology; review and re-assessment, *Progress in Geography*, **1**, 160–222.
Strahkov, N. M., 1967, *Principles of lithogenesis*, (S. I. Tomkeieff and J. E. Hemingway (Eds.)), Vol. 1., Oliver and Boyd, Edinburgh, 245 pp.
Strahler, A. N., 1965, *Introduction to Physical Geography*, Wiley, New York, 455 pp.
Strangeways, I. C., 1972, Automatic weather stations for network operation, *Weather*, **27**, 403–408.
Stringer, E. T., 1972, *Techniques of Climatology*, Freeman, San Francisco, 539 pp.
Tanner, W. F., 1961, An Alternate approach to morphogenetic climates, *Southeastern Geologist*, **2**, 251–257.
Tansley, A. G., 1929, Succession: the concept and its values, *Proc. Intnl. Congr. Plant Sciences (1926)*, 677–686.
Tansley, A. G., 1935, The use and abuse of vegetational concepts and terms, *Ecology*, **16**, 284–307.
Thomas, M. F., 1974, *Tropical Geomorphology*, Macmillan, London, 332 pp. (especially 258–276).
Thornes, J. B., 1971, State, environment and attribute in scree-slope studies, in *Slopes, Form and Process*, D. Brunsden (Ed.), Inst. Brit. Geogr. Sp. Pub. 3, 49–63.
Tricart, J., 1963, *Géomorphologie des régions froides*, Presses Univ. de France, Paris, 289 pp.
Tricart, J., 1965, *Principes et Méthodes de la Géomorphologie*, Masson, Paris, 496 pp.
Tricart, J., 1969, *Geomorphology of cold environments*, (Trans. E. Watson), Macmillan, London, 320 pp.
Tricart, J., 1972, *Landforms of the Humid Tropics*, (transl. C. J. K. de Jonge), Longman, London, 306 pp.
Tricart, J. and A. Cailleux, 1965, *Introduction à la géomorphologie climatique*, S.E.D.E.S., Paris, 306 pp.
Tricart, J. and A. Cailleux, 1972, *Introduction to Climatic Geomorphology*, (transl. C. J. K. de Jonge), Longman, London, 295 pp.
Tricart, J. and H. Vogt, 1967, Présentation des cartes détaillées de la France. Commentaire sur les cartes géomorphologiques Parignarques et Obernai au 1:25000, in *Progress made in geomorphological mapping*, J. Demek (Ed.), Proc. I. G. U. Comm. Applied Geom. (Subcomm. Geom.Mapping), Geogr. Cesko Akad., Brno.
Troll, C., 1948, Der subnivale oder periglaziale Zyklus der Denudation, *Erdkunde*, **2**, 1–21.

Troll, C., 1958, Structure soils, solifluction, and frost climates of the earth, *U.S. Army Snow, Ice and Permafrost Res. Establ. Translation 43*, 121 pp.

US Department of Agriculture, 1960, *Soil classification—a comprehensive system*, United States Dep. Agric., Washington DC.

Visher, S. S., 1937, Regional contrasts in erosion in Indiana, with especial attention to the climatic factor in causation, *Bull. Geol. Soc. Am.*, **48**, 897–929.

Visher, S. S., 1945, Climatic maps of geologic interest, *Bull. Geol. Soc. Am.*, **56**, 713–736.

Washburn, A. L., 1973, *Periglacial Processes and Environments*, Edward Arnold, London, 320 pp.

Water Resources Board and Scottish Development Department, 1974, *Surface Water Yearbook 1966–1970*, HMSO, London, 156 pp.

Waters, R. S., 1958, Morphological Mapping, *Geography*, **43**, 10–17.

Webster, R., 1968, Fundamental objections to the 7th approximation, *J. Soil. Sci.*, **19**, 354–366.

Whittaker, R. H., 1953, A consideration of climax theory: the climax as a population and pattern, *Ecol. Monoger.*, **23**, 41–78.

Wilhelmy, H., 1958, *Klimamorphologie der Massengesteine*, Westermann, Braunschweig, 238 pp.

Williams, P. J. 1961, Climatic factors controlling the distribution of certain frozen ground phenomena, *Geogr. Annlr.*, **43**, 339–347.

Wilson, L., 1969, Les relations entre les processus géomorphologiques et le climat moderne comme méthode de paléoclimatologie, *Rev. Géog. phys. Géol. dynam.*, **11**, 303–314.

Wilson, L., 1972, Seasonal sediment yield patterns of US rivers, *Water Resources Res.*, **8**, 1470–1479.

Wilson, L., 1973a, Relationships between geomorphic processes and modern climates as a method in paleoclimatology in *Climatic Geomorphology*, E. Derbyshire (Ed.), Macmillan, London, 269–284.

Wilson, L., 1973b, Variations in mean annual sediment yield as a function of mean annual precipitation, *Am. J. Sci.*, **273**, 335–349.

Wischmeier, W. H. and D. D. Smith, 1958, Rainfall energy and its relationship to soil loss, *Trans. Am. Geophys. Union*, **39**, 285–291.

Wolman, M. G. and J. P. Miller, 1960, Magnitude and frequency of forces in geomorphic processes, *J. Geol.*, **68**, 54–74.

Yaalon, D. H., (Ed.), 1971, *Paleopedology*, Israel University Press, Jerusalem, 350 pp.

Young, A., 1974, The rate of slope retreat in *Progress in Geomorphology*, E. H. Brown and R. S. Waters (Eds.), *Inst. Brit. Geogr., Sp. Pub.*, **7**, 65–78.

Zakrzewska, B., 1971, Nature of land form geography, *Prof. Geogr.*, **23**, 351–354.

CHAPTER TWO

# Chemistry of Rock Weathering: Fundamental Reactions and Controls

Charles D. Curtis

## 2.1 Introduction

This essay considers weathering from what might appear to be a specialist viewpoint: a chemical one. In another way, however, it can be argued that it is particularly general. Observations and conclusions apply equally to weathering across the total range of climatic environment, starting material, site and organic involvement. Furthermore, a brief analysis of denudation data demonstrates that chemical processes are very much more important on a worldwide scale than is generally appreciated.

Essential to the comprehension of chemical reaction is the concept of stability. Whether or not a particular mineral or mineral assemblage is unstable (and to what extent) obviously is the key to alteration in weathering. Unfortunately, however, quantitative assessment of stability and equilibrium is not easy. This is why a very considerable section of the essay is devoted to defining terms and discussing concepts.

Having laid the foundation of chemical thermodynamics it is possible to make the all-important step linking mineral assemblages with chemical environment. The influence of climate (the objective of the entire volume) is reflected in the aqueous chemical environment of the soil. Different geomorphic situations dictate different chemical environments and it is possible to predict the style of chemical reactions from the environments.

The final theoretical input is concerned with kinetic factors. The relationship between thermodynamic and kinetic stability is discussed. Thereafter, the more common types of natural weathering reactions are discussed and some comments on fundamental mechanisms and controls offered.

### 2.1.1 The general picture

The products and processes of weathering are familiar to everyone. In most descriptive accounts the tendency has been to concentrate on the more obvious physical processes of disaggregation, grain diminution, erosion and deposition of particulate material. It has been recognized, of course, (and most diligently researched) that limestone landforms are modified by solution processes. Every elementary earth science text explains how carbon dioxide from the air is carried down by rainwater to dissolve away calcium carbonate as soluble calcium bicarbonate. This 'special case' has been noted by geomorphologists studying landforms and by geologists thinking in terms of provenance.

Two fairly recent developments have underlined the inadequacy of the classic approach. Quantitative denudation rate studies have become increasingly popular and with them the recording of 'total dissolved solids'. Some work has gone further to report quantitative chemical analyses of stream and river waters. Sedimentary geochemists have been using similar data to establish the nature of the 'geochemical cycles' of different chemical elements. Great interest has developed in piecing together the early history and chemical evolution of the oceans. This obviously necessitates comparison of the chemical composition of seawater with that of continental runoff.

The net result of these investigations is to demonstrate very clearly that

Table 2.1. Some basic (if approximate) data

(a) Estimates of annual denudation (from Garrels and Mackenzine (1971, p. 120)).

| Continent | Total dissolved load (E, Figure 2·1) $\times 10^{14}$ g | Solution denudation tons. $km^{-2}$ | Total suspended load (F) $\times 10^{14}$ g | Suspension denudation tons. $km^{-2}$ | Ratio E/F |
|---|---|---|---|---|---|
| N. America | 7·0 | 33 | 17·8 | 86 | 0·4 |
| S. America | 5·5 | 28 | 11·0 | 56 | 0·5 |
| Asia | 14·9 | 32 | 145·0 | 310 | 0·1 |
| Africa | 7·1 | 24 | 4·9 | 17 | 1·4 |
| Europe | 4·6 | 42 | 2·5 | 27 | 1·8 |
| Australia | 0·2 | 2 | 2·1 | 27 | 0·1 |
| World | 39·3 | | 183·3 | | 0·21 |

(b) Composition of solution runoff (units p.p.m.) (simplified from Livingstone (1963, p. 41)).

| Continent | $HCO_3^-$ | $SO_4^{2-}$ | $Cl^-$ | $Ca^{2+}$ | $Mg^{2+}$ | $Na^+$ | $K^+$ | $SiO_2$ | Total |
|---|---|---|---|---|---|---|---|---|---|
| S. America | 68 | 20 | 8 | 21 | 5 | 9 | 1 | 9 | 141 |
| N. America | 31 | 5 | 5 | 7 | 2 | 4 | 2 | 12 | 68 |
| Asia | 95 | 24 | 7 | 31 | 6 | 5 | 2 | 8 | 178 |
| Africa | 79 | 8 | 9 | 18 | 6 | 7 | 2 | 12 | 141 |
| Europe | 43 | 14 | 12 | 13 | 4 | 11 | ? | 23 | 120 |
| Australia | 32 | 3 | 10 | 4 | 3 | 3 | 1 | 4 | 60 |
| World | 58 | 12 | 9 | 16 | 4 | 7 | 2 | 11 | 119 |

(c) Estimates of the percentage occurrence of the three major sediment lithologies in sediments and sedimentary rocks of the lithosphere.

|  | Holmes (1937) | Wickman (1954) | Horn (1966) | Garrels and Mackenzie (1971) |
|---|---|---|---|---|
| Sandstone | 16 | 8 | 20 | 11 |
| Shale | 70 | 83 | 73 | 74 |
| Carbonate | 14 | 9 | 7 | 15 |

present-day continental denudation occurs in significant part by removal in true solution. On a worldwide basis the contributions of particulate and solution material to total denudation have been estimated to be comparable (Table 2.1): different climatic/topographic situations can lead to the dominance of either fraction. This point has been made before but few authors seem to have identified the obvious corollary. Much of the particulate material owes its origin to chemical processes, or has at least been dramatically modified by them. It follows that general discussions of weathering processes or landform development cannot ignore chemical processes.

*2.1.2 The chemist's viewpoint*

Before proceeding to describe chemical processes in more detail it seems sensible to state in general terms what is the chemist's viewpoint or, perhaps more correctly, one chemist's viewpoint. Chemistry is concerned with substances described in terms of their fundamental (elemental) composition and with the processes whereby one substance is changed to another. The questions investigated boil down to: what are the structures of different substances (solids, liquids, gases) in terms of which elements are present therein and what is the elemental configuration in time and space? Furthermore, why do the particular elements in question combine together and why do they adopt a particular physical state? Supposing these questions to be satisfactorily answered, the final logical step is to enquire: what is the fundamental driving force which makes certain chemical reactions 'go' whilst others (or the reverse reactions) do not and what controls the rate at which reaction proceeds.

Chemical theory has evolved sufficiently far to be able to state unequivocably that fundamental chemical processes and properties stem from electronic reactions and configurations (between and within atoms). Bonds between atoms reflect the spatial configurations of the outer (negatively charged) electron shells which surrounded the positively charged atomic nucleus. These shells are modified by chemical reaction to adopt new states with different energies. The chemist's basic concern is with the particular types of electronic configuration that different chemical elements are known to adopt and the energy changes that accompany their modification. Although this may sound very academic, the relevance to geomorphology of the questions cited in the previous paragraph cannot be doubted and any approach that might give an answer must be considered.

### 2.1.3 Interplay between chemical and physical processes in weathering

Having made a case for considering chemical weathering reactions more carefully, it seems reasonable to attempt to locate them more precisely within the overall complex of geomorphological activity. Figure 2.1 is an example of a 'cycle diagram' much beloved of geochemists. The various geochemical 'spheres' are self-explanatory terms for the sum total of each constituent in the earth's crust. The first part of the sequence also is obvious. Weathering represents the interaction between rocks and the atmosphere, hydrosphere and biosphere. (The last mentioned is dominantly derived from the hydrosphere and atmosphere via photosynthesis. This 'externally driven' reaction (solar energy) actually stores up most of the chemical energy used up in other parts of the surface cycle. By separating carbon and oxygen, photosynthesis stores energy which can be liberated in combustion reactions.) Four kinds of products result (A to D). Some material goes into solution as 'silicic acid' and salts of sodium, potassium, magnesium and calcium. Significant solid organic material accumulates, mostly as plant debris much modified by biochemical reactions. Very often this material is so intimately mixed with inorganic material that physical separation is quite impossible.

Two inorganic 'rock' derivatives can be recognized, although sometimes the distinction between them may not be altogether clear. The first group consists of new silicates (clay minerals) and various hydrated oxides of aluminium, silicon and iron. Less obvious to those living in humid climates are the salts (especially calcium carbonate) which form in many soil profiles. None of these minerals represents the physical breakdown of primary rock materials: chemical reactions have been involved.

The second group consists of true physical weathering products (group D in Figure 2.1) and is made up of minerals present in the source rocks but liberated as free grains during weathering. Quartz is quantitatively by far the most important mineral in this class.

Weathering is the process of breakdown of the rock to form new products $(A + B + C)$ and a residuum of original material (D). The ratio of chemical to physical products obviously is given by $(A + B + C)/D$. This is much greater than the ratio of dissolved to suspended load $A/(B + C + D)$ or $E/F$ which is usually quoted in this context. Nor are the two ratios very simply related. Certain regions of very high solid runoff (South East Asia, for example) contribute mostly materials of group C. The relationship between these two lies in the intensity of chemical weathering in the source area. Clearly chemical weathering is qualitatively important and there are very obvious links with major climatic and topographic variables.

One of the difficulties with representations such as Figure 2.1 is that processes in different parts of the cycle may occur at the same time or quite separately. No

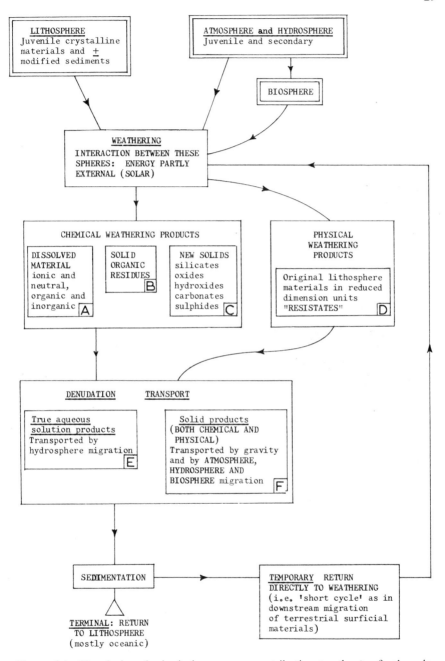

Figure 2.1. Chemical and physical processes contributing to the 'surface' cycle.

time concept can be included. Perhaps this is best illustrated by considering the development of a soil profile at some location within a drainage system. *In situ* weathering continues for a relatively long time-period accompanied by slight transport as slow downslope profile movement. At the same time, however, chemical products in solution are being continuously removed to the terminal sedimentation basin. This effects complete time separation of A from B, C and D. Lateral stream erosion of the soil profile would then cause transport and deposition of most of the solid residues downstream. This would occur rapidly compared with subsequent soil profile development before further erosion (short cycle). The solid products of weathering would thus expect to be subjected to weathering several times over before appearing in the continental denudation equation as runoff.

The balance in Figure 2.1 is modified further by climate. In drier situations, the soluble material fraction A may precipitate to give salts, especially calcium carbonate. This would increase component C at the expense of A. Whilst the relative importance of chemical and physical weathering products remains the same, the ratio of dissolved to suspended load in any runoff is much reduced.

## 2.1.4. Quantitative Considerations

Geologists have been interested in worldwide sedimentation estimates for many years. The recent text by Garrels and Mackenzie (1971) summarizes more recent developments in this field. Table 2.1 (a) enumerates data on dissolved and suspended materials presently being delivered each year to the oceans. Overall, the ratio of dissolved to suspended load is about 1/4.

Table 2.1 (b) shows that the most common anion in dissolved load is bicarbonate. This in considerable part is derived by photosynthesis. A significant fraction of the total $Na^+$ and $Cl^-$ is recycled directly as aerosols from the ocean. Neither component is derived directly from parent rock material in weathering so that the ratio E/F is a slight overestimate of the importance of solution products.

Tables 2.1 (a) and 2.1 (b) give some idea of the magnitude of the weathering processes and (as will be shown below) the style of weathering reactions. Some estimate of the relative importance of dissolved and suspended load also is given. Surprisingly, however, there are no comparable estimates of the composition of suspended load in terms of different mineral constituents (at least, none is known to me). About the only way of deciding the relative importance of fractions B, C and D in Figure 2.1 seems to be to estimate the percentages of the three major sediment lithologies both now (very poor estimates available, although these should improve dramatically as results emerge from the Deep-Sea Drilling Project) and in the geological column. Table 2.1 (c) indicates that true physical weathering products (sands) are minor in relation to chemical (carbonate) and chemically modified (clay) products.

All of these deliberations seem to suggest that chemical products of weathering (items A, B and C, Figure 2.1) far outweigh physical products in terms of mass. Again the importance of chemical processes is underlined, although it might be wise to point out at this early stage that involvement of the biosphere is very great indeed.

## 2.2 Chemical reactions

### 2.2.1 Preamble

The basic approach of a physical scientist towards an understanding of specific chemical reactions may be resolved into two essentially different components. In the first, systematic study of the properties of atoms of individual chemical elements leads to a general theoretical framework for bonding and reaction mechanisms at the atomic or molecular scale. This allows for prediction of unknown reactions or the response of given reactions to environmental modification.

The second treatment relies on measuring bulk physicochemical properties of chemical compounds and systems. From here, reactions are analysed in terms of energy changes. Energy is the key to stability. Once the energetics of reactions are worked out, both direction and rate may be predicted and environmental controls identified and quantified.

### 2.2.2 Types of chemical bonding

All substances are made of atoms; different atoms and/or different atomic arrangements define different substances and their physical properties. The three basic states of matter (solid, liquid, gas) illustrate the direct link between macroscopic property and atomic arrangement. In gases, the bonds between atoms (or small groups of atoms bound strongly together: molecules) are very weak indeed and random thermal energy available at normal temperatures keeps them widely separated in space and in random and violent motion relative to one another. In solids, by contrast, each atom is bonded firmly to all its neighbours and thermal energy is confined to oscillations of very limited magnitude. In crystalline solids (as opposed to glassy substances) order within the structure is very great with similar atoms or atomic groups occupying identical sites throughout the lattice. Liquids are more like solids than gases (comparable densities) and possess definite structure. This has only been appreciated relatively recently and the structure of liquids is the subject of intensive research at the present time. Ordering, however, is localized and of a transient nature with frequent rupture of fairly weak bonds. Consequently most liquids have little physical strength. The boundary between liquids of very high viscosity and glassy solids, however, is transitional.

Not all chemical bonds are alike, even those of comparable strength. Four principal types are recognized (ionic, covalent, dispersive and metallic) although intermediate types are common. In the earth sciences, metallic bonds are quantitatively insignificant. This leaves three: ionic, covalent (and all intermediates) bonds which are strong, and dispersive which are weak or very weak. The latter should not be dismissed as unimportant, however, for they are responsible for the structuring of all living tissue.

Much the simplest way of explaining strong chemical bonding is to state at the outset that all atoms appear to possess particularly stable states wherein the outermost electronic shell (the 'valence' shell) contains eight electrons. Since only very few neutral atoms (i.e. atoms where the basic protonic positive charge is countered by an exactly equivalent number of electrons) have this state, most free atoms are chemically very unstable. This is neatly confirmed by the observation that the only neutral atoms with this stable electronic configuration form no chemical bonds whatsoever (even with each other) and are appropriately called the inert gases. Elements with atoms containing just one more than the stable electronic complement tend to lose one:

$$Na \rightarrow Na^+ + e$$

this leaves the stable electronic configuration but the atom now has a net positive charge equivalent to one electron (likewise $Li^+$, $K^+$, $Rb^+$, etc). Elements with atoms short of the stable complement similarly tend to accept electrons and take on a negative charge ($F^-$, $Cl^-$, $Br^-$, $I^-$, etc.). The same process leads to doubly charged ions (charged atomic species) such as $Mg^{2+}$, $Ca^{2+}$, $Sr^{2+}$, $O^{2-}$, $S^{2-}$. All these are stable as atomic species, but form stable macroscopic materials only when their very large electrostatic charges are neutralized. Chemical compounds form (MgO, CaO, etc.) in which the atoms are always present in strict numerical ratios (thereby allowing for exact charge neutrality) as indicated by written formulae. Wherever the atomic structure is such that ions with one or two electronic charges can form, compounds are found in which the bonding is essentially electrostatic. This is referred to as ionic bonding.

With three or more electrons required to modify an atomic structure to the stable configuration, a different result is observed (or, at least, inferred). Adjacent atoms 'share' electrons. These become localized between the atoms in question and form covalent bonds. Covalent bonding can also be adopted, in special cases, by atoms which, in other situations, form ionic bonds. Oxygen is the classic example. This is particularly important in the earth sciences since oxygen is far and away the most abundant element in the earth's crust and mantle.

There are thus two distinctly different types of strong chemical bond that can form between atoms to yield compounds: covalent and ionic. As might perhaps have been anticipated by the cynical, however, virtually all strong bonds in nature are somewhere intermediate in character between these two 'end members'. Fortunately it is possible to predict, with fair certainty, whether a bond between two given atoms will be relatively more ionic or covalent. Such predictions will be shown to be a very useful guide to the behaviour of many substances in the natural environment. Table 2.2 lists estimates of percentage ionic character for element–oxygen bonds together with ionic radii as determined directly from X-ray crystallographic measurements. This shows that bonds between metals such as potassium, sodium, calcium and magnesium and oxygen can be reasonably thought of as being electrostatic in character. Perhaps surprisingly, the same table shows that carbonate and silicate linkages

Table 2.2. Crystal ionic radii and percentage ionic character estimates for bonds between oxygen and some common elements

| Element | Ion | Ionic radius (angstroms) basis $O^{2-} = 1.40$ | Percentage ionic bond character |
|---|---|---|---|
| Aluminium | $Al^{3+}$ | 0·53 (VI) | 60 |
| Barium | $Ba^{2+}$ | 1·60 (XII) | 84 |
| Calcium | $Ca^{2+}$ | 1·12 (VIII) | 79 |
| Iron | $Fe^{2+}$ | 0·77 (VI) | 69 |
| Iron | $Fe^{3+}$ | 0·65 (VI) | 54 |
| Magnesium | $Mg^{2+}$ | 0·72 (VI) | 71 |
| Manganese | $Mn^{2+}$ | 0·82 (VI) | 72 |
| Potassium | $K^+$ | 1·60 (XII) | 87 |
| Sodium | $Na^+$ | 1·16 (VIII) | 83 |
| Strontium | $Sr^{2+}$ | 1·39 (VIII) | 82 |
| Titanium | $Ti^{4+}$ | 0·61 (VI) | 51 |
| Carbon | * | | 23 |
| Phosphorus | * | | 35 |
| Silicon | * | | 48 |
| Sulphur | * | | 20 |

*These bonds are more covalent than ionic.
Sources: Radii from Shannon and Prewitt (1969). Effective ionic radii in oxides and fluorides, *Acta Crystallographica*, **B25**, 925–946. Parentheses include coordination number of the ion in common structures, i.e. number of nearest neighbouring ions; Percentage ionic character estimates from Krauskopf (1967, Appendix IV).

are much more covalent with definite bonds (as opposed to electrostatic attraction) between the atoms.

Elements such as titanium, aluminium and iron (ferric) form intermediate (hybrid) bonds. These exhibit distinctive behaviour in the hydrological cycle.

### 2.2.3 Crystal structures

Chemical compounds between atoms which readily form ions have structures which should simply reflect close packing (i.e. closest approach of positively and negatively charged spheres). Very simple mathematical analysis shows that specific structures can be predicted on the basis of the relative dimensions of positively and negatively charged ions (hence interest in ionic radii, as given in Table 2.2). These predictions are very successful for elements forming largely ionic bonds. All the predicted structures are simple; for example cubic rock salt (NaCl) and fluorite ($CaF_2$).

Conversely, silicate structures are very complex and predictions based on ionic bonding get absolutely nowhere. It is worth noting here that many mineralogy and geochemistry textbooks are very misleading in this context. The observed coordination numbers of many ions in complex silicates (coordination numbers give the number of ions immediately surrounding an ion in a structure) correspond with those predicted on the basis of ionic radius ratios. It is unreasonable, however, to conclude from this that ionic (that is electrostatic) forces are responsible for the overall structure. Most silicates are grossly anisotropic (i.e. they have crystal structures and hence physical properties which vary with direction). The charge on an ion, however, is spherically symmetrical. Complicated structures must reflect covalent bonding where individual atoms adopt definite spatial configurations of bonds which certainly are not spherically symmetrical.

A realistic and very useful picture of most naturally occurring silicates is to assume that the crystal structure is made up of two distinct parts. Silicon (and aluminium) atoms linked by bridging oxygens form a covalent skeleton which is negatively charged. Metals such as sodium, potassium, calcium and magnesium are present as positively charged ions (cations) which neutralize the negative skeletal charge. Bonding between them and the silicate framework is essentially electrostatic and close-packing principles apply.

Figure 2.2 attempts to illustrate these ideas. Ionic structures result simply

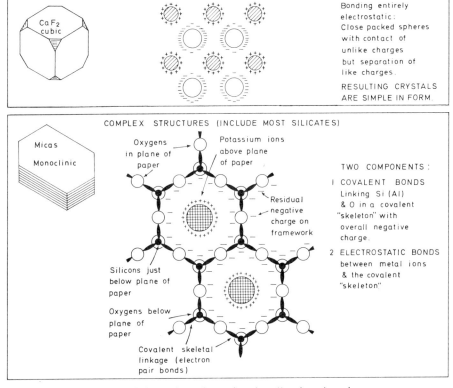

Figure 2.2. Ionic and covalent bonding in minerals.

from closest-packing of spherical charge of bodies with unlike charges in contact and like charges separated. Salt (NaCl) has this type of structure but in mica, covalent silicon–oxygen bonds (four at 109°28′) and silicon–oxygen–silicon bridging bonds (at an angle greater than 109°28′) dictate the complex structure of the negatively charged framework. Structures adopted by most silicate minerals are due to the rather few ways in which it is possible to link silicons and oxygens yet satisfy the rather specific geometrical requirements of individual covalent bonds. Various silicate classes (chain silicates, sheet silicates, etc.) differ in the number of oxygens that form two covalent bonds linking adjacent silicons (maximum four per silicon atom, as in framework silicates).

### 2.2.4 Reactions: rupture and reforming of bonds

Consider a very simple reaction such as the combustion of magnesium metal in oxygen. Classically this is written:

$$2Mg + O_2 = 2MgO$$

In fact, a great deal of heat is given out which accounts for the brilliant white flame. Better, therefore is the formulation:

$$2Mg + O_2 = 2MgO + \text{energy (many joules)}$$

The equation is now correctly balanced for both mass and energy. What has happened is that the covalent bonds which formerly linked oxygen atoms in pairs to form molecules ($O_2$) and metallic bonds in the magnesium have been broken. In the process ions have been formed by electron donation and acceptance ($Mg^{2+}$ and $O^{2-}$) and an electrostatic solid formed by close packing of the ions with opposite charges in contact. More energy obviously was stored up in the covalent and metallic bonds of the reactants than in the ionic bonds of the product since the difference was released as heat. The product clearly is more stable than the reactants since energy would have to be supplied to split it up again. This kind of argument is fundamental to chemical thermodynamics (Section 2.3.2).

Defining chemical reactions as occurring when chemical bonds are broken and reform in new ways leads to some less obvious, although perfectly valid examples. When water boils, for example, one type of between-molecule bond is broken and replaced by another, weaker one. Like the example given above, this reaction (phase change) is also accompanied by an energy change: the latent heat of vaporization.

Very similar to melting (another phase-change reaction) is dissolution. When sodium chloride dissolves in water, electrostatic bonds in the ionic solid are replaced by polar (also electrostatic) bonds between individual ions and water molecules. In dilute solutions considerable energy is released but as the solution gets more concentrated, so the energy liberated by further dissolution of solid gets less until a balance is reached. This equilibrium situation is a saturated solution.

There are thus many very different kinds of chemical reaction but all are characterized by the breaking of one set of chemical bonds and the formation of new ones. A knowledge of the particular type of bonds preferred by any given element is a useful predictor of behaviour. This will be analysed below as also will be the energy changes that accompany reaction.

*2.2.5 Links between chemical bonding and aqueous solubility*

Water is a polar solvent. This means that $H_2O$ molecules, although possessing zero net electrostatic charge, have small negative charges in the centre and positive charges at each end. This has the effect of making the liquid more structured than would be the case for liquids made up of non-polar molecules. Such liquids interact readily with charged atomic species (ions) to form stable ionic solutions. Polar solvents readily dissolve ionic solids, as in the case of sodium chloride referred to above.

The ions behave as free charged species in solution although they are surrounded by closely associated water molecules—the ions are said to be solvated. Naturally each positive charge must be countered by the presence of a negatively charged ion (anion) to preserve overall electrostatic neutrality (just as in solids). Positive and negatively charged solvated ions do not, however, become firmly bound one to another as in the case of crystalline solids because associated solvent molecules much reduce the electrostatic charge density at their surface. An interesting macroscopic analogy is the inverse dependence of electrostatic force upon the dielectric constant of the medium. Water has a very high value: attractive forces would be much reduced.

These reasons lie behind the dominance of charged (as opposed to uncharged) species in all natural aqueous solutions. The bulk of the hydrosphere resides in the ocean. Seawater is a fairly concentrated ionic solution with typical properties such as high conductivity and high corrosive capability. Less obvious perhaps is the fact that freshwaters are also typical dilute ionic solutions.

Amongst metals, those which form the most ionic bonds with oxygen (potassium, sodium) are not surprisingly most soluble. Those metals forming hybrid bonds (titanium, ferric iron and aluminium) do not form simple charged species and are much less soluble. Non-metals (carbon, phosphorous, nitrogen and sulphur) form strong covalent bonds with oxygen but the resulting 'molecule' is negatively charged and hence soluble (carbonate, phosphate, nitrate and sulphate).

When silicate minerals are brought into contact with water, therefore, those metals held by largely electrostatic forces (Na, K, etc.) might well be expected to go into solution. Certainly water is capable of holding high concentrations. Conversely, Si, Al, $Fe^{3+}$, Ti would be expected to be much less soluble. The former group of metals may be transported most efficiently by water in true solution whereas the latter group would tend to be left behind. This differentiation is one of the most important in weathering.

The data in Table 2.2 tally with another well known observation. The oxidized form of iron (ferric, $Fe^{3+}$) forms hybrid bonds much as does titanium and silicon. The reduced form, on the other hand, behaves much like magnesium. In oxygenated environments iron should be immobile and in waterlogged situations (oxygen poor) much more mobile. The presence of other substances in aqueous solutions, however, can alter this picture considerably. Hydrogen sulphide in waterlogged soils causes ferrous iron to be fixed as an insoluble sulphide. In contrast, the presence of organic molecules may enhance the solubility of otherwise highly insoluble oxides. In both cases these results are logical: metal sulphides are much less ionic than metal oxides whereas certain organic molecules react with poorly soluble metals to form organometallic complex ions which would be expected to be soluble in a polar solvent.

## 2.3 Thermodynamics and kinetics: quantification

### 2.3.1 Preamble

Centrally positioned within the logical framework of all experimental and observational science is the concept of stability. Yet, as many authors have pointed out, its definition is beset with difficulties. At the purely qualitative level things are not so bad: a substance or system may be said to be stable if it does not appear to change with time. Quantification of this theme, however, has been so difficult to achieve that two quite different approaches have been developed. Neither is sufficient, both must be considered in each unknown situation.

Chemical themodynamics deal with ultimate stability via analysis of energy changes. Less than 'ultimate stability' situations are treated by kinetic theory. Both disciplines are outlined below and links between them identified.

### 2.3.2 Thermodynamic analysis

Over the years, numerous texts have been devoted to the subject of chemical thermodynamics. Almost without exception, the better elementary treatises introduce the topic by considering physical analogies. The approach is illustrated in Figure 2.3(a), where a brick is shown in various hypothetical positions relative to a flat surface. The most stable position is clearly D, where the centre of gravity most closely approaches the surface. Position B is also stable, but less so than D because less effort is required to tip it over. Position A, C and F are all unstable (defined, perhaps, as having a spontaneous tendency to move, either by dropping or falling over). Stability, conversely, is the property of resistance to change. A stable state has no spontaneous tendency to change with time.

The continuous line joining centres of gravity in different positions A to F in Figure 2.3(a) can be considered to represent all possible states of orientation of the brick relative to the flat surface. Another physical analogy which can be better linked with a stability 'continuum' is given by Figure 2.3(b). This

(a)

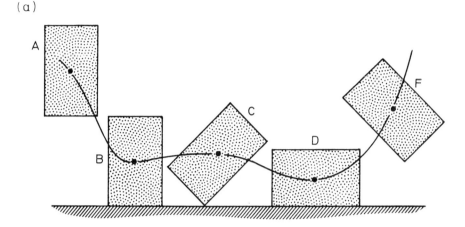

(b)

(c)

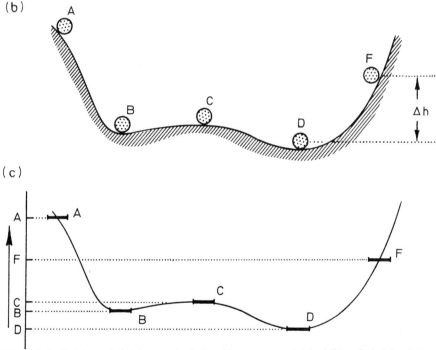

Figure 2.3. Stability: definitions and relationship to energy: (a) Stability of a brick relative to the floor: positions A, F and C are unstable; positions B and D are both stable but D more so than B since more work is needed to disturb it (i.e. D—stable, B—metastable). (b) Stability of a spherical boulder relative to smooth hillsides—precisely as in (a), the continuous line representing all possible energy states. The stable state is that of minimum height. Metastable states are local minima. *Force* on boulder of mass $m$ is $m.g$, where $g$ is the gravitational acceleration. *Work* needed to raise boulder from D to F is $m.g.\Delta h$. (c) Stability and energy. Energy is the capacity to do work. Position F has $m.g.\Delta h$ more energy than position D (potential energy). The stable state is that with minimum potential energy. The vertical scale is thus an energy scale.

could represent a boulder at various positions in a valley profile (double valley). Again all possible energy states are represented. B and D are stable states (the only two, as before) whilst all the others are unstable. Position D is more stable than B because the boulder would have to be rolled further uphill to push it into the second, higher valley. This means more work and that is obviously related to height difference. The amount of work is easily quantified: it is

mass × acceleration ( = force) × distance of application (of force)
= $m \times g \times \Delta h$ (height change)

Energy is defined as capacity to do work. The units of both work and energy are the same. The 'potential energy' of boulder mass $m$ at position F is $m.g. \Delta h$ since this is the maximum amount of energy that can be liberated in the system (i.e. by rolling down to position D, the lowest point). Obviously this is equal to the work that would have to be done pushing the boulder uphill from D to F.

This leads to a quantification of all the states shown (and all intermediate states) and also to a definition of maximum stability. This is when the potential energy of a state is zero (absolute basis) or a minimum (an arbitrarily defined system). The boulder at E has $m.g. \Delta h$ more energy than the boulder at D:

energy of boulder at F = energy of boulder at D + $m.g. \Delta h$ (joules)
or, for the change of state from F to D (downhill)
$\Delta E$ (energy change) = $-m.g. \Delta h$ (joules)

The negative sign denotes a loss of potential energy.

Thus, for any hypothetical movement, $\Delta E$ is negative for spontaneous motion towards a more stable state. $\Delta E$ must always be positive for any movement away from a stable state (away from position D, Figure 2.3). The same is true of motion away from position B although this is not such a stable state as position D. The distinction is recognized by referring to positions such as B as metastable states. True equilibrium is maintained in the stable state. Equilibria may also be defined in terms of the differential of energy with respect to the motion (or reaction) coordinate. Obviously the first differential is zero in stable states but the second differential is required to distinguish between unstable state C and states B and D.

The discussion has defined a whole series of terms (stable, unstable, mestastable, equilibrium) which obviously are reasonable, in so far as the physical models are concerned. The whole point of this analysis, of course, is that precisely the same kind of treatment suffices to describe chemical reactions. All substances possess energy tied up in their structures which is not unlike the potential energy tied up in a boulder sitting high on a hillside. A chemical system comprising several different substances has a total energy equal to the arithmetic sum of its individual chemical parts. In a chemical reaction energy may be liberated (or absorbed) exactly as in the case of boulder movement. Energy is often (although not exclusively) liberated as heat (much like the boulder again). In fact the boulder story is not really an analogue: mechanical work, heat and chemical energy are precisely interrelated by the laws of thermodynamics.

For the purpose of the present discussion, the important expression of this is that, in any spontaneous chemical reaction between reactant substances M and N to give product substances P and Q, there is a net release of free energy $\Delta G_r$:

$$M + N = P + Q + x \text{ (joules)}$$
$$\Delta G_r = -x \text{ (joules)}$$

Note the similarity between these equations and those representing movements of physical systems. The basic calculation of chemical thermodynamics is therefore to compare the total free energy of reactants with that of the products in any hypothetical reaction. Free energy values associated with specific substances are referred to as free energy of formation values, $\Delta G_f$:

$$\text{i.e. } \Delta G_r = \Sigma \Delta G_f \text{(products)} - \Sigma \Delta G_f \text{(reactants)}$$

Should the free energy of the products turn out to be less than that of the reactants, energy must be liberated ($\Delta G_r < 0$, negative). This condition demonstrates that the products would be stable relative to the reactants and that the forward reaction would be spontaneous. Free energy values for different substances can be measured directly and are tabulated in reference works of physical and chemical constants.

The free energy (in joules or, in older notation, calories or kilocalories) of any substance clearly depends on the physical state of that substance and the prevailing environmental conditions. The free energy of solid sodium chloride, for example, could hardly be anticipated to be the same as that for a saturated salt solution or for a dilute solution. The chemical bonds are different in each case. Dependence upon temperature and pressure must also be anticipated.

The framework of chemical thermodynamics thus envelops the measurement of different substances and determination of their dependence upon various environmental parameters (total pressure, temperature, component partial pressure in gas mixtures and solution concentration—note that pH is one such solution concentration variable). Sediment geochemists continue to gather together free energy values for a wide range of natural mineral substances. Calculations using these basic data provide complete descriptions of hypothetical systems under defined environmental conditions. The work of R. M. Garrels is quite outstanding in bringing this whole approach to the attention of earth scientists. Literally hundreds of research papers have since used the approach as an aid to the interpretation of mineral assemblages in terms of coexisting aqueous environments, particularly in the case of ancient or transient environments. All seriously interested workers should consult the text by Garrels and Christ (1965) which is not simply an exposition of theory but is a working manual wherein the reader is taken through a carefully selected series of calculations relating to commonly occurring natural environmental situations. Quantitative considerations of this kind are the substance of applied chemical thermodynamics. Unfortunately, space does not permit examples to be given here. Nevertheless some indication of the method can be given by qualitative reference to systems which are relevant to weathering.

Free energy data are available for gibbsite ($Al(OH)_3$), kaolinite ($Al_2Si_2O_5(OH)_4$), quartz ($SiO_2$), muscovite ($K_2Al_4Si_6Al_2O_{20}(OH)_4$), amorphous silica and dissolved species of Si, Al, K in water. The equilibrium concentration of dissolved species (including $H^+$ pH) with each of the above solids can be calculated at, say, 25 °C and 1 atmosphere total pressure. Different values are obtained in each case and also for the various combinations (such as kaolinite plus gibbsite, quartz plus muscovite). Some interesting results follow immediately: the assemblages gibbsite plus quartz and gibbsite plus muscovite cannot coexist stably with water: kaolinite would form by reaction in both cases.

But what of the relevance to weathering? The calculations show that for gibbsite to be stable (i.e. conditions under which gibbsite could form), the silica concentration in the coexisting aqueous solution must be extremely low and the pH must be near 7·0. Such waters are found in tropical leaching soil environments where, of course, gibbsite is found to occur. At slightly higher silica concentrations (as maintained in slightly less well-drained sites or, perhaps, lower in a given profile) kaolinite becomes stable relative to gibbsite. A numerical solution of this case is given in Curtis and Spears (1971).

Now should the pH be maintained at some very low value (2 to 4—as, for example, in high-latitude peat soils with continuous production of organic acid anions), the relative solubility of silica and alumina is reversed. The residuum of leaching would then be anticipated to be quartz-rich. This, of course, is found and constitutes a good example of environmental control on weathering process and products.

One quite separate point is worth making. Entropy, like energy, is a fundamental property of matter. It is measured directly in the laboratory and all thermochemical calculations of free-energy change and stability incorporate experimentally determined entropy values. The units of entropy are joules per degree. The definition and first appreciation of entropy came from heat flow measurements. More recently entropy has been linked with order–disorder relationships through the discipline of statistical thermodynamics. The link is quantitative and order–disorder relationships apply to atomic-scale phenomena.

In relatively recent times it has become fashionable to take logical frameworks of analysis and apply them to very different fields from those in which they were developed. Such a procedure may be useful and therefore must be acceptable to science as a whole. On the other hand, great care should be taken to define terms rigorously or to adopt new terminology, otherwise confusion will surely result amongst succeeding generations of workers. Entropy and energy have been clearly defined in the literature of physics and chemistry. Units are joules and joules per degree respectively. It is not reasonable to define entropy in terms of macroscopic disorder that cannot be quantified in real units. I have been given some alarming definitions of entropy by students who had no appreciation whatsoever that the term had been used in a quantitative sense to describe everyday phenomena for a very long time past.

## 2.3.3 Kinetics

Chemical thermodynamic analysis is concerned with equilibrium relationships and the identification of stable (or unstable) materials or systems. In natural situations, however, reacting systems are being observed in which chemical reactions have not proceeded to completion. The state of such systems will depend upon the various rates at which different reactions are proceeding. In most cool, humid climatic environments, for example, quartz tends to be concentrated in the soil profile indicating its 'resistate' character. Thermodynamic arguments, however, suggest that quartz is unstable and should dissolve.

The real situation is that quartz is unstable and dissolving but less rapidly than most other minerals in the same environmental situation. With time, therefore, quartz will be relatively enriched in the residuum. Another way of looking at the situation is to consider that different mineral grains take different times to dissolve completely, i.e. they have different residence times. For a bedrock with three minerals present in equal amounts but with different soil residence times, weathering will start with unequal rates of removal and a trend of relative enrichment of minerals in proportion to their residence times. Eventually a 'steady state' situation will be reached in which the input of disaggregated bedrock minerals (solid) into the base of the profile will equal the net removal of the same materials (in solution) from the whole profile. The slowest dissolving minerals will then be present in greatest amounts (assuming zero soil removal).

From this analysis it must be clear that kinetic factors alone can account for development of soil profiles markedly different in total composition from their source materials. Fortunately, the study of rates of chemical reactions is far advanced and it is fairly easy to draw general conclusions concerning factors which affect rates and, more important still, to see how these relate to the natural environment.

Chemical kinetics is concerned with the rate and mechanism of chemical change (Stevens, 1965). Reactions are generally classified as being homogeneous, that is occurring within a single chemical phase, or heterogeneous, where more than one chemical phase is involved. Weathering reactions almost exclusively occur via an aqueous phase and involve reactions at mineral surfaces. They are therefore heterogeneous. Rates are measured as the change of concentration of either products or reactants with time ($g.m^{-3}.s^{-1}$).

The rate of a chemical reaction depends upon a number of things. Perhaps most important is the inherent instability of the reacting system. In addition, however, all reaction rates depend upon the concentration of reactants. This amounts to the surface area of solids in mixed solid/solution systems and the partial pressure of individual gases within a gaseous phase. Reaction rates are also markedly dependent upon temperature. Dramatic changes in reaction rates can be accomplished by the addition of catalysts or inhibitors. These are defined as substances which take part in reactions but are not themselves altered. They are effective by altering the reaction mechanism.

Some of these factors are illustrated in Figure 2.4 which also attempts to identify the all-important link between thermodynamic and kinetic reasoning. Movement of the boulder (compare Figure 2.3(b)) from metastable state B to stable state D releases potential energy $y$ joules. Before that energy can be

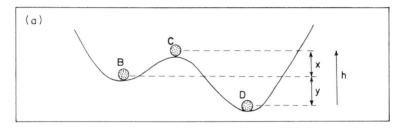

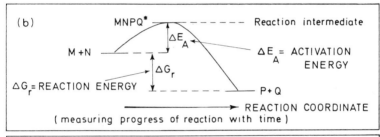

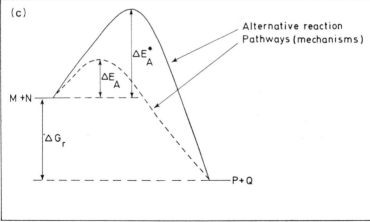

Figure 2.4. Reaction mechanisms and energy changes: (a) Energy profile as in Figure 2.3; (b) Activation energy diagram for reaction from metastable state (B) to stable state (D); (c) Alternative reaction mechanisms; rate dependence. $\Delta E_A^*$ = difficult pathway slower reaction—as in DRY reactions. $\Delta E_A$ = easy pathway faster reaction—as in WET reactions. Note zero dependence of rate on free energy of reaction $\Delta G_r$.

released, however, a certain amount of potential energy ($x$ joules) must be given to the boulder by pushing it up to the crest (position C). This amount is released again, of course, on the other side so that the net change at the end of reaction will be $y$ joules. It is also obvious that this value is unrelated to and gives no indication of the height of the barrier separating states B and D. The ease with which the reaction can proceed is reflected in the potential energy (height) barrier whilst the relative stability of the two states is given by the overall potential energy difference $y$ joules.

The second diagram in Figure 2.4 is a typical reaction diagram with the ordinate expressing progress of chemical reaction. For reactants M and N to produce stable products P and Q, an activation energy barrier $\Delta E_A$ must be overcome. The net energy change is $\Delta G_r$ as before. Thermodynamic arguments relating to stability and equilibrium involve $\Delta G$ measurements and calculations. They establish whether a particular reaction will occur spontaneously or not. Activation energy measurements, on the other hand, are the basic data of chemical kinetics. The rate of any chemical reaction is a function of $\Delta E_A$ for any given mechanism. Catalysts speed up reactions by providing an alternative reaction mechanism with lower $\Delta E_A$ value.

Numerous experimental investigations of the nature of the dependence of reaction rate on temperature have been undertaken and a single general relationship has been obtained:

$$\text{rate} = A . \exp(-\Delta E_A / RT)$$

where $A$ is a rate constant, $T$ is the absolute temperature, $E_A$ the activation energy required to overcome the energy barrier (Figure 2.4) and $R$ the gas constant. This is the Arrhenius equation. It is the basis for the approximate rule that the rates of chemical reactions tend to double for each ten degrees Celsius rise in temperature.

Here, then, is a very simple link between environment and reaction. It is to be anticipated that chemical weathering reactions will proceed something like ten times faster in hot regions ($\sim 30\ °C$) than in cold regions ($< 0\ °C$), simply on the basis of temperature difference.

The Arrhenius equation includes another important variable. In Figure 2.4(c) two alternative pathways are shown for a hypothetical reaction. If both were available, reaction would proceed by the lower energy pathway. The rate, according to the Arrhenius equation (a small $\Delta E$ value) would be greater. Modification of pathway in this sense is called catalysis.

A simple example of catalysis is provided by the oxidation of iron:

$$4Fe + 3O_2 = 2Fe_2O_3 + x \text{ joules}$$

Although spontaneous, this reaction proceeds very slowly in a totally dry atmosphere. The presence of moisture, however, dramatically accelerates reaction with very obvious and detrimental results.

Water plays an equally important role in weathering. Rapid chemical alteration occurs only when water is present. Continuously humid climate zones should (and, of course, do) show extensive weathering. Arid climates show very little.

For essentially the same reason, vegetation cover contributes enormously to weathering simply by providing a soil cover which holds water. Chemical weathering on exposed granite domes must be very slight compared with that on surrounding vegetated slopes.

Water, however only acts in the liquid state. Ice is not an active medium for interaction. Humid zones with extensive permafrost or extensive surface freezing will behave as arid zones: an aqueous phase is effectively absent. This factor may be more important than the general slowing effect of low temperatures. All these factors come under the general heading of catalytic effects: water present as a liquid phase is an enormously effective catalyst.

*2.3.4 Summary*

The preceding sections have dealt, albeit in a very general way, with the chemist's two most important theoretical approaches to the study of chemical reactions. The thermodynamic approach assesses the stability of a given mineral in a given environment relative to possible reaction products. In weathering reactions these tend to be hydrous solids and aqueous (dissolved) species. The precise nature of these reactions will be outlined in the following sections but suffice it to say that rapid removal of soluble products (maintained low concentrations of ions in soil waters) will stabilize hydrated aluminosilicates (clays, etc.) relative to minerals of igneous or metamorphic origin.

Kinetic considerations identify high temperatures and the presence of water as favouring rapid weathering reactions. Figure 2.5 demonstrates that these two separate theoretical approaches to chemical reaction mechanisms and controls may be linked simply and logically with well-accepted views of factors affecting soil formation based on field observations and classification. This diagram will be referred to again after detailed discussion of reaction mechanisms.

**2.4 Solids: a simple approach**

*2.4.1 Persistence of primary solids in weathering*

One aspect of weathering that has received a lot of attention is the resistance of particular minerals to chemical breakdown. The classic paper in the field was that of Goldich (1938) who studied soils on granite, diabase, amphibolite etc. and found that olivine and calcium plagioclase (high temperature and pressure minerals) weathered much more rapidly than low temperature minerals such as muscovite and quartz. There appeared, in fact, to be an almost perfect negative correlation between mineral persistence in soil profiles and the temperature and pressure of the inferred environment of formation of the mineral.

In qualitative terms, it is not difficult to elucidate the general principle:

Table 2.3. Energy released in some weathering reactions. The very high energies associated with oxidation ($Fe^{2+}$, C and H) should be noted

Standard free energy of reaction values: $\triangle G_r^0$

| Starting mineral | | Products | Energy change kJ g atom$^{-1}$ |
|---|---|---|---|
| (a) Simple 'rearrangement' reactions | | | |
| Diopside | $CaMgSi_2O_6$ | $Ca^{2+}$, $Mg^{2+}$, quartz | −2·72 |
| Anorthite | $CaAl_2Si_2O_8$ | $Ca^{2+}$, kaolinite | −1·32 |
| Albite | $NaAlSi_3O_8$ | $Na^+$, kaolinite, quartz | −0·51 |
| Kyanite | $Al_2SiO_5$ | Kaolinite, gibbsite | −0·54 |
| (b) Silicate oxidation plus rearrangement | | | |
| Fayalite | $Fe_2SiO_4$ | Haematite, quartz | −6·58 |
| (c) Highly reduced phases | | | |
| Pyrite | $FeS_2$ | Haematite, $SO_4^{2-}$ | −17·68 |
| Methane | $CH_4$ | $CO_2$, $H_2O$ | −20·54 |

the more similar the environments of formation and weathering, the less susceptible to weathering will the minerals be. Many workers, however, have tried to quantify this relationship in terms of the crystal structures of different rock-forming silicate minerals. Loughnan (1969), at the end (p. 60) of an excellent survey of such attempts, concluded 'the precise role played by the crystal structure in determining the rate of decay of a specific mineral is still far from understood'.

One way of assessing (on a comparative basis) the relative stability of minerals in weathering reactions is to calculate the energy change involved when different minerals alter to stable, or relatively stable, weathering products. A few examples are listed in Table 2.3. The calculations are based on standard state conditions (25 °C, 1 atmosphere total pressure and unit activity solutions) which are clearly not too realistic for many soil situations. Nevertheless, they are adequate for a comparative study. The energy changes are quoted as kilojoules per degree per gram atom: again a reasonable basis for comparison amongst silicates.

In simple 'arrangement' reactions of primary silicates to quartz, clays and aqueous solutions, small negative energy changes correlate with fairly stable minerals, larger values with less stable minerals. This is as it should be since a zero value would indicate equilibrium and zero tendency to weather. Reactions involving oxidation of iron, however (fayalite) liberate more energy. This accords with observations of universal instability of iron minerals in the surface environment. The very high energy associated with oxidation of pyrite (comparable with hydrocarbon combustion) is in keeping with a spontaneous ignition tendency in coal tips rich in pyritic shales.

There is thus a general correlation between resistance to weathering and calculated energy changes for typical reactions. Evaluation of the true significance of this correlation must await better thermochemical data and more detailed calculations.

*2.4.2 New mineral species*

There is no space here to undertake a description of minerals found in soils which are stable in the weathering environment. Clay mineralogy alone is an extremely complicated field. It is possible, however, to make some generalizations about the kinds of materials formed under earth surface conditions, namely low total pressure (1 bar), low temperatures ($-40$ to $+40$ °C) and high partial pressures of oxygen and water vapour. There are environments where oxygen is excluded (waterlogged, organic-rich) but, even in the most arid climates, water is present in significant amounts.

Therefore, soil minerals can be expected to be more hydrous than rock forming minerals. As a group they tend to be mixed oxide–hydroxides of elements forming intermediate ionic–covalent bonds: silicon, aluminium, iron (in the oxidized ferric form) and titanium (Table 2.2). Clay minerals such as kaolinite, $Al_2Si_2O_5(OH)_4$, hydroxides such as goethite $FeO.OH$ and gibbsite $Al(OH)_3$ are typical. These tend to be concentrated in soils where there is a large aqueous throughput to carry away soluble metal cations and anions and prevent their concentrations from building up in the soil solutions.

In less leaching environments, where cation concentrations do build up, cation bearing clays such as illite and montmorillonite will form. In arid situations where soil waters are maintained as highly concentrated ionic solutions, cation clays will form together with precipitated salts. Caliche soils are the most widespread examples of this condition wherein porewaters are calcite saturated.

## 2.5 Chemical formulation of weathering reactions

*2.5.1 Subtractive calculations from soil profiles*

Most early approaches to the problem of formulating weathering reactions relied on identification of *in situ* soil profiles and comparing the soil 'residuum' with parent rock material. Particularly clear examples often turned out to be cases of rapid tropical (humid) weathering of recent to sub-recent basic lavas and ashes. A good example is that depicted in Table 2.4 which is taken from Bunting (1965).

The first thing to notice is the very great extent of overall chemical change. No other process in the whole field of earth sciences (including fractional crystallization of magmas) achieves such dramatic differentiation. Water is

Table 2.4. Tropical weathering—Kaui, Hawaii (22°N) (Bunting, 1965)

|  | $SiO_2$ | $Al_2O_3$ | $Fe_2O_3$ | CaO | MgO | $K_2O$ | $Na_2O$ | $TiO_2$ | $H_2O$ |
|---|---|---|---|---|---|---|---|---|---|
| Soil | 9·2 | 24·4 | 35·8 | 0·3 | 0·3 | 0·21 | — | 6·89 | 15·0 |
| 'Subsoil' | 9·9 | 28·9 | 35·4 | 0·2 | 0·2 | 0·06 | — | 5·54 | 17·1 |
| 'Mantle' | 32·8 | 24·0 | 21·0 | 3·8 | 2·4 | 0·21 | 0·54 | 3·54 | 10·3 |
| Basalt | 49·0 | 13·7 | 13·2 | 7·3 | 13·5 | 0·27 | 1·62 | 1·73 | 0·4 |

added to the system and aluminium, iron (in the oxidized form) and titanium are much concentrated relative to the starting material. On the other side of the equation sodium is effectively removed and magnesium, calcium, potassium and silica much depleted.

The overall process is leaching. Kaolinite is formed together with hydrated oxides of aluminium and iron at the expense of metal silicates such as plagioclase feldspar ($CaAl_2Si_2O_8$) with metal cations lost in solution. Since water is added to the system continuously it is tempting to formulate weathering reactions as hydration reactions:

$$CaAl_2Si_2O_8 + 3H_2O \rightarrow Al_2Si_2O_5(OH)_4 + Ca^{2+} + 2OH^-$$
anorthite + water → kaolinite + solution

or

$$2NaAl_2Si_3O_8 + 3H_2O \rightarrow Al_2Si_2O_5(OH)_4 + 4SiO_2 + 2Na^+ + 2OH^-$$
albite + water → kaolinite + quartz + solution

These reactions would cause all soil waters to be alkaline. Since they most certainly are not, it is reasonable to enquire just how realistic such equations are. The obvious experiment to perform is to go out and analyse the surface waters in soils or those that drain out from them.

*2.5.2 Restraints imposed by the composition of natural aqueous solutions*

It is the usual practice to assess the dissolved load of streams and porewaters by measuring electrical conductivity. As has already been pointed out, this assumes that all dissolved species are charged, which is not entirely correct. It is more important, however, to repeat and emphasize that ionic solutions, just like solids, must exactly preserve electrostatic neutrality. In other words, every cation removed in weathering must be accompanied by an anion. If hydration were the key process, the anion would be hydroxyl and the solution alkaline. Dissolved silica is uncharged and could not affect the charge situation in neutral or acid solutions.

Most weathering solutions, however, are neutral to acid or even very acid. In these cases there must be an excess of anions over metal cations such that charge compensation is achieved by hydrogen ions ($H^+$). The inevitable conclusion to be drawn from this very simple argument is that anions play an extremely important part in weathering and may even be the dominant control.

The composition of rivers draining the world's continents (Table 2.1(b) shows that, in those waters at least, bicarbonate ions ($HCO_3^-$) dominate the anion fraction. Sulphate ($SO_4^{2-}$) ions are next most important and chloride ions next ($Cl^-$). These three account overwhelmingly for the negative charge fraction in natural waters. This is not quite the case in soil waters, but discussion of this is held over until later.

Starting with the least important anion, a significant fraction of the total $Cl^-$ in world runoff is contributed by precipitation or dissolution of evaporite beds in weathering. Neither component contributes to weathering since both are already matched by cations. Some sulphate is likewise introduced by dissolution of gypsum and anhydrite. The greater part of the sulphate, however, enters the hydrosphere as a result of oxidation, in weathering, of metal sulphides, especially pyrite:

$$4FeS_2 + 15O_2 + 8H_2O \rightarrow 2Fe_2O_3 + 16H^+ + 8SO_4^{2-}$$
$$\text{sulphuric acid}$$

In this case, the anions are not accompanied by cations and liberation of metals in weathering reactions is an obvious possibility. Black pyritic shales, therefore, carry within themselves the seeds of their own destruction. That this is realized is evidenced by the extreme alteration of shales in tips or mine workings.

Bicarbonate is always present in weathering solutions. The bicarbonate ion therefore is the primary agent of weathering on all types of rock, silicate as well as carbonate. It is formed partly by respiration of plant roots and partly as a result of bacterial degradation of plant debris. Photosynthetic fixation from atmospheric carbon dioxide is the precursor stage of the overall reaction. Soil atmospheres may be very rich in carbon dioxide. It dissolves and dissociates:

$$H_2O + CO_2 \rightleftharpoons H_2CO_3 \rightleftharpoons H^+ + HCO_3^-$$

Some carbon dioxide also many dissolve directly from the atmosphere in precipitation. The anion is produced without an accompanying metal cation and is capable of supporting such cations in neutral bicarbonate solutions.

### 2.5.3 Realistic weathering equations

From the preceding section it follows that destructive weathering of metal silicates in typical profiles maintaining neutral to acid soil porewaters is controlled to a great extent by the rate of supply of acids (carbonic and sulphuric in ionized form) to the system. In the more general and important case of carbonic acid, this will depend mostly upon the rate of botanical and microbiological productivity in the soil system (in addition to the other factors illustrated in Figure 2.5).

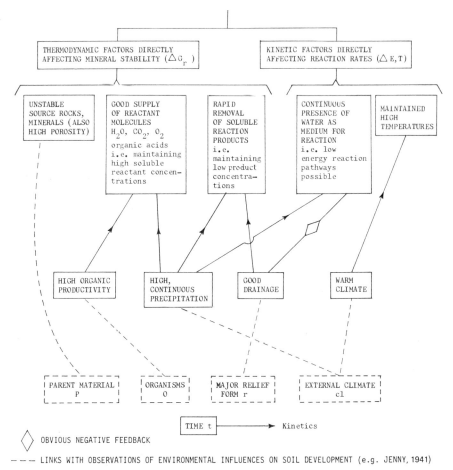

Figure 2.5. Environmental factors which directly influence the nature, extent and rate of chemical weathering.

It follows that realistic weathering equations should take into account the generation of anions and their subsequent involvement in the overall weathering process. The weathering of plagioclase feldspars may thus be represented:

$$\begin{cases} CaAl_2Si_2O_8 + 3H_2O + CO_2 \rightarrow Al_2Si_2O_5(OH)_4 + Ca^{2+} + 2HCO_3^- \\ \text{Plagioclase} \qquad\qquad\qquad\qquad\quad \text{Kaolinite} \qquad\quad \text{ionic components} \\ \text{feldspar end} \qquad\qquad\qquad\qquad\qquad\qquad\qquad\qquad \text{of resulting} \\ \text{compositional} \qquad\qquad\qquad\qquad\qquad\qquad\qquad\quad \text{solutions} \\ \text{members} \\ 2NaAlSi_3O_8 + 3H_2O + CO_2 \rightarrow 4SiO_2 + Al_2Si_2O_5(OH)_4 \\ \qquad\qquad\qquad\qquad\qquad\qquad\quad \text{quartz} \quad \text{Kaolinite} \\ \qquad\qquad\qquad\qquad\qquad\qquad\qquad\qquad\qquad\qquad + 2Na^+ + 2HCO_3^- \end{cases}$$

where $H_2O + CO_2 \rightleftharpoons H_2CO_3 \rightleftharpoons H^- + HCO_3$ equilibrium is maintained and hydrogen ions are removed from the system to form hydrated silicates with metal cations liberated therefrom. As hydrogen ions are removed, the carbonic acid/bicarbonate dissociation reaction is continuously displaced to the right.

An example of intermediate intensity weathering (something less than strongly leaching) would be the conversion of potassium feldspar to illite (muscovite formula assumed)

$$6KAlSi_3O_8 + 4H_2O + 4CO_2 \rightarrow K_2Al_4(Si_6Al_2O_{20})(OH)_4$$
feldspar + water + carbon dioxide → illite

$$+ 12SiO_2 + 4K^+ + 4HCO_3^-$$
+ quartz + solution

In humid, tropical weathering environments, however, quartz is rapidly leached away. This renders the feldspar equation given above less realistic. The problem is resolved by writing the equation to take account of complete hydration to give soluble 'silicic acid'. The transported species is the neutral $Si(OH)_4^0$ molecule. Dissolution of quartz may be written:

$$SiO_2 + H_2O \rightarrow Si(OH)_4^0$$
quartz water solution

Albite weathering under similar conditions could be formulated:

$$2NaAlSi_3O_8 + 11H_2O + 2CO_2 \rightarrow$$
albite + water + carbon dioxide →

$$Al_2Si_2O_5(OH)_4 + 2Na + 2HCO_3^- + 8Si(OH)_4^0$$
Kaolinite + Solution

The dissolution of limestone and dolomite follows the same pattern: only half the bicarbonate in weathering solutions is derived from the mineral phase:

$$CaCO_3 + H_2O + CO_2 \rightarrow Ca^{2+} + 2HCO_3^-$$
calcite + water + carbon dioxide → solution

One apparent complication is the role played by iron minerals during weathering. Pyrite weathering has already been cited as an example of a weathering reaction producing a net excess of anions. If iron is present as a silicate, oxidation by atmospheric oxygen and simple hydration will produce stable weathering products without any anion generation:

$$2FeSiO_3 + (O) + 4H_2O \rightarrow Fe_2O_3 + 2Si(OH)_4^0$$
iron silicate + oxygen + water → Iron oxide + solution

In this case, of course, no cations are removed from the system and iron accumulates in the oxidized form. If this kind of reaction is quantitatively important, the end product is a red laterite or bauxite.

### 2.5.4 Organic acids: a double contribution

It has been argued that anion generation is the key to decomposition and partial dissolution of silicates during soil formation. Bacterial oxidation of organic matter may not proceed to the ultimate stage of bicarbonate production. Organic acids represent intermediate oxidation products and are found in soils. The anions of such acids undoubtedly play an important role in mobilizing silicate cations in soil horizons. Detailed discussion of organic involvement is included in chapter 3. During surface transport however, especially in aerated rivers, it is likely that most organic anions are oxidized completely to bicarbonate. The overall weathering reactions are thus well approximated to by equations involving bicarbonate alone.

In very acid or in stagnant water masses, organic molecules may have a much lengthier existence and may contribute significantly to both mobilization and transport of metals. In these cases a new factor is introduced, that of metal complexing. Metal atoms form covalent-type bonds with single organic molecules and, as such, may have considerable solubility. Whether or not such a complex will form depends upon the relative stabilities of complex and original solid. Aluminium, which is extremely insoluble in ionic form, complexes readily with many organic molecules. Aluminium, therefore, may be leached and transported as an organic complex, but not in simple ionic solution. Movement of metals within soil profiles as complexes must be anticipated to be an important process but mass transport within surface water masses is probably limited to slow-moving organic-rich river systems.

### 2.5.5 Organic productivity: a major rate control

An interesting extension of the hypotheses outlined in the previous two sections concerns controls on the rate of chemical weathering. Leaching of metal cations demands the availability of acid anions: these are mostly generated within the soil profile. Bicarbonate (and organic acid) input must depend upon organic productivity. Environmental factors favouring high organic productivity, therefore, will also favour rapid chemical weathering. That physical weathering is seen to dominate over chemical weathering on steep alpine slopes is a consequence of low organic productivity retarding chemical weathering. Steep slopes in climates supporting high productivity often exhibit extreme chemical weathering. In this context it is essential to appreciate the very complex interplay between chemical, physical and biological factors in weathering and soil formation.

### 2.5.6 The special case of pyritic shales

Something like threequarters of the earth's land surface is covered by sedimentary rocks, of which shales constitute the most common lithology. Pyrite is a common mineral constituent of such materials and is extremely susceptible to oxidative weathering when sulphuric acid is added to the carbonic and organic acid spectrum.

Surface streams draining such areas often are particularly acid and capable of further leaching. In Britain, Namurian shales rich in pyrite frequently overlie the Carboniferous Limestone. Streams draining on to the limestone are then extremely effective at dissolving away calcite. Major cavern systems lying near to geological contacts of this type attest to the efficiency of the process.

### 2.5.7 The ultimate nature of weathering reactions

It has been argued that hydration and acid reactions (dominantly carbonation)

Table 2.5. Principal weathering reactions: summary

| | |
|---|---|
| (a) Simple hydration: solid and solution products $SiO_2 + 2H_2O \rightleftharpoons Si(OH)_4^\circ$ | Reaction of quartz with water to give silicic acid—removed in solution. |
| $Al_2SiO_5 + 5H_2O \rightleftharpoons Si(OH)_4^\circ + 2Al(OH)_3$ | Gibbsite residue from kyanite, silica lost. |
| (b) Anion generation reaction $H_2O + CO_2 \rightleftharpoons H_2CO_3 \rightleftharpoons H^+ + HCO_3^-$ | Complete oxidation (biological) of organic matter to give carbon dioxide: dissociation of carbonic acid. |
| $2S^{2-} + 4O_2 \rightleftharpoons 2SO_4^{2-}$ | Oxidation of metal sulphides: especially iron sulphides. |
| $2CH_2O + 2O_2 \rightarrow H_2C_2O_4 + H_2O \rightleftharpoons 2H^+ + C_2O_4^{2-}$ | Incomplete biological oxidation of organic matter (carbohydrate) to give organic acids (oxalic acid) |
| (c) Hydration and leaching (bicarbonate case) $CaAl_2Si_2O_8 + H_2O + 2H^+ + 2HCO_3^- = Al_2Si_2O_5(OH)_4 + Ca^{2+} + 2HCO_3^-$ | Anorthite feldspar altered to kaolinite: calcium leached away as calcium and bicarbonate ions. |
| $K_2Mg_6(Si_6Al_2O_{20})(OH)_4 + 14H^+ + 14HCO_3^- + H_2O =$ $Al_2Si_2O_5(OH)_4 + 4Si(OH)_4^\circ + 2K^+ + 6Mg^{2+} + 14HCO_3^-$ | Typical strong leaching reaction: biotite (Magnesian variety) leached to leave kaolinite soil residue; all cations and some silica removed in solution. |

are the most important weathering reactions with oxidation playing a significant supporting role. All the reactions listed in Table 2.5 involve hydrogen, hydration reactions by addition of water and carbonation reactions by exchange of hydrogen ions for metal cations (subtract bicarbonate ions from both sides of the equations).

Hydrogen is the first element in the Periodic Table and has the smallest and simplest atom of all elements. The cation of hydrogen, $H^+$ is not really an ion at all but a subatomic particle (proton). This is very much smaller indeed ($\sim 10^{-15}$ m) than typical ionic dimensions ($\sim 10^{-10}$ m). In consequence, hydrogen ions never occur 'free' but always associate with other ions or molecules. As well as simple bonds, hydrogen possesses a unique ability to form bridging bonds between electronegative atoms (such as oxygen). All this may be qualitatively summarized by stating that hydrogen has a peculiar ability to migrate in chemical systems involving oxygen. Interconversion of oxide and hydroxide ions and the dissociation of water are two important examples:

$$O^{2-} + H^+ \rightleftharpoons OH^-$$
$$\text{oxide} + \text{proton} \rightleftharpoons \text{hydroxide}$$

$$H_2O \rightleftharpoons OH^- + H^+$$
$$\text{water} \rightleftharpoons \text{hydroxide} + \text{proton}$$

In aqueous solutions, protons never exist in the free state. The dissociation of water into hydroxide and hydrogen ions (hydroxonium) is better written:

$$2H_2O \rightleftharpoons OH^- + H_3O^+$$
$$\text{water} \rightleftharpoons \text{hydroxide} + \text{hydroxonium}$$

In each case the reaction mechanism probably involves a first-step hydrogen bridge followed by electron readjustment. All the reactions listed in Table 2.5 can be thought of in these terms. Silicate hydration probably goes thus:

Close approach  Hydrogen bond  Electron migration  Hydroxide

The three steps involves close approach of a water molecule, formation of hydrogen 'bridging' bonds and then electron transfer to form two hydroxyl groups. Note that the Si–O–Si bridge, which is the fundamental structural unit of silicate skeletons, is broken.

The same process, of course, may be extended to rupture all bridging oxygen bonds such that a silicon atom is detached as a tetra-hydrate. This is the process whereby silicates dissolve to give silicic acid.

Similar reactions are also responsible for metal (M) cation liberation in carbonation reactions. Hydrogen bonding leads to hydration and cancellation of the residual ionic charge on the silicate framework, thus:

<pre>
   \|                                      \|
   Si —O⁻··                                 Si —O ⁻ H
  /        ⁻·H⁺ HCO₃⁻                      /           HCO₃⁻
 O    M²⁺                         →       O     M²⁺
  \                                        \
   Si —O⁻··                                 Si —O
  / |      ⁻·H⁺ HCO₃⁻                      / |    \H      HCO₃⁻

  Protonation (hydrogen bonds) of          Hydrate plus metal salt solution
  metal silicate by acid
</pre>

Cations formerly held by electrostatic attraction then would be released from the structure and pass into ionic solution, their charges compensated by the anions already present. These examples illustrate very clearly the unique part played by hydrogen in weathering.

The only other basic weathering reaction is oxidation.

Strictly speaking all such reactions should be referred to as oxidation-reduction or 'redox' reactions since oxidation of any substance must be compensated by reduction of another. Just as one subatomic particle can be considered basic to hydration and leaching reactions, so another, the electron, is responsible for oxidation-reduction. It is convenient to think of these reactions as two half-reactions involving electrons. Oxygen is reduced to oxide ions by addition of electrons (for which it has a great affinity) and iron is oxidized from metal to ferric (rust) by losing electrons:

$$O_2 + 4e^- \rightarrow 2O^{2-}$$
$$Fe \rightarrow Fe^{3+} + 3e^-$$

The overall reaction for oxidation of iron is obtained by adding the two half-reactions to balance electron gains with losses:

$$4Fe + 3O_2 \rightarrow 2Fe_2O_3$$

Oxidation of iron within a silicate structure therefore requires no more than removal of electrons. Iron-bearing minerals often conduct electricity fairly well showing that electrons can migrate freely. A ferrous mineral coming into contact with the atmosphere would tend to lose electrons. Schematically, this could be represented:

$$Fe(OH)_2 \rightarrow Fe(OH)_2^+ + e$$

Ejection of a proton from the charged hydroxide, however, would produce a stable mineral (goethite):

$$Fe(OH)_2^+ \rightarrow FeO.OH + H^+$$

as well as a proton which immediately might become involved in hydration reactions. The overall reaction is:

$$2Fe(OH)_2 + O \rightarrow 2FeO.OH + H_2O$$

but such a formulation gives no clue as to mechanism.

This section has been speculative. It had previously been suggested (based on reasonable evidence) that weathering, in all its complexity, really involves no more than three basic reactions: hydration, carbonation (or other acid involvement) and oxidation-reduction. It is here suggested that the fundamental reaction mechanisms may be even simpler. Protons and electrons seem to play very special roles.

## 2.6 Conclusion

This essay started with a grand overview of weathering processes in an attempt to demonstrate the immense importance of chemical reactions in surface processes. An attempt was made to clarify the rather complicated relationships between chemical and physical processes on the one hand and chemical and physical denudation on the other. Chemical processes are much more important than comparisons of solid and solution denudation rates might imply (not that they fail to underline their significance).

The various theoretical and experimental approaches applied by chemists to the understanding of chemical reactions in general were then reviewed. Two broad fields of investigation were outlined: that of considerations at the atomic scale and that of measurement and analysis of properties on a macroscopic scale. In the latter case a careful distinction was drawn between thermodynamic and kinetic stability. Examples from surface weathering and sedimentation environments were cited to illustrate the very general validity of each theoretical approach.

The specific nature of weathering reactions was then investigated by various methods. Resistance of different mineral structures was linked with energies of reaction. After a brief survey of the new minerals formed in weathering, alteration mechanisms were considered.

It was demonstrated that many (if not most) soil weathering reactions involve acids and that leaching necessitates net production of free acids. The generation of anions may well be a limiting control on soil formation.

Further analysis indicated that a small number of specific reactions (hydration, carbonation and oxidation) must be responsible for most weathering and further that these may be considered simply in terms of proton and electron transfer.

In the final analysis, therefore, this essay started on a global scale and ended up by considering subatomic particles. Global scale data collections were necessary to elucidate which chemical elements were involved and in what way. On the other hand, no chemical reaction can be explained or understood without reference to electron transfer, for electrons are the stuff of chemistry. All chemical reactions involve electron rearrangement. Knowledge is best increased by gathering data and ideas from all sources: the specialist following but one line of research lives in grave danger of misinterpreting observations. This is particularly true when complicated systems with delicate environmental balance are the subject of study. Chemical aspects of geomorphology should not be neglected in any general analysis of process or products.

## References

GENERAL

Blatt, H., G. V. Middleton and R. C. Murray, 1972, *Origin of Sedimentary Rocks*, Prentice-Hall, New Jersey.
Bunting, B. T., 1965, *The Geography of Soil*, Hutchinson, London.
Curtis, C. D. and D. A. Spears, 1971, Diagenetic development of kaolinite, *Clays and Clay Minerals*, **19**, 219–227.
Garrels, R. M. and F. T. Mackenzie, 1971, *Evolution of Sedimentary Rocks*, W. W. Norton, New York.
Goldich, S. S., 1938, A study in rock weathering, *J. Geol.*, **46**, 17–58.
Holmes, A., 1937. *The Age of the Earth*, Thomas Nelson, New Jersey.
Horn, M. K., 1966, see Garrels and Mackenzie (1971 p. 222), written personal communication.
Jenny, H., 1941, *Factors of Soil Formation*, McGraw-Hill, New York, 281 pp.
Livingstone, D. A., 1963, *Chemical Composition of Rivers and Lakes*, United States Geological Survey Professional Paper 440-G, Government Printing Office, Washington.
Loughnan, F. C., 1969, *Chemical Weathering of the Silicate Minerals*, Elsevier, New York.
Wickman, F. E., 1954, The total amount of sediment and the composition of the average igneous rock, *Geochim. Cosmochim. Acta*, **5**, 97–110.

SPECIFIC CHEMICAL OR GEOCHEMICAL TEXTS FOR FURTHER READING

Berner, R. A., 1971, *Principles of Chemical Sedimentation*, McGraw-Hill, New York.
Eglinton, G. and M. T. J. Murphy (Eds.), 1969, *Organic Geochemistry*, Springer-Verlag, Berlin.
Everett, D. H., 1959, *An Introduction to the Study of Chemical Thermodynamics*, Longmans, London.
Garrels, R. M. and C. L. Christ, 1965, *Solutions, Minerals and Equilibria*, Harper and row, New York.
Krauskopf, K. B., 1967, *Introduction to Geochemistry*, McGraw-Hill, New York.
Stevens, B., 1965, *Chemical Kinetics*, Chapman and Hall, London.
Stumm, W. (Ed.), 1967, *Equilibrium Concepts in Natural Water Systems*, American Chemical Societ Advances in Chemistry Series No. 67, Washington, DC.
Stumm, W. and J. J. Morgan, 1970, *Aquatic Chemistry*, Interscience, New York.

CHAPTER THREE

# Rock Weathering and Climate: Quantitative and Experimental Aspects

STEPHEN T. TRUDGILL

## 3.1 Introduction

Weathering systems are characteristically multifactorial and dynamic. The fact that they are dynamic means that they are amenable to experimentation in terms of input, outputs and internal processes. Furthermore, they can be characterized in terms of initial states, available energy for reactions and residual states. The multifactorial nature of weathering systems means that they must be analysed in terms of isolation of control and noise variables before the dynamics can be understood. An ideal programme of experimentation might adopt the procedure outlined below:

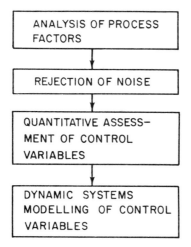

It is salutary for experiments to have clear objectives and for them to be designed to fulfil these objectives (Cochran and Cox, 1957, pp. 9–14). These objectives can only be clearly specified within an overall *a priori* framework. Energy models are becoming increasingly useful as they provide a unifying dynamic theme applicable at all levels of investigation. Such energy models have been elaborated by Chorley and Kennedy (1971) and Carson and Kirkby

(1972). Similar equilibrium models have been used in geology and are usefully outlined by Broecker and Oversby (1971). Recently, Runge (1973) has described the use of energy models which provide frameworks for the investigation of soil development sequences. He maintains that the dynamics of the soil system can be understood in terms of the factors of organic matter production ($o$), the amount of water available for leaching ($w$) and time ($t$):

$$s = f(o, w, t)$$
where $s$ = soil development. (3.1)

Furthermore, $w$ is determined by rainfall, time of rainfall during the year, runoff and relief. Thus climate ($c$) and relief ($r$) have their influence as expressed in Equation 3.2:

$$w = f(c, r) \qquad (3.2)$$

Moreover, organic matter production and accumulation may also be expressed in terms of climate and relief, thus:

$$o = f(c, r) \qquad (3.3)$$

The principal factors supplying the system with energy are organic matter production and water available for leaching. These factors, in turn, are governed by climate and relief.

This approach of Runge can be adopted for use in the rock weathering system by the addition of lithological factors ($l$), such as solubility, porosity and joint frequency. Then weathering (W) may be expressed as follows:

$$W = f(o, w, l, t) \qquad (3.4)$$

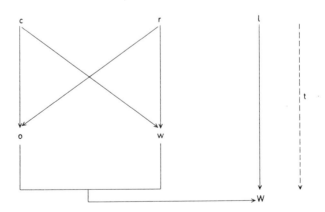

KEY
c climate  o organic matter production  l lithology
r relief   w water available for leaching  W weathering
           t time

Figure 3.1. The relationships between the weathering factors.

The expression of the climatic factor can be sought in terms of the importance of climate relative to relief and lithology in influencing the dynamic workings of the weathering system. The system can be modelled in terms of the factors discussed as shown in Figure 3.1.

Possessing such an *a priori* model of weathering systems, experiments can be designed to evaluate the importance of the various factors in controlling the system dynamics. If the models produced are of low explanatory power then greater elaboration of the models may increase the degree of explanation achieved.

While ideals, models and frameworks are important for the results of experiments to be interpretable in an overall fashion, the actual operations of the experiments are naturally controlled by feasibility. Each section that follows is thus a discussion of ideals and frameworks and also of feasibility. The reactions in weathering systems composed of porous media and those on planar surfaces are sufficiently distinct for a differentiated treatment to be given. The discussion is therefore divided into:

(1) *Leaching systems*, where alterations, transfers, removals and residua in rocks, soils and sediments, or other porous media, are considered.

(2) *Rock surface erosion systems*, where the dynamics of rock surface lowering are considered in situations where superficial covers do not occur.

## 3.2 Leaching systems

These systems are essentially ones of selective transformations and transportation within porous media. They are amenable to study in terms of climatic factors since they represent a lithological system which is tending toward equilibrium with the environmental forces acting upon it; these forces are primarily gravity, organisms, water and air. As Chorley and Kennedy (1971, p. 202) have illustrated there are many different forms which equilibrium may take but the leaching systems can be characterized in terms of adjustment to water inputs, water quality, organic factors, rock weatherability and water output. These factors adjust mutually over time tending towards a state of dynamic equilibrium. A steady state is achieved only if surface lowering and alteration by weathering proceed at the same rate so that the same weathering profile is visible through time.

Weathering is an equilibrium process in that the form of the materials involved tends to change so as to be more stable under the conditions occurring in the weathering environment. On the microscopic scale this involves chemical equilibrium. On the macroscopic scale it involves the organization of the constituents into a characteristic pattern known as the weathering profile.

Where minerals weather in the presence of gases, water and life, the equilibrium product is the clay–humus complex. Steps in the progression towards the formation of this complex can be interpreted in terms of the presence or absence of the various causal processes devolving from the presence of gases, water and life.

Equilibrium modelling can be elucidated by factor modelling. How many factors does the weathering material adjust to and what is their order of importance? Furthermore, how may weathering, and its products, vary over time? How may equilibrium reactions change as external conditions change and what transformations are rapid enough to adjust to external changes? What are the effects of feedbacks that occur within the system as it progressively alters?

Lines of analysis have proceeded along various routes, involving inductive reasoning from basic data, e.g. from chemical thermodynamic data (Garrels and Christ, 1965; Broecker and Oversby, 1971), or deductive reasoning from testing theories in the field. Simulation experiments with leaching columns are also common. This discussion will deal, first, with ideas and experiments on energy available for weathering (organic and rainwater inputs) and, second, with reactions within systems as correlated with inputs and outputs. To expand these themes extracted from literature and from theory some of the author's work on leaching systems in limestone soils will be discussed. To conclude this section, the investigation of residua will be examined.

### 3.2.1 Energy input and the potential for weathering

Forces motivating the weathering systems in porous media where leaching is dominant are the nature of the water input, gravity, organic products with chelatory or dissolving capacities and energy from nutrients stored in rocks which may be used directly by chemoautotrophic organisms and indirectly by higher plants. The motivations are thus essentially hydrological and organic. Climate is important in that the organic processes are strongly influenced by temperature and moisture and in that water input will be controlled by climate and topography.

Recent experiments continue to emphasize the organic factors in rock weathering. Processes include cation exchange, chelation and dissolution by organic acids. It appears that the capacity for such processes generally increases as decay of organic matter proceeds. The cation exchange capacity of organic appears to be related to the degree of decomposition and humic acid content (Marshall, 1964, pp. 163–164; Kononova, 1966). Schnitzer and Skinner (1963) show how stable organo-metallic compounds of Fe, Al, Ca, Mg, Cu and Ni are formed during podzolic weathering under decaying humus. Davies (1971) demonstrates how augite, hornblende and olivine can be weathered by polyphenols produced during organic matter decomposition. Fe and Mg chelates are released during the process. These processes depend upon water carrying the chelatory compounds down from the organic decomposition layer to the minerals. A process where the plants provide a more active motivating force is demonstrated by Keller and Frederickson (1952). Cation exchange between root systems and minerals establishes gradients along which the ions migrate. Hydrogen ions move from the root to the mineral, enhancing weathering by hydrolysis, and nutrient cations move back to the roots from the mineral.

In these ways the organic matter and organic processes provide the dynamic potential for weathering. Water provides the transport medium.

Further active processes have been discovered by experiments on bacteria and fungi. Kononova (1966, p. 184) lists examples of bacterial products (e.g. lactic and butyric acids) which may attack magnesium-rich minerals, chalk and other carbonates. Bacteria and fungi (Duff, Webley and Scott, 1963; Henderson and Duff, 1963), lichens (Syers, 1964; Iskander and Syers, 1972) and microbial metabolic products in general (Boyle and Voigt, 1973; Berthelini and Dommergues, 1972) have all been identified as involved in the weathering of silicate minerals.

The most recent experiments by Huang and Kiang (1972) and Huang and Keller (1973) have shown that weathering may be increased by as much as ten times where organic acids are involved. Huang and Keller shook 12 minerals, including augite, muscovite, microcline and clay minerals, in organic solvents, and showed that cations (especially Si, Al, Fe and Mg) dissolve far more rapidly than in water. Weakly complexing acids included the acetic and aspartic acid groups. The strongly complexing acids included citric, tartaric, salicylic and tannic acid groups. Organic acids in general tend to show the reverse order of dissolving powers that water does. Thus, they dissolve Al preferentially to Si whereas water containing only dissolved carbon dioxide tends to dissolve Si preferentially to Al.

Using this information on organic processes it is possible to begin to model the weathering system in terms of the organic potential provided. The two key factors which have been isolated by experimentation are:

(1) The population of micro-organisms present in porous media, their type and the reactions that they are capable of.
(2) The nature of the humus present, its cation exchange and chelatory properties.

It is relatively easy to model both of these in terms of climate as, where nutrients are available, both organic productivity and amount or organic matter decay in porous media are controlled primarily by moisture and temperature, as shown in Figure 3.2, following Kononova (1966, p. 248).

Using further information from Kononova on the nature of humus in different climatic zones it is possible to suggest models relating to how weathering may vary with climate. Tundra humus has a characteristically low humus acid content, low microbiological activity and low exchange capacity. These are highest in the chernozem zone and decrease again towards the serozem zone. Humus acid content in podzol soils is higher than in tundra soils and there is a very high fungal activity (Rode, 1970). It is known from the work of Henderson and Duff (1963) that fungi dominantly produce oxalic and citric acid. These, from the work of Huang and Keller (1973), encourage the preferential solution of aluminium, leaving a silica-rich residue behind. It can be further argued that the type of weathering can be influenced by the type of humus as

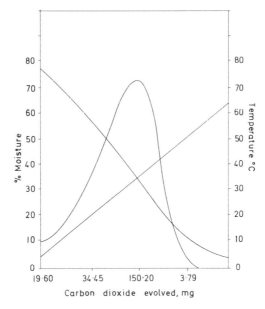

Figure 3.2. The control of organic matter decay (as measured by carbon dioxide production) by temperature and moisture (after Kononova (1966); reproduced by permission of Pergamon Press Ltd.).

organic matter decay first produces fulvic acid and then humic acid. A moderate moisture regime, a neutral pH and a temperature suitable for fairly intense microbiological activity are necessary for the progression of decay to the more complex humic acid molecules. Where these conditions are lacking, water with dissolved carbon dioxide and weakly complexing acids will be the main weathering forces, tending to dissolve Si preferentially over Al. As decay proceeds, where the conditions are suitable, increasing amounts of Al could be dissolved.

From the evidence outlined above it can be argued that the potential for weathering and the adjustment of weathering to environmental factors may operate primarily through the degree of organic matter decay and the amount of soil water. These will be controlled climatically by temperature and the precipitation/evaporation balance. While weathering type has been known to be broadly correlated with soil and vegetation conditions for some time the possibility that the system may be further understood by investigations of the composition of the humus type is in need of further substantiation.

Rainfall produces both a motivating force and a medium through which other agencies, especially the organic ones, act upon weathering material. Rainfall amount has been correlated with soil type from the time of Jenny and Leonard (1934) and Jenny (1941) through to Bryan's (1967) climosequences and, more recently, in conjunction with temperature, by Scrivner, Baker and Brees (1973). Of these works the last is perhaps the most thorough. Scrivner and coworkers have developed a model for converting long-term daily records of temperature and precipitation into daily depths of moistening and drying of the soil. These data are converted into a predictive equation for water available for weathering reactions.

The correlation of soil type with such climatic data is attractive in that it employs process mechanisms in the model. Useful overall models using the water balance of a soil profile are provided in Bridges (1971, pp. 17–25). Crompton (1960) outlines a well known model of weathering versus leaching. Rainfall, temperature, relief, runoff, microclimate (Southard and Dirmhirn 1972), freezing and drying (Hinman and Bisal, 1973) and many other factors influence weathering reactions and soil development. But the key process remains the rate at which chemical reactions occur as compared with the rate of water movement through the porous media. As such it will now be pertinent to consider weathering in terms of throughput and internal processes instead of just in simple terms of the potential provided by rainfall and the accumulation of organic matter.

### 3.2.2 System throughput—input, output and internal processes

If water flow rate is slow enough for chemical equilibrium to be achieved then the saturation value will be the controlling factor on solutional weathering in porous media. If the water flow rate is more rapid than the rate of solution then rate of flow will be the controlling factor. The importance of interaction of solution velocity (rate of achievement of equilibrium) with water flow-through time increases as flow rate increases. Thus in regions where highly transmissible media, steep relief and high rainfall occur the rate of flow and solution velocity are the governing factors on weathering. In conditions of very slow (near stagnant) water flow, reactions tend to proceed to equilibrium, chemical saturation is reached and these saturation conditions govern the weathering.

Solution velocity is the rate at which the equilibrium value is achieved, irrespective of the actual equilibrium value, and an illustrative example is given by Priesnitz (1972). Gypsum has a lower solubility than anhydrite (2500 mg/l at 15 °C as compared with 3500 mg/l) but it has a higher solution velocity, reaching equilibrium in 12 days as compared with anhydrite which had not reached equilibrium in 30 days in the experiments of Priesnitz (Figure 3.3).

In this example, if water flowed through a porous medium containing both gypsum and anhydrite at a rate of less than 20 days residence time then gypsum would be dissolved faster than anhydrite. For example, for a residence time of ten days the water would contain 3400 mg/l Ca derived from gypsum and only about 2200 mg/l from anhydrite solution. If, however, water residence time was longer than 20 days, and equilibrium values were reached, then more anhydrite would dissolve than gypsum. This has its expression in nature under humid conditions where in a mixed gypsum/anhydrite system the water flow-

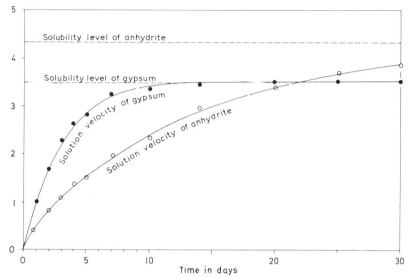

Figure 3.3. The principle of the difference between solution velocity and solubility illustrated by anhydrite and gypsum (after Priesnitz (1972); reproduced by permission of British Cave Research Association).

through rate is such that the less soluble mineral (gypsum) is the most eroded because flow-through rates are of the order 1 to 20 days.

The experiments of Wollast (1967) quoted in Berner (1971, pp. 170–175) also demonstrate the importance of water flow-through time to the products of weathering. Using the example of feldspar (albite) it is shown that the reactions proceed further as water flow rate increases. At low flow rates equilibrium is quickly reached. Thus, at flow rates approaching stagnancy, montmorillonite can be formed as a reaction between cations (especially Mg and Na) and Si and Al is possible because none of these is dissolved or flushed out. The reaction can be expressed as in Equation 3.3, assuming that albite, Mg and water are present.

$$Mg^{++} + 3NaAlSi_3O_{8\,albite} + 4H_2O \rightarrow$$
$$2Na_{0.5}Al_{1.5}Mg_{0.5}Si_4O_{10}(OH)_{2\,montmorillonite} + 2Na^+ + H_4SiO_{4\,aq} \quad (3.5)$$

If flow rates are faster the Mg and Na are quickly dissolved and removed. The dissolution of Al and Si is slower and so they remain to leave kaolinite as the weathering product, thus:

$$2H^+ + 2NaAlSi_3O_{8\,albite} + 9H_2O \rightarrow$$
$$Al_2Si_2O_5(OH)_{4\,kaolinite} + 2Na^+ + 4H_4SiO_{4\,a} \quad (3.6)$$

When water flow rate is fast relative to the rate of dissolution, or solution velocity, then silica is lost as well as other cations (Mg, Ca, Na). Aluminium oxide is produced in a bauxitic type of weathering:

$$H^+ + NaAlSi_3O_{8\,albite} + 7H_2O \rightarrow \\ Al(OH)_{3\,gibbsite} + Na^+ + 3H_4SiO_{4aq} \qquad (3.7)$$

In 1941 Jenny (p. 155) discussed the possibility that the ration of $Sio_2 : Al_2O_3$ in clays changed with temperature. The evidence was inconclusive. From the results of Wollast's experiments it is possible to state that the basic factor appears to be local conditions at the place of weathering. Factors such as flow conditions are important in the first place, rather than overall climatic factors like temperature. Both relief and climate will be important in controlling flow rate. Climate will determine the amount of input water. Relief will determine runoff. Factors such as infiltration rate and the transmissibility of the porous media will also control flow rate and these are complex factors adjusted in part to climate and vegetation, in part to relief and in part to the parent material and weathering system. Thus in steep areas with high rainfall, bauxitic weathering is possible and in flat, boggy areas montmorillonite may accumulate (Berner, 1971, p. 194). Relief will therefore operate within climatic zones to condition the exact weathering environment.

Laboratory experiments continue to contribute to the understanding of field situations but the two factors discussed above, organic processes and water flow rate, are the most important to replicate if the results of experiments, such as those in leaching columns, are to be interpretable in terms of natural processes.

Flow rate may be controlled by adjustments of input or, in confined systems, by adjustment of throughput. A controlled simulated rainfall input is possible using the method of Bryan (1970). His device controls intensities of water delivery from 1·27 to 91 cm/h. A constant head apparatus, such as those used in soil mechanics for measuring soil transmissibility (Lambe and Whitman, 1969, pp. 281–283) is useful for controlling flow rate in leaching columns. A pumping apparatus is described by Wierenga, Black and Manz (1973) for maintaining a steady state flow by use of multichannel syringe pumps attached to soil columns. They controlled outflow at 173 ml/day with a mean day to day variation of less than 0·2 ml. A variable flow apparatus was employed by Lai and Jurinak (1972) which is possibly more useful. Williams (1968) also describes a more modest apparatus where reverse (upwards) flow is used. This disperses the soil, so permitting free flow round the particles. Earlier references on leaching columns are given by Williams, and they include Black (1974), Gillingham (1965) and Legg (1963).

Simple open leaching columns are widely used, for example Kerpen and Scharpenseel (1967) and Atkinson and Wright (1957) but the use of Soxhlet extractors (Pedro, 1964; Henin and Pedro, 1965; Priesnitz, 1972) and of the perfusion apparatus has perhaps been neglected. It is difficult to control rate of flow precisely using a Soxhlet extractor but temperature may be controlled using the design of Priesnitz (Figure 3.4). Flow rate may be controlled more exactly in a perfusion apparatus by the control of the rate of flow of air supply

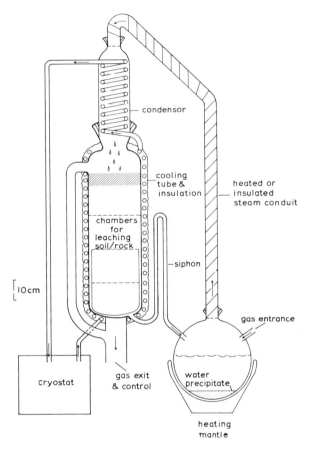

Figure 3.4. The Soxhlet extractor (after Priesnitz (1972); reproduced by permission of British Cave Research Association).

(Figure 3.5, after Alexander (1961) and P.H.T. Beckett (personal communication)).

The use of such experiments permits the control of solute type, rate of flow and often of temperature. It is difficult for them to replicate microbiological conditions, unless temperature is closely controlled. Moreover, some disturbance of pores is frequently unavoidable. Most methods of soil sterilization (in order to eliminate microbiological effects) may affect the soil chemically (e.g. mercury or other poisons) or physically (e.g. oven heating). Steam heating is perhaps the least objectionable. Pore disturbance, especially compaction on initial sampling, can be very detrimental to the replication of natural flow rates. The rate is influenced by pore geometry, especially tortuosity, rather than simple pore size (Childs, 1969). The tortuosity of interconnecting pores and rate of water flow frequently determine whether solutional weathering

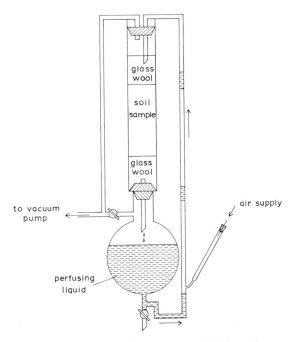

Figure 3. 5. Soil perfusion apparatus (redrawn from Alexander (1961)).

proceeds by simple diffusion or by diffusion and then displacement or by continual water displacement where solution velocity will be important. If the pores are compacted and the amounts of interconnecting channels reduced then the water may stand more chance of coming into equilibrium with the chemical constituents of the soil. Thus the use of core samples which do not compact the soil when the initial sample is being taken is essential if natural flow-through conditions are to be preserved. Moreover, sealing of the side of the soil column with resin or other suitable compound is necessary (see Curtis and Trudgill (1975) for sampling and soil moisture information).

Thus it seems that the experimenter has two alternatives: either to experiment in artificial situations, where at least the conditions can be measured even if they may not replicate natural conditions; or to experiment in essentially natural field situations where the processes are harder to define and the conditions more difficult to measure. However, developments of mathematical theory and improvements in field instrumentation have tended to minimize the dominance of these Heisenberg—type alternatives.

The evaluation of theoretical models for the calculation of diffusion rates in porous media and the valid applications of these models to real world situations is a challenging field. Multifactorial mathematical models have been attempted by Lindstrom and Boersma, (1971 and 1973), Lindstrom and coworkers (1971), Boast (1973) with Kurtz and Melstead (1973) and Gardner and Brooks (1957).

The biggest challenge comes from representing leaching under non-steady flows and allowing for feedback as solution proceeds and flow rate varies.

Geomorphologists have tended to adopt a rather more empirical approach and have attempted to study weathering in soil and bedrock in the field as much as they have by the use of experimentation and theory. Methods used for catching water to evaluate throughflow hydrology (Carson and Kirkby, 1972; Kirkby, 1969; Knapp, 1973; Weyman, 1970; Arnett, 1972) can be used to catch water for studies of dissolved content. Soil drainage waters have been collected, from field drains for example, by Williams (1970), Perrin (1965) and Massey and Jackson (1952). The problem with these studies is pointed out by Henin in the discussion in Hallsworth and Crawford (1965, p. 96): the results must be reviewed critically as it is difficult to collect drainage water whose origin (catchment area, tensions and flowthrough time) is precisely known. It is not known, for instance, whether the water comes from any particular size of pore, though Bourgeois and Lavkulich (1972) have developed a tension lysimeter which helps to overcome this problem in leaching studies. The same basic problem of process interpretation occurs in the analysis of stream waters (Davis, 1964; Edwards, 1971; Janda, 1971; Meade, 1969). The seat of weathering is a black box. Pedological processes do not necessarily find their expression in terms of solute load of a stream. In fact, it is common for chemical precipitation to occur at the base of the soil profile. For this reason attempts to understand variations in the solute load of streams in terms of climate may omit crucial soil and groundwater steps in the chain of events from rainfall input to streamflow output.

In the future weathering studies may look more to hydrological studies, especially where the estimation of residence time of water is involved. The investigation of this crucial factor should be improved by the use of water tracing (Corey, 1968; Kurtz and Melstead, 1973) as new methods of tracing, where dyes which are not absorbed on to the soil colloids are evolved or where a radioactive method is used. In this context the dye Rhodamine WT appears to be a useful tracer (Smart, 1972; Atkinson, personal communication; Laidlaw, personal communication). Initial experiments by the author and Smart have successfully indicated the residence times of water in the soil of a slope plot 10 m long and work is in progress on perfecting the technique. Such tracing would provide a final answer to flow-through time. This data could then be related to amount of rainfall, solution and chemical transport in the soil in order to understand chemical weathering and leaching in porous media in terms of climatic and other factors.

*3.2.3 A case study of a limestone leaching system*

The weathering processes that occur in the vegetation–soil–rock system on limestone epitomize in many ways the reactions where a climatically induced process, leaching, strives to reach equilibrium with the lithologically controlled factors (Trudgill 1972a, 1972b). The situation is very dynamic: rainfall and leaching encourage acidity in the soil and this acidity encourages rock solution but rock solution encourages soil alkalinity. In situations where leaching and solution are balanced a steady state occurs. Where one is dominant the soil progressively becomes alkaline or acid depending upon the balance of the factors, thus:

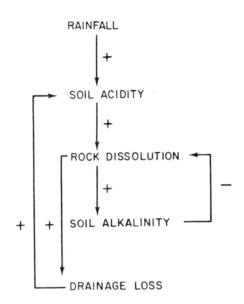

Under pluvial conditions the system is controlled by the drainage losses which occur through open joints in the rock. The losses determine whether or not alkaline solutes, the products of weathering, are lost from the system. If they are lost then the erosional trend is reinforced. Thus the progression of the system depends upon the balance between the negative and the positive feedback loops. Drainage loss is reinforced by solutional opening of joints and this reinforces soil acidity. Thus the equilibrium is dynamic with fluctuations from acidity and alkalinity in the soil as leaching and solution alternate around an erosional trend through time. If drainage loss is low the negative feedback loop dominates and decreases rock dissolution. Each leaching reaction is offset by reprecipitation and no progressive solution occurs.

The dynamics of the system were evaluated using field and laboratory experimentation. Initially, daily soil pH measurements, taken at various depths in the soil profile, were used as an indicator of the leaching process. Later, field studies were extended to include measurements of solute load of percolation water. Laboratory studies on both pH variations and solute load were made in leaching columns.

In the field study it was found that after heavy rainfall the soil was markedly acidified by leaching. After dry periods of weather the soil was alkaline. The actual pH values and the extent of the fluctuations depend upon the amount of rainfall, the vegetation type and the drainage conditions.

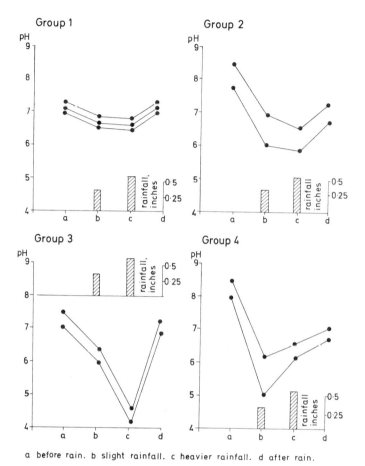

a before rain. b slight rainfall. c heavier rainfall. d after rain.

Figure 3.6. Daily pH fluctuations in thin soils. Group 1—Thin humus, calcicoles; Group 2—Thick humus, calcicoles; Group 3—Thick humus, calcifuges; Group 4—Same as Group 3, but on a slope.

All sites studied showed an increase in acidity with an increase in rainfall. The most marked changes were on well drained sites supporting a calcifuge vegetation capable of acidifying the soil (Grubb and coworkers, 1969) where comparatively small increases in rainfall lead to the largest pH changes. The least marked pH changes were under calcicole vegetation on poorly drained sites. More detailed information about variation from site to site is given in Figure 3.6. In this diagram the Group 1 sites are the least well drained and the Group 4 sites are the most freely draining. Groups 3 and 4 both support a calcifuge vegetation.

Thus the systems may be characterized as having an oscillatory nature. Under poor drainage the oscillations are restricted and under free drainage the oscillations are large. Where the amplitude of the oscillations is large the

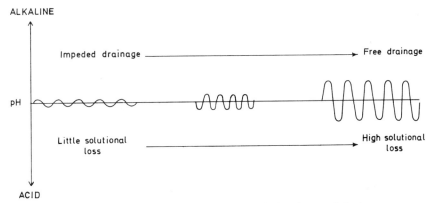

Figure 3.7. Model of pH fluctuations, solution loss and drainage.

system is prone to progressive change, i.e. progressive acidification (until the buffer capacity or reserve acidity value is reached). This is linked with a correspondingly high amount of solutional removal, the amplitude of the fluctuation being a measure of solute removal (Figure 3.7).

This study suggests that the operation of the climatic factor, leaching, is reinforced by calcifuge vegetation and good drainage but is inhibited by poor drainage. Thus, under conditions of good drainage, there exists a climatically dominated system: under conditions of poor drainage there exists a lithologically dominated system. In the first case the nature of the weathering and the nature of the soil can be explained with reference to climate as the dominant factor and lithology as the subordinate factor but the converse is true in the second case. Again, experimentation has shown how the factor of water flow conditions at a site is an important constraint on the expression of climatically motivated forces.

Slightly more sophisticated observations on the significance of water flow rate come from experiments on dolomite soil/rock systems. These are of experimental interest because the calcium and magnesium components of dolomite display different solution velocities. Chippings of the dolomite limestone studied were dissolved in the laboratory in a solution of weak carbonic acid. Sampling of water over time showed that magnesium tended to to into solution faster than calcium (Statham, personal communication). This laboratory experiment may, however, give a misleading picture of dolomite solution in nature. According to Douglas (1965) congruent dissolution occurs at specific partial pressures of carbon dioxide. By the term congruent dissolution it is meant that the rock constituents go into solution in the same ratio in which they are present in the rock. Incongruent solution, of the type shown in the initial stages of the laboratory experiment, occurs only at low concentrations of carbon dioxide. It follows that interpretation of the field results should include consideration not only of flow-through time but also of carbon dioxide production.

In the area studied (Inchnadamph, N. W. Scotland) the percolation springs that drain soil-covered slopes show marked fluctuations in Ca:Mg ratio with time. Daily sampling suggested that these fluctuations could be correlated with rainfall. At times of high rainfall a high Mg content was recorded (a low Ca:Mg ratio) but as rainfall decreased the Ca content, and the Ca:Mg ratio, rose. A high Ca:Mg ratio was also observed at the start of more intense rainfall but this appeared to fall off with time. This was attributed to the displacement of water previously stored in the soil. These results accord with those of Smart (in press) for variations in stream solutes. The detailed story of fluctuations varied from spring to spring, some showing very little chemical variation over time and others being more responsive to rainfall. The soils behind each spring were mapped. It was found that the Ca:Mg fluctuations were greatest where soils occurred which could be expected either to have the greatest flow-through time or where the routing of water through the soil would change, according to intensity of rainfall, and present differential opportunities for contact with limestone. Dilution of solute concentration occurred when overland flow took place and where the soils were very porous. Thus waters draining from peats and clays showed little variation unless overland flow occurred and then the solute content dropped. Waters flowing through gravel rendzinas, with finely divided limestone distributed throughout the soil profile, presenting large surface areas for dissolution, showed little variation except at very high flows. Percolation waters draining areas of freely draining open structured loams showed most fluctuations in Ca:Mg ratios.

When a dry soil is wetted by rainfall carbon dioxide production will increase but when a wet soil is moistened there will be less increase and there may even be a decrease, due to waterlogging. According to Douglas (1965) at low carbon dioxide partial pressures dolomite dissolution is incongruent, with magnesium going into solution before calcium. At higher pressures, around 1 atm, the dissolution is congruent but above this pressure dissolution again becomes incongruent but with calcium going into solution before magnesium. Thus it can be argued that when a dry soil is wetted initially magnesium will be dissolved before calcium; as wetting proceeds calcium dissolution will increase. In an already wet soil further additions of moisture will not radically alter the picture. Thus again, loose, porous soils of sandy texture and low organic matter which drain readily will show most fluctuations in Ca:Mg ratio. Compact soils which drain slowly will be least responsive. Figure 3.8 shows some examples of the field variations in Ca:Mg ratio for percolation springs.

Following the initial wetting (Day 2–3) the Ca:Mg ratio drops as more magnesium appears in solution. During days 3–5 the soil is gradually drying and the water which is present in the soil is coming into chemical equilibrium with the dolomite. During days 5–8 rainfall occurs. This is followed by a rise in calcium content, following on from incongruent (calcium before magnesium) solution at higher carbon dioxide levels. Also a displacement effect is seen by an increase in total dissolved load.

The rendzina soil and podzol soil springs show the weakest response in

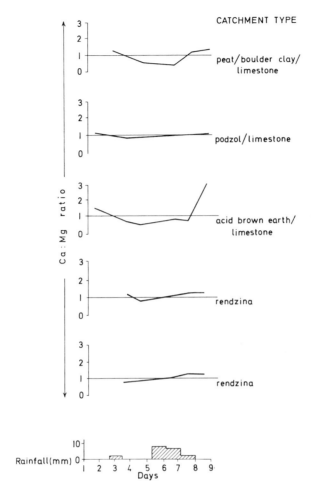

Figure 3.8. Ca:Mg fluctuations in percolation spring water, Inchnadamph, Sutherland.

terms of Ca:Mg ratios. In the former the water is always in intimate contact with finely divided limestone fragments. In the latter, flow-through is impeded by the iron rich layer (Bfe horizon). In the peat soil the pattern is confused by overland flow. The acid brown earth is a very open-structured, porous soil, rich in limestone. It will take up water, drain and dry easily and show the widest fluctuations.

It may be concluded that, under conditions of free drainage, carbon dioxide production and water residence time will be variable, generally with relatively rapid flow-through times. Because of these factors both calcium and magnesium will tend to be dissolved, on average, about equally. Magnesium dissolution will dominate at times of high flow and initial wetting and calcium dissolution will dominate at times of relatively high carbon dioxide production. Moister

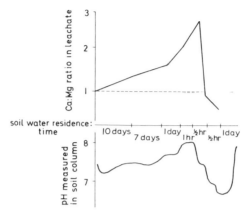

Figure 3.9. Ca : Mg ratio in leachate from experimental leaching column and variations in soil column pH.

soils, with less variable flow-through regimes and higher carbon dioxide production will tend to favour the dissolution of calcium.

Experimentation in the laboratory has attempted to link the pH fluctuations recorded in earlier work with the studies described above. A continuous recording pH meter was used with simple leaching columns of dolomite soils. The traces from the pH meter showed a familiar pattern of depression (acidification) upon the addition of distilled water acidified with carbon dioxide to pH 6·0. The drainage was free and the leachate was collected and then analysed for calcium and magnesium in solution.

In the experiment 100 ml of leaching water was added at increasingly rapid time intervals. The first addition caused a displacement of calcium and magnesium rich water (Figure 3.9) and little change in pH. The second displacement and subsequent replacements considerably moistened the soil, pH rose and calcium greatly dominated the leachate. This is interpreted as increased carbon dioxide production leading to the release of calcium which alkalized the soil. The addition of further leachate then reduced the acidity of the soil and as water flow-through was increased pH dropped and magnesium begn to dominate the leachate. This is interpreted as rapid dissolution of magnesium in a fast flowing, low carbon dioxide system.

The use of continuous recording of pH demonstrated an interesting response pattern. The rate of recovery, or restoration to a pre-leaching pH, seemed to be proportional to the degree of displacement, or acidification. Thus it would appear that a Le Chatelier-type effect takes place with equal and opposite reactions occurring to restore the effects of leaching. It can be argued that such an effect would not take place in a continually, or repeatedly, flushed system. Where rainfall is intermittent pH recovery can take place and the next rainfall event will effect less leaching. Where such events are closely spaced in time, recovery is prevented and the cumulative effect of rainfall is greater. Thus a dynamic climatic model of leaching can be proposed, based on the intensity and time distribution of rainfall.

*3.2.4 A dynamic model*

In these discussions of input potential and throughput the conclusion that can be advanced is that, complex and multifactorial as the systems are, in the cases studied they could be reduced to the one vector of rate of water flow. This vector has a high degree of explanatory power in the investigation of the dynamics of the system. Perhaps it cannot be emphasized too much that if the effects of climate are to be assessed then it is necessary first to understand the operation of the key factors in the processes and second to rate or assess the effect of climatic variables on these key factors. Neither the input nor the output alone provide the whole answer because they necessarily invoke inference of basic process. Experimentation on internal system process illuminates the interaction of solution velocity and water flow rate as the key factor.

From the study of limestone systems it can be suggested that the weathering dynamics in porous media may be best understood in terms of a pulsating system made up of alternately low and high throughflow rates (Figures 3.10, 3.11, 3.12, 3.13).

Consider two minerals A and B with solution velocities and equilibrium values Ae and Be as shown in Figure 3.10. If a rainfall pattern occurs such as that shown in Figure 3.11 then discharge from a downslope soil throughflow trough can be modelled as shown. The base flow water will have a long residence time and the peak flow, once the displaced water is past, will have a low residence time. If the flow rates at which Ae and Be are achieved are known

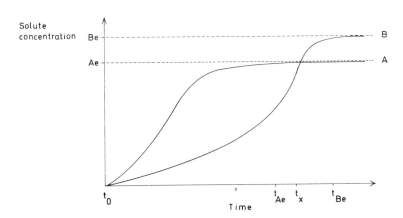

$t_{Ae}$  time taken for mineral A to come to equilibrium.   $t_{Be}$  time taken for mineral B to come to equilibrium.   $t_x$  cross over point

Figure 3.10. Solution velocities of minerals A and B, B having the greater solubility but the lower solution velocity.

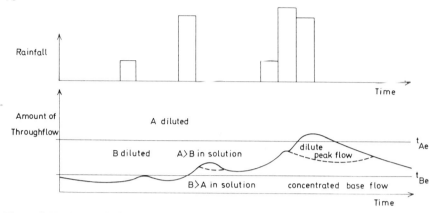

Figure 3.11. A model of the influence of rainfall on throughflow and its relationship with solution velocity.

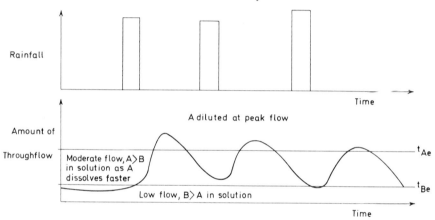

Figure 3.12. Model of throughflow and solution velocity under Climate 1: intense rainfall, widely spaced over time.

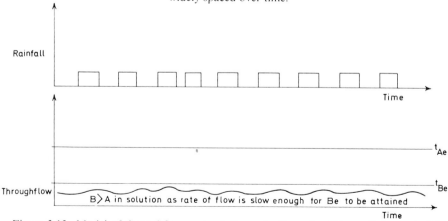

Figure 3.13. Model of throughflow and solution velocity under Climate 2: low rainfall, more or less continually distributed.

and plotted on the graph it can be seen that dilution of A is less than that of B, simply because A dissolves faster. In the upper portion of the hydrograph at rapid, high flows A is diluted; in the central portion the amount of A is greater than B and in the lower portion of slow, low flow the amount of B is greater than A.

Now consider Climate 1 defined here as one of intense rainfall widely spaced over time and where conditions of free drainage occur (Figure 3.12). Rapid throughflow interspersed with low flow encourages an overall approximately equal dissolution of both A and B in a pulsating system. B is dissolved more in low flow but dissolution of A approximates it in the long term because of its high solution velocity at times of high flow. Compare this with Climate 2 (Figure 3.13) where low rainfall is more or less continually distributed and where conditions of free drainage occur. In the absence of intense rainfall events, percolation rates are slow enough for equilibrium to occur. Solution velocity is not important but equilibrium values are and the dissolution of B is greater than A most, or perhaps all, of the time.

In these two cases the dissolution of mineral A is rainfall intensity dependent because it is flow rate dependent. The dissolution of mineral B is dependent only upon rainfall amount.

Variations of this model might include similar rainfall patterns in less freely draining situations. The differences between the two types of weathering would be far less marked as slower flow-through times would permit equilibration. The question of how the properties of the weathering systems influence the equilibration processes is of paramount importance. A fundamental distinction is that between porous media displaying dominantly colloidal properties and those where non-colloidal properties occur (Marshall, 1964, p. 2). The former include clay and humus-rich soils, the latter sand, silt, rock debris and porous rocks. The physical chemistry of the latter type of system can be approximately described by that of ordinary solution chemistry. Where colloids are the dominant constituent the solution kinetics will be influenced by the strongly adsorptive properties of colloids. The processes can be thought of more in terms of exchange, requiring some motivating force (see under 3.2.1) rather than in simpler mass transfer from solid to liquid in the case of larger mineral crystals.

Colloidal systems tend to display low throughflow times, unless they are structurally aggregated so as to permit rapid flow via structural interfaces. It can be argued that there may exist feedback relationships between the initial weathering environment, the type of weathering product (i.e. whether it is strongly colloidal or not) and the way in which the product may influence flow-through times and subsequently the local weathering environment. Certainly the incorporation of humus and clays into an open porous material will tend to slow down percolation rates and to encourage chemical equilibration.

It is to be hoped that attempts will be made to construct models of systems where climatic factors such as rainfall and temperature influence leaching, vegetation, and organic matter accumulation. The itemization of the subsequent pattern of events and the interplay between relief and parent material in influencing feedback relationships concerning flow-through time and equilibriation would constitute a fruitful study.

### 3.2.5 Residua

If dynamic models work it should be possible to test them by a study of the residual products of weathering or residua: that which has been involved in weathering but is left at or near the seat of weathering (Chesworth, 1973). It is not really possible to do this in any comprehensive way at the moment and studies in the literature are mainly those confined to an evaluation of the residua and inferences about the processes that induced them.

X-ray diffraction studies of weathering profiles often reveal diagnostic patterns of selective removal of elements and the formation of characteristic clay minerals. Interpretations concerning weathering sequences from the fresh rock to the soil are valid if the weathering profile is sedentary and no foreign material is included as a result of the action of any sedimentary process.

Like Wollast (1967), Ismail (1970) stresses the role of local environmental factors in the formation of clay types. Emphasis is given to the pH of the weathering environment, as controlled by climatic factors. By the use of mineral identification by X-ray diffraction, Ismail concluded that in some humid areas of California an acid, leached environment is encouraged, with loss of Fe and formation of vermiculite. In arid and semi-arid conditions leaching is minimal, leading to a neutral-to-alkaline environment and the formation of montmorillonite.

On granitic gneiss in the Texas central basin Stahnke and coworkers (1969) record the formation of illite as weathering proceeds. They suggest that the recycling of K by vegetation keeps the K content of the weathering profile high, especially near the surface of the soil. Vegetation recycling acts to offset leaching and it may be possible in the future to relate climatically controlled biological productivity to the amount of recycling.

Working in Yugoslavia, Gjems (1970) records a decrease in illite and chlorite and an increase in kaolinite up the weathering profile. Smectite (montmorillonite) was also discovered by X-ray diffraction, formed from illite and hydroxy-aluminium interlayered minerals. It is maintained that the well crystallized smectite is characteristic of temperate–humid weathering. An increase in kaolinite as weathering proceeds is also recorded by Syers and coworkers (1970) on greywacke in a humid environment.

None of these studies contradicts the idea that the more rapid the throughflow of water, the more strongly leached the weathering zone and the more advanced the residual system is along the montmorillonite–kaolinite–aluminium oxide sequence of Wollast. A good test of the dynamic model suggested above (3.2.4) would be to study data on site characteristics and rainfall type to see if there is any relationship between these and the clay type; but more comprehensive data are needed before this can be carried out.

A general model for the study of weathering residua is proposed by Chesworth (1973). The system is defined in terms of $SiO_2$–$Al_2O_3$–$Fe_2O_3$–$H_2O$ and, assuming water to be present always as a separate phase, the system can be prescribed by a SAF (Si–Al–Fe) triangular diagram. A model is presented in terms of climate (Chesworth 1973, Figure 3, p. 75) with halloysite (kaolin group) occurring in regions of high, year-round humidity, kaolinite in temperate areas with seasonally high humidity and bauxites and laterites in the tropics, where relatively high temperatures occur.

While clay mineralogy and the conditions of formation are large, polemical subjects (for instance there are extended discussions of tropical areas in UNESCO (1971)), one generalization at least may be offered in the present context, namely that the clay chemical composition tends to become simpler as leaching proceeds. Thus the illites and vermiculites of the hydrous mica group tend to be composed of Si, Al, Fe, Mg and K. Likewise the montmorillonites (or smectites) contain Si, Al, Mg and Na. The kaolin group (including halloysite and kaolinite) tend to be simpler, with just Si and Al (Marshall, 1964; Berner, 1971). Collectively, the evidence points to the significance of the degree of leaching in the clay forming microenvironment.

Before any attempt to correlate clay type with conditions of formation, the investigator must determine whether the clays are actively forming in the present environment. Dasch and Hills (1972) claim that it may be possible to distinguish modern from relict weathering profiles because Sr is removed at earlier stages than Rb. In studies on dolerite in N. Carolina it has been estimated that weathering ceased some 25 million years ago. Such information and the development of allied techniques will usefully avoid spurious correlations between rock weathering and climate, but the application of such studies is limited at the present time.

Studies of residua systems under experimental conditions are useful since it is possible to specify the conditions of formation, as for instance in the works of Farmer and Wilson (1970) on hydrobiotite formation, White (1962) on muscovite and Rausell-Colom and coworkers (1965) on mica. It may then be possible to translate these environmental conditions into real world situations in order to understand the evolution of weathering profiles.

In carbonate systems it is possible to employ carbonate staining techniques (Freidman, 1959; Wolf and coworkers, 1967) to detect selective removal. In the case study on dolomite, for instance, unweathered rock can be readily differentiated by staining from any high or low magnesium calcite left as a result of long-term incongruent solution.

The study of residua is perhaps a final test of dynamic process models. This section (3.2.5) has been necessarily piecemeal and eclectic. Studies are awaited that will interrelate weathering potential, lithology, leaching, flow-through time, solution loss and residua. Given these, more complete, comprehensive models may be constructed which largely avoid intuitive inferences concerning the links between input, internal process and output.

## 3.3 Surface systems

Unlike porous leaching systems, surface lowering is the only erosion that occurs in surface systems. Selective removal and residua do not occur, unless the system displays some porosity. Surface systems, like porous systems, can be characterized in terms of input, output and internal process. These can be monitored in terms of erosion potential, load output and surface lowering. Surface lowering proceeds by diffusion into aqueous media and, like porous systems, the amount lost depends upon the rate of water flow relative to the diffusion rate of the solid into the liquid. Where a gaseous phase is important, as in limestone solution, the diffusion of the gas into the liquid is also an important limiting factor. Physical removal, as opposed to chemical removal, is achieved by the incorporation of abrasive particles in the system input. Strictly speaking, if the system displays no porosity, the system can be characterized by erosion (surface removal) without previous weathering, the converse being true for porous systems which can be characterized by diffusion followed by erosional transport in many cases.

### 3.3.1 Simulation experiments

The evolution of morphology in solid rocks, especially limestones, has frequently been studied by hardware model experimentation. For example MacNeil (1954) experimented on atoll evolution using blocks of limestone treated by hydrochloric acid. As Watson (1972) has pointed out there are some limitations on these experiments, mainly in terms of solution kinetics. In particular, it is difficult to be certain whether the simulating study is analogous to the natural process and whether any direct time-scale substitution can be made if hydrochloric acid is used to accelerate the process. Roques (1969) has carefully documented the kinetics of the mass transfer in the solid–liquid–gas systems involved in limestone solution and his descriptions can be used to discuss the validity of experiments using a rapid rate of solution.

The general equation for the kinetics of solution is given by Roques as:

$$\frac{dm}{dt} = SE(c - c^*) \qquad (3.8)$$

where $m$ = mass of substance ($Ca^{++}$), $t$ = time, $S$ = surface area of contact (solid/liquid), $E$ = permeability coefficient, a measure of the ease of transfer of material across a boundary, $c$ = concentration ($Ca^{++}$), and $c^*$ = concentration at saturation ($Ca^{++}$).

It can be argued that when hydrochloric acid is substituted for weak carbonic acid, not only does the value of $c^*$ increase but the value of $E$ can also be affected. The solution velocity and the equilibrium value are raised. In natural situations the rate of dissolution is limited by the rate of mass transfer of the solid to the liquid, the rate of motion of the liquid and, often in practice the governing factor, the rate of transfer of carbon dioxide across the air/liquid interface (see also Coetzee and Ritchie (1969)). The reaction rates in a closed system with

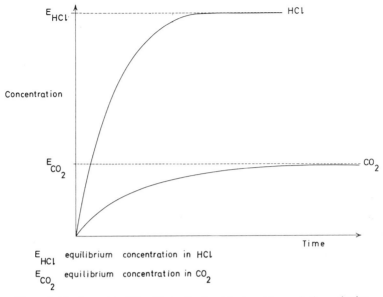

Figure 3.14. A model of the effect of hydrochloric acid on solution velocity.

an initial, but discontinued, supply of acid or carbon dioxide can be compared on a generalized diagram (Figure 3.14).

Since experiments are often used to simulate the evolution of morphology, and since reaction rates and the value of $c^*$ have been altered, the case of differential dissolution should be considered. Under situations of rapid (hydrochloric acid) dissolution differences in solution velocity could become exaggerated. Simply, if mineral A dissolves three times as fast as mineral B, then in an experiment where a weak carbonic acid is used the result could be, for the sake of argument, a loss of 3 mm in the case of mineral A and of 9 mm in the case of mineral B. If hydrochloric acid were poured onto the limestone the result could be 12 mm and 36 mm respectively. While the threefold ratio between losses from minerals A and B is preserved, it can be seen from Figure 3.15 and 3.16 that, from the morphological point of view, the results are considerably different, with the appearance of more fretted relief under conditions of more rapid solution. Of course, it cannot be argued that the situations are necessarily analogous in that I will eventually evolve into II given time in an open system with continued supply of carbon dioxide. This is because if solution proceeds at a slower rate, under carbon dioxide solution, as differential solution occurs the value of $S$ will increase but microsaturation can occur in the indented parts of the surface. Thus the projecting parts of the surface can be attacked preferentially and, because the system will, in this way, be partly self-limiting and oscillatory, a generally more uniform surface will evolve (Figure 3.16). This means that the system can be simulated by supplying fresh hydrochloric acid at a rate similar to that of natural carbon dioxide supply. In other words if

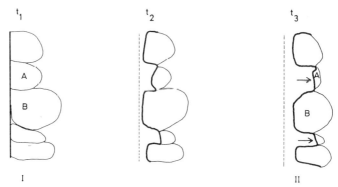

Figure 3.15. Microsolution experiments: form under rapid dissolution.

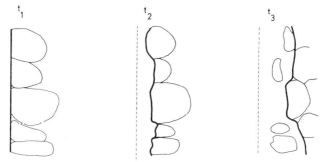

Figure 3.16. Microsolution experiments: form under slower dissolution.

carbon dioxide diffusion is sufficient to keep the pH of a solution actively dissolving limestone at pH 7·5 (whereas in a closed system with an initial but discontinued supply of carbon dioxide it may reach pH 9 at equilibrium) then hydrochloric acid should be supplied at a rate sufficient to keept the pH at 7·5. The dissolution may then proceed in an oscillatory fashion which would replicate natural conditions. In this way projecting grains may be removed. This would not tend to occur if the acid is simply poured on to the limestone and left to equilibriate.

Simulation of the process may be important for it is clear that solutional disintegration is common, especially amongst Pleistocene and Recent carbonate rocks. The cements are preferentially dissolved (aragonite and high magnesium calcite) with the physical dislodgement of grains occurring subsequently (Trudgill, 1972a). If rapid solution under hydrochloric acid occurs the differential solution may be exaggerated if acid is simply poured on to a block: if extremely strong acid is used, however, the danger is that differential solution may be lost altogether with the grains dissolving as fast as the cements. The strength of the acid is therefore critical in determining the degree of replication. In practice it has been found that with Pleistocene carbonate rocks (from

Aldabra Atoll, Indian Ocean) etching for 45 seconds to 30 seconds in a fresh solution of 1·75% HCl (50 ml concentrated + 750 ml distilled water) gives reasonable results. With acid concentrations weaker than this, or shorter periods of etching, replication failed to simulate natural surfaces. Differential solution became exaggerated with increasing acid concentrations until it was lost altogether.

It may be concluded that if dissolution conditions are to be accelerated then the strength of the acid has to be carefully monitored and the situation can be improved by use of weak acids in open systems, such as Priesnitz (1972) describes. Moreover, caution is demanded in that natural rock systems always have some degree of biological involvement, whether it is a thin covering of bacteria or fungi or some epilithic or endolithic algae. The darker, often greyer, colour of naturally weathering limestone surfaces is often due to the presence of algae (Syers, 1964). These biological factors can create conditions in interstitial microenvironments which may be important in naturally occurring dissolution and which may not necessarily be replicated under dissolution by artificially introduced acids.

### 3.3.2 Erosion potential

Chemical weathering occurs by the interaction of water and solid. It proceeds until equilibrium has been reached and thus measurements of departure from equilibrium can indicate the potential for further chemical weathering.

If the reaction involves the combination of aqueous anions with solid phase cations then departures from equilibrium, and thus erosion potential, can be assessed in terms of anion concentration in the liquid phase. $NO_3^-$, $SO_4^{2-}$, $Cl^-$ and $HCO_3^-$ are common anions capable of weathering reactions and the $OH^-$ ion is the motivating force in cation hydrolysis. The hydrogen ion, $H^+$, is the most usual motivating force when the reaction involves the combination of aqueous cations with anions of the solid phase, especially in limestone solution (Smith and Mead, 1962, Equation 8, p. 192) and in anion hydrolysis. Measurements of the concentrations of various ions in different aqueous media will indicate the general potential of the media for weathering. Rainwater is weakly charged with the anions mentioned above, as well as cations, hydrogen ions being evolved from the dissociation of carbonic acid. The $NO_3$ ion is common in soils, as is the hydrogen ion, derived from organic porcesses.

pH measurements, in combination with Eh measurements, can also indicate the stability of various metals (Broecker and Oversby, 1971; Garrels and Christ, 1965; Baas-Becking and coworkers, 1960). They may indicate, for instance, whether ferrous iron present in a rock-forming mineral is stable in its environment or not. Under aerobic conditions the ferrous iron spontaneously oxidizes to the ferric iron form (Iron II to Iron III compounds) which is then more stable.

While such measurements are valid in themselves they may suffer in the first case (of cation and anion measurement) in that they are non-specific, i.e.

they may not be interpretable in terms of a particular mineral or rock; and, in the second case, theoretical interpretation may only be possible in terms of pure phases of minerals. Rock-forming minerals may contain impurities which may increase or depress reaction rates. Therefore it may be better to experiment with samples of the actual rock in question in reactions in naturally occurring waters, or at least in controlled experiments with rocks and waters of known composition in the laboratory.

The technique for the measurement of solution potential, or aggressiveness, of karst waters is described by Stenner (1969, 1970a, 1970b, 1971). This technique, also described as the marble technique, involves the measurement of calcium content of two aliquots of water from the same sample, one of which has been treated by the addition of 'Analar' calcium carbonate. If the treated sample has more calcium in it than the untreated sample this indicates a potential for the solution of calcium carbonate.

As an adaptation of this principle, virtually any rock can be ground up and added as a powder to sub-samples of natural, or experimental, waters and relevant chemical attributes of the waters measured. The technique of Stenner has been extended by the author (Trudgill, 1972a, 1972c) using ground up limestones and by the measurement of both calcium and magnesium going into solution. The danger of using recent limestones is that aragonite may be altered to calcite by grinding so it is not only necessary to analyse the rock powder for calcium and magnesium (Bisque, 1962) and for trace elements by X-ray fluorescent spectrometry (Leake and coworkers, 1969) but also for mineralogy by X-ray diffraction and carbonate staining (Friedman, 1959; Wolf and coworker, 1967) before interpretation of results is possible.

Using this technique it has been possible to evaluate the view that the selective dissolution of magnesium carbonates is possible in many waters, even if the potential for calcium carbonate dissolution has been used up. For instance, studies of rainwater dissolution on bare rock surfaces have been undertaken by the author in tropical and temperate climates. It was found that the potential for dissolution of calcite was, on average, $+100$ mg/l for calcium carbonate and $+180$ mg/l for magnesium carbonates on Aldabra Atoll ($9°$S, $46°$E). In Co. Clare, Eire ($53°5'$N, $9°15'$W) the potential averaged around 70–75 mg/l $CaCO_3$ and up to 120 mg/l $MgCO_3$. No measurements were taken of the carbon dioxide content of the air in the two regions but from the data of Bolin and Keeling (1963) it can be deduced that, taken on an average annual basis, the figures are unlikely to differ significantly on a global scale to account for regional variations in the potential of rainwater for carbonate dissolution, though there are some seasonal differences. It is more likely that variations in the concentration of salt (sodium chloride) in the rainwater may account for the differences. Hem (1970, pp. 132–140) describes how sodium chloride increases the solubility of calcium carbonate; some of the calcium going into solution in combination with the chloride ion to form calcium chloride. Furthermore, Harned and Bonner (1945) show that small traces of salt may increase the solubility of carbon dioxide in water.

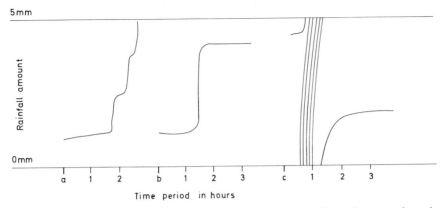

Figure 3.17. Rainfall type: Aldabra. Traces from a Dines recording rain gauge show the intense nature of the rainfall.

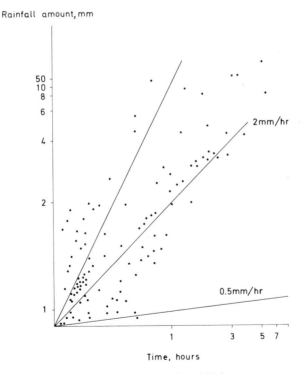

3.18. Rainfall intensity: Aldabra.

An important climatic difference between the two regions exists, however, in the intensity of the rainfall. Typical traces from an autographic rain gauge are shown in Figure 3.17. The scatter in Figure 3.18 shows that short heavy showers were dominant during the study seasons on Aldabra. Hot drying conditions occur between showers. In Co. Clare prolonged rain and drizzles of the type that keep the rock moist are common. These differences are important because the intensity of the rainfall and the nature of the surface will interact in the following way. It was found, in both climates, that the rainwater potential for calcium carbonate dissolution was used up very rapidly. Analysis of temporary freshwater pools showed that by the time the water had run over the surface of the rock, further calcium carbonate could not be dissolved but that the potential for magnesium carbonate solution remained for some time. However, in permanent or semi-permanent pools the respiration of blue-green algae (*Nostoc* spp.) could rejuvenate the potential for calcium carbonate dissolution. Thus the presence of standing water, under climates where storm pools persist and are replenished, greater dissolution of both calcium and magnesium will be encouraged. Where pools are transitory the calcium potential may be used up but the magnesium potential may not be. However, remembering the figures for potential dissolution (100 mg/l $CaCO_3$ and 180 mg/l $MgCO_3$) this does not necessarily mean that less magnesium will go into solution. What appears to happen is that, under short sharp showers, while 100 mg/l $CaCO_3$ may go into solution, the figure for $MgCO_3$ may only be 130–150 mg/l. Where topography permits the collection of waters, the $MgCO_3$ content may rise to about 180 mg/l (and both $CaCO_3$ and $MgCO_3$ may rise if *Nostoc* is present). Thus conditions of rapid runoff (short intense rainfall and dissected relief) encourage differential solution of minerals with high solution velocities. Conditions of prolonged wetting, where equilibriation with films of water is possible, encourages congruent solution. It can be concluded that rainfall intensity and evaporation are important climatic factors and that topography of the surface will be important, encouraging differential solution in the upstanding areas and congruent solution in the basins and pools.

Experiments are possible using silicate minerals and natural waters. The results of these may be interpreted in terms of the model suggested by Deju (1971) which are based on the experiments of Deju and Bhappu (1965, 1966) where minerals (including forsterite, quartz, hornblende and albite) were dissolved in acidified solutions. The process was monitored in terms of pH change manifested as an increase in alkalinity as hydrogen ions were exchanged at the surface for metal ions. The total surface area (number of exchange sites) and the oxygen: silicon ratio of the silicate mineral were the governing factors.

Since surface area for reaction is obviously a key factor in any chemical reaction the grinding of rocks into powders to increase reaction rate into measurable time can be criticized. The advantage of a high reaction rate is that external, or aqueous, conditions are not likely to change, especially by loss of carbon dioxide. But the situation when a mineral is ground up is essentially artificial even if a naturally weathering surface is ground up. Experiments using ground up rocks indicate dissolution potential and are comparable in themselves but the validity of the results can only really be assessed if predictions based on the assessment of potential are tested by measurement of actual erosion.

## 3.3.3 Surface lowering

The measurement of rock surface lowering may be effected by the measurement of differential erosion, as for example by Dahl (1967), where the erosion of rock adjacent to unweathered quartz veins was assessed. The time-scale was provided by assuming that the erosion began post-glacially. On a slightly larger scale microtopography can be measured by the use of a topograph (Boorman and Woodell, 1966), but recently the invention of a micro erosion meter (Hanna, 1966; Hanna and High, 1968; High and Hanna, 1969, 1970) has permitted the experimental investigation of rock surface erosion in the field (Newson, 1970, 1971b; Trudgill, 1972a, 1972c; Ford, 1971; Sweeting, 1970, p. 232).

The micro erosion meter consists of a dial gauge mounted in a framework constituting a three-point reference system. It can be placed upon three reference studs inserted into the rock. The height of the studs is taken as the datum and the level of the rock measured (usually to the nearest 0·01 mm) with reference to the studs. Repeated over intervals of time the micrometer measurements give an indication of surface lowering.

Micro erosion measurements can be used to test models of differing potential for erosion in various environments. A limitation is that while they may give a valid estimate for erosion in one small sub-system, say limestone pavement erosion at a point, they may not necessarily accord with measurements involving erosion in the whole system. This may be because reprecipitation may occur below the surface before the solute reaches the groundwater. For this reason, measurements of overall solution load, such as those discussed in Chapter 15 of this volume, may be less than average values for micro erosion measurements. The value of the micro erosion meter lies in small-scale experimental situations, rather than in estimating regional values of denudation.

## 3.3.4 Eroded load

As Keller (1957) points out, climate (the reactants and energy of weathering) is reflected in the soluble products of weathering. Such a tenet forms the basis of the discussion in Chapter 15. A shortcoming of this approach is that it is possible to be in a position where only backward inference about process is possible: the validity of the approach is greatly increased if measurements of weathering potential, processes, surface lowering and eroded load are performed together.

## 3.4 Climate as a factor

There is a fundamental weakness in the unqualified tenet of climatic geomorphology which states simply that different landforms occur in different climatic zones. Without measurements of the processes involved in weathering and erosion it is difficult to substantiate genetic arguments. The main objective of the quantitative and experimental work discussed so far has been to elucidate the genetic link in the chain of reasoning from landform to climate. In order to work out the relationship between rock weathering and climate as a factor it is necessary first to measure the relationship between rock weathering and process factors and secondly to see how these process factors vary in each climate.

It may be difficult to perform controlled experiments that seek to isolate the various contributory factors. This is because local factors, such as topography and lithology, frequently vary simultaneously with climate. Furthermore, it may be difficult to perform measurements which seek to relate present climates to landforms because it is difficult to separate the influence of past processes from those of present influences on the form of the land. It is perhaps because of these problems, which may be identified as those of covariance of factors and of polygenetic evolution, that small-scale field experimentation may be more successul than large-scale multivariate analyses.

### 3.4.1 Multivariate approaches

The multivariate approaches, such as that of Harmon and coworkers (1972), have outstanding potential and marked limitations. Harmon and his coworkers concluded that the collection of a large body of reliable, comparable data would be necessary over several years before a comprehensive analysis was possible. In the analysis of presently available data they found that the factor of climate had a relatively low explanatory power, when compared with the local factors of hydrology and geology, in accounting for data variations in water quality attributes. While such multivariate approaches are statistically sound and logically powerful, and will perhaps become the most useful tool of the future, small-scale studies have their own particular value.

### 3.4.2 The scale of the study

The advantage of studying microscale features is that they stand a better chance of being in equilibrium with present day climatic factors than do large-scale features. This is useful since it is only the present day climate that we can measure. Measurements and experiments with micro erosion meters and with micro weight loss of experimental rock tablets (Newson, 1970, 1971a, 1971b; Trudgill, 1970, 1972a, 1972c, 1975) last for about one or two years. Over this period climatic factors can be measured and climate–process–form relationships evaluated. The results may then be extrapolated in terms of the duration of particular conditions which lead to particular microforms. The duration of each process-form time unit will eventually determine the macroform, assuming that the time unit is long enough for the form to make the equilibrium adjustment.

The basic principle of the microscale argument is that the equilibrium form of erosion will adjust on the small scale to short-term processes. In a system of high lability (where erosion force is powerful or the resistance of the material being eroded is low) response to short-term events will be marked. Thus, the form of loose earth responds to individual storms: the form of a granite

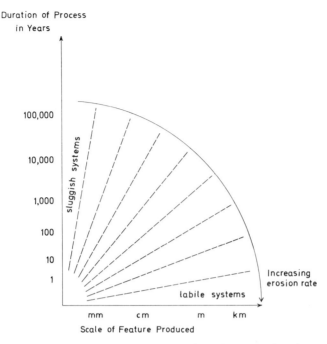

Figure 3.19. A model of erosion scale and process duration.

surface does not. Figure 3.19 illustrates the relationship between the duration of a particular process, the scale of form produced and the lability of the system.

An equilibrium erosional form is never a final one but it can be interpreted in terms of the forces acting upon it, including climate. A form in disequilibrium cannot be interpreted in terms of the forces presently acting upon it. Because of some inertia in the system, i.e. slow exchange across system boundaries relative to the speed of external change, the form will show some legacy of past processes.

The wavelengths of climatic fluctuation and the landform response times are illustrated in Figure 3.20. The solid line represents the motive force provided by climatic change. This perturbation of the system is followed by a fairly immediate response by labile systems. These are either small-scale rock systems or larger-scale systems of unconsolidated deposits. The exact magnitude of the fluctuations and the time lags will depend upon the fluidity within the system, i.e. the rapidity with which erosion and deposition rates can change to adjust to climatic conditions. Where weathering rates are slow the equilibrium response is delayed so that large-scale rock forms may be partly fossil unless climate is stable for a long time relative to the reaction rates. Near equilibrium forms will be more readily produced on the microscale.

To illustrate this model and the theme of the adaptation of small-scale features to climatic changes the observations of Waltham and Ede (1973) can be used. They observed in the Kuh-E-Parau limestone area of Iran how the

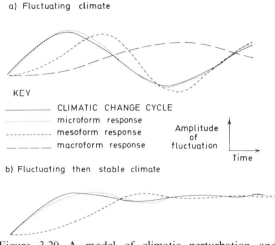

Figure 3.20. A model of climatic perturbation and landform response.

overall form on a large scale is solutional in origin, with rounded hills, dolines and ponors. However, it is to be concluded that these forms are in disequilibrium with the present climate. The larger forms show a legacy from a wetter period, probably during the Pleistocene, when solution was the dominant process. The crucial observation is that small-scale solutional forms are now being broken up by frost action. Inherited sharp-edged karren forms are being shattered and degraded; spalling flakes and granular disintegration are common. Measurement of present day climatic factors reveals a picture of a freeze–thaw climate (80 mm precipitation per annum, mostly as snow, summer maximum 37 °C and with diurnal ranges of temperature throughout the year showing fluctuations around freezing). The present day climate is not expressed in the larger-scale forms (where perhaps screes with headwalls may be expected) but only in the small-scale forms.

Similarly, in Ireland (Co. Clare) limestone pavement forms are present which are glacio-karstic in origin (Williams, 1966) and are not in equilibrium with the present climate. Present erosion rates, measured with a micro erosion meter vary from 0·003 to 6·3 mm/year. Such differential erosion is leading to very dissected landscapes, which are being actively formed near and beneath acid soils and peats, (Trudgill, 1972b). The flat pavement landscape is a legacy of glacial processes and is potentially unstable in the present climate. If lichens and mosses can establish a foothold and enhance weathering (Syers, 1964) rapid differential weathering can be expected. Thus the small-scale features, of which solution cups may be the largest, and (more likely) the rate of individual crystal erosion, may be interpreted in terms of the present climate. Larger-scale features, such as the pavements, and even the runnels (which may have been initiated by snowmelt) cannot necessarily be interpreted in terms of the present climate.

It may be concluded, then, that the present climate expresses itself in rock weathering by influencing processes which can be most readily seen on the small scale because small-scale rock features can adjust most readily to present processes. Valid interpretations of the relationships between the present rate of crystal erosion or grain erosion, present microenvironments and present climates are possible. Above this a hierarchy of systems occurs. The higher the order of scale the more polygenetic the landscape becomes.

Unconsolidated deposits are more labile and thus are more quick to adjust to climatic change. The same principle exists, but the scale of forms responsive to relatively rapid changes is much larger.

In summary, in systems of low lability (hard rock, low erosive force) adjustment to climatic factors is slow. Macroforms show legacies from the past and cannot be interpreted in terms of the present climate. Conditions of rapid weathering, such as occur on a microscopic scale, thus provide the best experimental situations for the evaluation of form–process–climate relationships.

Experimentation with micro erosion meters and on the micro weight loss of pre-weighed limestone tablets has been undertaken in and under soils and on bare limestone pavements. From the work on soils it has been possible to evaluate soil pH as the key factor in influencing erosion (Trudgill, in press; and see pp. 70–76). Factors which influence this are acidification by vegetation, organic matter and leaching as opposed by the factor of the original content of calcium carbonate left in any glacial drift present. Erosion has been worked out experimentally by the study of micro weight loss of identical tablets placed under different soils. The soils presented varying conditions of pH, calcium carbonate content, depth, permeability and vegetation type (Figure 3.21). Climate can be substantially established as an important factor in the way it influences leaching.

Using micro erosion meters, erosion rates have been measured for rock surfaces. In this case the work established lithology (cementation) as the prime factor in influencing rate of erosion. The conclusion is that the younger carbonate is more erodible than the older rock. An experiment has also been under-

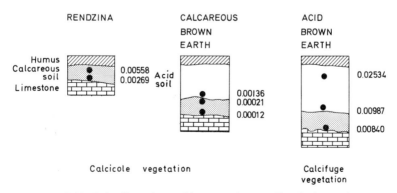

Figure 3.21. Subsoil erosion: tablets experiments. Results in mm/year

Table 3.1. Weight loss of limestone tablets exposed to sub-aerial weathering, mg/year

| Tablets | Temperate climate | Tropical climate |
|---|---|---|
| Carboniferous limestone | 0·3 mg | 0·1 mg |
| Aldabra limestone | 9·2 mg | 2·7 mg |

Conclusion: the effect of varying lithology is more marked than the effect of varying climate in determining erosion rate in the climates and lithologies selected.

*Micro erosion meter results*

| Aldabra limestone surface lowering | mm/year |
|---|---|
| Low magnesium calcite rocks | 0·35 |
| High magnesium calcite and aragonite rocks | 0·51 |
| Well cemented algal limestone | 0·09 |
| Poorly cemented algal limestone | 0·11 |

Conclusion: the results support the conslusions of aggressiveness measurements in that magnesium calcites are preferenctially eroded. Cementation is an important control on erosion rate.

taken subjecting tablets of two limestones to two contrasting climates and measuring weight losses. The results are shown in Table 3.1 together with micro erosion meter results.

In each case experimentation established the importance of the various factors that influenced weathering and erosion and then rated the climatically controlled factors against the other factors. This was possible because it was believed that the small-scale processes and features were interpretable in terms of the present climate.

The fact that the last experiment mentioned above stresses lithology more than climate is an important one. Much of the material presented in this chapter could equally be rewritten under the heading 'Rock weathering and lithology'. The role of climate is one which has to take its place amongst other factors. A last word on the weathering system is that it is mutually reactive. Climate, lithology, slope, soil type, vegetation type and hydrology interact considerably. It is perhaps vain to attempt to establish the role of climate as a factor *per se* by experimentation. It will certainly be concluded that climate is the most important factor if that is the dominant variable built into the experimental model. Comparing erosion rates, or any other single attribute or set of attributes, from one climatic region to the next begs the question of the role of climate as a factor. This is because such an experiment is not designed to test whether the role of lithology or relief or other factors should be discarded or included. One should not compare and contrast some attribute of the weathering system that it is believed to be important on *a priori* grounds. Rather, one should take each climate and, by experimentation, discover the processes in operation and ask, 'How does the system work here?' When the role of climate in influencing process dynamics has been assessed and the role of climate has been weighted relative to the other factors, it may then be possible to make some valid comparisons about differences in rock weathering in different areas. This is especially so when the dominant factor is truly climate and not some other factor or combination of factors. The rating of climate, relief and lithology in controlling the opportunity for chemical equilibrium of minerals with water is a very interesting topic worthy of attention.

## References

Alexander, M., 1961, *An introduction to soil microbiology*, Wiley.
Arnett, R. R., 1972, The field measurement of lateral soil water movement *Rev. Geom. Dyn.*, **4**, XXIe, 177–181.
Atkinson, H. J. and 1957, J. R. Wright, Chelation and vertical movement of soil constituents, *Soil Sci.*, **84**, 1–11.
Baas-Becking, L. G. M., I. R. Kaplan and D. Moore, 1960, Limits of the natural environment in terms of pH and oxidation-reduction potential, *J. Geol.*, **68**, 243–284.
Berner, R. A., 1971, *Principles of chemical sedimentology*, McGraw-Hill.
Berthelini, I. J. and Y. Dommergues, 1972, Effect of microbial metabolic products on the solubilisation of minerals in a granite sand, *Rev. d'Ecol. Biol. du Sol.*, **9**, 397–406: *Soils Fert. Abstr.*, **36**, 1973, Abstr. 678, p. 74.
Bisque, R. E., 1962, Analysis of carbonate rocks for calcium, magnesium, iron and aluminium with E.D.T.A., *J. Sedium. Petrol.*, **31**, 113–122.
Black, I. A., 1947, A modified apparatus for leaching soils, *Soil Sci.*, **63**, 337–340.
Boast, C. W., 1973, Modelling the movement of chemicals in soils by water, *Soil Sci.*, **115**, 224–229.
Bolin, B. and C. D. Keeling, 1963, Large scale atmospheric mixing as deduced from the seasonal and meridional variations of carbon dioxide, *J. Geophys. Res.*, **68**, 3899–3920.
Boorman, L. A. and S. R. J. Woodell, 1966, The topograph, an instrument for measuring microtopography, *Ecology*, **47**, 869–870.
Bourgeois, W. W. and L. M. Lavkulich, 1972, A study of forest soils and leachates on sloping topography using a tension lysimeter, *Canad. J. Soil Sci.*, **52**, 375–391.
Boyle, J. R. and G. K. Voigt, 1973, Biological weathering of silicate minerals, *Plant and Soil*, **38**, 191–201.
Bridges, E. M., 1971, *World Soils*, Cambridge University Press.
Broecker, W. S. and V. M. Oversby, 1971, *Chemical equilibria in the earth*, McGraw-Hill.
Bryan, R., 1967, Climosequences of soil development in the Peak District of Derbyshire, *East Midl. Geog.*, **4**, 251–261.
Bryan, R. B., 1970, An improved rainfall simulator for use in erosion research, *Canad. J. Earth Sci.*, **7**, 1552–1561.
Carson, M. A. and M. J. Kirkby, 1972, *Hillslope form and process*, Cambridge University Press, 475, pp.
Chesworth, W., 1973, The residua system of chemical weathering: a model for the chemical breakdown of silicate rocks at the surface of the earth, *J. Soil Sci.*, **24**, 69–81.
Childs, E. C., 1969, *An introduction to the physical basis of soil water phenomena*, Wiley.
Chorley, R. J., and B. A. Kennedy, 1971, *Physical Geography. A systems approach*, Prentice-Hall.
Cochran, W. G. and G. M. Cox, 1957, *Experimental designs*, Wiley.
Coetzee, J. F. and C. D. Ritchie (Eds.), 1969, *Solute–solvent interactions*, Marcel Dekker.

Corey, J. C., 1968, Evaluation of dyes for tracing water movement in acid soils, *Soil Sci.*, **106**, 182–187.

Crompton, E., 1960, The significance of the weathering/leaching ratio in the differentiation of major soil groups with particular reference to some very leached brown earths on the hills of Britain, *Rep. 7th Int. Congr. Soil Sci.*, **IV**, 406–412.

Curtis, L. F. and S. T. Trudgill, 1975, The measurement of soil moisture, *Brit. Geom. Res. Grp., Tech. Bull.*, **13**.

Dahl, R., 1967, Post-glacial micro-weathering of bedrock surfaces in the Narvik district of Norway, *Geografiska Annalar*, **49A**, 155–166.

Dasch, E. J. and F. A. Hills, 1972, Distinguishing modern from relict weathering profiles, *Nature Phys. Sci.*, **236**, 66, 73,

Davies, R. I., 1971, Relation of polyphenols to decomposition of organic matter and to pedogenetic processes, *Soil Sci.*, **111**, 80–85.

Davis, S. N., 1964, Silica in streams and ground water, *Amer. J. Sci.*, **262**, 870–891.

Deju, R. A., 1971, A model of chemical weathering of silicate minerals, *Geol. Soc. Amer. Bull.*, **85**, 1055–1062.

Deju, R. A. and R. B. Bhappu, 1965, Surface properties of silicate minerals, *N. Mex. Inst. Min. Tech., State Bur. Mines & Min. Res., Circ.*, *82*, 67–70.

Deju, R. A. and R. B. Bhappu, 1966, A chemical interpretation of surface phenomena in silicate minerals, *N. Mex. Inst. Min. & Tech., State Bur. Mines & Min. Res., Circ.*, *89*, 1–13.

Douglas, I., 1965, Calcium and magnesium in karst waters, *Helictite*, January, 23–36.

Duff, R. B., D. M. Webley and R. O. Scott, 1963, Solubilization of minerals and related minerals by 2-ketogluconic acid—producing bacteria, *Soil Sci.*, **95**, 105–114.

Edwards, A. M. C., 1971, *Aspects of the chemistry of four East Anglian rivers*, Unpublished Ph.D. Thesis, University of East Anglia (also published in *J. Hydrol.*, **18** (1973), 201–242).

Farmer, V. C. and M. J. Wilson, 1970, Experimental conversion of biotite to hydrobiotite, *Nature* **226**, (5248), 841–842.

Ford, D. C., 1971, Research methods in karst geomorphology, *Res. Meth. Geomorph., Proc. 1st Guelph symp. Geomorph. 1969, Dept. Geog., Univ. Guelph, Geog. Publ. 1*, 23–47.

Friedman, G. M., 1959, Identification of carbonate minerals by staining methods, *J. Sedim. Petrol.*, **29**, 87–97.

Gardner, W. R. and R. H. Brooks, 1957, A descriptive theory of leaching, *Soil Sci.*, **83**, 295–304.

Garrels, R. M. and C. L. Christ, 1965, *Solutions, Minerals and Equilibria*, Harper & Row.

Gillingham, J. T., 1965, A new soil-extracting apparatus, *Canad. J. Soil Sci.*, **45**, 102–104.

Gjems, O., 1970, Mineralogical composition and pedogenic weathering of the clay fraction in podzol soil profiles in Zalesine, Yugoslavia, *Soil Sci.*, **110**, 237–243.

Grubb, P. J., H. E. Green and R. J. C. Merrifield, 1969, The ecology of chalk heath: its relevance to the calcicole–calcifuge and soil acidification problem, *J. Ecol.*, **57**, 175–212.

Hallsworth, E. G. and D. V. Crawford (Eds.), 1965, *Experimental pedology*, Butterworths.

Hanna, F. K., 1966, A technique for measuring the rate of erosion of cave passages, *Proc. Univ. Bristol Speleol. Soc.*, **11**, 83–86.

Hanna, F. K. and C. High, 1968, Direct measurement of the rate of erosion of cave passages, *Proc. Brit. Speleol. Assoc.*, **6**, 29.

Harmon, R. S., J. W. Hess, R. W. Jacobson, E. T. Shuster, C. Haygood and W. B. White, 1972, Chemistry of carbonate denudation in northern America, *Trans. Cave Res. Grp. Gt. Brit.*, **14**, 96–103.

Harned, H. S. and F. T. Bonner, 1945, The first ionisation constant of carbonic acid in aqueous solutions of sodium chloride, *J. Amer. Chem. Soc.*, **57**, 1026–1031.

Hem, J. D., 1970, Study and interpretation of the chemical characteristics of natural water, *Geol. Surv. Water-Supply Paper 1473*, US Govt.

Henderson, M. E. K. and R. B. Duff, 1963, The release of metallic and silicate ions from minerals, rocks and soils by fungal activity, *J. Soil Sci.*, **14**, 236–246.

Henin, S. and G. Pedro, 1965, The laboratory weathering of rocks, in Hallsworth and Crawford (1965), q.v., 23–29.

High, C., 1970, *Aspects of the solutional erosion of limestone: with a special consideration of lithological factors*, Unpublished Ph.D. Thesis, University of Bristol.

High, C. and F. K. Hanna, 1969, Micro erosion measurements made in Jamaica in 1967, *J. Brit. Speleol. Assoc.*, **6**, 149–150.

High, C. and E. K. Hanna, 1970, A method for the direct measurement of erosion of rock surfaces, *Brit. Geom. Res. Grp., Tech. Bull.*, 5.

Hinman, W. C. and F. Bisal, 1973, Percolation rate as affected by the interaction of freezing and drying processes of soils, *Soil Sci.*, **115**, 102–106.

Huang, W. H. and W. D. Keller, 1973, Organic acids as agents of chemical weathering of silicate minerals, *Nature Phys. Sci.*, **239**, 149–151.

Huang, W. H. and W. C. Kiang, 1972, Laboratory dissolution of plagioclase feldspars in water and organic acids at room temperature, *Amer. Mineralogist*, **57**, 1849–1859.

Iskandar, I. K. and J. K. Syers, 1972, Metal—complex formation by lichen compounds, *J. Soil Sci.*, **23**, 255–265.

Ismail, F. T., 1970, Biotite weathering and clay formation in arid and humid regions, California, *Soil Sci.*, **109**, 257–261.

Janda, R. J., 1971, An evaluation of procedures used in computing chemical denudation rates, *Geol. Soc. Amer. Bull.*, **82**, 67–80.

Jenny, H., 1941, *Factors of soil formation*, McGraw-Hill, New York, 281 pp.

Jenny, H. and C. D. Leonard, 1934, Functional relationships between soil properties and rainfall, *Soil Sci.*, **38**, 363–381.

Keller, F., 1957, *Principles of chemical weathering*, Lucas Bros., Columbia, Missouri.

Keller, W. D. and A. F. Frederickson. 1952, Role of plants and coloidal acids in the mechanisms of weathering, *Amer. J. Sci.*, **250**, 594–608.

Kerpn, W. and H. W. Scharpenseel, 1967, Movements of ions and colloids in undisturbed soil and parent material columns, in *Isotope and radiation techniques in soil physics and irrigation studies. Proc. Symp. F. A. O./I. A. E. A., Istanbul, 1967*, 213–226.

Kirkby, M. J., 1969, Infiltration, throughflow and overland flow, in Chorley, R. J., *Water, Earth and Man*, Methuen, 215–227.

Knapp, B. J., 1973, A system for the field measurement of soil water movement, *Brit. Geom. Res. Grp., Tech. Bull.*, 9.

Kononova, M. M., 1966, *Soil organic matter*, Pergamon, Oxford.

Kurtz, L. T. and S. W. Melstead, 1973, Movement of chemicals in soils by water, *Soil Sci.*, **115**, 231–239.

Lai, S.-H. and J. J. Jurinak, 1972, The transport of cations in soil columns at different pore velocities, *Soil Sci. Soc. Amer. Proc.*, **36**, 730–732.

Lambe, T. W. and R. V. Whitman, 1969, *Soil mechanics*, Wiley.

Leake, B. E. and coworkers, 1969, The chemical analysis of rock powders by automatic X-ray fluorescence, *Chemical Geology*, **5**, 7–86.

Legg, W. T., 1963, An improved form of leaching apparatus, *Soil Sci.*, **95**, 214–215.

Lindstrom, F. T. and L. Boersma, 1971, A theory on the mass transport of previously distributed chemicals in a water saturated sorbing porous medium, *Soil Sci.*, **111**, 192–199.

Lindstrom, F. T. and L. Boersma, 1973, A theory on the mass transport of previously distributed chemicals in a water saturated sorbing porous medium: III. Exact solution for first order kinetic sorbtion, *Soil Sci.*, **115**, 5–10.

Lindstrom, F. T., L. Boersma and D. Stockard, 1971, A theory on the mass transport of previously distributed chemicals in a water saturated sorbing porous medium: II. Isothermal cases. *Soil Sci.*, **112**, 291–300.

Loughnan, F. C., 1969, *Chemical weathering of the silicate minerals*, Elsevier.

Lukashev, K. I., 1970, *Lithology and geochemistry of the weathering crust*, Israel Prog. Sci. Transl.

MacNeil, F. S., 1954, The shape of atolls: an inheritance from subaerial erosion forms, *Amer. J. Sci.*, **252**, 402–427.

Marshall, C. G., 1964, *The physical chemistry and mineralogy of soils, Vol. 1., Soil materials*, Wiley.

Massey, H. F. and M. L. Jackson, 1952, Selective erosion of soil fertility constituents, *Soil Sci. Soc. Amer. Proc.*, **16**, 353–356.

Meade, R. H., 1969, Errors using modern stream load data to estimate natural rates of denudation, *Geol. Soc. Amer. Bull.*, **80**, 1265–1274.

Newson, M. D., 1970, *Studies in chemical and mechanical erosion by streams in limestone terrains*, Unpublished Ph.D. Thesis, University of Bristol.

Newson, M. D., 1971a, The role of abrasion in cavern development, *Trans. Cave Res. Grp. Gt. Brit.*, **13**, 101–107.

Newson, M. D., 1971b, A model of subterranean limestone erosion in the British Isles based on hydrology, *Inst. Brit. Geogr. Trans.*, **54**, 55–70.

Ollier, C. D., 1969, *Weathering* Oliver & Boyd.

Pedro, G., 1964, *Contribution a l'étude Expériméntale de l'Altération géochimique des roches cristallines*, Inst. Natl. Rech. Agron., Paris.

Perrin, R. M. S., 1965, The use of drainage water analyses in soil studies, in Hallsworth and Crawford (1965), q.v., 73–92.

Picknett, R. G., 1964, A study of calcite solutions at 10 °C. *Trans. Cave. Res. Grp. Gt. Brit.*, **7**, 41–62.

Priesnitz, K., 1972, Methods of isolating and quantifying solution factors in the laboratory, *Trans. Cave. Res. Grp. Gt. Brit.*, **14**, 153–158.

Rauscll-Colom, J. A., T. R. Sweatman, C. B. Wells and K. Norrish, 1965, Studies in the artificial weathering of mica, in Hallsworth and Crawford (1965), q.v., 40–72.

Rode, A. A., 1970, *Podzol-forming process*, Israel Prog. Sci. Transl.

Roques, H., 1969, A review of present-day problems in the physical chemistry of carbonates in solution, *Trans. Cave. Res. Grp. Gt. Brit.*, **11**, 139–163.

Runge, E. G. A., 1973, Soil development sequences and energy models, *Soil Sci.*, **115**, 183–193.

Schnitzer, M. and S. I. M. Skinner, 1963, Organo-metallic interactions in soils. I. Reactions between a number of metal ions and the organic matter of a podzol Bh horizon, *Soil Sci.*, **96**, 86–93.

Scrivner, C. L., J. C. Baker and D. R. Brees, 1973, Combined daily climatic data and dilute solution chemistry in studies of soil profile formation, *Soil Sci.*, **115**, 213–223.

Shuster, E. T. and W. B. White, 1971, Seasonal fluctuations in the chemistry of limestone springs: a possible means for characterising carbonate aquifers, *J. Hydrol.*, **14**, 93.

Smart, P. 1972, *The use of activated carbon for the detection of the tracer dye Rhodamine W. T.* Unpublished M.Sc. Thesis, University of Alberta.

Smart, P. (in press) Solute variations during floods in the Traligill Basin, Sutherland, Scotland, *Caves and Karst*.

Smith, D. I. and D. G. Mead, 1962, The solution of limestone with special reference to Mendip, *Proc. Univ. Bristol Speleol. Soc.*, **9**, 188–211.

Southard, A. R. and I. Dirmhirn, 1972, Some observations on soils–microclimate interactions, *Soil. Sci. Soc. Amer. Proc.*, **36**, 843–845.

Stahnke, C. R., J. R. Rogers and B. L. Allen, 1969, A genetic and mineralogical study of a soil developed from granitic gneiss in the Texas Central Basin, *Soil Sci.*, **108**, 313–320.

Stenner, R. D., 1969, The measurement of aggressiveness of water towards calcium carbonate, *Trans. Cave Res. Grp. Gt. Brit.*, **11**, 175–200.

Stenner, R. D., 1970a, *An investigation into the measurement of aggressiveness of water towards calcium carbonate*, Unpublished M.Sc. Thesis, University of Wales.

Stenner, R. D., 1970b, Preliminary results of an application of the procedure for the measurement of aggressiveness of water to calcium carbonate, *Trans. Cave Res. Grp. Gt. Brit.*, **12**, 283–289.

Stenner, R. D., 1971, The measurement of aggressiveness of water to calcium carbonate, Parts II & III. *Trans. Cave Res. Grp. Gt. Brit.*, **13**, 283–295.

Sweeting, M. M., 1970, Recent developments and techniques in the study of karst landforms in the British Isles, *Geographica Polonica*, **18**, 227–241.

Syers, J. K., 1964, *A study of soil formation on Carboniferous Limestone with particular reference to lichens as pedogenic agents*, Unpublished Ph.D. Thesis, University School of Agriculture, King's College, University of Durham.

Syers, J. K., J. D. H. Williams, T. W. Walker and S. L. Chapman, 1970, Mineralogy and forms of inorganic phosphorous in a greywacke soil–rock weathering sequence, *Soil Sci.*, **110**, 100–106.

Trudgill, S. T., 1970, Micro erosion measurement of exposed bedrock, *Area*, **3**, 61.

Trudgill, S. T., 1972a, *Process studies of limestone erosion in littoral and terrestrial environments, with special reference to Aldabra Atoll, Indian Ocean*, Unpublished Ph.D. Thesis, University of Bristol.

Trudgill, S. T., 1972b, The influence of drifts and soils on limestone weathering, N. W. Co. Clare, Eire, *Proc. Univ. Bristol Speleol. Soc.*, **13**, 113–118.

Trudgill, S. T., 1972c, Quantification of limestone erosion in intertidal, suberial and subsoil environments, with special reference to Aldabra Atoll, Indian Ocean, *Trans. Cave Res. Grp. Gt. Brit.*, **14**, 176–179.

Trudgill, S. T., in press, Limestone erosion under soil, *Proc. Int. Speleol. Un. Conf.*, Oloumouc, Cezech., Sept. 1973.

Trudgill, S. T., 1975, *Brit. Geom. Res. Grp.*, Tech. Bull, 17.

Unesco, 1971, Soils and tropical weathering, *Natural Resources Research*, **XI**.

Waltham, A. C. and D. P. Ede, 1973, The karst of Kuh-E-Parau, Iran, *Trans. Cave Res. Grp. Gt. Brit.*, **15**, 27–40.

Watson, R. A., 1972, Limitations on substituting chemical reactions in model experiments, *Zeits. Geomorph.*, **16**, 103–108.

Weyman, D. R., 1970, Throughflow on hillslopes and its relation to the stream hydrograph, *Publ. Int. Assoc. Sci. Hydrol.*, **XV**, 25–32.

White, J. L., 1962, X-ray diffraction studies on weathering of muscovite, *Soil Sci.*, **93**, 16–21.

Wierenga, P. J., R. J. Black and P. Manz, 1973, A multichannel syringe pump for steady state flow in soil columns, *Soil Sci. Soc. Amer. Proc.*, **37**, 133–135.

Williams, B. G., 1968, An apparatus for leaching soil samples, *Soil Sci.*, **105**, 376–377.

Williams, P. W., 1966, Limestone pavements with special reference to Western Ireland, *Inst. Brit. Geogr. Trans.*, **40**, 155–172.

Williams, R. J. B., 1970, The chemical composition of water from land drains at Saxmundham and Woburn, and the influence of rainfall upon nutrient losses, *Rept. Rothamsted Exptl. Sta.*, Pt. 2 (1970), 36–67.

Wolf, G. and coworkers, 1967, Techniques of examination and analysis of carbonate skeletons, minerals and rocks, in Chilingar, G. V., H. J. Bissell and R. W. Fairbridge (Eds.), *Carbonate Rocks*, Developments in Sedimentology, 9B, Elsevier.

Wollast, R., 1967, Kinetics of the alteration of K-feldspar in buffered solutions at low temperature, *Geochim. Cosmochim. Acta.*, **31**, 635–648.

CHAPTER FOUR

# Mass-Wasting, Slope Development and Climate

MICHAEL A. CARSON

### 4.1 Varieties of mass-movement

Mass-movement processes, that is the downslope travel of earth or rock material under the influence of the gravitational body force of the mass, without the transportational aid of other moving media such as water, air or ice, comprise a diversity of denudational agencies. Classifications abound in the literature, those of Sharpe (1938), Varnes (1958) and, in the restricted context of clay slopes, Skempton and Hutchinson (1969) being, perhaps, the best-known. As Terzaghi (1950, p. 88) remarked: 'A phenomenon [landsliding] involving such a multitude of combinations between materials and disturbing agents opens unlimited vistas for the classification enthusiast'. Criteria used in distinguishing among different types of movement are many: velocity; type of material; consistency of the material; mode of deformation, i.e. rigid body versus internal strain; geometry of the moving mass, e.g. sheets, lobes, streams etc; and others. Indeed, the terms used may relate to the process itself (e.g. soil-creep), the event involving the process (e.g. landslide) or to morphological features (e.g. mud-flow, terracette). The classification used here (Table 4.1) involves many of these criteria, but is primarily a binary combination of mode of deformation and character of the material.

Creep was defined by Sharpe (1938, p. 21) as 'the slow downslope movement of superficial soil or rock debris, usually imperceptible except to observations of long duration'. The process, as envisaged by Sharpe, and by most geomorphologists, is dependent upon alternate heave and settling movements of the regolith (associated with freeze–thaw or cyclic moisture changes), shifting of soil by microfaunal activity, and other mixing mechanisms. The term is also used in a rheological sense by engineers, however, to denote small strains in earth material (and also in some rocks, e.g. rock-salt (Hendron, 1968)) at constant shear stresses which are smaller than the critical value to cause failure (e.g. a slide). Terzaghi (1950) recognized this distinction and differentiated between 'seasonal' and 'continuous' creep. The most significant feature of the latter process, however, is its ability to produce mass-movement of earth material at depths far below the influence of weathering and most biological

Table 4.1. A classification of mass movements. The symbols attached to those movements classified as 'frequent movements at same place' indicate the mode of remoulding or fragmentation of the material according to the legend in the second column

| Frequency of movement | Condition of material | Type of material: Rock | Type of material: Earth | Diagnostic characteristic | Name refers to |
|---|---|---|---|---|---|
| Continuing | Intact | Depth-creep | Depth-creep | Slow, shear strain | Process |
| Infrequent movements at the same place | Primary phase — failure: Initially intact | Slab-fall<br>Rock-slump<br>Rock-slide → <br>Rock-avalanche | Earth-fall<br>Earth-slump ⌐<br>Earth-slide ⌐→<br>Earth-avalanche<br>Earth-flow | Free-fall<br>Rigid body rotation<br>Translational slip<br>fragmented viscous | Event |
|  | Possible secondary phase: Remoulded during failure |  |  |  |  |
| Frequent movements at same place | Remoulded or fragmented by: Weathering*<br>or<br>Previous mass wasting upslope+ |  | Solifluction*<br>Mudslide+<br>Mudflow+ | Slow-flowage<br>Slip surface<br>Flowage | Process feature |
|  |  | Rock-fall* |  | Free-fall of fragmented rock | Process |
| Continuing | Remoulded | Talus-creep | Soil-creep | Heave/settling or biotic mixing | Process |

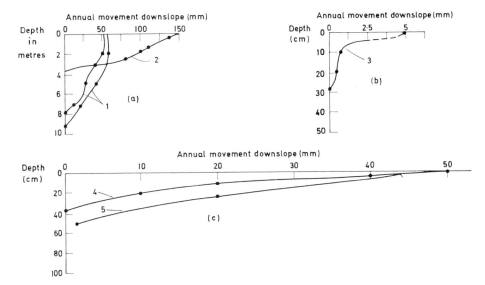

Figure 4.1 Velocity profiles for different types of slow mass-wasting: (a) depth-creep; (b) soil-creep (scale = (a) × 20); (c) gelifluction (scale = (a) × 10). (Redrawn after (1) Kojan (1967); (2) Ter-Stepanian (1965); (3) Carson and Kirkby (1972); (4) Williams (1966); (5) Rudberg (1962).)

activity. The terms *surficial-creep* and *depth-creep* (following Ter-Stepanian (1965)) are therefore, perhaps, preferable; the former is commonly separated into *talus-creep* and *soil-creep* according to the type of debris. Velocity profiles for depth-creep and surficial-creep, from the few available sources, are shown in Figure 4.1.

Catastrophic movements may take the form of falls, slumps, or slides and, if considerable structural disturbance (remoulding) of the mass occurs during failure, may develop into avalanches or flows.

Falls are defined (following Varnes, 1958, p. 23) as movements in which the mass 'travels mostly through the air by free fall, leaping, bounding or rolling, with little or no interaction between one moving unit and another'. Table 4.1 distinguishes between isolated events, involving the fall of an initially intact mass (slab-falls and earth-falls), and the intermittent process of release and fall of rock rubble. The character of the two types of movement, and the mechanisms involved are quite different. *Slab-falls* and *earth-falls* commonly originate from the removal of lateral pressure on cliffs and bluffs. They are characteristic of massive, cohesive formations and are preceded by the development of deep tension cracks within the mass. An extreme example of a slab-fall is the collapse of Threatening Rock, a huge monolith of sandstone in the Colorado Plateau, documented by Schumm and Chorley (1964); similar, but smaller, earth-falls have been described in loess by Lohnes and Handy (1968) and in clay by Hutchinson (1970). The term *rock-fall* is usually applied to the detachment and fall of fragmented rock material from cliffs in well-fractured rock masses. It is a surficial process dependent upon rock-weakening by weathering. Rockfall is particularly frequent from cliffs in cold areas (Rapp, 1960; Bjerrum and Jorstad, 1968) where freeze–thaw of water in fractures is a major mechanism of breaking off fragments from the main rock mass. Rock-fall often produces talus or scree at the cliff base, but this may also happen after slab-falls or earth-falls (e.g. Hutchinson, 1970), if fragmentation of the mass occurs on impact with the ground.

*Slides* denote rigid body translations of slabs of intact earth or rock along a distinct, approximately planar, slip surface. They should be distinguished from earth or rock avalanches in which, after failure, the mass becomes considerably remoulded or fragmented, and continues to travel, often at high velocity, far past the site of the initial slide.

The term *rock-avalanche* has been previously used (Carson and Kirkby, 1972, pp. 120–125) to describe a rapid movement in fractured rock masses which, beginning as a slide, involves fragmentation of the mass as it travels downslope. If they occur high upslope, rock-avalanches may attain very high velocities and, in extreme cases, e.g. Frank, Alberta (Kent, 1966), Elm, Switzerland (Howe, 1909), may continue travelling up the opposite valley slope. These rare occurrences may be dependent upon unusually high pore (air) pressures in and under the mass (see Section 4.2) as suggested by Shreve (1968b).

A well-documented example of an *earth-avalanche* (synonymous with Sharpe's (1938) debris-avalanche, a term widely used in North America) is the Caneleira failure, Brazil (Figure 4.2(b)) described by Vargas and Pichler (1957). The failure involved a sandy-clay regolith overlying parent gneiss; the mass rapidly disintegrated, overriding the ground downslope, and left behind a broad, but shallow scar of exposed bare rock. *Earth-slide* masses may, if they occur upslope of a cliff or bluff, travel far downslope but, more usually, they will encounter stable ground immediately downslope and, because the mass remains intact, movement is relatively small. The Jackfield slide (Figure 4.2(a)), for instance, involved movements of only 10–20 m (5–10% of the length of the slide mass), a marked contrast to the Caneleira failure. Why some earth-slides develop into avalanches and others do not, is not fully clear. Steepness of the hillside is presumably a factor. Vegetation may also be important in binding regolith mantles and reducing disintegration. The natural cohesiveness of the slide mass, the rate and amount of strength loss on remoulding, the consistency of the material, and drainage conditions during shear may all be relevant.

Whatever the reason for the different behaviour of earth-slides and earth-avalanches, the distinction between them is sufficiently important geomorphologically to warrant emphasis. The latter type of movement entails complete stripping of material from hillsides, e.g. the scars of earth avalanches on Exmoor during the 1953 storm (Gifford, 1953), and, with repeated movements during the course of geological time, leads to slope retreat. Earth-slides occur-

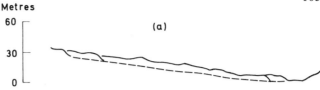
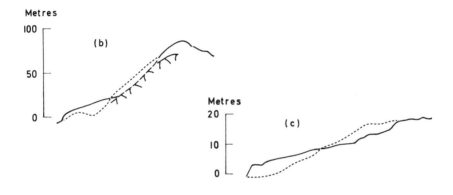
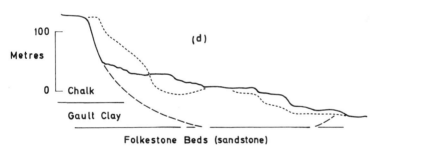

Figure 4.2. Varieties of slope failure: (a) slide (Jackfield, England) in stiff, fissured clay; (b) earth-avalanche (Caneleira, Brazil) of sandy clay regolith over gneiss; (c) slump (Lodalen, Norway) in clay; (d) slump-slide (Folkestone Warren, England). Dotted lines denote pre-failure slope profile; solid lines denote post-failure profile; dashed lines denote failure surface. (Redrawn after (a) Henkel and Skempton, 1954; (b) Vargas and Pichler, 1957; (c) Sevaldson, 1956; (d) Hutchinson, 1969.)

ing intermittently on the same hillside do not lead to retreat but, as discussed in Section 4.4, possibly produce slope profile changes similar to soil-creep.

The terms slump, slide and slip are often used interchangeably to describe rigid body movements, but a *slump* is quite distinct from a slide in being a rotational, rather than a translational, movement. Slumps, *sensu stricto*, just like slides, involve limited displacements; the motion itself, by reducing the surface slope of the mass (Figure 4.2(c)) in this case helps to restore stability. The occurrence of slumps, *vis-à-vis* slides depends primarily on the uniformity (in strength) of the slope: slides usually occur on slopes where there is an abrupt change in strength between surficial and underlying material; deep-seated slumps are restricted to uniform masses, although shallow slumps may occur in the regolith. In certain situations, where an underlying competent material interferes with the full development of a rotational movement (Figure 4.2(d)), a *slump-slide* may occur, combining both processes. Slumps are found only in cohesive material; rock-slumps tend to be more irregular than earth-slumps, because of pre-existing planes of discontinuity in the mass, and often occur in combination with slides. If the movement of a slumped mass is excessive, lateral support of the mass behind may be reduced sufficiently to cause, either immediately or much later, further slumps progressing retrogressively upslope.

Slumps are commonly subdivided using the geometry of the slope surface, e.g. circular, shallow, non-circular (Skempton and Hutchinson, 1969); equally important is the post-failure behaviour of earth-slumps. If a slumped earth mass becomes broken up, it may, especially if wet, travel by flowage beyond the site of the initial failure. Such *slump-earthflows* are recognized in most classifications of mass-movements. In soft earth masses, remoulding of the material may produce very rapid flowage and sudden release of supporting pressure on the mass upslope; this, possibly in conjunction with structural collapse of the mass due to shock (Section 4.2), may lead to rapid retrogression of the unstable area by slumping. The viscous behaviour of the mass, after failure, is so striking in these cases that the term *earthflow* (with the slump prefix omitted) is often used, e.g. Sharpe (1938), Peck and coworkers (1951) and Carson and Kirkby (1972), to describe these catastrophic movements. Such earth-flows are especially common in the St. Lawrence Lowlands of eastern Canada and in western Norway, creating huge bowl-shaped depressions.

Slow earth flowage may take place in front of slumping cliffs and bluffs for long periods of time in the form of large sheets and lobes of muddy debris, e.g. Gros Ventre, Wyoming (Blackwelder, 1912), and continual slump-flowage merges into *mud-flow* movements. Orginally (as defined by Blackwelder (1928) and used by Sharpe (1938)) mud-flows were distinguished from earth-flows in that they were movements of previously remoulded debris accumulated in narrow, steeply sloping valleys and ravines, i.e. they resembled streams. They were believed to be particularly characteristic of mountain terrain in semi-arid areas, producing debris fans (comparable to alluvial fans formed by fluvial action) at the piedmont junction. More recently, however, the term mud-flow has been applied to the much slower (maximum rates of 10–30 cm day$^{-1}$) movement of debris lobes and streams in front of unstable clay cliffs along the coast of Britain, e.g. Hutchinson (1970), Prior, Stephens and Douglas (1971) and Brunsden and Jones (1972), although (see below) some of these are in fact *mudslides* rather than flows. Blackwelder's usage of the term mudflow did not, apparently, carry any implication regarding the mode of deformation of the debris and, indeed, he himself seemed to believe that sliding occurred: 'Observers are not wholly agreed as to how the mass moves, but the various phenomena indicate that it slides or glides over the surface without that internal churning that characterizes a rapid stream of water' (Blackwelder, 1928, p. 473). Johnson (1970) suggests that the movement of debris flows at Wrightwood, California, corresponds to Bingham (viscoplastic) deformation: the flows are stagnant until a critical condition is attained, at which time shear occurs within the lower layers, carrying a dead crust of intact debris. The mechanics of mudflow movement are still poorly understood.

Slow-flowage is not restricted to accumulation slopes beneath cliffs, however,

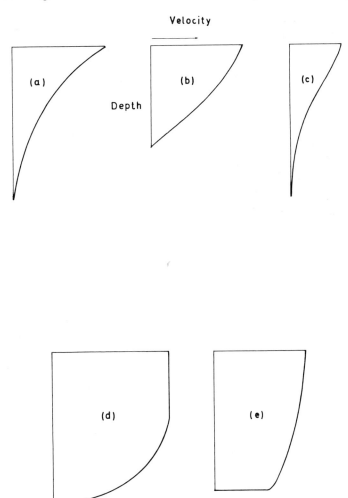

Figure 4.3. Theoretical velocity profiles for different modes of slow mass-movement: (a) heave and settling; (b) viscous flow; (c) combination of (a) and (b); (d) viscoplastic flow (Bingham flow); (e) basal sliding.

according to many observers, and the term *solifluction* has long been in use (Andersson, 1906) to describe the process. Unfortunately, the whole topic of earth flowage is badly confused by casual and inconsistent use of terminology. Solifluction, for instance, is often defined as a periglacial process but, although (believed to be) widespread in tundra environments, Andersson did not claim that it was restricted to cold areas. It has therefore been proposed (Baulig, 1956; Washburn, 1973) that the term *gelifluction* be used in reference to the special case of solifluction in the active layer of tundra areas. More seriously, under what conditions slow earth flowage, i.e. involving internal shear strain rather than basal slip (Figure 4.3), really exists, is a matter of some uncertainty. Hutchinson (1970), for instance, after investigating slowly moving (14m year$^{-1}$) mud lobes in the Thames Estuary, has shown that movement, in fact, is a sliding motion. Possibly other lobes, believed to be solifluction phenomena, are really *mudslides*. Chandler (1972b), after examining gelifluction-like mud lobes in Vestspitsbergen, inferred (from the abrupt discontinuity between surficial and underlying material) that they were mudslide features; unfortunately no velocity profiles were obtained. Nonetheless, Williams (1966), Rudberg (1962), Benedict (1970) and others have all demonstrated curved, rather than vertical, velocity profiles in moving debris mantles on tundra slopes. These profiles are concave downslope (Figure 4.1(c)), rather than upslope (as might be expected from true viscous flow (Carson, 1971, p. 89)), and are comparable in shape with soil-creep profiles. Indeed, as emphasized by Washburn (1973, p. 173), it is difficult to distinguish between the separate effects of gelifluction and frost-induced soil-creep. The shear strain displayed by gelifluction velocity profiles is, however, considerably greater than that produced by soil-creep (Figure 4.1) and it is believed (e.g. Williams (1966)) that annual frost heave and subsequent settling alone could not account for the magnitude of these movements. Again, the topic of slow earth flowage awaits a great deal more quantitative research.

The question why some mass-movements occur along a well-defined shear surface (slumps and slides) while others begin as, or develop into, flows, is one which has attracted attention among geomorphologists for some time (e.g. Yatsu (1966, p. 103)). The necessary evidence, e.g. velocity profiles, data on consistency, mineralogy etc. of the moving mass, is unfortunately meagre. Available observations do suggest, however, that the Liquidity Index of the material involved (a measure of the actual moisture content relative to the liquid limit) is a useful diagnostic. The liquidity chart of Figure 4.4 indicates a clear distinction between earthflows, gelifluction, the Caneleira earth-avalanche and other flow features (above the LI = 1 line) and slumps and slides. High Liquidity Indices are not common among consolidated argillaceous sediments. In over-consolidated clays, the index is typically very low, often approaching zero, and even for normally-consolidated clays (Figure 4.5) high values are rare. Exceptions are deposits with open fabrics, e.g. the Champlain Clays (Leda Clays) of eastern Canada, and possibly radiolarian earths. Values should be higher in surficial debris; this is particularly true for silty regoliths (for which

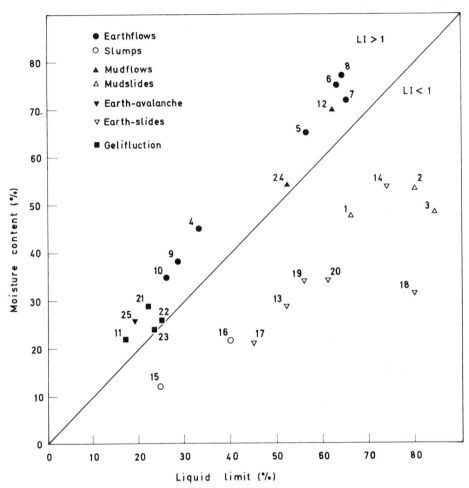

Figure 4.4. Types of mass-movement (flows versus slips) plotted in relation to Liquidity Index of material. 1. Isle of Sheppey; 2. Bouldner, Isle of Wight (both reported by Hutchinson and Bhandari (1971)); 3. Beltinge (Hutchinson, 1970); 4. St. Thuribe (Peck and coworkers 1951): 5. Nicolet (Hurtubise and Rochette, 1956); 6. Hawkesbury (Eden and Hamilton, 1957); 7. Green Creek; 8. Breckenridge (both reported by Crawford and Eden (1967)); 9. Manglerud (Bjerrum, 1954); 10. Bekkelaget (Eide and Bjerrum, 1954); 11. Mesters Vig, Greenland (Washburn, 1967); 12. Stockosick, Derbyshire; 13. and 14. Atlow Mill, Derbyshire (all from author's files); 15. Selset (Skempton and Brown, 1961); 16. Monte Serrate, Brazil (Vargas and Pichler, 1957); 17. Jackfield (Henkel and Skempton, 1954); 18. London Clay (Skempton, 1964); 19. Rockingham and 20. Uppingham (both in Lias Clay reported by Chandler (1970b)); 21., 22. and 23. Okstindan, Norway (Harris, 1972); 24. Melvin Hill, Barbados (author's files); 25. Caneleira, Brazil (Vargas and Pichler, 1957).

liquid limit values are small) in tundra areas, where frozen water accumulation can be large, and is probably the reason why solifluction is so common on such slopes.

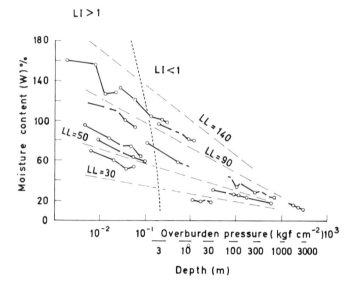

Figure 4.5. Sedimentation compression curves for normally consolidated argillaceous sediments (redrawn after Skempton, 1970).

Finally, it should perhaps be emphasized that, while mass-movements, particularly falls, avalanches, slides and creep, are commonly treated as processes of slope profile development, their role in geomorphology is much broader than this. Earth-falls and slumps are often active mechanisms of bank erosion in alluvial stream channels and may play a more important role in controlling hydraulic geometry than previously considered. Earth-slumps and slump-earthflows, together with mudflows and fluvial erosion of the broken up material, have also been observed to be major processes in the extension of drainage networks during periods of rejuvenation (e.g. Johnson and Rahn (1970)) and, certainly, warrant more study than presently afforded by geomorphologists concerned with stream network morphometry. Earth-flows, in contrast, directly fashion their own topography in areas of soft rock, quite differently from the normal combination of fluvial and mass-movement processess. And, finally, the deposits of mass-movements, including talus cones, piedmont 'alluvial' fans, block streams and infilled valleys (e.g. Slumgullion, Colorado (Howe, 1909)), may themselves constitute distinctive landforms. The role of mass-movements in geomorphology is indeed a varied one.

## 4.2 Mechanisms of mass-movement

Mass-movements are rarely attributable to a single cause; they stem from changes, some gradual, others sudden, in many factors controlling the stability

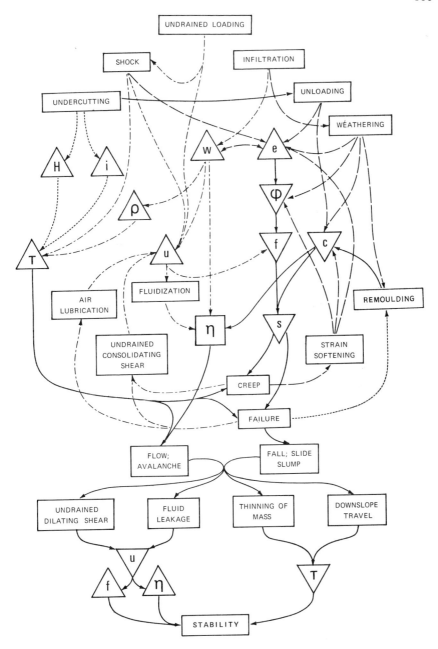

Figure 4.6. Mechanisms of mass-wasting (symbols are defined in text). Conventional triangle denotes increase in value of variable represented by symbol within triangle; inverted triangle denotes a decrease; dotted line denotes sudden, irreversible change; dashed line denotes slow, irreversible change; alternation of dots and dashes denotes sudden reversible change; solid line denotes change undifferentiated by type.

of a sloping mass. These changes and their ramifications are shown in flow-chart form in Figure 4.6; emphasis is directed to the distinction between short-term changes, e.g. in rainstorms, earthquakes and other short-lived events, and longer term changes, e.g. deconsolidation (see below) and weathering. While short-term changes are often the trigger for a rapid mass-movement, their effectiveness is strongly controlled by slow, but cumulative, changes over a longer period. Schumm and Chorley (1964), in their account of the fall of Threatening Rock, and Hutchinson (1961) in his analysis of the landslide at Furre, Norway, provide two of the few available accounts of progressive deterioration in slope stability leading to rapid movements.

The long-term changes in strength($s$), though poorly understood, may be conveniently grouped into (i) unloading and associated deconsolidation and (ii) alteration by the interrelated processes of biological, physical and chemical weathering.

*Unloading*, the reduction in confining pressure at points within an earth or rock mass, is an inevitable consequence of denudation, and the volume expansion, *deconsolidation*, which accompanies the release in pressure is a universal mechanism of strength loss, either by fissuring and fracturing the material (and thereby reducing the cohesion ($c'$)), or by swelling of earth masses, again reducing cohesion and, by increasing the void ratio ($e$), decreasing frictional resistance also.

Unloading and deconsolidation may take many forms depending on the material involved. Igneous domes, exposed by the erosion of surrounding sediments, commonly develop exfoliation joint surfaces parallel to the surface; if the surface is steep enough this may lead directly to mass-movement in the form of a slab-fall or a rock-slide; if not, weathering may be necessary to break down the surficial rock further. Dissection of massive sandstone formations by streams, e.g. in southwestern USA, leads to deep tension cracks (Carson, 1971, pp. 102–103) parallel to the canyon walls, and provides advance warning of slab-falls; indeed, provided that a mechanism exists for removing the accumulated debris, lateral unloading and formation of tension cracks may constitute a mechanism of scarp retreat far beyond the initial canyon wall. Koons (1955), for instance, describes such a process in New Mexico in which removal of talus and basal undercutting are maintained by gullying at the cliff base. Many of the rock-slides in Norway have been shown (Bjerrum and Jorstad, (1968)) to originate on joint surfaces parallel to the ground slope, attributable to unloading by glacial erosion. Aisenstein (1965) has described similar deconsolidation fissures in limestone in Israel and emphasized their role in promoting slides. Unloading of limestone is also believed (Kiersch, 1964) to have contributed to the failure of the Vaiont Dam in Italy. In other instances, rather than large, discrete joint surfaces, a network of random gaps and fractures may be produced by the release of pressure; as noted by Terzaghi (1962), intensification of such networks may gradually transform an intact rock mass into a cohesionless, though densely packed, aggregate of angular blocks, leading to rock avalanches. Skempton (1964) has outlined the compara-

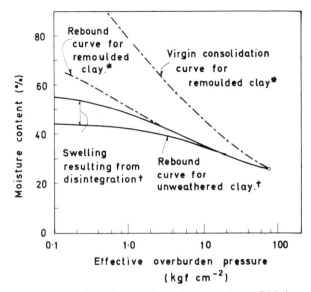

Figure 4.7. Changes in moisture content of Little Belt clay during consolidation, unloading and weathering (redrawn after Bjerrum, 1967). *Based on laboratory tests; †based on field observations.

ble role of fissures in over-consolidated clays, providing an explanation for the frictional, rather than cohesive, character of such clays at failure.

Strength reduction by deconsolidation is not confined to destruction of the cohesive component. Continued volume increase, i.e. decreased density of packing, by distingeration of intact blocks, reduces interlocking and thus the frictional component ($\phi'$) of strength. This appears to be particularly important in clays and clay shales (Bjerrum, 1967) where intake of water ($w$) by, and between, fissures accentuates swelling. The factors controlling the limit of this process are still unclear, although it may be inferred from the swelling (rebound) curves of undisturbed and unweathered clays (e.g. Figure 4.7) that there is a definite limit to strength reduction by deconsolidation-induced swelling alone.

Volume expansion of over-consolidated material, by both fissuring and swelling, may be considerably extended by *weathering*, i.e. *in situ* alteration, and by *strain-softening* and *remoulding* during mass-movement. Again, the processes are varied and, while many are well known, e.g. frost-action, cyclic temperature and moisture content changes and microfaunal and macrofaunal activity, their detailed effects on the strength of earth material are, as Kenney (1967a) has emphasized to engineers, not really known.

Swelling of clays as a result of *weathering* has received some attention in the last few years (Bajerrum, 1967; Nakano, 1967; Chandler, 1972a) and it has been shown that the increased water content (Figure 4.7) and concomitant strength reduction at a given overburden pressure, may be appreciable. The limits to weathering-induced swelling by moisture intake are not known, although it is unlikely that the process is sufficiently powerful to overcome all the bonds developed during consolidation and diagenesis. Moreover, swelling is not dependent upon moisture intake and, indeed, may be greater in aerobic environments. The reaction products of oxidation commonly have molar volumes appreciably greater than those of the unaltered components and can create large amounts of swelling and heave above the water table. Volume expansion of this type has been reported by Quigley and Vogan (1970) and Penner and coworkers (1970) in a weathering pyritiferous shale and is worthy of more attention than has so far been afforded it. Dessication is also believed to be effective in weakening earth material, under some circumstances, and may be a mechanism by which weathering extends fissures initiated by unloading; Eden and Mitchell (1970), for example, have recently attributed shallow slides in the Ottawa area to destruction of cohesion, by micro-fissuring, in the weathered crust of the clay slopes.

*Weathering* may also exert a strong control on the strength of surficial slope material through changes in grading (especially in coarse regoliths) and alterations of the solid and pore-fluid chemistry in fine-earth material.

Engineering tests have shown that the frictional strength ($\phi'$) of earth and rock debris, previously indicated as being dependent on density of packing, is also affected by changes in size composition. Vucetic (1958), in tests on clayey-schist material, observed that $\phi'$ (peak friction) may decrease discontinuously as fragment size decreases; similar changes were noted by Holtz (1960) during the testing of sand and gravel mixtures, with decreasing gravel fractions. Holtz and Gibbs (1955) concluded that friction values increase with gravel content (up to about 50–60%) and then show little, if any, increase and, in some cases, a decrease as the gravel fraction becomes larger. Such tests may be regarded as simulations of the weathering progress on slopes cut in jointed rock masses, and indicate a general, but non-linear, decrease in frictional strength with progressive conversion from rock rubble to fine earth. On a geological time-scale, slopes mantled by rock rubble may, therefore, be expected to experience several phases of instability (Carson, 1969) as weathering reduces the regolith to fine earth, in a manner anticipated, to some extent, long ago by W. M. Davis in his discussion of slope profile development.

The further weathering of fine-earth regoliths and deposits is less sensitive to changes in grain size and grading and far more dependent on chemical effects. Data by Kenney (1967b), for instance, has indicated the importance of minerology in determining residual (frictional) strength. Replacement, or dilution, of connate water by infiltrated surface water is believed to be a factor in the reduction of strength and increased sensitivity of the post-glacial clays of eastern Canada (La Rochelle and coworkers, 1970) and western Norway (Bjerrum, 1954). Soil scientists (e.g. Yong and Warkentin, 1966, pp. 63–66; Rose, 1966, pp. 98–115) have long been aware of the dependence of structural stability (of the soil fabric) on pore-fluid chemistry and ion-exchange on sorption surfaces and, while this has attracted attention primarily in the context of soil conservation, i.e. soil erosion, it is also relevant to mass-movement studies.

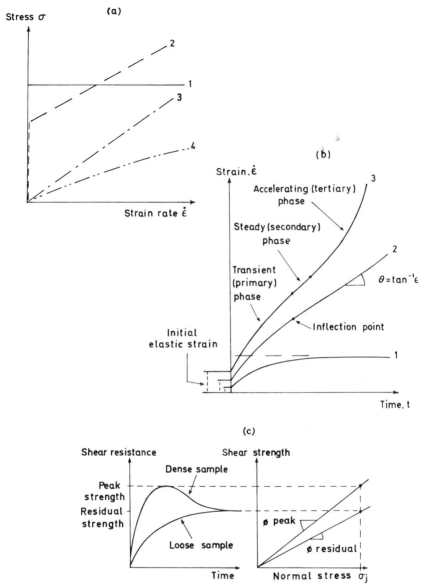

Figure 4.8. Stress–strain relationships for earth materials: (a) rheology of (1) ideal rigid, plastic material, (2) Bingham solid, (3) Newtonian fluid, (4) non-Newtonian fluid; (b) contrasting modes of creep: (1) attenuating creep, (2) steady creep, (3) accelerating creep; (c) strain-softening from peak strength to residual strength for cohesionless material.

If rock and regolith could be regarded as ideal rigid plastic materials, the stability of a slope would be a simple dichotomous affair: the slope is either stable (with a factor of safety in excess of unity) or is in the process of failure.

This follows immediately from the stress–strain curve of Figure 4.8(a). For at least two reasons, however, this approach is sometimes invalid. Firstly, some materials creep (deform very slowly) at stresses less than the so-called shear strength of the material (Figure 4.8(a)); in addition, after an initial period of decelerating creep (under constant stress), and a subsequent period of steady creep, deformation may begin to accelerate (Figure 4.8(b)) leading, eventually, to failure. This acceleration in the creep rate with increased strain (deformation) is one aspect of a process referred to as *strain-softening* by engineers. Under these conditions the stability of a slope is a function of time: without any change in external conditions, a rapid mass-movement will eventually occur arising from the strain-induced weakening of the mass. Secondly, once failure has occured, there is commonly an abrupt decrease in strength (Figure 4.8(c)), from peak to residual, associated with the early post-failure strain. This decrease in strength may originate in several sources: dilation (volume expansion) during shear is known to be a factor in initially dense materials; increased preferential orientation of platy particles during shear is another. The significance of this phenomenon has, however, only recently been appreciated (Skempton, 1964; Bjerrum, 1967). On an 'infinite' slope, with uniform stress and strength at a given depth, the only effect of this would be an acceleration of the post-failure movement, although this may extend the amount of travel of the slide mass. On an undercut slope, however, the shear stress ($\tau$) at the slope base may be greater than the gravity stress (assumed in the 'infinite' slope model) leading to *local* strain on that part of the slope. If this occurs in material which displays sufficient strain-softening (from peak to residual strength), it may, as Bjerrum (1967) and Wilson (1970) have shown, lead to a progressive (spatial) extension of strain upslope. The occurrence of a rapid mass-movement, in this case also, may therefore demand a long period of time.

The full destruction of the structure and associated cohesive bonds in a mass, whether by artificial or natural means, is termed *remoulding* by engineers. Full strain-softening throughout a mass will contribute to remoulding, but other natural processes, e.g. earthquake shocks, heave effects of weathering and post-failure disturbances of slide masses, are likely to be far more important. The significance of remoulding is that the entire rheological character of a mass may change from quasi-plastic to viscous if amounts of pore-fluid are large enough. Crawford (1968, p. 244) presents a photographs of two samples of Leda Clay (at the same moisture content) showing the undisturbed sample supporting a load in excess of 100 N and the remoulded sample flowing from a beaker. It is hardly surprising that slumps in this material degenerate into earthflows (Section 4.1) although the details of the remoulding process are not fully understood.

More than any other factor, *infiltration* of water during rainstorms and snowmelt periods acts as a trigger for most rapid mass-movements. In part this is due to the increased bulk unit weight ($\gamma$) of the mass and an associated increase in shear forces, but this is relatively minor. More important is the rise of water table levels, or the creation of perched water bodies, building up pore pressures ($u$) in the mass, reducing effective normal stresses (Section 4.3) and thus reducing friction ($f$) developed on potential shear surfaces. Reports by Henkel and Skempton (1954) on the Jeckfield slide, by Vargas and Pichler (1957) on the Caneleira slide avalanche, by Skempton and Brown (1961) on the Selset slump and by Chandler (1970a) on a renewed slide in Lias Clay, and many others, confirm the role of pore pressure build-ups in triggering rapid mass-movements. In addition, measurements by Mitchell and Eden (1972) on stable clay slopes in the Ottawa area have shown that rates of depth-creep increase with seasonal peaks in groundwater pressure. More recently, Hutchinson and Bhandari (1971) have demonstrated that movements of mud lobes on slopes beneath cliffs of London clay may be accelerated by transient artesian pressures caused by periodic falls of debris on to the lobes; this mechanism of rapid *undrained loading* may be far more significant than hitherto recognized in maintaining movement of wet debris downslope of unstable cliffs. Infiltration of water into remoulded masses is also possibly a common mechanism in the reactivation of stable mudflows and solifluction features.

Under certain circumstances, pore pressures may intensify considerably during, and immediately after, failure and may play a major role in determining the magnitude and character of post-failure deformation. Shear of loose or soft material, for instance, is typically accompanied by consolidation and expulsion of pore-fluid; if shear is rapid, relative to the drainage rate, pore pressures will increase leading to a reduction of strength. It has been suggested (Crawford and Eden, 1967) that such *undrained shear* occurs in certain sensitive clay deposits; it may also be important in the liquefaction of saturated sands during earthquakes. Excess pore-fluid (air) pressures are probably also important in the more catastrophic forms of rock slides and avalanches; Shreve (1968a, 1968b) believed that such slide masses, when they occur high up rock slopes, trap a cushion of compressed air beneath them and, using this *air lubrication*, accelerate rapidly downslope. Leakage of the entrapped air through flow may lead to *fluidization* (Reynolds, 1954) of the debris, converting the movement into an avalanche as postulated by Kent (1966) for the Frank catastrophe.

Not all mass-movements during rainstorms are dependent upon the build-up of pore pressure, however, and many are attributable directly to the *undercutting* of slopes by streams, with the concomitant increase in shear stresses ($\tau$) inside the valley slopes. Cohesionless masses show an immediate response to undercutting, being unable to support vertical faces; but cohesive masses will not fail until a critical combination of slope height (H) and angle ($i$), in relation to the strength and bulk density of the mass, is attained, and a deep-seated failure (wedge-slide or a slump) occurs. Stability charts (e.g. Figure 4.9) have long been used for predicting such failures.

Analysis of the dynamic aspects, i.e. *shock*, of slope instability, stemming from earthquakes and tremors, blasting, sonic booms, vehicle traffic on high-

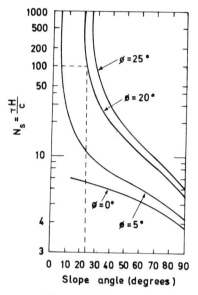

Figure 4.9. Stability chart for prediction of deep-seated slips (redrawn after Scott, 1963). $N_s$ = stability number; $\gamma$ = bulk unit weight; $H$ = slope height; $c$ = cohesion; $\phi$ denotes angle of internal friction.

ways and related phenomena, has lagged far behind the static aspects. Whatever the detailed mechanisms involved, it is evident that tectonically induced mass-movement is a significant process in many parts of the world. Simonett (1967), for example, demonstrated a strong spatial association between frequency of landslide scars and distance from the epicentre of the 1935 earthquake in northern New Guinea. The Alaska earthquake of 1964, and the catastrophic failures associated with it, has spurred research in the field, especially in North America; numerous articles have also been published in the Japanese geotechnical journals. Shock involves both transient and cumulative weakening of ground material (e.g. by pore pressure build-up and by fissuring, respectively) and cyclic fluctuations in stresses. Various workers (e.g. Newmark (1965) and Seed (1967)) have emphasized the need for a rational approach to the problem, but the dynamic aspects of mass-movements are far from being fully understood.

While much more is known about mass-movement than in 1950 when Terzaghi published his classic paper on 'The mechanism of landslides', topics such as strength loss on deconsolidation and weathering, progressive failure, the effects of pore-fluid chemistry and soil dynamics are likely to prove fruitful avenues of research for several years to come.

## 4.3 Characteristic and threshold angles of slopes in non-cohesive material

### 4.3.1 Threshold slope angles

Although few earth materials can be regarded, *sensu stricto*, as rigid plastic in their stress–strain behaviour, most are quasi-plastic; that is, there is, for any material in a given condition, a distinct threshold stress controlling its deformation. At values of the stress below this critical level, strain is very slow and, at larger values, strain is rapid. This threshold stress value is commonly called the strength of the material. In nature, the shear stresses within a mass beneath an inclined surface cannot always be predicted accurately but, in most cases, can be assumed equal to the stresses imposed by the downslope component of the gravitational body force. (see Carson, 1971, pp. 1–14 for a general discussion of stresses in earth materials). These stresses are thus a function of overburden weight and slope angle. As indicated in the previous discussion of undercutting of cohesive masses, the critical combination of slope angle and height can be predicted for a mass provided its engineering properties are known. For non-cohesive materials, slope height is of much less significance and the critical stress condition is represented by a particular *threshold slope angle*, which will vary with the frictional properties of the material.

The stability analysis involved in predicting this threshold angle (treating the slope mass as infinite in width and length, and bounded by a potential failure plane parallel to the ground surface) is quite simple. The threshold angle

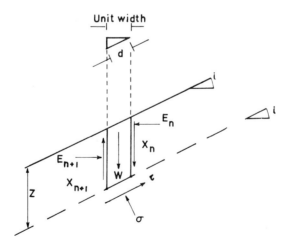

Figure 4.10. The infinite slope approach to shallow slide stability analysis. $E_n = E_{n+1}$; $X_n = X_{n+1}$ (assumed); shear stress $= \tau = \sin i \div d = \gamma z\ i,\ \cos i;$ normal stress $= \sigma W \cos i \div d = \gamma z\ \cos^2 i;$ strength $= s = \sigma'\tan \phi' = (\gamma z \cos^2 i - u) \tan \phi';$ $u =$ porewater pressure on failure surface.

($i_t$) is given by balancing shear stress ($\tau$) and strength ($s$) on any section of the failure plane:

$$(\tau) = \gamma z \sin i_t \cdot \cos i_t \, (=s) = (\gamma z \cos^2 i_t - u) \tan \phi'$$

using symbols defined in Figure 4.10. The threshold slope angle is thus a function of the angle of internal friction (also termed angle of shearing resistance) ($\phi'$) of the material comprising the failure zone, the maximum pore-fluid pressure ($u$) on the failure plan and, within much narrower limits, the bulk unit weight ($\gamma$) of the mass. The restriction of this equation to non-cohesive material is not as severe as it would appear because, over long time-periods (as indicated in Section 4.2) the cohesive component of strength, particularly near the ground surface, is destoryed by unloading and weathering. Skempton and DeLory (1957) have demonstrated the validity of this equation for the long-term stability of clay-mantled slopes which, by fissure development, have become essentially cohesionless. Terzaghi (1962) similarly suggested that the steepness of many rock slopes was controlled by the frictional strength of the fractured rock mass and was completely unrelated to the cohesive strength of intact rock samples from the mass. The concept of threshold slope angles should, therefore, be of widespread significance, applicable to various slope materials, and embrace slope angles ranging from over seventy to less than five degrees.

Two categories of threshold slope angle are particularly frequent: the $u = 0$ case, e.g. dry rock rubble, and the $u = \gamma_w z \cos^2 i$ case which can be shown (Skempton and Delory, 1957) to correspond to saturation of the mass (or merely of the regolith (Carson and Petley, 1970)), with the water table at the ground surface and seepage downslope parallel to it. In the former case, substitution for $u(=O)$ in Equation 4.1 yields a threshold angle given by

$$i_t = \phi' \tag{4.2}$$

This may be defined as a *frictional* (threshold) slope angle. In the latter case, substitution for $u$ in the general equation yields

$$\tan i_t = \left[\frac{\gamma - \gamma_w}{\gamma}\right] \tan \phi'$$

$$\tan i_t \doteq \frac{1}{2} \tan \phi' \tag{4.3}$$

This is defined here as a *semi-frictional* slope angle. Threshold slope angles in the London Clay (Skempton and DeLory, 1957) and in the Lias Clay of the East Midlands (Chandler, 1970a) have been shown to be semi-frictional. The development of pore-fluid pressures controlling semi-fractional slopes may occur quite frequently (e.g. seasonal rise of the water table to the ground surface on gentle slopes) or, where a more permeable regolith overlies less permeable bedrock, as a sporadic, transient development of a perched water system along the hillside.

Slope masses which never become saturated or dry out completely, i.e. $u$ remains negative, will possess threshold angles greater than the friction angle. Schumm (1956a) believed that slopes cut in silty clay landfill at Perth Amboy were so steep (49°) because of this extra strength imparted by (negative) capillary pressures. Unfortunately, without data on $\phi'$, this cannot be verified. Such *capillary* slope angles would, in any case, be expected to be short-lived phenomena: saturation of the surficial layer must occur on some occasion, raising pore pressures at least to zero. Similarly, slope angles intermediate between the frictional and semi-frictional values are likely to be temporary features: if water tables exist sufficiently close to the ground surface to reduce slope angles below the frictional value, then, on a geological time-scale, it must be anticipated that water tables will rise to the ground surface, producing instability on all slopes in excess of the semi-frictional angle.

Conditions in which pore-fluid pressures on hillsides can exceed those given by $u = \gamma_w z \cos^2 i$, i.e. artesian conditions occur, are not widely reported in the literature. Artesian pressures are known to occur in shallow, surficial aquifers in tundra areas, confined between surface frost and permafrost during spring thaw periods, and Washburn (1973, p. 89) has emphasized that they may play a significant, though previously overlooked, role in mass-movement on tundra slopes. Chandler's (1972b) findings confirm the occurrence of artesian pore pressure in the development of earth slides in such areas. Whether these are isolated occurrences, or whether *artesian* slope angles are the norm in tundra areas has yet to be established. Slides on slopes with angles less than the semi-frictional value have not been widely reported outside tundra areas except (Section 4.5) in relation to inferred former permafrost conditions. Artesian pressures are known to occur beneath clay regoliths overlying more permeable interbedded shales, sandstones, and limestones in certain areas of England (e.g. Rycroft, 1971), but have yet to be incorporated into slope stability analyses.

Implicit in the previous discussion is the notion that pore pressures, at their maximum, will be uniform downslope along a potential failure plane. In situations where perched water systems occur, this is probably valid. But with classic groundwater flow nets, involving descending flow in upslope areas and asending flow in the valley bottom, pore pressures will increase downslope from the divide. If the strength characteristics of the mass or regolith are uniform, this means that the threshold angle must decrease downslope. This offers a rather obvious hypothesis for the development of concave basal slopes, although it does not seem to have been considered previously.

*4.3.2 Characteristic and threshold slopes*

Whether or not the concept of threshold slope angles is important in slope development depends on the rate at which, and the manner in which, other processes alter slope form, once rapid mass-movements have reduced hillsides to a limiting stable angle. If flattening of hillside slope profiles is as significant a process as geomorphologists have believed in the past, slopes will stand at the threshold angle for only a brief period of time, after which the slope profile is dictated by other processes. In such circumstances, threshold slope angles have no lasting significance and actual slope angles are essentially a function of time. This was the conclusion reached by Young (1961, p. 130): 'the majority of the characteristic slope angles of an area are related to local morphological history, and are not intrinsic features of slope development'.

It is unfortunate that Young grouped together all measurements of slope angle in his study, irrespective of whether drawn from a convex crest slope, a straight slope or a concave basal slope. The concept of a limiting or threshold slope angle applies primarily to straight slopes and has little meaning in the context of curved lengths of profile. By combining measurements of slope angle for many short lengths, on only a few slope profiles, instead of focusing on the angle of the straight section of many hillslope profiles, it is possible that Young's data are misleading. The data presented in this section show, in marked contrast to Young's thesis, that the frequency distribution of slope angles for straight slopes is strongly controlled by the threshold angle (or angles) of the regolith.

These data are summarized in Figure 4.11. They refer to observations in six areas. Slope regoliths of differing strength, depending especially on weathering, may occur in the same area imparting varying degrees of bi- or tri-modality to the histograms. Six different categories of slope are indicated: rock slopes (highly fractured); rubble slopes (mantled by a regolith of loose rock debris derived from weathering of the fractured rock beneath); talus slopes (accumulations of similar rubble derived from falls and slides from upslope); earth rubble slopes (similar to rubble slopes, but in which weathering has converted part of the rubble to finer earth; comparable to Wentworth's (1943) taluvium on mobile hillsides); silty-sand slopes; clay slopes. Only limited discussion of each area is possible here, but attention is directed to the strong dependence of the histograms on the threshold slopes of the various regoliths.

The Mesa Verde *rock* slopes occur in fractured Mancos Shale, are actively undercut by highly ephemeral closely-spaced gullies, and are bare of vegetation. The region is semi-arid, water table levels are far below the ground, and the thin surface crust (5 cm) over the fractured, thinly bedded shale is not conducive to perched water systems. It is perhaps not surprising then that the slope angle histogram is confined entirely within the range of the frictional angle of the material and displays a mode almost identical to the modal $\phi'$ angle of the surficial shale. The Verdugo Hills slope data (surveyed by Strahler (1950)) refer to steep granite-gneissic *rock* slopes in ravines, cut by ephemeral streams, and to gentler *rubble* slopes and *talus* slopes, located where the slope base is protected from undercutting. Unpublished tests (author's files) on repacked samples from the rock slopes yielded $\phi'_p$ values from $38°$ to $55°$ (a function of the density of packing), again, enclosing the histogram of angles of undercut rock slopes. The modal $\phi'_p$ angle ($46°$) is close to the modal slope angle. Shear strength tests on the sandy rubble produced $\phi'$ values ranging from $45°$ ($\phi'_p$) to $33°$($\phi'_r$), again, almost coincident with the slope angle data. Chandler (1973) has suggested that angles of straight talus slopes are generally less than the $\phi'$ value of the

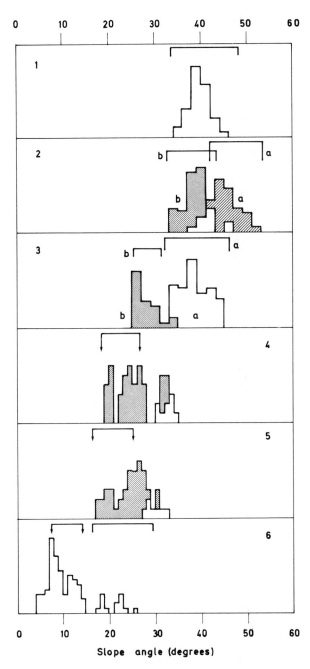

Figure 4.11. Histograms of slope angles in relation to range of threshold slopes for six areas. 1. Mesa Verde, Colorado (shale): straight slopes 105 angles; 2. Verdugo Hills, California: maximum angles (171 under cut slopes (a); 33 protected slopes (b)); 3. Queenston Shale, Ontario: straight slopes (114 angles) subdivided according to regolith ((a) rubble, (b) clay); 4. Exmoor and Pennines, England (sandstones): stable straight slopes (54 angles); 5. Laramie Mts., Wyoming: stable, straight slopes (46 angles; unshaded denotes dry scree); 6. Namurian clayshales, Derbyshire: stable straight slopes (93 angles). Sources: 1, 2, 3 and 6: author's files; 4: Carson and Petley, 1970; 5: Carson, 1971; slope angle data in 2 from Strahler (1950). Threshold angles are differentiated into frictional (without arrow head) and semi-frictional.

talus, but this does not appear to be true of the Verdugo Hills. The angle of repose of the gneiss debris (36°–38°) and the value of $\phi'_p$ for the debris in the loosest state ($\phi_{cv}$), found to be 37°, are quite close to the modal slope angle on protected slopes. The strength data from the Verdugo Hills thus emphasize the wide range in $\phi'$ possible according to the condition of the regolith. Data from valley slopes in southern Ontario cut in shale and mantled by regoliths in varying stages of weathering, display similar features: straight slopes stand at, or close to, the friction angle, but encompass a broad range (44°–25°) according to the degree of breakdown of the regolith. The data from *earth rubble* slopes and *silty-sandy* slopes in the southern Pennines, Exmoor and the Laramie Mountains, Wyoming, indicate that stable straight slopes stand at the semi-frictional angle. Observations on stable slopes cut in *clay-shale* in south Derbyshire, where winter piezometric pressures have been measured close to the surface, indicate that these also stand at the semi-frictional angle, unlike those in Ontario.

The role of climate in determining whether frictional or semi-frictional angles are likely to dominate a landscape is considered in Section 4.5. For the moment, the important point is the close association between the characteristic straight slope angles of these six areas and the threshold angles of the slope material. The data provide a mechanics explanation for the common geomorphological observation that 'strong rocks produce steep slopes and weak rocks yield gentle gradients', but, as anticipated by Chorley (1959), in relation to the weathered products rather than the parent rock.

## 4.4 Models of slope profile development by mass-movement

Slope profiles commonly possess rounded, convex upslope elements and lower (basal) concavities. In areas of low relief, the two may unite with no intervening slope unit; with greater relief, these two components are invariably separated. On many occasions the intervening slope unit, the main slope, may completely dominate the slope profile. The main slope may take many forms: a single straight segment; a double-segment profile; a rockwall-talus sequence; and others. The development of the main slope over time, i.e. whether it retreats, declines, shortens or undergoes replacement (Figure 4.12), will vary with the type of processes acting upon the slope. The effects of mass-movement processes are discussed below.

Falls, avalanches and rock-slides, by periodic stripping of material off the main slope, result in slope replacement or retreat, depending on the rapidity with which debris can be removed from the slope base. Often, the lateral (contour) extent of such movements is small, and complete stripping of the regolith sheet requires successive movements of the adjacent material. If for some reason, e.g. rapid weathering and related moisture accumulation along a fault zone, movements show preferential development in certain parts (laterally) of the hillside slope, the retreating slope surface, and the associated debris accumulations, will display a distinctly three-dimensional character, e.g. rockwall chutes

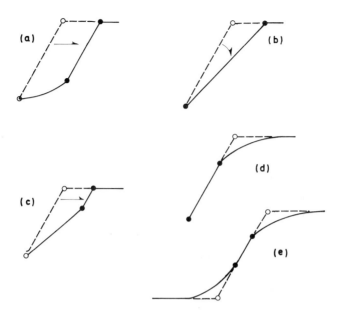

Figure 4.12. Modes of development of the main slope: (a) retreat; (b) decline; (c) replacement; (d) shortening.

and talus cones in high altitude areas (Rapp, 1960; Luckman, 1971) and coombes cut in coastal clay cliffs (Brunsden and Jones, 1972). Indeed, it should be emphasized that slope development is not entirely a matter of slope *profile* development, but also (Hack and Goodlett, 1960) a reflection of the rate of obliteration of side slope areas by the growth of hollows and noses. In extreme instances, such preferential weathering and mass-movement merge into processes of drainage network extension, as indicated earlier, but this is beyond the scope of the present discussion.

Slumps, in contrast, are mechanisms of slope decline. Circular slumps, so commonly emphasized in pre-1960 accounts of mass movements, seem to be restricted to slopes with definite cap rocks (capable of maintaining steep overall gradients), as reported by Hutchinson (1969) and by Brunsden and Jones (1972), or situations with high undercutting rates. Shallow slumps in contrast are extremely common, singly and in succession, on clay, and clay-shale slopes of moderate inclination.

Falls, avalanches, rock-slides and, in a different way, slumps all tend to produce slope profiles which tend to be dominantly straight and inclined at the limiting angle of stability of the frictional material. Talus slopes, admittedly, commonly show extensive basal concavities, but it seems probable that these reflect transportational processes (e.g. slush-avalanching), or subsequent mass-wasting, rather than the initial rockfall process.

Solifluction and debris-creep are alike in being point-to-point transfer processes. It is commonly believed (e.g. Kirkby (1971)) that debris movement by creep is essentially a distance-independent process and may be described by an expression of the form:

$$S = Cx^m(\partial y/\partial x)^n \tag{4.4}$$

where $S$ is the rate of debris transfer downslope (the area contained by the velocity profile curve), $x$ and $y$ are horizontal and vertical distances from the divide, $m = 0$, $n > 0$ and C is a coefficient. Processes of this type demand that the combination of velocity and thickness of the moving mantle remains fixed down a slope of constant gradient. Whether or not debris-creep is indeed a distance-independent process has yet to be verified, but this seems probable for both creep and solifluction. By definition, distance-independent processes are incapable of causing retreat or flattening of straight slopes: both of these changes require that, at any point on the profile, discharge of incoming debris is less than that of the outgoing debris and this is contrary to the $m = 0$ criterion. Such processes, therefore, lead to shortening of main slopes, by the downslope extension of the convexity and, where basal accumulation of debris is possible, by upslope extension of a concave element.

The role of earth-slides in slope profile development is uncertain. The limited movement involved in earth-slides, as distinct from earth-avalanches, is more comparable with point-to-point transfer, albeit rather intermittently, than with complete stripping of regoliths from a hillside. In this respect earth-slides resemble debris-creep, but differ in their dependence upon a threshold shear stress, and, for this reason, only very limited shortening of main slopes is likely. More detailed descriptions of earth-slides, not merely in terms of slide mass geometry, but in the context of slope profile development, are needed. Blong (1971) has described slides in New Zealand in which underthrusting occurs within the slidemass over most of its length. Instead of a rigid-body movement, thickening of the slidemass downslope results from the underthrusting and, as in the case of shallow slumps, produces a small flattening of the slope. The relief of underthrust slide masses shows small, sub-parallel, transverse ridges similar to terracettes, a feature long associated (in an uncertain manner) with mass-movement. Whether there is a genetic association between terracettes and underthrusting is unclear, but terracettes (which appear to be particularly characteristic of shallow regoliths) have been previously associated (Carson, 1969) with slope decline from one threshold angle to another.

In attempting to model slope profile development under any individual mass-movement process, much will depend on the relative basal erosion rate affecting the slope, i.e. the rate of undercutting (laterally or vertically) relative to the rate of rock disintegration by unloading and weathering. With continuing undercutting, slopes will steepen and increase in height until mass-movement processes attain sufficient rapidity to balance basal erosion and produce a state of quasi-equilibrium.

With very low relative erosion rates, slow mass-wasting could, theoretically, maintain equilibrium profiles, although because of the distance-independent nature of such processes, the profiles would have to be convex. This can be modelled quantitatively from the continuity equation (Kirkby, 1971):

$$\frac{\partial y}{\partial t} = \frac{\partial S}{\partial x} \qquad (4.5)$$

i.e. lowering of the ground, at a point, over a period of time $(\partial y/\partial t)$ requires that $S$ increase downslope (with $x$). Assuming that $S$ is purely a function of slope angle, as before, and substituting

$$S = f(\theta_s) = K(\theta_s) \qquad (4.6)$$

where $\theta$ is the sine of the slope angle in the previous equation, yields

$$\frac{\partial y}{\partial t} = K \cdot \frac{\partial \theta_s}{\partial x} \qquad (4.7)$$

i.e. the rapidity with which the slope profile is lowered is proportional to the rate of increase of slope gradient downslope. Provided that the rate of debris transfer (and hence $K$) is sufficiently large, a convex profile could, theoretically, be lowered at the same rate as stream downcutting. The data on $f(\theta_s)$ are unfortunately scant. Kirkby's (1967) data for soil-creep on grass-covered slopes in Scotland yield $K = 5$ cm$^2$ year$^{-1}$ which, with, for example, a rate of downcutting of 1 m in 10,000 years, would demand convex curvature of 0·2 (in sine form) per metre. Such curvature is extreme and, if it is assumed that the steepest slope angle attainable before creep merges into rapid mass movement is 45° ($\theta_s = 0.7$), this could apply to a convexity no greater than 3·5 m (0·7 ÷ 0·2) in length. Surface velocity data for sheet gelifluction obtained by Washburn (1973, p. 178), however, correspond to a value of $K \doteq 170$ cm$^2$ year$^{-1}$ (assuming it to be applicable to a 10 cm thick regolith) and would require a contour curvature of only 0·006 m$^{-1}$, equivalent to a convexity longer than 100 m. As downcutting proceeds, the length of the convex profile must increase (unless the position of the divide is fixed) and therefore mean convex curvature must decrease with time. With the concomitant decrease in the rate of lowering of the slope profile, mean slope steepness must increase leading, eventually, to more rapid forms of mass-movement.

At the other extreme, i.e. conditions of maximum relative erosion, slab-falls and rock-slides would, theoretically, occur at certain intervals during downcutting (at critical combinations of slope height and steepness). The early stages of such a sequence have been described in loess (Lohnes and Handy, 1968), with critical conditions predicted from stability formulae. In strong rocks, the threshold condition would never be attained (if rock weakening did not occur) and valleys would remain as canyons. Such conditions are rare in nature, however, because undercutting must result in strength loss by unloading.

If rock weakening is sufficiently rapid to destroy cohesion within the mass, slopes will be maintained at a threshold angle in a state of dynamic equilibrium, provided that debris does not accumulate and that downcutting continues. As indicated in the previous section, the steepness of such undercut V-shaped valleys may range from 70° to 5° depending upon the rock type, the degree of rock breakdown, i.e. the relative erosion rate, and the hydrological environment. Approached from the standpoint of mechanics, the V-shaped valley profile, comprising a pair of threshold slopes, is the typical product of stream incision.

If the basal slope is free from undercutting, e.g. fault scarps, or slopes protected by a floodplain or terrace, sequential profile changes will occur, their nature dependent upon antecedent erosional rates, and the extent to which accumulation of slope debris may occur. With impeded removal (Savigear, 1952), accumulation of the debris of mass-movement will extend the slope outwards. The amount of curvature of the basal slope will depend on the extent to which the processes involved are controlled by threshold slope conditions. Hutchinson (1967) has shown the Bakker–Le Heux (1946) model of accumulation to be just as applicable to depositional basal slopes in London Clay (formed by slides) as to talus accumulations from rockfall. Creep, solifluction and fluvial processes, in contrast, are likely to produce broad concave accumulation slopes. The case of unimpeded removal has also been treated quantitatively by Bakker and Le Heux (1952; their $c \to -\infty$ case), who demonstrate that the receding upper slope is gradually replaced by a straight basal slope, extending upslope, at the threshold angle of the debris mantling the slope.

The Bakker–Le Heux models refer to a single *phase of instability* (Carson, 1969), i.e. a sequence in which one threshold slope is gradually replaced by another. But in certain cases, slopes may undergo more than one such sequence. Observations on slopes on Exmoor and in the Pennines led Carson and Petley (1970) to infer that slopes experienced up to three phases of instability in those regions: retreat of rockwalls to produce scree-mantled slopes; weathering of the scree to earth rubble, and subsequent rapid mass-movement, leading to retreat and replacement by a taluvium-mantled slope at its threshold angle; weathering of the taluvium, sliding and retreat, or decline, producing a slope mantled by silty-sandy regolith at its threshold angle. The number of phases of instability is thus controlled by the changing shear strength of the regolith during weathering. It also depends on antecedent erosional rates: if previous down cutting has been slow, regoliths may be thoroughly weathered, and the equilibrium threshold angle during downcutting may be the lowest which the slope can attain; no new phase of instability will occur during free degradation of the slope. Recognition of these phases of instability poses the question whether the change occurs by retreat or flattening. Available evidence suggests that both may occur.

## 4.5 Mass-movement and climate

The effects of climate on mass-movement, i.e. the frequency and mode of sporadic movements, the threshold angles associated with these movements, and the absolute (and relative) importance of slow mass-wasting, have been touched upon in the previous sections. The following attempt to draw these observations together is inevitably speculative in places: detailed studies of process are scant and are localized in certain areas; observations on form, while abundant, have rarely been properly linked to the processes responsible: there is, moreover, evidence that present-day forms often reflect former climatic conditions rather than those of the present; and finally, present-day mass-wasting, just like fluvial activity, is commonly affected by landscape changes wrought by man.

Sporadic movements are characteristic of upland and mountain landscape, rejuvenated lowland areas and earthquake-prone regions throughout the world, and are not controlled by climate, although the type and frequency of movement can be affected by this factor. Rock-fall and talus production, for instance, have been observed in desert areas (Koons, 1955), temperate areas (Hack, 1960) and humid tropical areas, but appear to be especially prolific in cold areas, where freeze-thaw acts to create, and wedge apart, fractures in rockwalls. Indeed, almost all forms of mass-wasting appear to be especially effective in areas sufficiently cold to contain ground ice. Mackay (1966) has emphasized this in connection with slumping of steep banks in the Mackenzie Delta area; others (e.g. Isaacs and Code (1972) and Chyurlia (1973)) have noted that ice contents in fine-grained deposits may exceed the liquid limit, leading, on thaw, to rapid flowage of slope material. Mudflows of the rapid type are more commonly associated (Sharpe, 1938) with alpine and mountainous semi-arid areas, where the seasonal character of runoff production allows accumunation of debris on ravine bottoms, the sparsity of trees prevents anchorage of it to the underlying ground, and the abundance of runoff, when it occurs, leads to lubrication of the waste. Slow mudflows and mudslides, in contrast, seem to occur in most climates, in situations where debris accumulates in front of unstable cliffs in weakly consolidated material. Earth-flows also appear to be characteristic of particular deposits, rather than climate, and though particularly frequent in the marine clays exposed by post-glacial rebound in eastern Canada and coastal Norway, have also been described (Jordan, 1949) in a Miocene clayey sand deposit in Florida. Slides and avalanches are dependent on steep terrain (in relation to the strength of the regolith) and are not directly linked to climate.

In contrast, it might be expected that the threshold angles of slope regoliths would, because of their strong dependence on pore pressures and slope hydrology, show a definite association with climate. The limited data available (e.g. Figure 4.11) tend to support this view. Frictional slopes occur in the Mesa Verde, the Verdugo Hills (both marginally semi-arid) and in southern Ontario, whereas regoliths in the Pennines and on Exmoor (humid temperate) and in the Laramie Mountains are, with one exception, stable only at semi-frictional angles. The exception is rock rubble and talus (not shown in Figure 4.11) which, because of the very high permeability, is probably never saturated. The distinction between the frictional clay-shale slopes of southern Ontario and the semi-frictional slopes, in similar material, in Derbyshire is especially interesting. In terms of amount and distribution of annual precipitation and runoff, the two are similar. Groundwater levels are much lower in the Ontario shale, however, with well levels typically between 15 m and 30 m below the surface, in contrast to near-surface winter water tables in Derbyshire. Whether the low water table levels of southern Ontario are a product of plugging of surficial fissures by clay till, which mantles the inter-valley areas, or attributable to impeded infiltration of winter and spring precipitation by surface frost, or another factor, the contrast highlights the difficulty of generalizing on the effect of climate alone. The Laramie slopes present another problem; with an annual precipitation of about 320 mm the present climate is semi-arid, water table levels are deep, and yet semi-frictional slopes prevail. Whether these are the result of infrequent development of perched water systems during snowmelt or rainstorms at the present time, or the product of more pluvial periods during Pleistocene times, is uncertain.

Whether threshold slopes are frictional or semi-frictional is not a matter of climate directly but of hydrology, and is essentially dependent on the hydrologic storage of the slope. Thin regoliths may develop perched water systems frequently; deeply weathered rocks in contrast, are characterized by permanent water tables well below ground level (e.g. Lumb (1962) and Ruddock (1967)). Unfortunately few regional stability analyses on deeply weathered rocks are available. Lumb (1962) has provided some data for slopes on deeply weathered (30 m) granite in Hong Kong which support the idea of frictional threshold angles. Slopes are steep (reports state $30°-45°$, but frequency distributions are unavailable) and comprise an oxidized crust over the decomposed granite. The shear strength data of the latter is given as $\phi' = 33°-41°$, with $c'$ decreasing from 100 kN m$^{-2}$ to zero with saturation. While saturation of the surficial material must occur during rainstorms, the great depth of weathering, supplemented by the impedance to infiltration by the oxidized crust, renders the development of positive pore pressures under the slope surface most unlikely. Such conditions should be typical of slopes cut in deeply weathered tropical rocks.

Artesian slopes (i.e. threshold slopes gentler than the semi-frictional angle) are believed to be frequent in, but probably not confined to, tundra areas. Indeed, wherever clayey regoliths cover more permeable rock, it is possible that the threshold angle will be determined by artesian pressures. Evidence of mass-wasting during former tundra conditions in the Pleistocene abounds in many present-day temperate areas, and the possibility exists that some artesian threshold slopes in these areas are relicts of artesian conditions in earlier periods. Weeks (1969), in particular, has identified slip surfaces on very gentle ($4°$), presently-stable, clay slopes in southeast England and has attributed their origin to former tundra conditions. The role of climatic change, and not just climate, in controlling threshold slope angles must not be overlooked.

The influence of climate on the rate of slow mass-movement particularly in comparison with fluvial slope processes, has attracted much comment in recent years. Observations by Young (1960) and Kirkby (1967), on grass-covered slopes in England (Table 4.2) support the notion that creep is far

Table 4.2. Rates of slow mass wasting processes and comparison with soil erosion rates

| Surficial-creep[a] | cm² year⁻¹ | Surface velocities of mass-wasting processes[b] | mm year⁻¹ | Soil-wash[c] | mm year⁻¹ |
|---|---|---|---|---|---|
| Malaya (Eyles and Ho, 1970) | 12.4 | Depth-creep (Ter-Stepanian, 1965) | 110 | S. Dakota (Schumm, 1956a) | 13.4 |
| N. T., Australia: granite (Williams, 1973) | 7.3 | Gelifluction (Jahn, 1960) | 65 | New Mexico (Leopold and coworkers 1966)[d] | 4.7 |
| Alaska: till (Barr and Swanston, 1970) | 7.0 | Talus-creep (Caine, 1964) | 50 | France: bare soil (Gabert, 1964)[d] | 4.6 |
| Ohio (Everett, 1963) | 6.0 | Gelifluction (Williams, 1966) | 45 | Java (Coster, 1938)[d] | 0.7 |
| N. T., Australia: sandstone (Williams, 1973) | 4.4 | Gelifluction (Harris, 1972) | 35 | Madagascar: crops (Roche, 1954)[d] | 0.5 |
| N. S. W., Australia: sandstone (Williams, 1973) | 3.2 | Depth-creep (Kojan, 1967) | 30 | N. S. W., Australia (Williams, 1973) | 0.1–0.05 |
| S. Alps, N. Z. (Owens, 1969) | 3.2 | Gelifluction (Washburn, 1973) | 25 | N. T., Australia (Williams, 1973) | 0.05 |
| Scotland (Kirkby, 1967) | 2.1 | Gelifluction (Rudberg, 1962) | 24 | | |
| N. S. W., Australia: granite (Williams, 1973) | 1.5 | Soil-creep (Barr and Swanston, 1970) | 6 | | |
| England (Young, 1960) | 0.9 | Depth-creep (Mitchell and Eden, 1972) | 2 | | |
| | | Soil-creep (Kirkby, 1967) | 2 | | |
| | | soil-creep (Everett, 1963) | 1 | | |

[a] cm² year⁻¹ denotes area enclosed by annual velocity profile (also expressed as cm³/cm/year).
[b] mm year⁻¹ denotes downslope travel vector.
[c] mm year⁻¹ denotes vertical lowering of ground surface.
[d] Cited by Williams (1973).

more important than soil wash in humid areas, as suggested by Schumm (1956b), and Hack and Goodlett (1960) earlier. Data collected by Williams (1973), however, conflict with this view: in both tropical (savanna) and temperate (forest) areas of Australia, undifferentiated soil-wash (including rainsplash) was found to yield 5–7 times more sediment than creep. The possibility that soil-wash rates on grass-covered slopes in humid areas are unnaturally low, in relation to those under the indigenous forest cover, must therefore be considered. William's data also indicate higher creep rates in the savanna area, possibly a reflection of termite activity in the soil there. Few data are available on creep rates in humid tropical areas, but if depth-creep occurs, rates could be very high on deeply weathered slopes.

Slow mass-wasting in cold areas (combined gelifluction and frost-creep) appears to be (Table 4.2) of an order of magnitude higher than debris-creep in other areas. Surface movements of $10-100$ mm year$^{-1}$ (equivalent to $10-100$ m per thousand years) are typical; the higher values refer to lobate features and the lower to sheet movements. If these high rates were applicable to temperate areas during Pleistocene tundra conditions, they would provide quantitative support for the widely held view that mass-wasting on present-day stable slopes is in a relatively dormant period.

Overall, the role of climate as a control of mass-movement processes must be considered a significant, albeit complex, one. Any attempt to classify mass-movement type and frequency, and its role in landscape development, in terms of climatic parameters is certainly doomed to failure. While mass-movement processes are intimately related to ground hydrology, and while the latter is strongly linked to climatic conditions, it is not a univariate function of climate.

**Acknowledgment**

The author is indebted to Professor Eiju Yatsu for his criticism of the original script, and to Mr Abal Sen for his drafting of the diagrams. Some of the data presented here derive from research, previously unpublished, by the author over several years, supported by the National Research Council of Canada and the Geological Survey of Canada.

**References**

Aisenstein, B., 1965, Some observations on the deconsolidation of limey rocks on steep slopes, *Proc. 6th Int. Conf. Soil Mech. and Found. Eng.*, **2**, 439–441.

Andersson, J. G., 1906, Solifluction, a component of subaerial denudation, *J. Geol.*, **14**, 91–112.

Bakker, J. P. and J. W. N. Le Heux, 1946, Projective-geometric treatment of O. Lehmann's theory of the transformation of steep mountain slopes, *Kon. Ned. Akad. van Wetensch.*, Ser. B, **49**, 553–5547.

Bakker, J. P. and J. W. N. Le Heux, 1952, A remarkable new geomorphological law, *Kon. Ned. Akad. van Wetensch.*, Ser. B, **55**, 399–410 and 554–571.

Barr, D. J. and D. N. Swanston, 1970, Measurement of creep in a shallow, slide-prone till soil, *Am. J. Sci.*, **269**, 467–480.

Baulig, H., 1956, Pénéplaines et pédiplaines, *Soc. belge Études géog.*, **25**, 25–58.

Benedict, J. B., 1970, Downslope soil movement in a Colorado alpine region: rates, processes, and climatic significance, *Arctic and Alpine Research*, **2**, 165–226.

Bjerrum, L., 1954, Stability of natural slopes in quick clay, *Proc. European Conf. on Stability of Earth Slopes*, **3**, 101–119.

Bjerrum, L., 1967, Progressive failure in slopes of over-consolidated plastic clay and shales, *J. Soil Mech. and Found. Div., A. S. C. E.*, **93**, 1–49.

Bjerrum, L. and F. Jorstad, 1968, Stability of rock slopes in Norway, *Norwegian Geotechnical Publication*, **79**, 1–11.

Blackwelder, E., 1912, The Gros Ventre slide, an active earth-flow, *Bull. geol. Soc. Am.*, **23**, 487–492.

Blackwelder, E., 1928, Mudflow as a geologic agent in semi-arid mountains, *Bull. geol. Soc. Am.*, **39**, 465–483.

Blong, R. J., 1971, The underthrust slide—an unusual type of mass movement, *Geogr. Annlr*, **53**, Ser. A., 52–58.

Brunsden, D. and D. K. C. Jones, 1972, The morphology of degraded landslide slopes in south west Dorset, *Quart. J. eng. Geol.*, **5**, 205–222.

Caine, T. N., 1964, Movement of low angle scree slopes in the Lake District, northern England, *Rev. de Geom. dynam.*, **14**, 171–177.

Carson, M. A., 1969, Models of hillslope development under mass failure, *Geog. Analysis*, **1**, 76–100.

Carson, M. A., 1971, *The Mechanics of Erosion*, Pion, London.

Carson, M. A. and D. J. Petley, 1970, The existence of threshold hillslopes in the denudation of the landscape, *Trans. Inst. Brit. Geog.*, **49**, 71–95.

Carson, M. A. and M. J. Kirkby, 1972, *Hillslope Form and Process*, Cambridge University Press, Cambridge.

Chandler, R. J., 1970a, A shallow slab slide in the Lias Clay near Uppingham, Rutland, *Geotechnique*, **20**, 253–260.

Chandler, R. J., 1970b, The degradation of Lias Clay slopes in an area of the east Midlands, *Quart. J. eng. Geol.*, **2**, 161–181.

Chandler, R. J., 1972a, Lias Clay: weathering processes and their effect on shear strength, *Geotechnique*, **22**, 403–431.

Chandler, R. J., 1972b, Periglacial mudslides in Vestspitsbergen and their bearing on the origin of fossil 'solifluction' shears in low angled clay slopes, *Quart. J. eng. Geol.*, **5**, 223–241.

Chandler, R. J., 1973, The inclination of talus, Arctic talus terraces and other slopes composed of granular materials, *J. Geol.*, **81**, 1–14.

Chorley, R. J., 1959, The geomorphic significance of some Oxford soils, *Am. J. Sci.*, **257**, 503–515.

Chyurlia, J. P., 1973, Stability of river banks and slopes along the Liard River and Mackenzie River, Northwest Territories, in *Hydrologic Aspects of Northern Pipeline Development*, Dept. of the Environment, Ottawa, 113–152.

Crawford, C. B., 1968, Quick clays of Eastern Canada, *Eng. Geol.*, **2**, 239–265.

Crawford, C. B. and W. Eden, 1967, Stability of natural slopes in sensitive clay, *J. Soil Mech. and Found. Div., A. S. C. E.*, **93**, 419–436.

Eden, W. J. and J. J. Hamilton, 1957, The use of a field vane apparatus in sensitive clays, *ASTM Special Publication 193*, 41–53.

Eden, W. J. and R. J. Mitchell, 1970, The mechanics of landslides in Leda Clay, *Canad. Geotech. J.*, **7**, 285–296.

Eide, O. and L. Bjerrum, 1954, The slide at Bekkelaget, *Proc. European Conf. on Stability of Earth Slopes*, Stockholm, **3**, 88–100.

Everett, K., 1963, Slope movement, Neotoma Valley, southern Ohio, *Ohio State Univ., Inst. of Polar Studies, Rept. No. 6*.

Eyles, R. J. and R. Ho, 1970, Soil creep on a humid tropical slope, *J. trop. Geog.*, **31**, 40–42.
Gifford, J., 1953, Landslides on Exmoor caused by the storm of August 15, 1952, *Geography*, **38**, 9–17.
Hack, J. T., 1960, Origin of talus and scree in northern Virginia, *Bull. geol. Soc. Am.*, **71**, 1877–1878.
Hack, J. T. and J. C. Goodlett, 1960, Geomorphology and forest ecology of a mountain region in the Central Appalachians, *U. S. Geol. Surv. Prof. Pap. 347*.
Harris, C., 1972, Processes of soil movement in turf-banked solifluction lobes, Okstindan, northern Norway, *Polar Geomorphology, Inst. Brit. Geog. Spec. Publ. No. 4*, 155–174.
Hendron, A. J. Jr., 1968, Mechanical properties of rock, in *Rock Mechanics*, K. G. Stagg, and O. C. Zienkiewicz (Eds.), Wiley, New York.
Henkel, D. J. and A. W. Skempton, 1954, A landslide at Jackfield, Shropshire, in an over-consolidated clay, *Proc. Conf. on Stability of Earth Slopes*, **1**, 90–101.
Holtz, W. G., 1960, Effect of gravel particles on friction angle, *Proc. Amer. Soc. Civ. Engrs Research Conf. on Shear Strength*, 1000–1001.
Holtz, W. G. and H. J. Gibbs, 1955, Shear characteristics of pervious gravelly soils as determined by triaxial shear tests, *Paper presented to Convention of Amer. Soc. Civ. Engrs, San Diego, Calif.*
Howe, E., 1909, Landslides of the San Juan Mountains, Colorado, *U. S. Geol. Surv. Prof. Pap. 67*, 58 pp.
Hurtubise, J. E. and P. A. Rochette, 1956, The Nicolet slide, *Proc. 37th Canadian Good Roads Association*, 143–155.
Hutchinson, J. N., 1961, A landslide on a thin layer of quick clay at Furre, central Norway, *Geotechnique*, **11**, 69–94.
Hutchinson, J. N., 1967, The free degradation of London Clay cliffs, *Proc. Geotech. Conf., Oslo*, **1**, 113–118.
Hutchinson, J. N., 1969, A reconsideration of the coastal landslides at Folkestone Warren, Kent, *Geotechnique*, **19**, 6–38.
Hutchinson, J. N., 1970, A coastal mudflow on the London Clay cliffs at Beltinge, North Kent, *Geotechnique*, **20**, 412–438.
Hutchinson, J. N. and R. K. Bhandari, 1971, Undrained loading, a fundamental mechanism of mudflows and other mass movements, *Geotechnique*, **21**, 353–358.
Isaacs, R. M. and J. A. Code, 1972, Problems in engineering geology related to pipeline construction, *Proc. Canadian Northern Pipeline Research Conference*, 147–177.
Jahn, A., 1960, Some remarks on the evolution of slopes on Spitsbergen, *Zeits. für Geom., Supp. Bd.* **1**, 49–58.
Johnson, A. M., 1970, *Physical Processes in Geology*, Freeman, Cooper and Co., San Francisco, 577 pp.
Johnson, A. M., and P. Rahn, 1970, Mobilization of debris flows, *Zeits. für Geom., Supp. Bd.* **9**, 168–186.
Jordan, R. H., 1949, A Florida landslide, *J. Geol.* **57**, 418–419.
Kenney, T. C., 1967a, Shear strength of soft clay, *Proc. Geotech. Conf., Oslo*, 3–12.
Kenney, T. C., 1967b, The influence of mineral composition on the residual strength of natural soils. *Proc. Geotech. Conf., Oslo*, 123–129.
Kent, P. E., 1966, The transport mechanism in catastrophic rock falls, *J. Geol.*, **74**, 79–83.
Kiersch, G. A., 1964, Vaiont Reservoir disaster, *Civil Engineering*, **34**, 32–39.
Kirkby, M. J., 1967, Measurement and theory of soil creep, *J. Geol.*, **75**, 359–378.
Kirkby, M. J., 1971, Hillslope process-response models based on the continuity equation, *Inst. Brit. Geog. Spec. Publ. No. 3*, 15–30.
Kojan, E., 1967, Mechanics and rates of natural soil creep, *U. S. Forest Serv. Exp. Stn. (Berkeley, Calif.) Rept.*, 233–253.
Koons, D., 1955, Cliff retreat in southwest United States, *Am. J. Sci.*, **253**, 44–52.

La Rochelle, P., Chagnon, J. Y. and G. Lefebvre, 1970, Regional geology and landslides in the marine clay deposits of eastern Canada, *Can. Geotech. J.*, **7**, 145–156.
Lohnes, R. A. and R. L. Handy, 1968, Slope angles in friable loess, *J. Geol.*, **76**, 247–258.
Luckman, B. H., 1971, The role of snow avalanches in the evolution of alpine talus slopes, *Inst. of Brit. Geog. Spec. Publ. No. 3*, 93–110.
Lumb, P., 1962, The properties of decomposed granite, *Geotechnique*, **12**, 226–243.
Mackay, J. R., 1966, Segregated epigenetic ice and slumps in permafrost, Mackenzie Delta area, N. W. T., *Geogr. Bull.*, **8**, 59–80.
Mitchell, R. J. and W. J. Eden, 1972, Measured movements of clay slopes in the Ottawa area, *Canad. J. Earth Sciences*, **9**, 1001–1013.
Nakano, R., 1967, On weathering and change of properties of Tertiary mudstone related to landslide, *Soil and Foundation*, **7**, 1–14.
Newmark, N. M., 1965, Effects of earthquakes on dams and embankments, *Geotechnique*, **15**, 139–159.
Owens, I. F., 1969, Causes and rates of soil creep in the Chilton Valley, Cass, New Zealand, *Arctic and Alpine Research*, **1**, 213–220.
Peck, R. B., Ireland, H. O. and T. S. Fry, 1951, Studies of soil characteristics, the earthflows of St. Thuribe, Quebec, *Soil Mech. Series No. 1*, Univ. of Illinois.
Penner, E., Gillott, J. E. and W. J. Eden, 1970, Investigation of heave in Billings shale by mineralogical and biochemical methods, *Canad. Geotech. J.*, **7**, 333–338
Prior, D. B., Stephens, N. and G. R. Douglas, 1971, Some examples of mudflow and rockfall activity in north east Ireland, *Inst. Brit. Geog. Spec. Publ. No. 3*, 129–140.
Quigley, R. M. and R. W. Vogan, 1970, Black shale heaving at Ottawa, *Canad. Geotech. J.*, **7**, 106–112.
Rapp, A., 1960, Recent developments of mountain slope in Karkevagge and surroundings, northern Scandinavia, *Geogr. Annlr*, **42**, 71–200.
Reynolds, D. L., 1954, Fluidization as a geological process, and its bearing on the problem of intensive granites, *Am. J. Sci*, **252**, 577–614.
Rose, C. W., 1966, *Agricultural Physics*, Pergamon, Oxford.
Rudberg, S., 1962, A report on some field observations concerning periglacial geomorphology and mass movements on slopes in Sweden, *Biul. Peryglacjalny*, **11**, 311–323.
Ruddock, E. C., 1967, Residual soils of the Kumasi district in Ghana, *Geotechnique*, **17**, 359–377.
Rycroft, D., 1971, Drainage investigations in the Southwest, *Annual Report, Field Drainage Experimental Unit, Ministry of Agriculture, Fisheries and Food*, Cambridge, 7–15.
Savigear, R. A. G., 1952, Some observations on slope development in south Wales, *Trans. Inst. Brit. Geog.*, **18**, 31–52.
Schumm, S. A., 1956a, Evolution of drainage systems and slopes on badlands at Perth Amboy, New Jersey, *Bull. geol. Soc. Am.*, **67**, 597–646.
Schumm, S. A., 1956b, The role of creep and rain-wash on the retreat of badland slopes, *Am. J. Sci.*, **254**, 639–706.
Schumm, S. A. and R. J. Chorley, 1964, The fall of Threatening Rock, *Am. J. Sci.*, **262**, 1041–1054.
Scott, R. F., 1963, *Principles of Soil Mechanics*, Reading, Mass.
Seed, H. B., 1967, Slope stability during earthquakes, *J. Soil Mech. and Found. Div., A. S. C. E.*, **93**, 299–323.
Sevaldson, R. A., 1956, The slide in Lodalen, October 6th, 1954, *Geotechnique*, **6**, 167–182.
Sharpe, C. F. S., 1938, *Landslides and Related Phenomena*, Columbia University Press, New York.
Shreve, R. L., 1968a, The Blackhawk landslide, *Geol. Soc. Am., Spec. Pap. 108*, 47 pp.
Shreve, R. L., 1968b, Leakage and fluidization in air layer lubricated avalanches, *Bull. geol. Soc. Am.*, **79**, 653–658.
Simonett, D. S., 1967, Landslide distribution and earthquakes in the Dewani and Torri-

celli Mountains, New Guinea, in *Landform Studies from Australia and New Guinea,*. J. N. Jennings and J. A. Mabbutt (Eds.), Methuen, London.

Skempton, A. W., 1964, The long-term stability of clay slopes, *Geotechnique*, **14**, 75–102.

Skempton, A. W., 1970, The consolidation of clays by gravitational compaction, *Quart. J. Geol. Soc. London*, **125**, 373–411.

Skempton, A. W. and J. D. Brown, 1961, A landslide in boulder clay at Selset, Yorkshire, *Geotechnique*, **11**, 280–293.

Skempton, A. W. and F. A. DeLory, 1957, Stability of natural slopes in London Clay, *Proc. 4th Int. Conf. on Soil Mech. and Found. Eng.*, **2**, 90–108.

Skempton, A. W. and J. N. Hutchinson, 1969, Stability of natural slopes and embankment sections, *Proc. 7th Int. Conf. on Soil Mech. and Found. Eng., State of the Art Volume*, 291–340.

Strahler, A. N., 1950, Equilibrium theory of erosional slopes approached by frequency distribution analysis, *Am. J. Sci.*, **248**, 673–696 and 800–814.

Ter-Stepanian, G., 1965, *In-situ* determination of the rheological characteristics of soils on slopes, *Proc. 6th Int. Conf. on Soil Mech. and Found. Eng.*, **2**, 575–577.

Terzaghi, K., 1950, Mechanism of landslides, *Bull. geol. Soc. Am., Berkey Volume*, 83–122.

Terzaghi, K., 1962, Stability of steep slopes on hard unweathered rock, *Geotechnique*, **12**, 251–270.

Vargas, M. and E. Pichler, 1957, Residual soil and rock slides in Brazil, *Proc. 4th Int. Conf. on Soil Mech. and Found. Eng.*, **2**, 394–398.

Varnes, D., 1958, Landslide types and processes, in *Landslides And Engineering Practice* N. R. C. Publication 544, 20–47.

Vucetic, R., 1958, Determination of shear strength and other characteristics of coarse, clayey schist material compacted by pneumatic wheel roller, *Proc. 6th Int. Cong. on Large Dams*, **4**, 465–473.

Washburn, A. L., 1967, Instrumental observations of mass wasting in the Mesters Vig district, Northeast Greenland, *Medd. om Greenland*, **166**, 318 pp.

Washburn, A. L., 1973, *Periglacial Processes and Environments*, Edward Arnold, London.

Weeks, A. G., 1969, The stability of natural slopes in south-east England as affected by periglacial activity, *Quart. J. Eng. Geol.*, **2**, 49–61.

Wentworth, C. K., 1943, Soil avalanches on Oahu, Hawaii, *Bull. geol. Soc. Am.*, **54**, 53–64.

Williams, M. A. J., 1973, The efficacy of creep and slopewash in tropical and temperate Australia, *Australian Geographical Studies*, **11**, 62–78.

Williams, P. J., 1966, Downslope movement at a subarctic location with regard to variations with depth, *Can. Geotech. J.*, **3**, 191–203.

Wilson, S. D., 1970, Observational data on ground movement related to slope instability, *Journal of Soil Mechanics and Foundations Division, Proceedings of American Society of Civil Engineers*, **96**, SMS, 1521–1543.

Yatsu, E., 1966, *Rock Control in Geomorphology*, Sozosha, Tokyo.

Yong, R. N. and B. P. Warkentin, 1966, *Introduction to Soil Behavior*, Macmillan, New York.

Young, A., 1960, Soil movement by denudational processes on slopes, *Nature*, **188**, 120–122.

Young, A., 1961, Characteristic and limiting slope angles, *Zeits. fur Geomorph.*, **5**, 126–131.

# CHAPTER FIVE

# Catenas in Different Climates

CLIFFORD D. OLLIER

It is well known that soils vary with climate. Indeed the zonal concept of soils which still colours much discussion about soils on a world scale assumes from the beginning that climate is the dominant factor in soil formation. It is also believed that the processes of slope formation are likely to be climatically controlled. A lot of work has been put into measuring the relative and absolute rates of slopewash and creep over a wide range of climates, and theories of slope evolution are often climatically oriented. It seems possible, therefore, that there may be a range of regular associations of soils and slopes, each related to a different climate.

The regular association of distinctive soils with certain topographic slopes has been observed by soil scientists for many years, and the term *catena* was introduced by Milne (1935) for the slope and its associated soils. The definition of a catena has become somewhat confused (see Watson (1960) for further discussion) and toposequence is sometimes preferred. The term catena will be used here to mean a regular association of soils and slope.

In this chapter I shall examine the range of catenary variation and then try to relate the information to pedological and geomorphic theories concerning soils and slopes.

## 5.1 Case studies of soils and slopes

Soil surveys in many parts of the world provide descriptions of soils in relation to slope, but the details of soil–slope relationships seldom interest the investigator and there are usually gaps in the data that prevent use of the soil survey for slope studies. Fortunately there are many detailed studies of catenas to supplement survey data. Simple and complex geological situations offer different possibilities for investigation: both will be examined.

### 5.1.1 Simple catenas

A simple catena is one on simple parent material without complications from extraneous surficial material such as wind-blown sand and without inherited soil features. An example of a very simple catena is provided by the work of E. B. Joyce (personal communication) in Gippsland in temperate southeast Australia. Here, in a climate characterized by a rainfall of 40 in with common frosts, is found a region of dominantly convex slopes on Mesozoic mudstones and arkoses. Weathering is rapid, and the dominant soils are podzols containing fragmented rock including fresh feldspars. Soil cover is continuous, and of even depth, despite variation in slope. Slumping and terracettes are common. Joyce ascribes the slopes to rapid downcutting in a recently uplifted area of uniform lithology and rapid weathering, with mass-movement by creep predominating. Before 1900 the area was densely forested but for the past 70 years has had only a grass cover. Nevertheless the uniform depth of soils suggests that creep is the dominant movement of soils, and neither creep nor sheetwash are exposing bare rock because weathering and soil formation are fast enough to keep pace with downslope movement.

Sparrow (1966) describes a very simple catena on Pre-Cambrian schists in the subtropical high rainfall Sugar Belt of Natal. The slopes have a very large convex element of at least 100 m relief, with maximum angles of up to 70°, below which is a smaller concave element (Figure 5.1). Despite this extremely steep topography the soils remain remarkably uniform, with the A and B horizons making a layer which scarcely deviates from 40 cm thickness. Weathered rock below this is about a metre thick. Sparrow believes that the constant thickness of these soils is caused by tenacious vegetation hindering creep, but he points out that even on cleared and cultivated steep slopes there are no signs of soil movement even after a considerable period. An alternative explanation might be that rapid pedological processes are keeping pace with soil removal.

Some slopes suffer no erosion. They are not helpful in discussions of slope

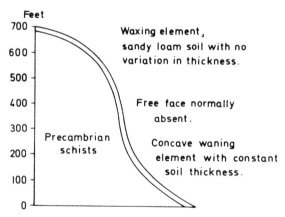

Figure 5.1. Simple catena on schist under coastal forest in a subtropical humid climate, Natal (after Sparrow (1966)).

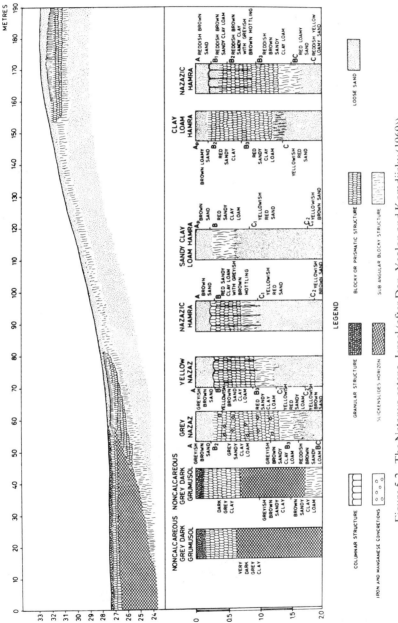

Figure 5.2. The Netanya catena, Israel (after Dan, Yaalon and Komdjisky (1969)).

evolution, but they facilitate the study of catenary variation of soil profiles where erosion and deposition are absent. Such slopes occur on sand dunes, beach ridges, and uneroded slopes on volcanic products such as scoria cones. An example is provided by the Netanya catena from Israel (although, in fact, slightly complicated by extraneous material) which has been very thoroughly described by Dan, Yaalon and Komdjisky (1969). The soils are Red Mediterranean soils, and the parent material is sand or sandstone on what were originally coastal dunes, with clay pans in depressions between dunes (Figure 5.2).

Textural differentiation in well developed profiles on the upper slope is pronounced, indicating a high degree of leaching and clay mobility. The leaching is also evident in pH values, clay mineral composition, and the exchange complex. Soils on the bottomland are poorly drained and gleyed. This study clearly distinguished pedological problems from slope problems, and considerable soil variation is shown to occur without erosion or deposition.

A catena in which the soil moisture regime is the chief variable may be called a hydromorphic catena. Ellis (1938) considered that upper slopes, losing water by runoff, are locally arid, whereas depressions or footslopes, receiving water from upslope as well as their own precipitation, are locally humid.

Another simple catena is the erosion catena, where simple erosion on upper slopes and deposition on lower slopes causes variation from a uniform soil cover. The example provided by Ellis (1938) from Manitoba (Figure 5.3)

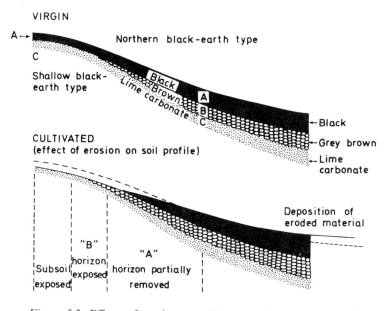

Figure 5.3. Effects of erosion on soil type in Manitoba (after Ellis (1938)).

is almost an example of an erosional catena, although even before erosion there were different soils in different parts of the slope.

In many parts of the tropics the soil catena is formed on thoroughly weathered bedrock which acts as a uniform parent material. An example of such a catena from Zambia is described by Webster (1965), and this catena is slightly more complicated than the previous examples because the soil varies downslope. Most of the slope is a gently convex plateau and there is a shorter concave section leading to the swampy valley soils (*dambo*). A representative sequence of soils ranges from red on the upper slopes through yellowish-brown to grey. Termites are active and are thought to be responsible for a stone line which extends to crest sites. By comparing the non-opaque heavy minerals above and below stone lines Webster was able to show that the soil profiles had developed in place. Even the sandy dambo does not appear to be predominantly colluvial material. It seems that soil material once mobilized on the surface by runoff water has little opportunity to become mixed with the lower slope soils over which it is washed. Sooner or later it reaches the headwater streams and is carried away. Webster regards surface wash as the main mechanism of mass-wastage as there is no evidence of creep but the activity of slopewash can be directly observed at times of heavy rainfall. In general, the sand content of the soils increases downslope. Webster believes the picture is one of increasing surface flow of water downslope with consequent increase in erosional intensity, and preferential removal of the final soil particles. Erosion is greatest on the lower and steeper part of the slope and it is near here that the coarsest-grained soil occurs.

In catenas in Malaysia, also, Morgan (1973) found that the clay fraction decreases downslope even across the gentle basal elements. This, and the

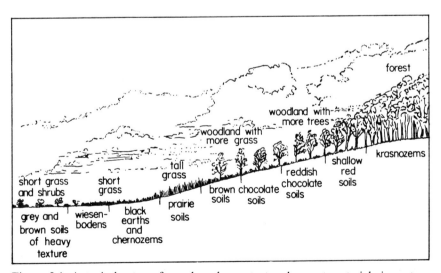

Figure 5.4. A typical catena formed on heavy textured parent materials in eastern Australia (after Corbett (1969)).

absence of colluvial deposits on footslopes, is evidence of free transit of material across the whole of the hillslope. Subsurface wash or throughflow is thought to be largely responsible.

Similar tropical catenas have been described in detail by many other workers including Nye (1954–55), Ollier (1959) and Moss (1968).

Catenary variations with distinct changes in soil down the slopes have been described from many areas of New South Wales, and they show interesting variation with parent material and with climate. A typical catena on weathered basalt in a warm temperate climate is shown in Figure 5.4, and might be summarized thus:

>     krasnozem
>       chocolate soil
>         prairie soil
>           black earth

This catena shows quite a remarkable dependence on rainfall, as shown in Figure 5.5.

Under a similar climate, but on quartz-bearing parent materials that provide sand, the following podzolic sequence is found:

>     red podzolic soil
>       yellow podzolic soil
>         grey podzolic soil

Under a colder climate the sequence is:

>     alpine humus soil
>       brown podzolic soil
>         grey-brown podzolic soil

In the Arctic, Cruickshank (1971) reports a crude catena type of variation with polar desert soils and lithosols on dry, moisture-shedding sites, whereas bog soils and tundra gleys occur in wet, poorly drained and plant-colonized sites. In these examples it is the soil forming processes that vary downslope.

Schumm (1964) described soils and slope evolution in Colorado where variation in slope processes is mainly responsible for catenary variation. On the crest of the slope the soil which is eroded is restored by weathering and swelling of the platy zone and bedrock, whereas on the lower part of the slope eroded material is largely replaced from upslope. The steep straight segment is eroding on an average about as much as the convex segment, but accumulation of debris from the straight and convex segments retards erosion on the concave segment. He concludes that the typical concave–convex hillslopes are fashioned primarily by creep and believes that if sheetwash became more important the upper convexity would be destroyed.

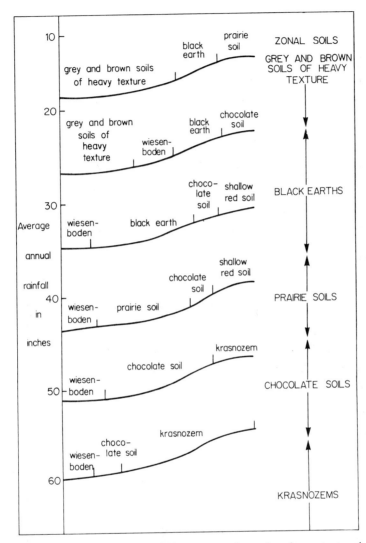

Figure 5.5. The effect of rainfall on catenas formed on heavy textured parent materials in eastern Australia.

## 5.1.2 Complex catenas

Complex catenas have structural, lithological or historical complications not present in simple catenas. Many African catenas, for example, are complicated by rock outcrops, or the presence of hard ferricrete (plinthite). Such catenas were classified by Ollier (1959) and the classification was elaborated by Moss (1968): Figure 5.6 shows the various types now recognized. Ollier (1959) used mineralogical evidence (as did Webster (1965)) to show that most soils were sedentary or very nearly so, but some workers believe that transport-

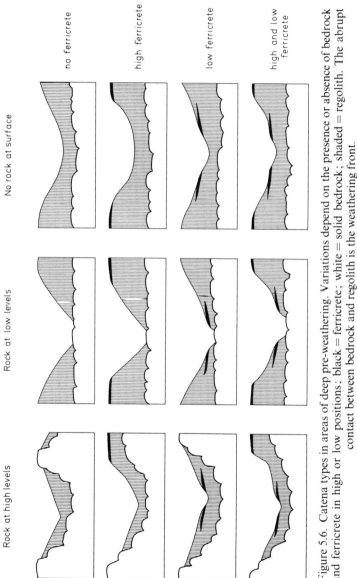

Figure 5.6. Catena types in areas of deep pre-weathering. Variations depend on the presence or absence of bedrock and ferricrete in high or low positions; black = ferricrete; white = solid bedrock; shaded = regolith. The abrupt contact between bedrock and regolith is the weathering front.

ed material may be present in significant, even dominant amounts (Charter, 1958: Clarke, 1971).

These tropical landscapes are controversial and there are arguments about the origin of the rock outcrops, especially the larger ones (inselbergs or bornhardts), about the nature of the general ground surfaces (pediplains, peneplains, or etchplains), about the depth of weathering and the age of rock weathering, about whether the landforms are formed under the present climate or are inherited from some past climate, and about the origin of the ferricrete.

Such controversies will not be resolved by looking at the ground surface alone, or by looking only at the top soil. The shape of the base of the regolith is also important, as is the location and variation in the water table. In this connection it should be added that such profiles on previously deeply weathered rock are not confined to the tropics. Landscapes produced by partial stripping of regolith from deeply weathered rocks have been described from southern Australia, the United States, Canada, England, Czechoslovakia and Scandinavia.

Ferricrete, when present, is usually responsible for a break of slope. If it is a caprock then it will outcrop around a plateau edge: if it is a lower slope feature there will usually be a break of slope coinciding with the ferricrete band. The interpretation of ferricrete landscapes is evidently difficult, for the many explanations are often so diverse that it is hard to see how different workers could have observed the same material. Some correlate ferricretes with distinct periods of planation, others believe they are diachronic, and others believe they form periodically but randomly, depending on local landscape evolution rather than continent-wide climatic changes. Some ferricretes appear to be detrital while others are formed in place and preserve rock structures in ironstone. There is some evidence that iron may be mobilized and remobilized in the tropical environment, so that it can be dissolved from one site and redeposited in another. Radwanski and Ollier (1959) suggest that as a hillside is lowered by erosion the ferricrete at depth in the profile is dissolved at the top and redeposited at the base of the ferricrete. In this way, even a profile with ferricrete can sink to keep pace with surface lowering. Maignien (1958) explains the evolution of stepped ferricrete residuals by successive downcutting of rivers, with new deposition of ferricretes as terraces are formed, the iron being derived from partial destruction of higher ferricretes and new weathering products.

In southwest Nigeria, Moss (1965) made a detailed study of slope development at a plinthite breakaway (ferricrete outcrop). As is usually the case, the hard ferricrete is restricted to a fairly narrow zone behind the actual outcrop and excavations further back from the hillside enter soft mottled clay which is capable of hardening if exposed on the valley side. The sandy clay horizon above the mottled clay gets gradually thinner towards the ferricrete outcrop, but the sandy topsoil and humus layers show little variation, suggesting that erosion (by slopewash?) is thinning the soil towards the break of slope, but

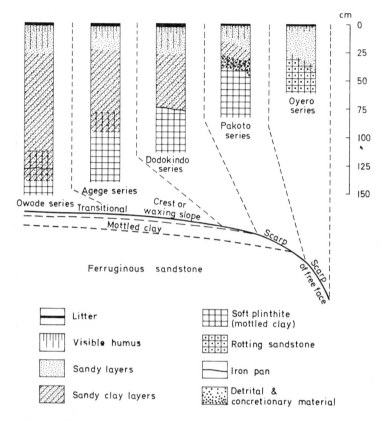

Figure 5.7. Soils associated with the slope above a high level ferricrete in southwest Nigeria (after Moss, 1965).

pedological processes (and winnowing of clay?) are keeping pace with surface lowering so far as the top 25 cm are concerned (Figure 5.7).

The ferricrete itself is seen to be cambered in cross-section, suggesting that there is subsurface erosion and spring sapping beneath the outcrop. Downslope from this there is a mass of detritus consisting of ferricrete and other mineral matter derived by slope processes from above, and partially from the washing out beneath the plinthite. The debris itself appears to be fairly rapidly disintegrated and altered, and the iron content appears to be rapidly reduced with distance from the breakaway.

An interesting complex profile is described by Sparrow (1966) from a cool temperate area of Natal at over 700 m, where a caprock of dolerite overlies Ecca (Permo-Carboniferous) Shales (Figure 5.8). On the caprock there is a well-rounded convex element with a relief of about 70 m. The red-brown ferruginous clays that cover this element maintain a constant thickness of 25 cm on the crest but thin to 3 cm on the steeper portions due to the effects

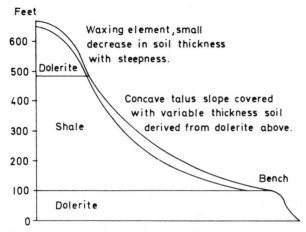

Figure 5.8. Complex catena formed on dolerite and shale under grassland in a cool subtropical humid climate, Natal (after Sparrow (1966)).

of creep. The base of the B horizon becomes progressively sharper as the angle increases, suggesting that the B horizon has been moved by creep. This is supported by the occurrence of small terracettes on the waxing element on the steepest slopes. On the shale the slope is concave and soils thicken from a few centimetres at the top to a maximum of about a metre in mid-slope. Much of this soil is clay loam derived from the dolerite, and frequently it overlies little-weathered shales.

Costin and Polach (1971) report that slopes between 5° and 25° in the Snowy Mountains of southeastern Australia appear to be well established by the existing forest vegetation. However, the slope deposits consist of fines and oriented gravels which are the product of periglacial conditions, commencing 31,000 to 34,000 years ago. Similar deposits are widespread in southeastern Australia.

A complex catena occurs on many Chalkland soils in England, and has been described by Avery (1964), Ollier and Thomasson (1957) and French (1972). The bedrock of Chalk is overlain by a layer of clay-with-flints up to several metres thick, and this is overlain in turn by brickearth (loess) which has been largely incorporated into the soil profile to form a silty topsoil. On the plateau the soil profile is complete, with silty topsoil over clay-with-flints over Chalk. At the plateau edges a convex slope element appears, the silty topsoil disappears, and soils are formed on fairly thin clay-with-flints directly over Chalk. The base of the clay is here topographically lower than on the plateau, so the Chalk must have been eroded away to some extent. Further downslope we can see how this happens, for here the clay is even thinner and and is incorporating pieces of chalk plucked from the bedrock. On the steepest slopes (which decline at only about 18°) the surficial clay is totally absent

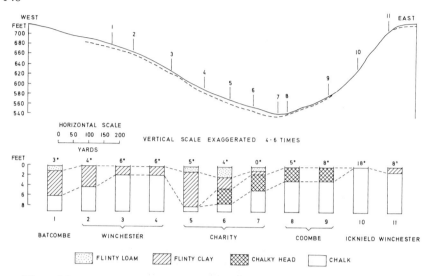

Figure 5.9. A typical catena formed on Chalk in the Chiltern Hills. Cross-section of the Little Hampden Valley (after Ollier and Thomasson (1957)).

and there are only shallow organic soils directly on Chalk (rendzina soil). At the base of this free face there is a debris slope of lower angle, upon which a mixture of chalk and debris from the plateau soils has been deposited. This increases towards the valley bottom, but also increasing towards the valley bottom is a decalcified horizon in the upper part of the profile from which chalk has been dissolved. This story seems to account satisfactorily for the topographic and soil relationships as seen in the field (Figure 5.9) and is further supported by mineralogical evidence. The silty topsoils contain distinctive minerals which are absent from the underlying clay-with-flints and absent from the soils on the convex slope. The base of the clay-with-flints contains distinctive minerals derived as an insoluble residue from the underlying Chalk, and not present in the topsoils or in the silty topsoils. The derived soils on the footslopes contain all the distinctive minerals from the plateau, thoroughly mixed up and dispersed throughout the profile.

A quite different kind of complex catena (Figure 5.10) is described by Huddleston and Riecken (1973) from an area of loess in Iowa. Deposition of loess ceased about 14,000 years ago, and soon afterwards leaching and oxidation created four weathering zones:

> oxidized and leached
> oxidized and unleached
> deoxidized and leached
> deoxidized and unleached

These four zones provided different parent materials for later soil formation and the situation is further complicated by slope processes.

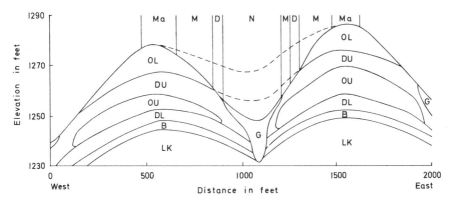

Figure 5.10 Relationship between surficial deposits, weathering zones and soil series in a cutting in western Iowa (after Huddleston and Riecken (1973)). Key to letter symbols: Ma—Marshall; M—Monona; D—Monona-Dow intergrade; N—Napier; OL—oxidized and leached Wisconsin loess; DU—deoxidized and unleached Wisconsin loess; OU—oxidized and unleached Wisconsin loess; DL—deoxidized and leached Wisconsin loess; B—basal soil Wisconsin loess; LK—Loveland loess and Kansan till; G—gully fill sediments.

In mid-slope locations, there is a thin layer of sediment in transport overlying various substrates and the soil on such sites 'has inherited the initial sorting of the loess deposit, the downslope sorting during erosive evolution, and the distribution of carbonates and iron produced by geological processes of weathering zone formation and subsequent truncation'.

The summit position has not been affected by subsequent periods of erosion. The shoulder position is a complex zone in which pedological and geological processes have been interacting simultaneously, and the toeslope is similarly complex.

## 5.2 Soil and slope elements

Geomorphologists have evolved a number of slope elements to help in discussion of slope evolution. The first description of slope elements in English was by Wood (1942) and additions have been made by writers such as King (1957) and Dalrymple, Blong and Conacher (1968). The work of Penck (1953) is frequently involved in discussion of slope elements, but his genetic terms will be avoided in this chapter. A sequence of slope units is shown in Figure 5.11. The convex element appears to be common in all classifications of slope elements. The free face refers to a slope free of debris, but genetically means a slope with the steepest gradient that a given rock can maintain; on some rocks this angle may be sufficiently low for some debris to be retained for a while. Below the free face there may be a concave slope, grading out in some

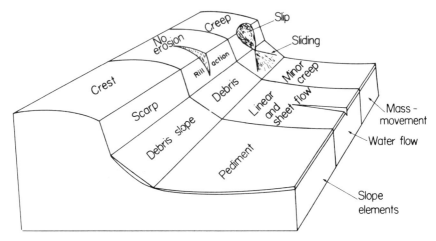

Figure 5.11. Slope elements in a fully developed hillslope profile (after King (1962), with permission).

situations to a pediment. In other hillslopes there may be another straight slope beneath the free face, built of debris of a characteristic angle of rest.

Explanations of the formation of these slope elements have been put forward by geomorphologists with varying degrees of factual restraint. Some hypotheses are based on simplified theories and involve only form; some are based on simple observations on slope deposits, which are extrapolated in a theoretical model; and some are based on assumed textural features of the deposits on slopes, which in reality are seldom found. Pedologists have rarely concerned themselves with models of slope evolution, and the number of soil/slope studies based on thorough soil profile description *and* detailed slope data is remarkably small, as the following résumé will make apparent.

*5.2.1 The convex slope*

Ideas on the origin of convex slopes derive from the work of Gilbert (1909). Gilbert argued that the regolith creeps as a continuous layer and the amount passing a given part of the slope is a product of cross-sectional area and the velocity. The further downslope the given part of the slope occurs, the more material it receives from above, and as the depth of the creep layer does not increase the velocity must increase. An increase in velocity requires a greater effect of gravity and hence a steeper slope, so slopes must steepen progressively in a downhill direction: that is, they must be convex.

Numerous objections have been raised to this hypothesis. The hypothesis fails to explain why the regolith should remain constant in thickness. This is not a deliberate decision made by an intelligent slope and it seems possible in theory for the regolith to thicken, thin, or remain constant. We now know in practice that while Gilbert's assumption has been verified in many places it is untrue in others. In the convex slopes of catenas considered earlier, variation of regolith varies from uniform (e.g. Gippsland Hills and Natal Hills) to slight thinning (Natal plateau) to thinning to zero (Chiltern Hills). In those places where the regolith thins, the explanation is probably that on steeper slopes velocity is even greater than is required on gentler slopes. The continuity equation can be maintained without necessarily invoking uniform thickness of regolith.

It should be mentioned that other methods of producing convex slopes have been suggested, including the dominance of overland sheet flow, while concave slopes are formed by flow in rills or gullies. In some climates solifluction may be important.

*5.2.2 The straight slope*

In simple geomorphic discussion the free face is regarded as free of debris. In more elaborate discussions it is part of a slope sequence and freedom from debris is not essential. In the Chiltern catena, for instance, the slope usually has a shallow organic soil (rendzina, the Icknield Series) rather than bare chalk.

In Uganda Pallister (1956) demonstrated by profile series the slope retreat of a free face with an angle of 17°. From the soils data of Radwanski (1960) it appears that this slope was not covered by mobile debris, but nevertheless on a parent material of rotted rock *in situ* (saprolite), a deep tropical red earth (Buganda Series) was present. In southeast Uganda, Radwanski and Ollier

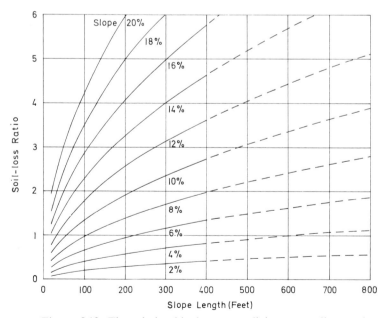

Figure 5.12. The relationship between soil losses, gradient and slope length (after Wischmeier and Smith (1965)).

(1959) proved the absence of drift layers on the free face by mineralogical studies.

Figure 5.12 shows how the ability of runoff to erode soils increases with an increase in either the length or the gradient of a straight slope. The data are derived from experiments and are related to standard plots 73 ft long and of 9% slope.

### 5.2.3 The debris slope

Debris slopes are essentially of two kinds. One is the Richter slope, with a thin veneer of debris actively passing over a bedrock slope of the same gradient (see Section 5.3.1), while the other has a variable thickness of debris which may or may not be in motion, may be undergoing pedogenesis, or may be eroded out of the system by sheetwash.

The catena described by Moss (Section 5.1.2) is an example of the latter. Both pedological and slope transport processes combine to produce the variation in debris type and thickness, and in soil type across the concave slope covered with debris. The example from the Chiltern Hills (Section 5.1.2) is another, but here there is a possibility that the slope deposit is inherited from a former climate and present pedological activity is leading to de-calcification of the upper part of an originally homogeneous mixed deposit of debris derived from upslope. The Nowra catenas (Section 5.3.6) show patchy deposition of debris and, in many slopes, even on fairly low angles, there may be no extensive accumulation of debris at all. The lower slopes at Coober Pedy (Ollier and Tuddenham, 1962) have essentially similar soils to the steeper, upper part of the slope, and there is no indication of a definite zone of sedimentation.

Some geomorphologists have been concerned to show that the shape of a slope in plan affects potential debris accumulation (Sparks, 1972). A concave plan means the various slopes are converging and this should lead to thicker sediments; a convex plan means the slopes are diverging and so should lead to thinning of soils. I have been unable to find any detailed investigations that support this theoretical picture. The simple two-dimensional model of slope and sediments seems to give a better picture of actual events than the theoretically superior three-dimensional model.

## 5.3 Some models of slope evolution in relation to soils

Besides being concerned with the mode of development of various slope elements, geomorphologists have evolved several theories of slope evolution to account for entire slope profiles. Some soil scientists have also developed models of landscape evolution, often remarkably dissimilar to those of the geomorphologists. The slope–soil relationships in a number of these models will be discussed below.

### 5.3.1 Richter slope

In some areas it is found that the angle of some slopes in bedrock is equal to that of scree slopes in the same area. Such slopes are known as Richter slopes. These are straight and steep slopes, and the soils are no more than lithosols. Pedogenesis is virtually absent, so in most areas these slopes are of no further concern to us.

The Richter slope is most likely to be encountered in areas of lithosols or polar desert soils. In the ice-free valleys of southern Victoria Land, Antarctica, Selby (1971) has suggested that salt weathering processes in the inland areas, and frost and solifluction near the coast, have produced straight bedrock slopes (Richter slopes) at a nearly constant angle of 35°.

*5.3.2 Parallel retreat of slopes*

Theories of parallel retreat of slopes during slope evolution have been put forward particularly by Wood (1942), Penck (1953) and King (1957). Essentially the theory is based on the idea that weathering may be uniform over a free face of constant angle. If the weathered debris is removed a new free face of the same gradient will appear a little further from the valley. The processes may be repeated many times and the slope retreats parallel to itself. The slope element above the free face may also undergo parallel retreat, though not by simple shedding of debris. The slope elements below the free face may undergo parallel retreat to a limited extent, but a footslope at a lower gradient must be developed as well. Most variations on this theme envisage a debris slope element on some part of the lower slope (as in Figure 5.11).

It might be expected that sedentary soils should be developed on the upper slopes: the profile thickness might be greater on flatter sites, thinning (sometimes to zero) on the steeper slopes. If a debris slope exists, debris is the parent material of the soils and the profiles will be of the type in which a balance between additions from upslope and losses downslope governs the profile morphology. Pedogenesis may keep pace with addition or removal of debris, or with softening of bedrock by weathering.

Such slopes should have very contrasting soils on the bedrock and the debris slope. Some examples are known, such as the steep slope of the Chiltern catena (Section 5.1.2) where the slope that appears on morphological grounds to be a debris slope turns out to have a cover of debris derived from upslope and to have a distinctive soil (Coombe series). In many other areas, however, especially in arid and semi arid regions, the geomorphic evidence indicates parallel retreat of slopes. However, there is neglible accumulation of debris and the soils are formed on bedrock or weathered bedrock, in various pedogenetic equilibrium relationships, e.g. examples described by Schumm (1967) and Ollier and Tuddenham (1962).

The regular association of different mechanisms on different parts of a slope, as envisaged by King (Figure 5.11) has not been verified by detailed studies to date.

*5.3.3 Slope decline*

The mechanism of landscape evolution postulated by W. M. Davis in his famous geographical cycle envisaged the gradual decline of slopes. No specific zone of debris accumulation is described and such slopes could be developed (in theory at least) by either creep (at varying rates on different gradients) or by sheetwash (affecting different parts of the slope in different ways). The soils on such slopes could therefore be in equilibrium, truncated, or otherwise related to the slope.

Demonstrable slope decline has been recorded by Savigear (1952) and Strahler (1950) who both found that an actively undercut slope had a steeper gradient than a protected one. As Carson and Kirkby (1972) point out, while these studies show that slope flattening has occurred, neither study offers any support to the idea that slope flattening is a continuing process. It is possible that the decline in slope angle is simply the replacement of one threshold slope by another (cf. Carson, 1969). Such a situation was recorded by Carson and Petley (1970) who found that most steep talus slopes on Exmoor at about 33° correspond to the angle of repose of the debris, and that gentler slopes at about 26° are probably newer stability slopes related to a more advanced stage of weathering of the mantle. Nevertheless Carson and Kirkby (1972, p. 321) suggest that there is probably no threshold slope angle for transport-limited processes on subdued concavo-convex slopes. Much appears to depend on variations in the shear strength of the slope material with weathering, especially in periods of instability when both retreat and decline may occur (Carson, Chapter 4).

The process of slope decline implies removal of soils over entire slopes, not merely the free face and upper slopes. With continuous renewal of soil profiles this may be hard to detect, especially if the erosion rates are slow compared with the speed of soil formation.

### 5.3.4 Landslide slopes

There are some areas where the gentle processes of creep and wash are replaced by landsliding as the dominant mechanism of slope development. Such areas, dominated by mass-movement, may be isolated patches within a region of 'normal' slope development, but in others the entire area is moulded by landslides or other forms of rapid mass-movement.

Landslides strip all the soil from the area they affect, and soil development begins afresh. Horizontally along the slope, beyond the limits of the landslide, well developed soil may persist. It is found that where large amounts of landsliding take place the soil varies little over the whole slope from crest to valley floor. It is only on the crests and valley floors that different soils may occur, and even this is not necessarily so. Mosaics of soils do occur, but the variations are *along* the slope and dependent on the frequency with which mass-movement occurs. There is thus a horizontal rather than a vertical variation in soil morphology.

An example is provided by the Hunua ranges of North Island, New Zealand, described by Pain (1969, 1971). Extreme climatic events in the form of high intensity rainstorms of tropical cyclonic origin in late summer trigger the movements, and vegetation has a marked effect on mass-movement processes and resulting landforms. Under forest the whole soil mass becomes saturated and flowage usually results. Under grass the A horizon of the soil remains coherent except where it cracks vertically, so that when slope failure occurs the A horizon slides on a wet shear-plane and the deposits take the form of a number of coherent blocks or rafts of A horizon material. Rapid mass-movement occurs about five times more frequently under grass than under forest. The soils affected are skeletal soils and semi-mature podzolic soils, both developed on greywacke, the main difference between the two being that the skeletal soils (Te Ranga) have no B horizon, while the podzolic soils (Marua) have a B horizon.

Another region where this mechanism has been investigated is the Adelbert Range, New Guinea (Pain, 1973; Pain and Bowler, 1973), where the aftermath of the Madang earthquake of 1970 was studied in some detail. Relative relief is between 100 m and 500 m and measured slope angles were generally found to be between 35° and 40°, with maximum slopes of 50° or more. Ridge crests are narrow. Landslides occurred on ground with all vegetation types, ranging from new gardens through regrowth to mature forest. The soils are of two types: lithosols, with little or no horizon differentiation, and earths with well-developed horizons and a sharp boundary between the B and C horizons. Some 240 km$^2$ were found to have been affected by dense landsliding and, beyond that area, there were scattered landslides in areas of generally undamaged forest. The landslides appear to be very shallow. No areas of debris accumulation were found on slopes and only minor and temporary accumulation occurred in channels. The whole landscape, with sharp divides and steep straight slopes, suggests a rate of denudation that is controlled by the availability of debris rather than the competence of the channel system to carry it. Where they occurred, the landslides removed nearly all the material available for transport. The shallow roots of the forest led to limited disturbance of regolith and usually the effect was restricted to less than 50 cm. However, soil depth does have an effect and slides on earths were significantly deeper than those on lithosols.

The complete removal of soil and forest cover from large areas of steep hillslopes is the most dramatic aspect of this extreme form of slope development. Approximately 25% of the slope areas in the 240 km$^2$ affected actually failed and, in some small headwater areas, more than 60% of the ground surface moved. This is an area where removal of 'skins' of debris results in parallel retreat of slopes, with no accumulation of debris at the base.

*5.3.5 Maignien's ferricrete model*

A special model of landscape and slope development with strict climatic confines was developed by Maignien (1958). In this model it is assumed that iron moves in solution downslope, but is re-precipitated as a footslope ferri-

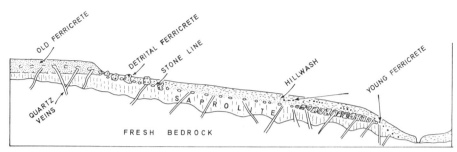

Figure 5.13. Typical soil–slope relationship in the African savanna, redrawn from Rohdenburg (1969). The old, high ferricrete is being destroyed, and a new ferricrete is froming on the lower slope.

crete. This ferricrete becomes sufficiently thick to be resistant to further erosion, so inversion of relief occurs and the footslope comes to occupy a ridge crest. The process may be repeated many times and ferricretes formed at many different elevations at many different times. If this process is occurring, great care needs be taken in reconstructing old erosion surfaces by casting an imaginary surface over laterite-capped residuals.

This process can be demonstrated through several cycles in Madagascar and several other places (Figure 5.13). However, many ferricrete areas do not have the right setting or relative relief to allow many cycles so that, although the existence of plateau-top and footslope ferricretes is acknowledged, the further landscape evolution with inversion of relief and the incidental development of new valley-side slopes is perhaps not especially common.

*5.3.6 Periodic soils*

The idea of periodic soils has been developed, especially in Australia by Butler and his colleagues.

A well-developed soil takes time to form, and so represents a period of stability at the ground surface. Alternating with such stable periods there may be periods of activity when slopes are eroded and eroded material is deposited on lower slopes or in valleys. The two phases together (stable and unstable, soil forming and erosion/deposition) make a cycle which has been called a K-cycle (Butler, 1959). A number of Australian studies, summarized by Butler (1967) indicate that there have been several successive K-cycles in parts of Australia and, furthermore, each stable period was characterized by a different dominant soil, as shown below:

        K1 (most recent)  minimal prairie soil
        K2                    red earth or red-brown earth
        K3                    red podzolic soil

On slopes, some K-cycle soils are preserved at the surface as palaeosols, others are truncated and some are buried.

An example is provided by Costin and Polach (1973) who investigated slope deposits in Canberra, Australia, which revealed clear evidence of former instability. Excavations revealed an upper layer of very stony, earthy material, up to 2 m deep, bleached near the surface and yellowish-red below. Beneath this was a less stony, compact, yellowish-red, mottled layer extending from 2 m to 3 m, containing a lens of transported topsoil with carbonized wood fragments. The lowest part of the profile consisted of heavy clay with few stones; pedogenic cracks extended through the middle and lowest layer. This sequence is interpreted as consisting of two age groups of slope deposits overlying the eroded stump of an older soil. The onset of unstable slope conditions, dated by the carbonized wood fragments, occurred about 27,000

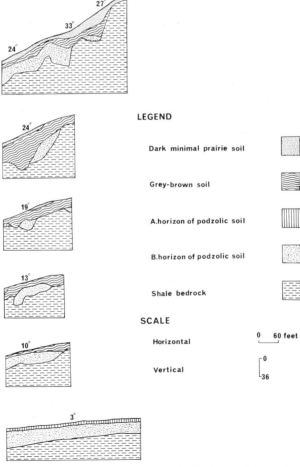

Figure 5.14. Soil layering on six shale hillslopes of decreasing slope angle at Nowra, New South Wales (redrawn from Walker (1962)).

years ago and is attributed to periglacial conditions with temperatures 8-10°C below present values.

Another example comes from the Nowra area of New South Wales (Walker, 1962) where three kinds of soil are found on the slopes: Minnamurra (minimal prairie soil), the Nowra (grey-brown soils) and the Wandandian (red podzolic on upper slopes, yellow podzolic on lower slopes). These soil 'layers' occur in various complex ways (Figure 5.14). On some slopes, any one of the three soils is present as the only layer. In such cases the soils are developed over materials ranging from *in situ* bedrock at the hill crest to deep hillside mantle deposits on the lower slopes. Usually, however, the soil layers occur on a variety of substrata over a wide range of slopes. The sequence is thought to result from alternating stable and unstable conditions in the past, with soil development phases alternating with periods of hillwash, creep and gully erosion.

Periodic soils are not restricted to the southeastern Australian environment. Mabbutt and Scott (1966) have described periodic soils from a region of savanna near port Moresby, Papua.

## 5.4 Asymmetrical valleys and soil distribution

Asymmetrical valleys are widely distributed throughout the world and very commonly show not only asymmetry of slope profile form and gradient but also asymmetry of soil type and distribution. If we take out of consideration those asymmetrical valleys that might be structurally controlled, the remainder are brought about by having different, climatically controlled, slope processes, or different rates of slope processes on opposite sides. Within any one region there is usually consistency in orientation of the steeper and gentler slopes, but there is no overall consistency within one hemisphere, or over large climatic ranges. It is usually assumed that different microclimates on opposing slopes are the fundamental cause of the asymmetry.

A good example is provided from the Beaufort Plain, Banks Island, in northern Canada (French, 1971). In most valleys aligned in a direction northeast-southwest, the steeper slope faces towards the southwest (Figure 5.15). The asymmetry is the result of differing microclimates, favouring the development of solifluction processes on the northeast-facing slopes and fluviothermal erosion on the southwest-facing slopes. The gentler slopes are particularly favourable localities for nivation processes since they are in the lee of the dominant winds. The steeper slopes may be regarded as boulder controlled slopes, since they are usually bare of vegetation, and the slope angles (25°–35°) probably represent repose angles for the sands and gravels. The soils, which are closely associated with the geomorphic units, include polar desert soils, tundra soils (differentiated into two groups on the basis of vegetation, wetness and topography) and regosols consisting of sands and gravels.

Kennedy and Melton (1972) studied valley asymmetry over a wide area of northwest Canada and found that in areas of severe climate the north-facing slopes are steeper, while in areas of milder climate the south-fading slopes

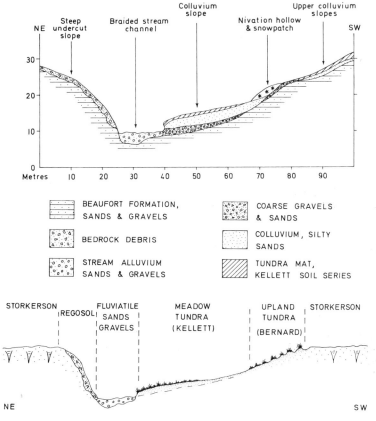

Figure 5.15. Generalized relationships between slopes and materials in asymmetrical valleys (above) and simplified soil catena (below) found in northwest Banks Island, Canada (after French (1971)) reproduced by permission of the National Research Council of Canada.

are steeper. This picture appears to be general. Reports of asymmetry with the steeper side facing north come from severe climates of Alaska (Hopkins and Taber, 1962; Kachadoorian, 1960; Currey, 1964), the Northwest Territories of Canada (Brönhofer, 1958) and Siberia (Presniakov, 1955; Surmach, 1962; Gravis, 1969). Asymmetry with steep south-facing slopes is reported from many areas in Europe, now temperate but where former periglacial conditions are postulated (Büdel, 1953; Maarlveld, 1951; Dylik, 1956; Thomasson and Avery, 1963). See Figure 5.16.

The situation can be more complicated than in the Banks Island area, described above, in areas where pedogenesis is not closely associated with geomorphic units. In the Chiltern catena, for instance, the gentle slopes are covered by clay-with-flints which may have been emplaced by periglacial processes, but the well developed soils are similar to those on the upper, convex

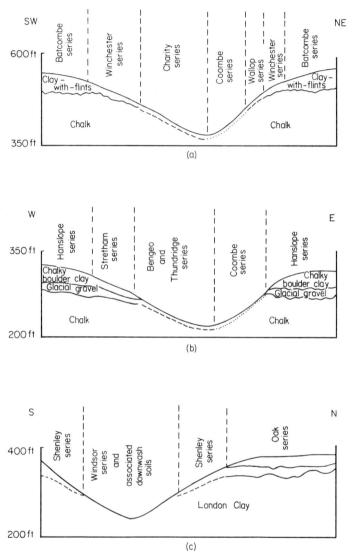

Figure 5.16. Asymmetrical catenas in Hertfordshire. (a) The Clay-with-Flints area of west Hertfordshire. (b) The Chalky Boulder Clay area of east Hertfordshire. (c) The London Clay area of south Hertfordshire. All sections are one mile long and there is great vertical exaggeration (after Thomasson and Avery (1963)).

slope element on the steep slope. In the same region there are some valleys which are morphologically symmetrical, but have a symmetry of soil type. It would appear that, even when microclimate is not sufficiently different on opposite slopes to cause morphological differences, it may still be sufficient to give rise to very different types of soil.

The phenomenon of asymmetry is not confined to periglacial or former periglacial areas, but is much more widespread. A warm temperate example comes from the Mount Crawford area of South Australia (R. Woods, personal communication). Here east–west valleys cross a belt of schist and quartzite striking north–south and dipping about 80° E (Figure 5.17). The south-facing slopes are steep and convex with shallow soils while the north-facing slopes are gentle and straight with deeper soils. Woods suggests that the form probably indicates slope retreat to the north concurrent with downcutting—an explanation similar to that proposed for the Chiltern valleys described above. Maximum retreat occurs on the soft schists, moderate retreat on the minor quartzites and little retreat on the thick quartzite. Woods also suggests that the asymmetry is a relict feature because the large embayments are abandoned and the present river has 'normal' meanders with undercuts on alternate sides.

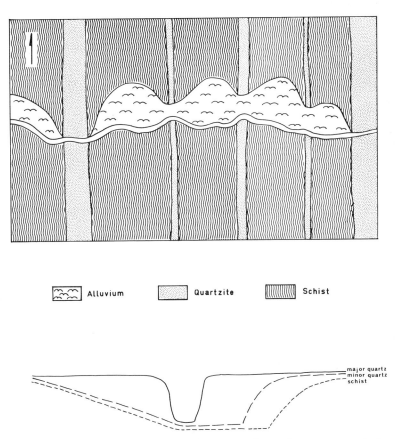

Figure 5.17. Diagrammatic map and cross sections of an asymmetrical valley developed in the Mount Crawford area, South Australia, (unpublished information from R. Woods).

The widespread existence of asymmetrical valleys suggests that they might provide a very fruitful field for the study of soil–slope relationships with respect to microclimate. Since opposite sides of the same valley have the same tectonic and geomorphic history, some of the variables present in comparing valleys from very different climatic areas are absent and firm determinations might eventually be drawn.

## 5.5 Some Principles of Methodology

In the preceding section we have seen various attempts to relate soils, slopes and climate. Hypotheses are in conflict, and at present the data are selective and inconclusive. Soil scientists may describe soils in great detail but ignore the geomorphology; geomorphologists describe slopes in great detail but describe the soils or surficial sediments inadequately. Both may relate catenas to climate on inadequate evidence.

The paper by Nettelton, Flach and Nelson (1970) is an example of an excellent pedogenic study with a naive approach to slope geomorphology. It is a study of soil development on tonalite (quartz diorite) in the mediterranean climate of Southern California. On the hilltops and upper slopes a leaching environment prevails. B horizons of soils here have base saturations of less than 90%, lack salts in soil solution and contain little exchangeable magnesium or sodium. Biotite weathers to vermiculite and kaolinite. On lower slopes soils have significant soluble salts and large amounts of extractable magnesium and sodium. Minor amounts of secondary carbonate and amorphous silica are found at the base of the profile. Biotite weathers to vermiculite and montmorillonite, and hornblende weathers to beidellite-montmorillonite.

This study presents valuable information on pedology, mineralogy and chemistry, but the geomorphology is reduced to a distinction between upper and lower slopes and, since we are told nothing of slope form, slope elements, gradient or length, we cannot use the information to study soil–slope relationships in detail.

Some early geomorphological theories of slope developments made unwarranted statements about the calibre of material being moved. Thus Baulig (1940) states that on concave slopes increasing fineness and impermeability (which results from chemical weathering) favour the concentration of rill wash. Reduction to fine grades is a very slow process and therefore shows its effects only after the waste has been a long time in transit, and especially when it is moving very slowly across the gentle lower slopes. The gravity-controlled creep and the unconcentrated wash become more effective in transportation and in wastage of the surface as the thickness of the waste sheet increases and it becomes progressively finer—but only until the point is reached at which so much debris is of clay grade that concentrated rill wash is brought into play. Thus Baulig invents a landscape-debris system in which particle size varies from the top to the bottom of the layer and from the crest to the foot of the slope. This type of slope does not in fact occur in nature.

Birot (1949) is another geomorphologist who makes unfounded statements such as 'the change from convexity to concavity is due to a progressive increase in impermeability'; 'the movement of waste becomes slower as the declivity diminishes, so that there is more time available for the breaking down of the debris to finer grades'; 'there is greater impermeability on the lower slopes'; and, finally, 'as the thickness of soil diminishes upslope, the rate at which the bedrock is wasting away increases' (translation by Cotton (1953)). Theories should not be based on such wild generalizations.

Finally we might consider the model-makers. It is possible to derive a whole range of possible slope forms and slope development sequences by mathematical modelling, as has been done for instance by Scheidegger (1961), Carson and Kirkby (1972) and by Kirkby in this volume (Chapter 8). Nikiforoff (1949) provides an example of a theoretical soils approach: he discusses what might happen to a soil under various conditions of pedogenesis and erosion. The trouble with most models is that they are not tested and the authors do not always feel constrained to devise tests for them.

One model that can claim to be tested is that of Beckett (1968) who derived a model for the relationship between soil formation and slope development related to the *Aufbereitung* concept of Penck in which soil maturity leads to increased detachability and erodibility. The model was tested by Furley (1968) who found that soil maturity can be related to gradient. He also claims that a *junction* can be recognized between the zone of erosion (upper slope) and the zone of accumulation (lower slope). However, Young (1969) has pointed out that the different soil properties found by Furley might be associated with soil moisture variation and do not necessarily indicate that accumulation is occurring below the junction. On a broader scale it must be noted that it is perfectly possible for concavities to be produced and maintained entirely by erosion, and many of the catenas described in this chapter have no zone of accumulation on lower slopes. Scientific progress only results when models are put to the test and rejected, modified or accepted in the light of the findings. At present model-making is far ahead of factual information and there is a real need for *crucial* tests that will provide a firm basis for theorizing.

Field problems do not normally permit adequate control of all the variables that an observer would like to measure, or provide all the information he would like to know; but we might at least consider the kind of study that seems desirable in the light of current knowledge to advance the study of soil/slope/climate relationships. The desiderata fall into several groups.

(1) *Consideration of the nature of the evidence*. So far as possible assumptions should be discarded and detailed evidence sought for the following:

(a) Evidence of creep; the depth to which it occurs; the rate of creep.
(b) Evidence of hillwash; the thickness of soil removed; the rate of hillwash loss.
(c) Evidence of downslope movement of solutions; gains in some sites from precipitation. These should be quantified if possible.

(d) Evidence of incorporation of material at the base of the soil profile by weathering. Comparison of the rate of incorporation with rates of soil loss by wash and creep will indicate whether erosion is dominant, pedogenesis is dominant, or whether the profile may be in equilibrium with current rates of slope erosion.

(e) Evidence of translocation within the profile and downslope. This is perhaps one of the most self-evident features of soil profiles which has been repeatedly disproved! Many B horizons are created by weathering of primary minerals to clay and not by illuviation.

(f) Evidence of deposited material: this is especially important on lower slopes and alternative means of producing thick horizons on lower slopes should be seriously considered.

(g) Evidence of rapid mass-movement: this will usually be obvious.

(2) *Quantification of the processes and materials involved.* Geomorphologists have long been familiar with the idea of a slope budget (Twidale, 1960) and many of the theories of slope development take into account the amount of debris produced under any given style of slope development. Some soil scientists have not been constrained by this idea, and some of them have ideas of 'drift' deposits which do not seem compatible with the size of source areas. This is exemplified by the ideas of Charter (1958) who envisaged a widespread drift in Africa, similar to the glacial drift of the northern hemisphere but of unknown origin. The diagram of Clarke (1971), however much we allow for the exaggeration for diagrammatic purposes, is erroneous and misleading (Figure 5.18). The thickness of drift is enormously exaggerated and it is most unlikely that so much drift could be derived from the slopes extending almost to the hilltop. In recent years many geomorphologists have believed in equilibrium

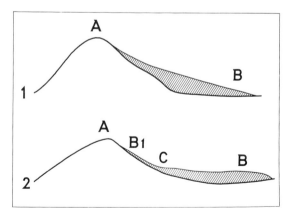

Figure 5.18. Stages in the weathering of a hill, A, redrawn from Clarke (1971) who describes it thus: 'The shaded portion B is a quartz drift derived from A; B1 is a small secondary drift. The main mass of B becomes separated from A by erosion of the drainage basin C.'

slopes almost as a matter of faith, but it is not scientifically permissible to *assume* that equilibrium conditions prevail at a site.

It has been pointed out that in many places soil forming processes are much faster than slope forming processes, so soil profiles may fail to indicate how the slopes were produced. For example, on granites in the Tenterfield area of New South Wales podzols cover all parts of the slope. Sandy topsoils mantle the entire slope, but are produced by pedogenesis—not by slope processes. Detailed studies of the relationship between slope and topsoil (e.g. Furley (1968)) are therefore of dubious value and any conclusions drawn from such relationships require many (usually unwarranted) assumptions.

It is in the handling and interpretation of good quantitative data from soil/slope studies that the model makers could make a most valuable contribution.
(3) *Detailed consideration of the climatic factor*. Climate is commonly treated in a very simple manner, though it seems from the studies of asymmetrical valleys that quite minor features of microclimatology may be very important in controlling processes. Furthermore, it is not always clear whether a given situation results from present-day processes or is inherited from some former set of processes and conditions. Any past climates or conditions that are postulated should be independently proved, and dated if possible (cf. Kennedy, chapter 6). If former conditions are merely postulated on the soil/slope data there is a danger of circular arguments and the interpretation is much less reliable.

Present-day climate should be considered in detail. What aspects of the climate are responsible for distinctive features of soil or slope: rainfall, temperature, evaporation/precipitation ratio or occasional violent storms? Is microclimate of great significance? A great many climatic factors or combinations of factors need to be considered and it will probably be a long time before the climatic factor can be assessed with confidence. What we can be sure is that the rather vague climatic descriptions commonly used in morphogenetic studies, such as 'maritime', 'savanna' or 'boreal' will almost certainly be inadequate for detailed studies.

## 5.6 Summary of Soil–Slope Relationships

It is not possible to draw up a table of climates and state what kind of catena will be found in each category. We can make a few general statements about certain situations, leaving a residue of areas for which generalization is merely speculative.

(1) In cold dry areas, frost action and salt weathering are dominant processes. Soil formation is limited to lithosol formation and Richter slopes may be formed.
(2) In cold wet (periglacial) areas, solifluction is dominant. Lobes of mobile debris accumulate on low areas, moving on even gentle slopes, and soils are mainly lithosols and bog soils.

(3) In hot dry areas wash (from occasional storms) appears to be a dominant erosional process and pedimentation a common slope process. Salt weathering, and various kinds of mechanical weathering such as slaking, may be important. Lithosols are the commonest soil type on steeper slopes.

(4) Duricrust areas. Ferricrete is not confined to a very narrow climatic range but is most often associated with landscapes of plateaux and shelves, and soils are usually organized in typical catenas. Wash appears to be the dominant slope process on upper slopes and there is great disagreement about the existence and importance of depositional slopes.

The silcrete variant of this landscape is found in more arid areas: the bauxite variant in areas of continuous high rainfall and high temperature. Parallel slope retreat appears to be general in such areas.

(5) In tropical areas there is a tendency for lower slopes to have sandy soils, these being created when throughflow washes out the clay fraction.

(6) Landslide-dominated landscapes are generally associated with wet areas, where accidents such as exceptional rainfall or earthquake shocks can trigger mass-movement. Landslides in such regions are usually shallow relative to the areal extent of the mobile layer. They are of course associated with steep, often straight, slopes in a region of feral relief (angular valleys and ridges). Soils reach various stages of maturity, depending on the frequency of landslides and the rates of pedogenesis, but variation in soil types is frequently more evident along the slope rather than up or down it.

(7) Non-eroded slopes. Certain rare sites, such as fixed sand dunes, may be sufficiently porous to suffer no erosion. The slopes are constructional and tell us nothing of slope erosion processes, but the catenas that form on non-eroding sites do provide a basis for comparison with other slopes where erosion is active.

(8) All other situations. Areas of non-extreme type, including humid and temperate zones, show a wide range of soils and slope morphology. Creep, wash, slope-sedimentation and pedogenesis interact in many different ways and historical changes of environment present a further complication. It is not yet possible to state any firm principles of catena formation and it is certainly premature to attempt to relate soil–slope relationships to climate.

## 5.7 Conclusion

On examining the catena–climate relationships of the world, we find a contrast between extreme and non-extreme situations. Distinctive soil–slope relationships occur in the extreme situations dominated by frigid or arid conditions, by landslides, by duricrust, or by highly porous parent materials. In all the rest of the world, under non-extreme conditions, the processes of slope erosion, slope deposition and pedogenesis are almost inextricably interwoven. The local studies that are soundly based are too few to allow serious generalization; and the many hypotheses that have been proposed to date appear to be limited in application and often demonstrably false.

Harris (1968) provided good reasons for doubting the 'Law' of soil zonality; Stoddart (1969) concluded that much of climatic geomorphology is not well established and is inadequate for setting up generalized worldwide schemes of climatic-morphological regions. These two findings add up to the same conclusion derived here from a consideration of case studies, namely that it is premature to relate climatic-geomorphological systems to soils.

## References

Avery, B. W., 1964, *The soils and land use of the district around Aylesbury and Hemel Hempstead*, Soil survey of Great Britain, HMSO.
Baulig, H., 1940, Le profil d'équilibre des versants, *Ann. Géogr.* **49**, 81–97.
Beckett, P. H. T., 1968, Soil formation and slope development; 1., *Z. Geomorph.*, **12**, 1–24.
Birot, P., 1949, *Essais sur quelques problemes de morphologie génerale*, Lisbon.
Bronhofer, M., 1958, Field investigations on Southampton Island and around Wager Bay, N. W. T., *The Rand Corp. Memo 1936*, Santa Monica.
Büdel, J., 1953, Die 'periglazial' morphologischen Wirkungen des Eiszietklimas auf des Ganzen Erde, *Erdkunde*, **7**, 297–314.
Butler, B. E., 1959, Periodic phenomena in landscapes as a basis for soil studies, *C. S. I. R. O. Soil Publ. 14*, Melbourne.
Butler, B. E., 1967, Soil periodicity in relation to landform development in south-eastern Australia, in *Landform Studies from Australia and New Guinea* (ed. J. N. Jennings Mabbutt), A. N. U., Canberra, 231–255.
Carson, M. A., 1969, Models of hillslope development under mass failure, *Geogrl. Analysis*, **1**, 76–100.
Carson, M. A. and D. J. Petley, 1970, The existence of threshold hillslopes in the denudation of the landscape, *Trans. Inst. Brit. Geogr.*, **49**, 71–95.
Carson, M. A. and M. J. Kirkby, 1972, *Hillslope form and process*, Cambridge University Press.
Charter, C. F., 1958, Report on the environmental conditions prevailing in Block 'A', Southern Province, Tanganyika Territory, *Ghana Dept. Agric. Occ. Pap. No. 1*.
Clarke, G. R., 1971, *The study of soil in the field*, Oxford.
Corbett, J., 1969, *The living soil*, Martindale Press, West Como.
Costin, A. B. and H. A. Polach, 1971, Slope deposits in the Snowy Mountains, south-eastern Australia, *Quatenary Research*, **1**, 228–235.
Costin, A. B. and H. A. Polach, 1973, Slope development at Black Mountain, *Aust. J. Soil Res.*, **2**, 13–25.
Cotton, C. A., 1953, Review and discussion, *J. Geol.*, **62**, 73–81.
Cruickshank, J. G., 1971, Soils and terrain units around Resolute, Cornwallis Island, *Arctic*, **24**, 195–209.
Currey, D. R., (1964), A preliminary study of valley asymmetry in the Ogtoruk Creek area, N. W. Alaska, *Arctic*, **17**, 85–98.
Dalrymple, J. B., R. J. Blong and A. J. Conacher, 1968, A hypothetical nine-unit land surface model, *Z. Geomorph.*, **12**, 60–76.
Dan, J., D. H. Yaalon and H. Komdjisky, 1969, Catenary soil relationships in Israel; 1, The Netanya Catena on coastal dunes of the Sharon, *Geoderma*, **2**, 95–120.
Dylik, J., 1956, Esquisse des problêmes periglaciaires en Pologne, *Biul. Peryglacjalny*, **4**, 57–71.
Ellis, J. H., 1938, *The soils of Manitoba*, Manitoba Econ, Survey Board, Winnipeg.
French, H. M., 1971, Slope asymmetry of the Beaufort Plane, Northwest Banks Island, N. W. T., Canada, *Canadian J. Earth Sci.*, **8**, 7, 717–731.

French, H. M., 1972, Asymmetrical slope development in the Chiltern Hills, *Biul. Peryglacjalny*, **21**, 51-73.
Furley, P. A., 1968, Soil formation and slope development; 2, The relationships between soil formation and gradient angle in the Oxford area, *Z. Geomorph.*, **12**, 24-42.
Gilbert, G. K., 1909, The convexity of hilltops, *J. Geol.*, **17**, 344-50.
Gravis, G. F., 1969, Fossil slope deposits in the northern arctic asymmetrical valleys, *Biul. Peryglacjalny*, **20**, 239-258.
Harris, S. A., 1968, Comments on the validity of the law of soil zonality, *Trans. 9th Int. Cong. Soil Sci.*, **IV**, 585-593.
Hopkins, D. M. and B. Taber, 1962, Asymmetrical valleys in Central Alaska, *Special Papers No. 68, Abstracts for 1961, Geol Soc. Am.*, **33**, 116.
Huddleston, J. H. and F. F. Riecken, 1973, Local soil-landscape relationships in Western Iowa, I, Distribution of selected chemical and physical properties, *Soil Sci. Soc. Amer. Proc.*, **37**, 264-270.
Kachadoorian, R., 1960, Geological investigations in support of Project Chariot in the vicinity of Cape Thompson, N. W. Alaska, Preliminary Report, *U. S. Geol. Surv. Trace Elements Invest. Report 753*.
Kennedy, B. A. and M A. Melton, 1972, Valley asymmetry and slope forms of a permafrost area in the Northwest Territories, Canada, *Polar Geomorphology*, R. J. Price and D. E. Sugden (Eds.), *Inst. Brit. Geogr. Spec. Pub. 4*.
King, L. C., 1957, The uniformitarian nature of hillslopes, *Trans. Edin. Geol. Soc.*, **17**, 81-102.
King, L. C., 1962, *The morphology of the earth*, Oliver and Boyd, Edinburgh and London.
Maarveld, G. C., 1951, De asymmetrie van de kleine dalen op het noordelijk half rond, *Tijdscjr. van het kon Neder. Aardijksundig Genootschap*, Ser. 2, **28**, 297-312.
Mabbutt, J. A. and R. M. Scott, 1966, Periodicity of morphogenesis and soil formation in a savannah landscape near Port Moresby, Papua, *Z. Geomorph.*, **10**, 69-89.
Maignien, R., 1958, Le cuirassement des sols en Guinee, *Mem. Carte Geol. Alsace-Lorraine, No. 16*.
Milne, G. M., 1935, Some suggested units of classifications and mapping particularly for east African soils, *Soil Res.*, **4**, 183-198.
Morgan, R. P. C., 1973, Soil-slope relationships in the lowlands of Selangor and Negri Sembilan, West Malaysia, *Z. Geomorph.*, **17**, 139-155.
Moss, R. P., 1965, Slope development and soil morphology in a part of south-west Nigeria, *J. Soil Sci.*, **16**, 192-209.
Moss, R. P., 1968, Soils, slopes and surfaces in tropical Africa, Ch. 2 in *The Soil Resources of Tropical Africa*, R. P. Moss (Ed.), Cambridge U. P., London, 29-60.
Nettleton, W. D., K. W. Flash and R. E. Nelson, 1970, Pedogenic weathering of tonalite in Southern California, *Geoderma*, **4**, 387-402.
Nikiforoff, C. C., 1949, Weathering and soil evolution, *Soil Sci.*, **67**, 219-30.
Nye, P. H., 1954-55, Some soil forming processes in the humid tropics, I-IV, *J. Soil. Sci.*, **5**, 7-21 and **6**, 51-83.
Ollier, C. D., 1959, A two cycle theory of tropical pedology, *J. Soil Sci.*, **10**, 137-148.
Ollier, C. D. and A. J. Thomasson, 1957, Asymmetrical valleys of the Chiltern Hills, *Geog. J.*, **123**, 71-80.
Ollier, C. D. and W. G. Tuddenham, 1962, Slope development at Coober Pedy, South Australia, *J. Geol. Soc. Austrl.*, **9**, 91-105.
Pain, C. F., 1969, The effect of some environmental factors on rapid mass movement in the Hunua Ranges, New Zealand, *Earth Sci. J.*, **3**, 101-107.
Pain, C. F., 1971, Rapid mass movement under forest and grass in the Hunua Ranges, New Zealand, *Austr. Geog. Studies*, **9**, 77-84.
Pain, C. F., 1973, Characteristics and geomorphic effects of earthquake-initiated landslides in the Adelbert Range, Papua, New Guinea, *Eng. Geol.*, **6**, 261-274.

Pain, C. F. and J. M. Bowler, 1973, Denudation following the November 1970 earthquake at Madang, Papua–New Guinea, *Z. Geomorph.*, Suppl. Bd., **18**, 92–104.
Pallister, J. W., 1956, Slope development in Buganda, *Geogr. J.*, **122**, 80–7.
Penck, W., 1953, *Morphological analysis of landforms* (trans. H. Czech and K. C. Boswell). Macmillan, London.
Presniakov, J. A., 1955, Les vallees asymmétriques en Siberie, *Questions de Géol. de l'Asie*, II, Moscow, 391–396.
Radwanski, S. A., 1960, The soils and land use of Buganda, Uganda *Dept. Agric. Mem. Res. Div. Ser. 1, No. 4*.
Radwanski, S. A. and C. D. Ollier, 1959, A study of an East African catena, *J. Soil Sci.*, **10**, 149–168.
Rohdenburg, H., 1969, Slope pedimentation and climatic change as principal factors of planation and scarp development in tropical Africa, *Giessener Geogr. Schrift*, **20**, 57–152.
Savigear, R. A. G., 1952, Some observations on slope development in South Wales, *Trans. Inst. Brit. Geogr.*, **18**, 31–52.
Scheidegger, A. E., 1961, *Theoretical Geomorphology*, Springer-Verlag, Berlin.
Schumm, S. A., 1964, Seasonal variations of erosion rates and processes on hillslopes in western Colorado, *Z. Geomorph.*, Supp. **5**, 215–138.
Schumm, S. A., 1967, Rates of surficial rock creep on hillslopes in Western Colorado, *Science*, **155**, 560–561.
Selby, M. J., 1971, Slopes and their development in an ice-free, arid area of Antarctica, *Geogr. Annlr*, **53A**, 235–245.
Sparks, B. W., 1972, *Geomorphology*, Longmans, London.
Sparrow, G. W. A., 1966, Some environmental factors in the formation of slopes, *Geogr. J.*, **132**, 390–395.
Stoddart, D. R., 1969, Climatic geomorphology: review and re-assessment, *Progress in Geography*, **1**, 161–222.
Strahler, A. N., 1950, Equilibrium theory of erosional slopes approached by frequency distribution analysis, *Am. J. Sci.*, **248**, 673–696 and 800–814.
Surmach, G. P., 1962, Water erosion study on the left bank of the Volga River in Kiybyshev, *Soviet Soil Sci.*, **2**, 187–193.
Thomasson, A. J. and B. W. Avery, 1963, The soils of Hertfordshire. *Trans. Hertfordshire Nat. Hist. Soc.*, **25**, 247–263.
Twidale, C. R., 1960, Some problems of slope development, *J. Geol. Soc. Austrl.*, **6**, 131–147.
Walker, P. H., 1962, Soil layers on hillslopes: a study at Nowra, N. S. W., Australia, *J. Soil Sci.*, **13**, 167–177.
Watson, J. P., 1960, Soil catenas, *Soil and Fertilizers*, **28**, 307–310.
Webster, R., 1965, A catena of soils on the Northern Rhodesia plateau, *J. Soil Sci.*, **16**, 31–43.
Wischmeier, W. H. and D. D. Smith, 1965, *U. S. D. A. Handbook 282*.
Wood, A., 1942, The development of hillside slopes, *Proc. Geol. Ass.*, **53**, 128–140.
Young, A., 1969, The accumulation zone on slopes, *Z. Geomorph.*, **13**, 231–233.

CHAPTER SIX

# Valley-side Slopes and Climate

BARBARA A. KENNEDY

## 6.1 Introduction

*6.1.1 General*

The primary concern of this chapter is with the form of slope profiles within valleys created by fluvial erosion. Such basins, whatever their size, have four distinctive components of form: a perimeter, an outlet, a channel network and a suite of valley-side slopes. It is obvious that any change in the geometry or position of any one of the form elements which comprise the basin will necessarily result in alterations to one or more of the rest, although there will be some time-lag involved in the adjustment. In consequence, one cannot realistically consider problems of the relationship between climate and slope form without an equal consideration of the influence of climatic factors upon the remaining three components of the drainage basin, of which the channel network is in many cases the most critical. Further, it may well be the case that variations in slope form reflect, not a direct response to the influence of climate, but rather adjustments required to counterbalance climatically-induced modifications of some other feature of the basin morphology: the alterations in slope form that may follow from a change in the location of the outlet after a post-glacial rise in sea level may serve as an example. It is therefore the basin as a whole, not simply the valley-side slopes, that must be considered.

Moreover, all drainage basins are oriented forms, in that their channels and slopes possess particular compass orientations. As a result, every basin lies at a particular angle to the prevailing winds and to incoming direct solar radiation. Furthermore, the orientation of each basin—as well as its location within the landmass—will also determine its relationship to the trend of geologic structure, to the passage of glacier or continental ice, and to the inland transmission of the effects of eustatic or isostatic movements in sea level. It is clear that the overall geometry of any valley will to a large extent reflect the degree to which the directions of operation of the controlling factors mentioned above have coincided or conflicted. Just as the slopes comprise but one aspect of the geometry of the valley as a whole, so the climatic variables of precipitation and insolation create but one of the set of factors operating in preferred directions upon those slopes.

The geomorphological importance of the climatic variables is in itself a major source of discussion. In the case of valley-side slopes, we know that few, if any, profiles can be considered to develop under the influence of a single geomorphological process. Further, it is clear that few, if any, of those processes can be related simply to specific climatic or meteorological events, although the case of rainsplash may prove the exception (de Ploey and Savat, 1968). In the overwhelming majority of situations we have to deal with slopes exhibiting a mixture of processes, each one of which is related in a complex fashion to the magnitude, frequency and relative timing of the incidence of sun, snow, rain and wind. Finally, the climate created by these meteorological events which is of geomorphological importance is that of the surface layers of the ground, rather than of the standard weather station. It is clear that it is the variation in temperature and moisture conditions at and immediately beneath the surface which exerts the most fundamental control over the majority of processes operating upon valley-side slopes, including soil-creep (Kirkby, 1967), slumping (Beatty, 1972) and block-falls (Schumm and Chorley, 1964), rill initiation (Kirkby and Chorley, 1967) and mud-flows (Prior, Stephens and Douglas, 1971). Our direct knowledge of these crucial climatic conditions is, at present, very limited indeed.

It is necessary to bear the complexity and uncertainty of the basic situation firmly in mind. Unambiguous associations between climatic and topographic features are likely to prove rathere rare. Nevertheless, we shall hope to show that some such associations can be made, for certain elements of slope forms, and it is to be hoped that this may provide encouragement for the pursuit of further and more complex interrelationships.

*6.1.2 The problem of scale*

As we have indicated above, the climatic conditions which are of greatest importance to the operation of slope processes are those within the upper layers of the slope mantle (or the bedrock). It has only become technically possible to measure these conditions within the last 15 years or so and there are few sites where recording has been continuous over this period. Moreover, once attention is focused upon the soil microclimate, it becomes apparent that the network of recording sites required within even the smallest river basin is very large indeed, since these microclimatic conditions will vary not only with the orientation of the ground surface, but also with type and degree of vegetation cover, soil or regolith depth and texture, position upon the hillside and slope angle.

In consequence, attempts to assess climatically-controlled variations in process (cf. Gerlach (1963) and Soons and Rayner (1968)) have generally been complicated by the fact that several features which presumably control the soil climate vary from one observation point to another. For example, if one controls all potential differences between two sites except their aspect (Gerlach, 1963), one nevertheless cannot necessarily distinguish between the role of sun and rain in the creation of the variations observed. Even such a degree of control is unusual, however: more usually, we can only infer that climatic factors are at least partially responsible for the differences concerned, for sites will differ by far more than orientation alone (Soon and Rayner, 1968).

Such small-scale investigations are undoubtedly crucial to our understanding of the manner in which climatic events, geomorphological processes and slope forms are functionally related. The difficulties of instrumentation and the period of time required to obtain acceptable limits of accuracy in measurements have led some workers to opt for a broader view in which regional climatic indices have been associated with more general measures of slope form, with the linking processes inserted by inference, rather than direct observation. A major study of this type is that of Melton (1957) which will be discussed in detail below.

Whatever the spatial scale on which climatic and topographic features are viewed, we are faced with the knotty and persistent problem of the temporal variation in climatic conditions. Slopes are, on the whole, features of the landscape which will only alter perceptibly over tens or hundreds of years, in the most unstable environments: low angle profiles may persist for thousands or tens of thousands of years virtually unchanged. Because the overall geometry (i.e. convex, concave) as well as the angle of slope of different sections of the profile will govern the rate of weathering and the passage of water through the surface layers, the soil microclimate and hence the nature and rate of operation of the slope processes may be highly dependent upon the slope form. As a direct consequence, 'climatically controlled' variations in processes may persist within a river basin purely because those processes at work are unable to alter the morphology of slopes created under past climatic conditions. We can, in practical terms, do no more than acknowledge that this position may exist, unless the slopes concerned are datable with an unusual degree of precision (cf. Strahler (1956)).

The problems of scale, both spatial and temporal, can be seen to pose severe difficulties in the theoretical and practical assessment of links between climate and slope form. The temporal considerations are undoubtedly the more acute. However, it is possible to overcome both groups of difficulties in some, though by no means all, instances.

## 6.2 Two contrasting studies

### 6.2.1 Introduction

Consideration of valley asymmetry has been an important component in the study of the relationship between climate and slope form. Before turning to the consideration of that highly complex topic, however, it is desirable to examine the results of studies that have not been directly concerned with asymmetry in order to establish a general frame of reference.

Although there are a number of examples which could be quoted at this juncture, only two will be used in illustration. The first, Bik's (1968) study of the form of prairie mounds in Alberta, may be taken as a good example of the manner in which variations in slope form may be inferentially related to microclimatic differences within a comparatively simple geomorphological environment. The second, taken from Melton's (1957) investigation of basin forms and processes in the southwestern United States, illustrates the difficulties which are experienced when attempting to isolate the linkage between generalized features of present-day climate and slope morphology.

*6.2.2 The prairie mounds of southern Alberta*

Bik's investigation of the nature and origin of the prairie mounds of southern Alberta (50° N, 111° W), provides interesting evidence of the influence of microclimatic variations upon slope form, despite a lack of direct information concerning the processes at work.

The prairie mounds concerned are low-angled features of subdued relief (the maximm angle recorded is 14° and the average 7°, while the relative relief is of the order of 11m/36 ft). The material composing the mounds is a sandy till which appears to be relatively homogeneous in composition and unaffected by structural trends. Not only does Bik consider that the mound topography is geologically very young (dating from approximately 10,500 B.P.), but it would also seem to have escaped subsequent modification by fluvial activity. The simplicity of the relief, structure and erosional history of the prairie mounds make them an excellent testing ground for climatic–geomorphological relationships. Against this, however, must be set the fact that climatic conditions have changed substantially since the creation of the mounds and the general lack of direct observation of the rates of present-day processes upon their slopes must also be borne in mind. Moreover, it must be assumed that if differences are to be detected on such low-angle forms, such observations must be conducted over a prolonged period.

Bik himself was not concerned directly with modern processes, but rather with the elucidation of the sequence of events leading to the formation of the mounds. As part of his investigation, however, he provides (1968, Figure 8) the angles of 305 profiles measured during traverses across the mound topography in an area to the east of Medicine Hat, subdivided according to the compass orientation of the slopes. These data have been re-plotted as Figure 6.1. It is clear that all eight sub-groups demonstrate the normal distribution of angles which, following Strahler (1950) may be taken to indicate homogeneity of geomorphologic environment (see below). The maximum difference between the average angles of the sub-groups is 1·3° (northeast-facing profiles average 7·3°, those facing west only 6·0°). When tested by the analysis of variance this difference is found to be statistically significant at the 95% level. A similar,

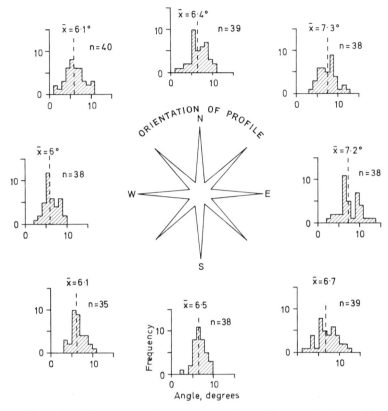

Figure 6.1. Distribution of slope angles of prairie mounds, by profile orientation, for the area to the east of Medicine Hat, Alberta (redrawn from Bik, 1968, Figure 8).

and apparently more pronounced, tendency for south-and west-facing profiles to be most subdued in angle is reported by Bik for the mound area to the south of Medicine Hat (1968, Table II, p. 421). In both locations, the low-angled slopes with southerly and westerly orientations are described as being distinct from other profiles in that they are characterized by large concentrations of stones at their surfaces. However, Bik does not proceed to any detailed consideration of the mechanisms which may have created the topographic variations described, beyond the following observation: 'The rate of mass-wasting or erosion appears to have varied for differing exposures' (1968, p. 421).

This inference of microclimatic control would seem to be readily acceptable, given the apparent absence of structural or fluvial controls, despite the obvious impossibility of specifying the details of the processes involved. We may consider, then, that the prairie mound topography provides a clear indication that the differential receipt of direct insolation and precipitation by low-angled slopes may, even within as short a period as 10,500 years and unassisted by differential fluvial erosion, serve to modify the action of slope processes sufficiently to produce significant variations in profile form. True, the maximum average variation in slope angle noted is rather small in absolute terms, but it nevertheless represents an 18% reduction from the steepest mean value observed.

While it would be of considerable interest to know more of the modern pattern of slope processes and microclimatic conditions within these (and similar) prairie mound areas, nonetheless it seems clear that this is one situation in which it is safe to infer that variations in slope form are directly related to climatic considerations.

*6.2.3 Small basins of the southwestern United States*

The study undertaken by Melton (1957) of the relationship between form and process in some 80 basins in Arizona, Colorado, New Mexico and Utah remains one of the most comprehensive sources of material for the investigation of climatic influences upon drainage basin features in general and profile form in particular. This information provides a valuable contrast to Bik's (1968) study, in that it relates to the regional rather than the local scale and, as an inevitable corollary, is concerned with topographical features which are influenced by structural and historical controls, in addition to recent climatic variations.

Melton considered as many as 40 quantitative aspects of drainage basin form (1957, Table 2), including the average maximum angle of valley-side slopes (according to Strahler's (1950) definition), the drainage density (in miles/square mile) and the relative relief (in feet). The principal description of the modern climatic conditions employed was the precipitation–effectiveness (P-E) index of Thornthwaite (1931). In addition, Melton described the general lithology, structure and vegetation cover of each basin studied.

Clearly, the detection of a general relationship between climatic factors and the slope forms of such a variety of drainage basins will be complicated first by variations in the lithologies concerned (for these will modify the manner in which weathering occurs) and, second, by differences in relative relief (in part a function of structure and lithology, in part a reflection of geologic history and, in turn, dependent to some extent upon climatic events). Finally, the general relationship will be complicated by the non-uniformity of fluvial dissection, as measured by the drainage density index (itself dependent upon lithology, structure, relief and climatic events, both past and present). Nevertheless, Melton's analysis (1957, Figure 15, p. 68) revealed a significant, positive correlation between the average angle of valley-side slope and the logarithm of the P-E index, for some 59 of the basins studied. The angles concerned range from $5°$ to $40°$ and the logarithms of P-E from 1·0792 (12) to 2·0414 (110). However, the level of this association is predictably rather weak ($0·05 < p < 0·025$). In fact, Melton considered that far more important controls over slope angle are exerted by variables such as infiltration capacity and relative relief which are influenced jointly by climatic and geologic factors.

It is of interest to look more closely at Melton's data. While few of the basins sampled are of absolutely identical lithology, it is possible to identify four groups cut into rocks of similar composition. Within each of these relatively homogeneous samples it is then instructive to consider in some detail the relationship between relief, drainage density and purely climatic factors (as expressed by the P-E index) on the one hand, and the observed variation in maximum slope angles on the other. The results of such an analysis are described below (data from Melton, 1957, pp. 15–21 and Table 2): all differences discussed have been tested by the Mann–Whitney U test or the Kruskal–Wallis one-way analysis of variance (Siegel, 1956), using two-tailed tests and the 95% confidence level.

*6.2.3.1 Basins on granite, granite-gneiss and schist.* Melton provides details for 23 basins on these rock types, located in Arizona, Colorado and New Mexico. The precipitation-effectiveness index varies from 40 to 82 and the vegetational Life Zones represented include the Upper Sonoran, Canadian and Hudsonian, together with the two intervening transitional zones. The observed climatic variation is largely a product of differences in elevation above sea level, which ranges from approximately 5350 to 9500 feet (1630–2886 m).

The first potential control of slope angles to be examined is relative relief and this is found to exert a pronounced effect. The six basins in this group with a relative relief of less than 750 feet (229 m) possess mean maximum slope angles which are significantly lower than those of the basins with more rugged relief. When the influence of varying drainage density is then assessed within the six low-relief basins, differences of 13–23 mile/mile$^2$ are found to have no detectable effect upon average slope angles. This allows the response to changing precipitation-effectiveness to be tested and, as P-E rises from 40 to approximately 47·5, there is found to be a corresponding and significant increase in the average maximum slope angle of some 4°.

Reverting to the sub-group of 17 basins with relative relief in excess of 750 feet, it is found that their maximum slope angles show significant variation as drainage density ($Dd$) rises. Basins with values of $Dd$ of less than 15 mile/mile$^2$ possess significantly lower mean maximum angles than those which are more heavily dissected (cf. Strahler (1950)). Neither of these new sub-groups indicates a statistically detectable response in slope forms to changing values of the P-E index over the range 40–82.

*6.2.3.2 Basins on sandstone, limestone and shale.* Melton describes 17 basins of this type, again located in Arizona, Colorado and New Mexico. Values of the precipitation-effectiveness index range from 12·8 for the shale basins of the Chinle badlands, Arizona, to 77 for the limestone and shale area of the Poleo basin in New Mexico. The majority of these valleys fall within the Upper Sonoran Life Zone, with a few representatives of the transitional zone to the Canadian: once again, the primary control over climatic variation is exerted by differences in elevation, from 4000 to 8600 feet above sea level (1220–2621 m).

As in the case of the granitic basins, the effect of varying relative relief upon slope angle is found to be significant. Here, too, basins with a relief of less than 750 feet (229 m) are found to possess significantly lower mean maximum angles of valley-side slope.

The larger of the two sub-groups identified is that with the lower relative relief and contains 11 basins. Rather surprisingly, the variations in drainage density within this class (from 11·2 to 349 mile/mile$^2$) are not found to exert any statistically detectable effect upon the slope forms. Accordingly, it is possible to examine directly the influence of P-E for the whole sub-group. It emerges that a rise in that index from 12·8 to 77 is accompanied by a significant *decline* in mean maximum slope angles, from approximately 24° to 18·5°.

The smaller of the two sub-groups defined on the basis of differences in relative relief (that containing 6 basins with a relief of more than 750 feet) is significantly influenced by changing drainage density. Values of $Dd$ in excess of 11 mile/mile$^2$ prove to be associated with steeper maximum slope angles (a situation comparable to that described by Strahler (1950) and noted in some of the granite valleys discussed above) and it is therefore impossible to assess the influence of variations in the P-E index upon the profile angles for this class.

*6.2.3.3 Basins on limestone, sandstone and quartzite.* It may appear illogical to separate the group of six drainage basins containing quartzites from the rather similar sedimentaries discussed in the preceding section. However, not only are the members of this sub-group drawn from a different region (Utah), they also possess significantly greater relative relief and steeper slope angles. Thus, it is justifiable to treat the six as representatives of a distinct population from the outset. The precipitation-effectiveness index within these valleys ranges from approximately 47 to 75; the vegetational Life Zones represented are the Upper Sonoran and the transitional zone to the Canadian; and the elevations above sea level vary between approximately 6000 and 7500 feet (1829–2286 m).

Relative relief amongst these six basins varies from 1289 to 4524 feet (393–1379 m), but has no statistically significant influence upon the values of maximum slope angles. However, the control apparently exerted by drainage density is pronounced and significant. In this case, valleys with levels of $Dd$ in excess of 11 mile/mile$^2$ are found to experience a significant *reduction* in the average steepness of their maximum slope angles; this is in complete contrast to the situation found in the other three sets of valleys discussed here and, while it conflicts with the general theory outlined by Strahler (1950), it is in agreement with Melton's generalized findings for 43 of his sample basins (1957, Figure 4, p. 65). It is clear that the pattern of association between drainage density and slope angles is a complex one. In this particular case, however, the strong influence of drainage density makes it impossible to test for the effect of precipitation-effectiveness.

*6.2.3.4 Basins on volcanic rocks: basalt, felsite, dacite, rhyolite and andesite.*
Information is available for 25 valleys cut into various combinations of these volcanic rocks and located in Arizona, Colorado and New Mexico. Precipitation-effectiveness ranges from 33 to 110 and all Life Zones from the Lower Sonoran to the Hudsonian are represented. Once more, the major source of climatic variation is the differing elevation of basins above sea level, with a range from approximately 5000 to 10,000 feet (1524–3048 m).

Within this sub-group the effect of relative relief upon maximum slope angles is found to be immediately significant, steeper slope profiles being located within those valleys which possess more than 1000 feet of relief (305 m).

When the set of 15 basins with the greater relief is examined, it emerges that average maximum angles exhibit a significant response to variations in drainage density: the steeper angles are associated with $Dd$ values in excess of 13 mile/mile$^2$. Neither the group of nine basins with high relief and high drainage density, nor that of the six with high relief and lower drainage density, is found to display any significant variation in slope angles as the P-E index rises from 33 to 60.

Turning to the remaining set of valleys with relative relief of less than 1000 feet, neither variations in $Dd$ from 4·5 to 18 mile/mile$^2$, nor in P-E from 60 to 110, have any statistically detectable effect upon the maximum angles of valley-side slopes.

From this extensive re-examination of some of Melton's 1957 data, two points of general interest emerge. First, the weak overall correlation obtained by Melton (1957) between the precipitation-effectiveness index and the values of maximum slope angles is hardly surprising. In only two of those sub-sets identified above as comparatively homogeneous (in terms of lithology and the range of both relative relief and drainage density values) does variation in P-E appear to be accompanied by significant changes in average slope steepness. Moreover, the direction of that significant association is different in each case (positive for the granitic valleys, negative for the group of sandstone, limestone and shale basins). This observation (together with the evidence for the non-uniform response of angles to increasing values of the drainage density index) would seem to emphasize that it is not only difficult but also probably highly misleading to endeavour to establish broad regional associations between selected climatic and morphological variables.

Secondly, the analysis has illustrated one of the major features of climatic geomorphology: whereas the degree of clear association between climate and landform is apparently slight, the *indirect* evidence for an element of climatic influence is far more substantial. Both relative relief and drainage density (factors found to be far more immediately linked to changes in maximum slope angles than the P-E index) are in themselves measures of the landscape that inevitably reflect climatic as well as structural and lithological influences. This is particularly true in the case of drainage density, which Melton has shown (1957, Figure 1, p. 65) to have a strong negative relationship to the level of precipitation-effectiveness.

In summary, a detailed examination of Melton's data provides a clear picture of the pattern of success that may be expected in any wide-ranging discussion of the influence of climate upon drainage basin forms in general and slope forms in particular. At the regional scale, the patterns of association will appear blurred, for non-climatic forces will intrude upon and modify the action of climatic elements. Only when structural considerations, in particular, are held relatively constant will it be possible to identify co-variation in climatic and morphological indices and it is to be expected that the nature of such relationships will vary with the precise location and composition of the basins concerned.

*6.2.4 Summary*

The contrasted evidence presented above has been selected in order to illustrate two different routes which may be employed to lead to an association of variations in certain features of valley-side slope forms with climatic differences. The problems of interpretation posed by temporal fluctuations in climate and by regional variety in structure, geologic history and climatic conditions should have emerged from both studies. As we shall see, these difficulties assume a more prominent role when we consider the question of valley asymmetry.

## 6.3 The vexed question of asymmetrical valleys

*6.3.1 Introduction*

'All valley cross-sections are asymmetrical, but some are more asymmetrical than others.' This apparently frivolous axiom provides a useful starting point for any discussion of valley asymmetry, since the crux of the problem lies in an acceptable definition of the term 'asymmetrical' (Section 6.3.2). Only when such a distinction can be made is it possible to proceed to an examination and explanation of the asymmetry observed (Section 6.3.3).

Undoubtedly, many valleys can be shown to be persistently asymmetrical in cross-section. Further, some of those valleys do not appear to owe their peculiar morphology to the vagaries of structure, lithology, glacial erosion or the operation of the Coriolis force: in these circumstances it is justifiable to infer that the differences in form are the result of the variable operation of geomorphological processes which, in their turn, have been (or may still be) controlled by the non-uniform distribution of precipitation and direct insolation. In fact, there is evidence to suggest that some examples of valley asymmetry are climatically (or, better, microclimatically) controlled (Section 6.3.4).

Nevertheless, it has proved very difficult to identify the role of climate in the creation of asymmetrical valleys with any degree of certainty, for many descriptions have either been vague in their basic definition of asymmetry or less than rigorous in their consideration of the role of non-climatic agencies.

Frequently this imprecision seems to have been rooted, in a rather curious fashion, in a desire to illustrate or identify a global pattern of asymmetrical valley forms. Most regrettably, this search for continent-wide or even worldwide regularities linked to major climatic zones, has tended to overshadow the far more basic discussion of the manner in which individual cases of valley asymmetry can be legitimately related to climatic factors.

The whole question of the role of climate in the creation of valley asymmetry is both complex and emotive. It is hoped that the following discussion will indicate that much of the emotion can be removed from the debate by a careful consideration of the nature and implications of the complexities involved.

*6.3.2 The definition of asymmetry*

In strict geometrical terms, an asymmetrical valley is one in which the opposing walls are not exact mirror-images of each other, about the valley axis. Few, if any, natural valleys fail to be asymmetrical in this sense.

When geomorphologists speak of asymmetry they refer to 'substantial' differences either in the overall form, or in the steepness of opposing valley sides and thus the identification of asymmetrical valleys turns upon an acceptable definition of 'substantial' differences. As it is difficult to arrive at clear descriptions of valley-sides as three-dimensional forms, definitions of asymmetry have come to be based upon comparisons of two-dimensional slope profiles. Here again, the accurate description of the geometry of an entire profile is often complex (cf. French (1971)), with the result that asymmetry is generally determined simply with reference to the angular values of comparable sections of the opposing slopes. The final question to be answered is, therefore, 'which sections should be compared?' The problem is trivial where the profiles are straight (or nearly so), since either the average angle between crest and thalweg, or the maximum angle (according to Strahler's (1950) definition: see below) may be used virtually interchangeably. The more complex the profile form, however, the less generally useful the mean (and to a certain extent the maximum) angle becomes.

It is necessary here to look more closely at the 'Strahler maximum angle'. This was defined by Strahler in his classic 1950 paper as the angle of the steepest section of a slope profile, other than a cliff-face, provided that the section was more than 5 feet (2 m) in length. (The minimum length selected was chosen with reference to the scale of the valleys in which Strahler was working and was not intended to be generally applicable.) The maximum angle identified can be regarded as corresponding to that of the 'constant-slope' sector of the classic four-segment profile, i.e. that portion lying below the free face and above the basal concavity or wash slope. Where the free face is absent (as on many vegetated profiles) the constant-slope section will be that on which the down-slope component of the gravitational force is at a maximum such that (other things being equal) subaerial processes will operate most efficiently on this sector. Intrinsically, then, the constant-slope segment and its angle are of

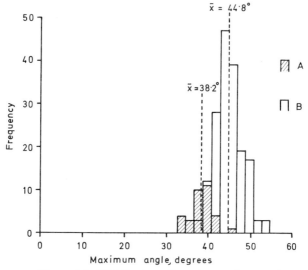

Figure 6.2. Distribution of maximum slope angles, Kline Canyon area, Verdugo Hills, California, decomposed into two samples representing A, slopes protected from stream attack and, B, basally corraded profiles (redrawn from Strahler, 1950, Figure 14).

geomorphologic interest. However, it is a further property of these slopes that is perhaps of greater importance to the definition of valley asymmetry. According to Strahler (1950, p. 685): 'Within an area of essentially uniform lithology, soils, vegetation and climate and stage of development, maximum slope angles tend to be normally distributed with low dispersion about a mean value determined by the combined factors of drainage density, relief and slope-profile curvature'. Two such distributions are illustrated in Figure 6.2.

Strahler's finding, so amply supported by later evidence (cf. Figure 6.1), is of major importance. First, it is possible to consider the average maximum angle in any uniform environment as characteristic or descriptive of the sum total of its geomorphological conditions. Second, it is then possible to compare these characteristic angles for environments which are thought to be non-uniform in any way: for example, in the amount of direct solar radiation or precipitation they receive. Third, the nature of the distribution followed by maximum angles makes it not only possible but necessary to define 'substantial' differences between groups of profiles, i.e. valley asymmetry, upon a statistical basis. The advantage of clarity gained by such a definition is substantial. (Further, the normal distribution characteristic of homogeneous samples of maximum angles is similarly exhibited by average slope angles, although the dispersion of the latter is considerably reduced: distributions of the angles of crestslopes and footslopes and of free face sections are, however, almost invariably skewed, thus reducing the ease with which statistical comparisons may be made.) The definition of asymmetry discussed above has been used by a number of workers (Strahler, 1950; Melton, 1960; Hack and Goodlett, 1960; Kennedy and Melton, 1972), with the result that their findings are directly and readily comparable. However, the majority of those who have studied the problem of asymmetry in the last twenty years have preferred to use other means of identifying asymmetrical valleys. This variety of practice creates difficulties when synthesis is attempted. The two alternative forms of definition most generally employed are those of slope mapping (cf. Ollier and Thomasson (1957)) and the use of asymmetry indices (cf. Gloriod and Tricart (1952) and Hadley (1961)). The advantages and disadvantages of each method will be outlined below.

Slope maps possess the obvious merit that they deal with the entire valley-side slopes. It is therefore possible to see at a glance whether all segments of the slopes on one side of the valley axis are consistently steeper than those opposing them, or whether this is true only of certain portions of the profile. However, the construction of such maps requires a classification of the angular values involved. A valley in which the mid-slope section of one side falls in the class $7.5°–9.9°$, while that of the opposite side has angles between $5.0°$ and $7.4°$ will certainly appear to be asymmetrical. However, the actual differences involved may range from $4.9°$ to $0.1°$ and there is no way, once the map has been constructed, in which either the magnitude or the constancy of the variation can be inferred. Of course, this problem is less severe if the two opposing sections differ by more than one class, but even so the construction of a slope map may still involve a loss of precision which renders comparison between different areas extremely difficult.

The use of asymmetry indices is rather more difficult to justify. In the simplest form, such indices represent the ratio between the steeper and the gentler angles of opposing profiles (Gloriod and Tricart, 1952): in this case, the ratio is always greater than unity, if any degree of asymmetry exists. A modification of this basic procedure is also found (Hadley, 1961), in which the observation from the right- or the left-hand valley side is always used as the numerator: this allows index values of less than one. In either case, although a valley, or groups of valleys, which consistently registers index numbers other than unity may safely be described as asymmetrical, there is little or no way of assessing whether the difference involved is truly 'substantial', although some workers (cf. French (1971)) have endeavoured to overcome this problem by taking an arbitrary value of $1.5$ (for the simple index) to represent 'genuine' asymmetry. However, this scarcely solves the basic problem. Consider, for example, a series of cross-profiles which yield the following pairs of angles for the constant-slope sectors: $3°$ and $6°$, $4°$ and $6°$, $30°$ and $45°$, $35°$ and $45°$, with corresponding indices of $2.0$, $1.5$, $1.5$ and $1.28$. Unless the original measurements are provided together with the index values (which rather destroys the point of calculating the latter), it appears that the first cross-section is the most strongly asymmetric and the fourth, the least. Whilst this is obviously true in relative terms, for most purposes the geomorphologist would surely consider a difference of

10° to be of equal or greater moment than one of 3°. The use of an asymmetry index, however, makes the former difference appear far less substantial than the latter and, in the example given, the former would not even be rated as a 'real' difference, if a critical index value of 1·5 were employed.

From the preceding discussion it should be clear that no consensus of opinion yet exists amongst those geomorphologists most interested in the problem of valley asymmetry concerning the manner in which asymmetry should be described and defined. This leads to practical difficulties when one seeks to collate the results of studies which have employed different methods of definition, for one of the most basic observations in the study of asymmetrical valleys is that cross-sections which are judged to be asymmetrical according to one scheme, fail to appear so when other criteria are employed. A good example of this general difficulty is provided by French's study (1971) of asymmetry in the forms of valley-side profiles in the Beaufort plain of Banks Island, N.W.T., Canada. The slopes concerned are undoubtedly asymmetrical in the sense that their overall geometries are quite different (1971, Figure 4). However, when statistically analysed, French's data (1971, Table 5) on the maximum angles of the profiles indicate that the only significant difference between the values from north- and south-facing profiles lies in the relative variability about the mean values, not in the mean values themselves. French himself employs an asymmetry index with a critical level of 1·5 in his examination of the data, but it might be the case that the use of average slope angles would, in this instance, have provided a clearer statement of the position. It is patently very difficult to describe all cases of asymmetry in precisely the same fashion, but it is equally impossible to compare different accounts if one cannot be sure that similar sorts of asymmetry are involved.

In the sequel, the statistical definition of asymmetry (with reference to the Strahler maximum or the average slope angle) is employed whenever possible and the results of studies which have employed other criteria are considered only when the reported differences in slope steepness are marked and persisent. Such a procedure will lead to a conservative view of the problem; even so, its true magnitude should become apparent.

*6.3.3 Forms and causes of asymmetry*

The search for relationships between climatic influences and valley asymmetry is complicated, first and foremost, by the fact that detectable asymmetry in cross-sections may occur in any one of four forms. Although one of these manifestations may be safely said to be *non*-climatic in origin, none of them may be linked to climatic causes alone. (All four forms of asymmetry may be considered as arising from an imbalance in the volumes of material transported down opposing profiles to the valley floor, variation in the rates of removal of debris and solid rock from the base of the slopes, or from some combination of these two.)

Ignoring, for the moment, the causes of such an imbalance, the four forms in which asymmetry may occur may be summarized as follows.

(a) If one valley side is prone either to greater overall denudation or to continuous basal erosion, then the orientation of relatively gentle or relatively steep slopes (and the degree of angular discrepancy between those slopes and the opposing profiles) will be comparatively constant throughout the basin. Such asymmetry may be considered 'valley-wide' (cf. Ollier and Thomasson (1957)).
(b) If either valley side may experience greater denudation or more efficient basal attack at the limited number of points within the basin and thus be reduced or steepened in angle to a degree which is completely independent of its azimuth, then the orientation of relatively steep and gentle slopes will be variable. Such asymmetry may be described as 'localized' (cf. Strahler (1950) and Figure 6.2).
(c) If there is a general tendency for greater or lesser erosion upon one valley side, coupled with localized steepening or reduction at certain points, the latter being independent of the orientation of the profile, we have a form of asymmetry which is a simple combination of types (a) and (b). This may be said to represent a simple addition, or superimposition, of the effects of at least two distinct and independent controlling agencies (cf. Melton (1960)).
(d) If, although there is no general tendency for one valley side to be relatively steepened or reduced in angle, there is nevertheless a tendency for the effect of localized erosion to vary in intensity with the orientation of the profile concerned, then there is a complex combination of types (a) and (b). This may be said to result from the interaction of the effects of at least two controlling factors which are not independent in their operation (cf. Kennedy (1969)).

Melton (1960) has observed that the immediate cause of valley asymmetry is the nature of the 'local erosional environment', or the position of the stream channel with respect to the two valley sides. Subsequent work (cf. Currey (1964), French (1971) and Kennedy and Melton (1972)) would seem to bear out this observation. (It is also possible that the role of valley glaciers may be viewed in much the same light, but see below.) If for any reason, the stream channel is held against one valley wall then that wall will be generally steepened (Currey, 1964). If, on the other hand, the channel meanders freely, then either strictly localized asymmetry associated with cut-banks will develop (Strahler, 1950), or the influence of basal sapping will simply be added to a more general pattern of asymmetry created by the differential denudation of opposing profiles (Melton, 1960). If the asymmetry is of the fourth type ((d) above), then the behaviour of the meandering stream may be viewed as a 'trigger' mechanism. However, the disparity in the slope angles created is really the product of the differing processes at work upon the two valley sides, once undercutting has occurred (Kennedy, 1969, pp. 75–76).

While the variation in fluvial efficiency is almost always to be found as the proximate agency in the creation of valley asymmetry, it can only be accepted as a satisfactory explanation for localized asymmetry (of type (a)). In all other circumstances it is necessary to enquire why the balance between linear and sheet erosion is as it is and this necessitates the search for underlying controls, of which climatic differences are only one possibility. It is essential to consider all of these factors in turn and it is proposed to examine them in an approximately decreasing order of generality.

(1) *Coriolis force*. In the preceding discussion of asymmetrical valleys it has been implicitly assumed that there is no *a priori* reason to consider that substantial differences in the form of opposing slope profiles should develop within any basin. This assumption would have been vigorously disputed by many workers at the turn of this century (cf. Davis (1908)) who were impressed with what Fairchild (1932, p. 423) has termed 'the singular and romantic fact that all moving bodies on the earth are subject to a deviating or deflecting force due to the rotation of the globe'. The operation of the Coriolis force, which has the effect of deflecting moving objects to the right in the northern hemisphere and to the left in the southern, is, of course, of considerable importance in governing the atmospheric and oceanic circulation.

What is important here is the frequently maintained assumption that the Coriolis force will cause sufficient deflection of bodies of water contained by earth banks to result in excessive erosion along the right banks of rivers in the northern hemisphere. Although Fairchild, in his 1932 paper, would appear to have shown conclusively that this hypothesis is untenable, the subject has been re-opened more recently both in an empirical study (Currey, 1964) and in a general discussion (Carson and Kirkby, 1972, pp. 385–386). These latter investigations, however, serve to reinforce Fairchild's view, rather than contradict it and it would seem to be eminently reasonable to state that the Coriolis force is of negligible importance in the creation of valley asymmetry.

(2) *Differences in insolation and precipitation receipts*. There are universal differences in the amounts of direct solar radiation received by slopes of different orientations (Geiger, 1965; Sellers, 1965, Figure 13). In the northern hemisphere, radiation totals are greatest on slopes which face due south, decreasing through west- and east-facing surfaces to reach a minimum in north-facing sites. In addition, the total received by any slope profile will also vary with its angle of inclination and its overall curvature. The width of the valley will also prove important, in that it governs the degree of mechanical shading of one side by the other.

Overlaid upon this fundamental pattern in all but the driest deserts are variations in the amount and temporal distribution of precipitation received by slopes oriented in different directions. As the distribution of rain and snow is highly dependent upon wind direction, it is generally more variable both in time and space than that of direct solar radiation. For example, Lacy (1951) has shown that, although southwest-facing surfaces in southern England will receive most moisture at all times of the year, the location of the driest sites will vary from season to season. Where substantial amounts of precipitation occur in the form of snow, the position becomes even more complex for, although drifts are likely to accumulate most deeply on leeward slopes, those which perist longest will be the best-shielded from direct insolation (cf. Evans (1972)). In areas of prevailing easterly or northerly winds in the northern hemisphere, the optimum locations for depth and persistence of snowdrifts will not coincide.

From the interaction of the three climatic elements of sun, moisture and wind emerges a pattern of soil moisture distribution within any valley, the details of which will vary from time to time, both on annual and longer timescales. It is reasonable to assume that these variations will determine the nature or speed of processes operating at or near the ground surface (cf. Gerlach (1963)) and, indeed, it has already been shown (Section 6.2) that differences in slope forms may result (Bik, 1968). Undoubtedly the most dramatic effects are those associated with the initiation and maintenance of cirque glaciers (cf. Chapter 15).

The potential importance of microclimatic variations as an agency in the creation of valley asymmetry is therefore clear, most especially because the differences described may be expected to exist to some degree in almost any valley (the exceptions being, perhaps, those very close to the equator). It should be easy to see why so much attention has been directed to this relationship by climatic geomorphologists, but there are other factors which may be at work and which considerably complicate the picture.

(3) *Differences in slope dimensions.* It is self-evident that the height and length of any valley side will exert some control over the rate of operation of subareal processes upon that slope and will also provide a limit to the volume of material which may be moved downslope.

Arguing from this starting point, it is possibe to consider that angular variations between opposing profiles may be explained essentially as the product of 'auto-asymmetry', i.e. resulting purely from the differences in slope dimensions (cf. Alexandre (1958)). On reflection, however, it can be seen that this is far from satisfactory as an explanatory mechanism for all but the most localized and variable asymmetry. In any other circumstance it is surely necessary to ask why it is that the valley sides themselves are consistently lacking in symmetry: the factors (4), (5), (6), (7) or (8) discussed below will usually be found to be responsible.

(4) *Variable lithology.* A valley located with its long axis parallel to the junction of two beds of different resistance will almost inevitably develop consistent asymmetry of cross-section. Similarly, the presence of lenses of more or less resistant material within any one bed may result in locally asymmetrical valley sections. Either of these effects, when persistent, is frequently the result of structural control (see (5), below), but not invariably, for they may also arise from the deposition of materials within pre-existing valleys. The most obvious examples (ignoring glacial effects) involve the emplacement of lava flows (cf. Ollier (1969, pp. 83–85)) and the deposition of loess (cf. Faucher (1931) and Guilcher and Cailleux, 1950).

Localized asymmetry created in this fashion may be very difficult to detect, but any valley-wide form should, at least in theory, be readily identifiable; however, subsequent partial removal of the infilling material will create practical problems. Much of the discussion of climatically-controlled valley asymmetry in Europe has been complicated by such uncertainty concerning the role of loess deposits (which are, of course, climatically-controlled events in themselves, cf. Taillefer (1944), Pinchemel (1954) and Brown (1965)) and similar problems have also been encountered by some North American workers (cf. Zakrzewska (1963)).

(5) *Geological structure.* Undoubtedly the most dramatic form of valley asymmetry is that created by faulting, where one valley side represents either a fault- or fault line-scarp. Less spectacularly, both faulting and folding, followed by erosion, frequently give rise to asymmetrical valleys by bringing beds of varying resistance into juxtaposition at the surface of the earth. The form of scarpland valleys, produced by the uniclinal, down-dip shifting of the courses of strike streams, is almost certainly the best-known example of asymmetry.

Whereas long-term fluvial dissection of tilted strata of differing resistance is well-documented as a cause of large-scale asymmetry, there is a great deal of uncertainty about the more detailed effects of all aspects of structural control, particularly in basins developed upon a single largely homogeneous lithology. It is known (cf. Judson and Andrews (1955)) that jointing may exert a very strong control over the location of river meanders, resulting in localized valley asymmetry, but further information on the precise role of joints is sparse. It is quite clear that the attitude of strata cannot be expected to influence valley form when the beds concerned are absolutely horizontal or vertical, yet we possess extraordinarily little data concerning the upper and lower limiting angles at which uniclinal shifting—or other structural effects—may occur and it is, in consequence, scarcely surprising that these few observations are conflicting. For example, French (1967) has argued that the asymmetrical form of certain Chalk valleys in southern England may be the product of uniclinal shifting by streams which ran sub-parallel to dips of as little as $1°$. Kennedy (1969), on the other hand, has found that dips of less than $4°$ in the Galena dolomite of the Driftless Area of Wisconsin apparently exert no influence over the average angles of valley-side slopes. Work undertaken by Crowther (1973) on the forms of river-cliffs in the shales of the southern Pennines seems to confirm that dip angles of less than $4°$ have no detectable influence upon profile forms, whereas Hack and Goodlett (1960) find that 'gentle' dips of sandstones and shales in the Appalachians of Virginia are associated with significant steepening of the maximum angles of profiles which slope in the same direction as the beds (i.e. a situation the reverse of uniclinal shifting). Information concerning the upper limits for structural control is even more scanty than that related to the lower, the only major study being that of Macar and Lambert (1960) on the limestone slopes of the Condroz, Belgium. These workers find that very small differences in slope angles develop where strata dip at angles of between $30°$ and $65°$ from the horizontal (the steeper slopes being those inverse to the direction of dip), but that these differences disappear on higher-angled structures.

It may be that dipping strata will only influence slope form in valleys cut within one lithology over the range 4°–65°, but much more information is needed to substantiate this suggestion. More disturbing are the indications that there is no consistent relation between the direction of dip and the orientation of the steeper slopes, for the findings of Hack and Goodlett (1960) run counter not only to those of Macar and Lambert (1960), but also to general views on the dominance of uniclinal shifting as the means whereby structure becomes reflected in valley cross-profile. A contributing factor, which again is very poorly understood, may be the degree of small-scale lithological variation, with particular reference to the differential water-transmitting capacities of alternating facies (cf. Hack and Goodlett (1960)). The unpublished study by Crowther (1973) is relevant here, as he finds that the relationship between dip and cliff angles varies not only between sandy- and clay-shale facies, but also according to the precise mixture and relative location of the two lithologies within the river-cliff faces.

The ubiquity of tilted beds of non-homogeneous composition makes the role of structure in the creation of valley asymmetry potentially enormous. The current lack of quantitative studies of the extent and nature of structural control, in all its aspects, is not only inherently unfortunate, but also serves to complicate any discussion of the role of climatic factors.

(6) *Warping*. The effects of gentle differential movements of the land surface upon valley forms are patently difficult to distinguish from those of low-angle folding, in the majority of cases. However, where warping is of recent geological date, effects comparable to those produced by uniclinal shifting may result, despite the lack of lithological variations. Coleman (1952) and Clayton (1957) have both proposed this mechanism to explain the asymmetry of valleys in southern England.

(7) *Evolution of the drainage net*. Although the form of the channel network within any basin will generally represent a response to one or more of the controls already outlined, there appear to be certain circumstances in which the direction of flow and the alignment of channels may be substantially altered after the initiation of the network, resulting in the creation of asymmetrical valleys. This 'secondary' asymmetry is clearly akin to that produced by warping, but may result from one of several agencies.

For example, an imbalance in the rate of downcutting between major and minor streams flowing sub-parallel to each other down a gently sloping surface may result in the capture of the latter by the former and the creation of a pinnate drainage pattern which may involve the development of asymmetrical cross-sections in the lower reaches of the tributaries. In other situations, it is possible that the growth of bars or spits across the mouth of a trunk stream may force the main channel to shift laterally, creating asymmetry. White (1966) has invoked this explanation for the asymmetry of several of the major rivers of the Atlantic Coastal Plain of the United States.

On the regional scale, these causes of asymmetry may well become obscured by structural or climatic influences: nevertheless, they would seem to represent possible causes of asymmetrical valley forms.

(8) *Glaciation.* Dramatic examples of valley asymmetry have been described from glaciated areas (cf. Tuck (1935), Malaurie (1952) and Ragan (1962)) and it is also characteristic for glaciated peaks themselves to be strongly asymmetrical (cf. Gilbert (1904) and Evans (1972)). Since the development of cirque glaciers and ice caps is a climatically controlled process, it is plain that the resulting asymmetry of valleys can be said to be climatically determined. However, when one considers the present-day form of glacial troughs, it becomes apparent that the role of glaciation in the creation of asymmetry is highly complex. In the first place, it is almost always impossible to separate the effects of the ice itself from those of pre- or supraglacial frost action (these latter processes being primarily controlled by microclimate). Secondly, the majority of glaciated valleys in upland areas must have been fluvial in origin and, therefore, initially as susceptible to climatic or structural controls over their form as any other basins. Finally, since the retreat of the ice, these valleys have undergone tremendous changes in both macro- and microclimates and the processes which are currently at work within them will be as susceptible as those elsewhere to the complex influences of pre-existing forms and contemporary conditions of precipitation and insolation (cf. Beatty (1962)).

In brief, the asymmetry of glaciated areas, particularly glaciated highlands, would seem to pose far more problems for the analyst than that of non-glaciated regions. The most promising approach to this particular field of climatic geomorphology would seem to lie in studies such as those of Rapp (1960) or Kirkby (1967), in which attention is focused upon the operation of modern processes, rather than on the geomorphic history of the valley-side slope forms.

There are, then, at least eight mechanisms which have been considered responsible for the creation of asymmetrical valleys. It seems reasonable to eliminate one (the Coriolis force) from further serious consideration, but microclimatic variations remain only one of seven possible agencies that may be invoked to explain the asymmetrical cross-section of any valley. Furthermore, it is important to recall that all these potential controls operate in specific directions upon the oriented surfaces which are valley-side slopes and that two or more agencies may either reinforce or conflict with each other's operation.

Neither the forms in which asymmetry may be manifested nor the factors which may be responsible for those forms are simple. If one is looking for a single, major cause of difficulties in any attempt to elucidate the role of climatic factors, then the uncertain influence of structure would seem to be paramount. As a potential control of form, structural variation is almost as widespread as climatic variation and its manner of operation is even less well known.

*6.3.4 The asymmetry of small valleys*

Valley asymmetry is clearly a complex phenomenon, even in theoretical terms, for it can appear in one of four forms and can be the product of at least seven major influences, working separately or in combination. Nevertheless, there has been a persistent belief that climatically-controlled variations in slope processes are the fundamental agency producing asymmetrical valley cross-profiles in many different locations. One problem which is encountered when endeavouring to substantiate this belief is that studies of asymmetry have been conducted by a multitude of workers, using a wide variety of definitions of the phenomenon. Moreover, there has been a fairly sharp difference in emphasis between European and American workers, in particular: the former have concentrated upon what they consider to be 'fossil'asymmetry produced under conditions of 'periglacial' climate, whilst the latter have generally emphasized 'modern' cases that are apparently produced by the forces at work in the present landscape. It is clearly impossible to decide how far the European examples are entirely fossil and how far the North American ones are entirely contemporary. Both sets of studies are also bedevilled by lack of precise information on the role of structure and a further complication is introduced by the fact that valleys of very different depths are involved. This last point is frequently critical, for there is evidence (cf. Gregory (1966) and Karrasch (1972)) that the orientation of steeper profiles varies either with elevation above sea level or with the overall depth of the valley concerned, or both (cf. Gloriod and Tricart (1952) and Kennedy and Melton (1972)). There does not seem, at this point in time, to be any really adequate explanation for these observations, which simply add one more dimension to the whole asymmetry problem.

There are clearly various ways out of this dilemma. In fact there have been almost as many solutions proposed as there have been studies of asymmetry. As each worker tends to make a particular definition of the problem, it might be fair to say that we have been going round in circles. It must be accepted that no starting point will meet with anything approaching universal approval, yet it is clear that an attempt has to be made. What follows is a highly personal view of one potential line of attack.

The following, quite possibly unwarranted, assumptions are made:

(a) That asymmetry may be defined in statistical terms, with reference to either the Strahler maximum or the average angles of opposing slope profiles.
(b) That the Coriolis force can be ignored as an effective agency in the creation of asymmetry.
(c) That any manner or structural influence upon the cross-profiles of valleys cut in essentially homogeneous rocks will cease to operate at dip angles of 1° or less.
(d) That the asymmetry of glaciated valleys in upland areas is too complex for analysis in the present state of the art.
(e) That asymmetry is generally linked, in immediate terms, to the action and position of stream channels.
(f) That any climatically-linked asymmetry will be most straightforward to

identify and interpret in rather shallow valleys, particularly if these are post-glacial in origin.

(g) That the nature of climatic control over asymmetry may vary between valleys which run east-west (where differences in direct radiation receipts are likely to be most pronounced) and those trending north–south (where differential exposure to wind may prove more critical).

(h) That if such a simplified situation cannot be resolved, the prospects for unravelling worldwide truths about the role of climatic factors in the development of valley asymmetry must be regarded as poor.

Tables 6.1 and 6.2 provide information from three sources (Melton, 1960; Kennedy, 1969; Kennedy and Melton, 1972) concerning the asymmetry observed in 11 sets of small valleys in North America, seven of which (Table 6.1) trend east–west and four of which (Table 6.2) are aligned north–south. The observations from the Northwest Territories, Manitoba and Wyoming come from field measurements made with abney levels and tapes; those from Wisconsin, Kentucky and Texas represent data drawn from the 1:24,000 topographic sheets of the United States Geological Survey. All valleys sampled contain stream channels which oscillate from side to side, so that slopes of each orientation are equally likely to suffer from basal attack and steepening (cf. Strahler (1950)). Moreover, the random samples of cross-profiles used in each study have been stratified so that there are equal numbers of observations of undercut north- and south-facing profiles in the east–west valleys, or of east- and west-facing slopes in those which trend north–south. It is arguable that, if there is no climatic control over the asymmetry of these 11 sets of basins, then the only form in which that phenomenon should be manifested should be the purely localized variety, i.e. that characterized by an alternation of steeper and gentler profiles which is dependent solely upon the location of the stream channel with reference to the base of the valley side.

Table 6.1 indicates that all seven sets of east–west valleys exhibit a statistically detectable degree of asymmetry in cross-section and that the purely localized form exists only in the Grand Prairie area of Texas. The remaining groups of basins all demonstrate some component of asymmetry linked to the orientation of the profiles: this may be considered climatic in origin. It is of interest to note that in each case, the steeper profiles have a northerly aspect, from the permafrost region of northern Canada as far south as humid temperate Kentucky (Section 6.3.5).

From Table 6.2 it appears that the position in north–south trending basins is rather different. Here only one group of valleys (that from the Eden Shale belt of Kentucky) demonstrates what can be considered climatically-controlled asymmetry. The two semi-arid regions (Wyoming and Texas) show only the localized, stream-dominated form and the classic periglacial area of Wisconsin demonstrates none that is detectable. (It should be noted that the Kentucky asymmetry could equally well be ascribed a structural rather than a climatic origin, for the rapidly alternating sequences of thin limestone and shale beds

Table 6.1. The asymmetry of selected sets of small east–west trending valleys in North America

| Location and source | Geology and structure | Age of valleys | Present climatic characteristics | | | | Average values maximum[*] or mean[+] angles | | | | Sample size | Type of asymmetry statistically detectable |
|---|---|---|---|---|---|---|---|---|---|---|---|---|
| | | | Precipitation (mm/year) | Mean January temperature (°C) | Mean July temperature (°C) | Relative relief (m) | N-facing undercut | S-facing slip-off | S-facing undercut | N-facing slip-off | | |
| Caribou Hills, N.W.T. (Canada) 69° N, 134° W (Kennedy and Melton, 1972) | Sands and claystones (?) Cretaceous: apparent dip northerly ≤ 1° | c. 12,000 | 250 | −30.0° | 14.4° | 15 | *27° | *22° | *22° | *24° | 60 | Valley-wide: north-facing steeper |
| Pembina Basin, S. Manitoba (Canada) 49° N, 100° W (Kennedy, 1969) | Sandy till : no definable structure | c. 12,000 | 490 | −15.6° | 20.0° | 15 | *24° | *13° | *19° | *14° | 100 | Addition of localized and valley-wide: north-facing steeper |
| Military Ridge Driftless Area, Wisconsin (USA) 43° N, 90° W (Kennedy, 1969) | Galena dolomite (Ordovician): dip southerly < 1° | (?) Post-Pliocene | 760 | −3.9° | 21.1° | 46 | +7.5° | +6.0° | +6.5° | +6.5° | 200 | north-facing steeper. |
| Laramie Mountains, Wyoming (USA) 41° N, 105° W (Melton, 1960) | Deeply weathered granites: 'No influence of structural planes on valley-side slope angles' | (?) 12,000 | 270 | −3.3° | 17.2° | 30 | *21° | *14° | *16° | *17° | c.60 | Addition of localized and valley-wide: north-facing steeper |
| Gangplank, Wyoming (USA) 41° N, 105° W (Kennedy, 1969) | Ogallala sandy conglomerate (Pliocene): dip easterly < 1° | c. 8000 | 380 | −2.8° | 21.1° | 30 | *48° | *15° | *33° | *17° | 64 | Interaction of localized and valley-wide: north-wide: north-facing undercut steepest |
| Eden Shale Belt, Kentucky (USA) 38° N, 85° W (Kennedy, 1969) | Ordovician limestones and shales: dip westerly < 1° | (?) Pre-Pleistocene | 1090 | −1.1° | 24.4° | 46 | +12.5° | +9.5° | +10.0° | +12.5° | 200 | Valley-wide: north-facing steeper |
| Grand Prairie, Southwest Texas 31° N, 98° W (Kennedy, 1969) | Edwards and Comanche Peak limestones (Cretaceous): dip easterly < 1° | (?) Pre-Pleistocene | 860 | 7.2° | 29.4° | 15 | +7.5° | +6.0° | +8.0° | +6.5° | 200 | Localized : undercut steeper |

Table 6.2. The asymmetry of selected sets of small north–south trending valleys in North America. For locations, see Table 6.1.

| Location | Average values, maximum* or mean† angles | | | | Sample size | Type of asymmetry statistically detectable |
|---|---|---|---|---|---|---|
| | E-facing undercut | W-facing slip-off | W-facing undercut | E-facing slip-off | | |
| Military Ridge, Wisconsin (USA) | †5·0° | †4·5° | †5·5° | †5·0° | 200 | None |
| Gangplank, Wyoming (USA) | *32° | *17° | *29° | *16° | 36 | Localized : undercut steeper |
| Eden Shale Belt, Kentucky | †13·0° | †10·5° | †12·0° | †12·5° | 200 | Valley-wide: east-facing steeper |
| Grand Prairie, Southwest Texas (USA) | †7·0° | †6·5° | †8·0° | †6·5° | 200 | Localized: undercut steeper |

which comprise the Eden Shale dip to the west and *might* produce a degree of uniclinal shifting, although the dip is only approximately 0.4°)

These studies suggest two relationships. First, climatic asymmetry can be clearly identified in a wide range of North American environments and it is both more frequent and better-developed in east–west trending valleys than in those which run north–south (the converse of what appears to be the position in Europe). This last observation implies that differential direct radiation receipts are of greater importance in controlling variations in soil moisture conditions and subaerial processes than are variations in precipitation amounts linked to wind direction. Second, it is extremely interesting to note the different reactions of north- and south-facing profiles to basal corrasion. Table 6.1 demonstrates that the north-facing valley sides in all seven areas become relatively steeper than the opposing south-facing profiles when stream channels are located close to their bases. However, in four of the regions (the Caribou Hills, Military Ridge, the Laramie Mountains and the Eden Shale area) south-facing slopes which are similarly subject to basal attack fail to become the steeper valley sides. Hence, development of valley-wide, climatically controlled asymmetry can be assumed to depend directly upon the differential response of slopes of varying orientation to episodes of fluvial corrasion and, possibly more crucially, upon the manner in which the undercut and steepened profiles are subsequently degraded by slope processes. Where slopes of opposing orientations respond in a comparable fashion (e.g. both groups of Texas valleys in the sample) no valley-wide asymmetry can develop. Where one category of profiles fails to respond detectably to corrasion, the opposing slopes must inevitably become and remain steeper, creating persistently asymmetrical valleys (as in the Caribou Hills, Military Ridge and the Eden Shale belt).

Most interesting, however, are the intermediate cases (the Pembina Basin, Laramie Mountains and the Gangplank valleys), where the relationships between stream and slope processes are more delicately balanced. In the two former areas, where the asymmetry is of the additive type, it would appear that the crucial factor in establishing the generally greater steepness of north-facing profiles is their lesser decline after direct corrasion has ceased. In the Gangplank area, on the other hand, it seems to be the response to undercutting which is climatically controlled, with the development of exceptionally steep constant slopes above north-facing meander scars.

The foregoing is simply descriptive, however. What is still in doubt is the explanation of the variable response of valley-side slopes to fluvial attack. Why should the streams of the Grand Prairie area of Texas apparently exert complete control over the development of asymmetrical valley cross-profiles (which are purely localized), while those in the Caribou Hills, the Military Ridge region or the Eden Shale belt seem to be virtually powerless to modify the form of one side of their valleys? This central question is by no means original: it has frequently been assumed that genuine climatic asymmetry occurs either because stream channels are 'forced' up against one valley wall by excessive debris production from the opposing hillside (cf. Pinchemel (1954)), or because the basal corrasion of one profile is inhibited, particularly by the existence of permafrost very close to the slope surface (cf. Smith (1949)). However, as the studies outlined in Tables 6.1 and 6.2 illustrate, the foregoing explanations deal only with the end member of a series of situations which result from the joint activities of stream and slope processes. Since the effect upon profile form of a meandering stream is potentially similar everywhere, the crucial factor must obviously be the variable operation of the slope processes themselves (and hence the nature of the slope microclimates).

It must be recalled that valley sides are subject to the operation of a great variety of geomorphological controls; however, it is possible to divide the processes at work into two general categories—those which produce a decline in gradients (notably creep and mass-movements) and those which are capable of maintaining or even locally steepening sections of the profile (principally rainwash, rill action and the early stages of gully development, cf. Schumm (1956)). Although both groups of processes are controlled by the moisture conditions of the slope mantle, it is possible to make a rather important distinction between the ways in which that control is primarily exerted. The 'reducing' agencies are particularly influenced by moisture and temperature conditions and changes *within* the mantle and it is for this reason that they are the dominant processes on vegetated surfaces. The 'maintaining' agencies, on the other hand, are more straitly governed by surface conditions and, in consequence, are inhibited in their action by any absorbent covering of that surface, whether it be bare (cf. Schumm (1964)) or vegetated.

If this simple, perhaps naive, distinction between the major categories of slope processes be accepted, the nature of the opposing slope surfaces becomes the critical factor and some pattern in climatic asymmetry such as that outlined in Table 6.1 and 6.2 may emerge. Where either valley side becomes stripped of vegetation after basal fluvial attack, rill wash and perhaps even gullying will take over, however temporarily: if greater moisture is available to profiles of one orientation, then these may be expected to become unusually steep. (Hack and Goodlett's evidence (1960) is crucial here, for the surplus moisture concerned in the case they describe is provided by the dip of the strata.) Where both valley sides remain largely vegetated, even during fluvial attack, the critical factor in the creation of asymmetry will be the speed with which slope angles are reduced when corrasion ceases and this, in turn, implies that the frequency of moisture-change and/or freeze–thaw cycles will be critical (cf. Kirkby (1967)). The less-affected profiles (which will probably be those which are moister and retain the most dense vegetation mat, cf. Gerlach (1963)) will reduce in angle more slowly and therefore remain comparatively steep. Patently, the most dramatic asymmetry is likely to be produced where opposing slopes have such different microclimates that, on being undercut, they respond in the one case by the complete loss of vegetation and in the other, by almost complete retention of the surface cover. This will inevitably lead (cf. Pomerol (1964) and French (1971)) to valley-wide asymmetry with the bare slope, maintained in angle by wash and rilling, being opposed by a declining, vegetated hillside subject to creep and small mass-movements (or solifluction).

This interpretation of the diverse nature of asymmetry in shallow valleys is inevitably open to challenge, yet it endeavours to re-state the nature of the relationship between slopes and streams in the creation of 'climatic' asymmetry in the most general of terms.

In sum, it is suggested that 'climatic' elements in valley asymmetry can be seen not simply as the products of differential fluvial attack, but rather as the outcome of microclimatically controlled variations in the responses of opposing valley sides to such basal corrasion. The development of any valley-wide asymmetry depends upon the creation, first of relatively steep slopes by stream activity and, second, upon the existence of microclimatic variations that are sufficiently strong to allow some degree of difference in the nature or rate of operation of the subaerial processes which then set to work upon those steepened slopes. A comparison of Tables 6.1. and 6.2 indicates that the second criterion appears to be rarer in the north–south valleys of several North American areas than it is in those which run east–west. Table 6.1 further suggests that sufficiently strong microclimatic differences are less common in areas of semi-arid climate. These final observations, however, cannot be explored without some general discussion of the thorny question of the link between climatic asymmetry and the major climatic zones.

*6.3.5 Asymmetry and macroclimate*

Many of the early observations of asymmetrical valleys which could be assumed to owe their origin to climatic controls, came from those portions of Europe that were known to have suffered 'periglacial' conditions during the Pleistocene glaciations. The valleys concerned were almost universally observed to contain steeper south- or west-facing profiles and it rapidly became conventional to regard this situation as uniquely 'periglacial'. Since the 1950s, however, evidence has accumulated from many areas of the northern hemisphere which suggests that this early assumption was grossly oversimplified. Valleys containing steeper north- or east-facing profiles have been described from areas of both past and present 'periglaciation' (cf. Guilcher and Cailleux (1950), Currey (1964), Kerney, Brown and Chandler (1964) and Kennedy and Melton (1972)) and valleys with steeper south- and west-facing slopes have been located in areas beyond the spatial or temporal reach of 'periglacial' zones (cf. Bryan and Mason (1932) and Tricart and Michel (1965)). The former evidence led Tricart (1963) to propose a modification of the original theory and to suggest that there are two forms of periglacial asymmetry: 'warm' or 'marine' being characterized by steeper south- or west-facing slopes, 'cold' or 'continental' producing steeper north- or east-facing profiles. Whereas this suggestion may go some way toward separating the categories of periglacial asymmetry, it does not assist with the problem of temperate examples.

More recently Karrasch (1972) has endeavoured to resolve the question of the distribution of asymmetry in Europe by reference to a latitudinal zonation of the two major types. Although this classification may hold for the areas concerned, it breaks down in England (Kennedy, 1969) and can be seen not to apply to North America (Tables 6.1 and 6.2). Moreover, Karrasch explicitly assumes that all asymmetry is periglacial and, further, that asymmetry characterized by steeper north- or east-facing profiles is peculiar to situations in which there is no 'direct' fluvial erosion: Table 6.1, in particular, should serve to refute both suggestions.

At this point, it is necessary to pose a very basic question. It is reasonable to assume, first that there should be a major distinction (in the northern hemisphere) between those asymmetrical valleys in which north- or east-facing profiles are the steeper and those in which south- or west-facing slopes possess the greater gradients; and, even if this should prove to be the case, is it feasible to envisage a mechanism which will create clear-cut zones of asymmetry of each type?

In the writer's opinion neither part of this question can be answered in the affirmative. The first suggestion is acceptable only if it can be shown that differential heating of the ground surface is the dominant factor in the separation of microclimatic zones in all latitudes and longitudes and at all altitudes. In view of the obvious importance of variable receipts of precipitation and of differing exposure to wind, it would seem most unwise to make this assumption at the present time. However, as has been shown above, the greater prevalence of 'climatic' elements in the asymmetry of east–west valleys of North America does suggest that insolation receipts may outweigh the variable effects of other factors. Nevertheless, that remains an inference rather than a demonstrable truth and until a good deal more is known about the roles of the different climatic elements in controlling the operation of slope processes, some scepticism about the universal equation of south- with west-facing profiles, or north- with east-facing is recommended.

Clearly, the absence of certainty about this fundamental distinction renders the second part of the question redundant. It has been suggested above that much of the discussion of asymmetry in the literature has been ill-advised precisely for this reason. Schemes of global patterns of asymmetry are obviously a nonsense when basic understanding of the mechanisms involved is so precarious.

Despite the negative tone of the latter part of this essay, there remains room for some positive thinking on this question. To the writer's knowledge, reports from the northern hemisphere of asymmetric valleys which are apparently in some kind of equilibrium with prevailing climatic conditions, and in which south- or west-facing valley sides are the steeper, are limited to areas where those profiles are very sparsely vegetated (Bryan and Mason, 1932; Tricart and Michel, 1965; French, 1971; Kennedy and Melton, 1972) where unusually large amounts of surface runoff are supplied by the substratum (Hack and Goodlett, 1960; Kennedy and Melton, 1972), or both. It may be that there exist general locations where macroclimatic conditions are such that the weakening of the hold of the vegetation cover by stream action against the base of south- or west-facing valley sides promotes or triggers rill action and the maintenance of the steepened profiles. In other words, it is simply suggested that the macroclimate predisposes, that the stream action initiates and that the microclimatic, process and form variations are all, in a sense, the product of those two more general controls. This would not, of course, mean that the role of the range of the climate would necessarily be the same in all valleys within one area. This would go some way to explain the variable fashion in which valley depth seems to enter into the asymmetry question.

It seems clear that climatically-created valley asymmetry exists and that it is capable of geomorphological explanation. However, it is equally plain that a universally satisfactory statement cannot be derived from the present data.

First and foremost, an adequate general definition of asymmetry must be agreed upon. (It need not, of course, be that employed in this duscussion.) Second, the role played by structure in the control of moisture conditions, in particular, within slope mantles, must be clarified; this would serve to eliminate what remains a major source of confusion in asymmetry studies. Third, but possibly most pressing, is the need for far more detailed studies of the relationships between contemporary climatic events and subaerial processes: the studies by Rapp (1960) and Kirkby (1967) have remained isolated milestones for too long.

The whole field of asymmetry studies, although it may have come to be regarded by some as a fruitless quest, can nevertheless be seen to contain many of the major keys to a balanced evaluation of the role of climatic factors in shaping the landscape.

## 6.4 Conclusion

Much of the discussion of climatic influences upon valley-side slope forms in the past has, since the time of W. M. Davis, been concerned with the search for broad links between major classes of features and major zones of climate. It may be that there are some profound but simple truths awaiting discovery at this scale of investigation. At the medium, or regional level, however, it seems clear that slope forms cannot usefully be viewed as responses to climatic variables alone. The most desirable focus for the climatic geomorphologist interested in hillslopes would undoubtedly appear to be at the local level, with attention given to studies of the interactions between microclimate and process. Sufficient evidence surely exists to suggest that this should prove an extremely fruitful field for investigation.

## References

Alexandre, J., 1958, Le modelé quaternaire de l'Ardenne Centrale, *Annales, Geol. Soc. Belg.*, **81**, 213–332.

Beatty, C. B., 1962, Asymmetry of stream patterns and topography in the Bitterroot Range, Montana, *J. Geol.*, **70**, 347–54.

Beatty, C. B., 1972, The effect of moisture on slope stability: a classic example from southern Alberta, Canada, *J. Geol.*, **80**, 362–366.
Bik, M. J., 1968, Morphoclimatic observations on prairie mounds, *Z. Geomorph.*, **12**, 409–69.
Brown, E. H., 1965, Glacial and periglacial landscapes in Poland, *Geography*, **50**, 31–44.
Bryan, K. and S. Mason, 1932, Asymmetric valleys and climatic boundaries, *Science*, **75**, 215–216.
Carson, M. A. and M. J. Kirkby, 1972, *Hillslope form and process*, Cambridge University Press, 475 pp.
Clayton, K. M., 1957, Some aspects of the glacial deposits of Essex, *Proc. Geol. Assoc.*, **68**, 1–21.
Coleman, A., 1952, Some aspects of the development of the lower Stour, Kent, *Proc. Geol. Assoc.*, **63**, 63–86.
Crowther, J., 1973, *A study of the influence of dip of strata in a basally undercut slope system*, Unpublished B. A. dissertation, Dept. of Geography, University of Cambridge, 110 pp.
Currey, D. R., 1964, A preliminary study of valley asymmetry in the Ogotoruk Creek area, N. W. Alaska, *Arctic*, **17**, 84–98.
Davis, W. M., 1908, Deflection of rivers by the earth's rotation, *Science*, **27**, 32–33.
De Ploey, J. and J. Savat, 1968, Contribution à l'étude de l'érosion par le splash, *Z. Geomorph.*, **12**, 174–193.
Evans, I. S., 1972, Inferring process from form: the asymmetry of glaciated mountains, in W. P. Adams and F. M. Helleiner (Eds.), *International Geography, 1972*, **1**, 17–19.
Fairchild, H. L., 1932, Earth rotation and river erosion, *Science*, **76**, 423–427.
Faucher, D., 1931, Note sur la dissymetrie des vallees de l'Armagnac, *Bull. Soc. Hist. Naturelle de Toulouse*, **3**, 262–268.
French, H. M., 1967, *The asymmetrical nature of chalk dry valleys in southern England*, Unpublished Ph.D. dissertation, University of Southampton, 269 pp.
French, H. M., 1971, Slope asymmetry of the Beaufort Plain, northwest Banks Island, *Can. J. Earth Sci.*, **8**, 717–721.
Geiger, R., 1965, *The climate near the ground*, Harvard University Press, 611 pp.
Gerlach, T., 1963, Extension des transformations des versants meridionaux du Haut Beskide a l'epoque actuelle, *Report VIth INQUA Congress*, **III**, 101–104.
Gilbert, G. K., 1904, Systematic asymmetry of crest lines in the high Sierra of California, *J. Geol.*, **2**, 579–588.
Gloriod, A. and J. Tricart, 1952, Etude statistique de vallees asymetriques sur la feuille St. Pol au 1/50,000, *Rev. Géomorph. Dyn.*, **3**, 88–98.
Gregory, K. J., 1966, Aspect and landforms in north east Yorkshire, *Biul. Peryglac.*, **15**, 115–120.
Guilcher, A. and A. Cailleux, 1950, Reliefs et formations quaternaires du centre-est des Pays-Bas, *Rev. Geomorph. Dyn.*, **1**, 128–143.
Hack, J. T. and J. C. Goodlett, 1960, Geomorphology and forest ecology of a mountain region in the central Appalachians, *U. S. Geol. Survey, Prof. Paper 347*.
Hadley, R. F., 1961, Some effects of microclimate on slope morphology and drainage basin development, *U. S. Geol. Survey, Prof. Paper*, *424B*, 832–833.
Judson, S. and G. W. Andrews, 1955, Pattern and form of some valleys in the Driftless Area, Wisconsin *J. Geol.*, **63**, 328–336.
Karrasch, H., 1972, The planetary and hypsometric variation of valley asymmetry, in, W. P. Adams and F. M. Helleiner (Eds.), *International Geography, 1972*, **1**, 31–34.
Kennedy, B. A. 1969, *Studies of erosional valley-side asymmetry*, Unpublished Ph.D. dissertation, University of Cambridge, 289 pp.
Kennedy, B. A. and M. A. Melton, 1972, Valley asymmetry and slope forms of a permafrost area in the Northwest Territories, Canada, *Spec. Pub. Inst. Brit. Geog.*, **4**, 107–121.
Kerney, M. P., E. H. Brown and T. J. Chandler, 1964, The late-glacial and post-glacial history of the Chalk escarpment near Brook, Kent, *Phil. Trans. Roy. Soc. B.*, **248**, 745, 135–204.

Kirkby, M. J. 1967, Measurement and theory of soil creep, *J. Geol.*, **75**, 359–78.
Kirkby, M. J. and R. J. Chorley, 1967, Throughflow, overland flow and erosion, *Bull. Int. Assoc. Sci. Hydrol.*, **12**, 3, 5–21.
Lacy, R. E., 1951, Observations with a directional rain gauge, *Q. J. Roy. Met Soc.*, **79**, 283–292.
Macar, P. and J. Lambert, 1960, Relations entre pentes de couches et pentes des versants dans le Condroz (Belgique), *Slope Comm. Rep.*, **2**, 129–132.
Malaurie, J. N., 1952, Sur l'asymétrie des versants dans l'île de Diskö, Groenland, *C. R. Acad. Sci. Paris*, **234**, 1461–1462.
Melton, M. A. 1957, An analysis of the relation among elements of climate, surface properties and geomorphology *O. N. R. Tech. Rep. 11.*
Melton, M. A., 1960, Intravalley variation in slope angles related to microclimate and erosional environment, *Bull. Geol. Soc. Am.*, **71**, 133–144.
Ollier, C. D., 1969, *Weathering*, Oliver and Boyd 304 pp.
Ollier, C. D. and A. J. Thomasson, 1957, Asymmetric valleys of the Chiltern Hills, *Geog. J.*, **123**, 71–80.
Pinchemel, P., 1954, *Les plaines de craie*, Armand Colin, 502 pp.
Pomerol, C, 1964, Influence du climat periglaciaire sur le modele des versants crayeux de la vallee de la Seine a l'aval de Mantes, *Ann. Géogr.*, **73**, 704–707.
Prior, D. B. N. Stephens and G. R. Douglas, 1971, Some examples of mudflows and rockfall activity in northeast Ireland, *Spec. Pub. Inst. Brit. Geog.*, **3**, 129–140.
Ragan, D. M., 1962, Valley asymmetry in the Twin Sisters Range, northern Cascades, Washington, *Spec. Paper, Geol. Soc. Am.*, **68**, 118–119.
Rapp, A., 1960, Recent development of mountain slopes in Kärkevagge and surroundings, northern Scandinavia, *Geogr. Annlr*, **42**, 65–200.
Schumm, S. A., 1956, The role of creep and rainwash on the retreat of badland slopes, *Am. J. Sci.*, **254**, 693–706.
Schumm, S. A. 1964, Seasonal variations of erosional rates and processes on hillslopes in western Colorado, *Z. Geomorph.*, Supp. Band 5, 215–238.
Schumm, S. A. and R. J. Chorley, 1964, The fall of Threatening Rock, *Am. J. Sci.*, **262**, 1041–1054.
Sellers, W.D., 1965, *Physical climatology*, University of Chicago Press, 272 pp.
Siegel, S., 1956, *Nonparametric statistics for the behavioural sciences*, McGraw-Hill, 312 pp.
Smith, H. T. U., 1949, Physical effects of Pleistocene climatic changes in non-glaciated areas, *Bull. geol. Soc. Am.*, **60**, 1485–1516.
Soons, J. M. and J. N. Rayner, 1968, Microclimate and erosion processes in the Southern Alps, New Zealand, *Geogr. Annlr.*, **50A**, 1–15.
Strahler, A. N., 1950, Equilibrium theory of erosional slopes, approached by frequency distribution analysis, *Am. J. Sci*, **248**, 673–696 and 800–814.
Strahler, A. N., 1956, The nature of induced erosion and aggradation, in W. L. Thomas (Ed.), *Man's Role in changing the face of the earth*, University of Chicago Press, 621–638.
Taillefer, F., 1944, La dissymétrie des vallées gasconnes, *Rev. Geog. des Pyrénées et du Sud-Ouest*, **15**, 153–181.
Thornthwaite, C. W., 1931, The climates of North America according to a new classification, *Geogr. Rev.*, **21**, 633–655
Tricart J., 1963, *Geomorphologie des regions froides*, Presses Univ. de France, 289 pp.
Tricart, J. and M. Michel, 1965, Monographie et carte geomorphologique de la région de Lagunillas, (Andes Vénezueliennes), *Rev. Geomorph. Dyn.*, **15**, 1–33.
Tuck, R., 1935, Asymmetrical topography in high latitudes resulting from alpine glacial erosion, *J. Geol.*, **43**, 530–538.
White, W. A., 1966, Drainage asymmetry and the Carolina Capes, *Bull. geol. Soc. Am.*, **77**, 223–240.
Zakrzewska, B., 1963, An analysis of landforms in a part of the central Great Plains, *Ann. Assoc. Am. Geogr.*, **53**, 536–568.

CHAPTER SEVEN

# The Role of Extreme (Catastrophic) Meteorological events in Contemporary Evolution of Slopes

Leszek Starkel

## 7.1 Introduction

The influence of processes of extremely high intensity on landform change and on the morphogenetic balance in general is an appropriate current concern of geomorphology. The function of these extreme phenomena may only be explained on the basis of processes occurring each year. The frequency of extreme events has been increasing in recent centuries together with the transformation of the natural environment by man.

There is a diversity of viewpoints on the role of extreme meteorological events in the shaping of relief. Rapp (1963), Tricart and coworkers (1962), Mortensen (1963), Emmett (1968) and Lamarche (1968) underline their leading role in relief development. The work of Soviet geomorphologists, e.g. Rajonirovanije SSSR (1965), stress their importance in particular climatic zones. Mieshtcheriakov (1970), however, ascribes only a local role to them, while King (1957), in discussing the rules of slope evolution and the uniformitarian outcome of processes, omits reference to the essential contribution of catastrophic events to slope evolution.

Wolman and Miller (1960), basing their conclusions on their own observations of fluvial activity and on Bagnold's aeolian studies, stress events of medium intensity rather than those of extreme character both in transportation (with as much as 90% of suspended load carried by rivers during floods with recurrence periods of less than 5 years) and in the shaping of landforms (assuming the bankfull stage as critical). They fail to consider slopes on which displacements are much larger although they may occur over short distances. Moreover, the duration of even the 'normal' processes may be limited in such cases to a few days or even to a few hours in any one year.

Dynamic equilibrium in morphological systems resulting from a set of endogenous and exogenous factors is to be expected in most, if not all, climatic zones (Gierasimov, 1970). The exogenous factors are often grouped together as the climate–soil–vegetation climax system (Tricart and Cailleux, 1972) which determines slope equilibrium. In each system there exist limiting values of intensity for particular processes, e.g. on solifluction slopes or wash-pediments, below which the equilibrium remains undisturbed. Nevertheless, meteorological phenomena do occur which, owing to their intensity exceed even the threshold values established in the course of prolonged denudation. They consist either of an exceptional element affecting, for example, the trend of slope evolution (Rapp, 1963) or a recurring component of the morphogenetic system as a whole.

In considerations of extreme events causing slope transformation, it is most often the processes due to gravitation, slope-wash, linear erosion and wind action that are discussed. Apart from seismic or volcanic events, these processes may be triggered by extreme meteorological phenomena the most important of which are:

(a) rapid downpours of high intensity, mainly local in range and connected with convection currents or the passage of cyclones;
(b) long-lasting continuous rains of lesser intensity and regional extent due to advection of oceanic air-masses;
(c) periods of rapid warming, bringing about quick melting of snow and ice as well as ground-thaw, often accompanied by precipitation.

Ignoring the effect of strong winds, the action of extreme morphogenetic processes is controlled by water circulation on slopes, i.e. by the ratio of surface runoff to infiltration as well as by the mutual relationship between gravitation forces and the 'resistance' of slope masses. In actual cases, the type and intensity of the phenomena are influenced by endogenous processes reflected in the slope form and inclination, the lithology of the substratum, the past history of the slope (landforms and debris mantles may be inherited from periods characterized by quite a different equilibrium), the present-day climate–soil–vegetation system and its disturbance by human activity. Variations in the long profiles of slopes will involve changes in the intensity and function of extreme processes.

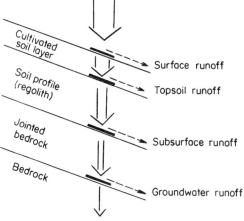

Figure 7.1. The effect of varying permeabilities in different near-surface layers in the formation of surface and subsurface runoff.

In analysing the slope infiltration model, it is possible to distinguish a few boundary surfaces at which changes in infiltration may allow critical, limiting values to be exceeded during rare meteorological events (Figure 7·1). The first boundary surface is that of the topsoil. When the infiltration capacity of this material is surpassed during high intensity rains, surface runoff begins and slope-wash and linear erosion result. The second boundary surface is the base of the arable layer (or of the A horizon), a surface separating two layers of different water holding capacity. Surplus water may either be drained through soil channels or it may be impeded, in which case the soil may become liquefied. Rapid flows of topsoil may result. The third boundary surface is that between the regolith (waste-mantle or gravitation cover) and the bedrock. Water which fails to soak into the bedrock (because of high water input rates or low bedrock permeability, or both) flows down this interface under hydrostatic head. Suffosion channels, mud-flows, debris-flows or landslides and slumps commonly result. Finally, the fourth boundary surface is formed by the, frequently irregular, base of well developed rock-jointing which, in many rocks, determines the limit of shallow groundwater aquifers. The presence of an overburden, and the plasticity of this horizon, often leads to landslides and rock-falls which tend to occur late in the sequence initiated by the torrential rains.

Slope inclinations and varying regolith patterns, as well as the parameters of precipitation, decide whether the surfaces in question, and which of them, will instigate the disturbance in slope equilibrium. Human activity has brought about changes in limiting values of particular processes, especially through decreasing or increasing infiltration rates (change of soil structure, deforestation, ground reclamation, road building) and in altering the coherence of the ground surface by the destruction of plants, artificial slope undercutting and the like.

*7.1.1 The terms 'extreme' and 'catastrophic' event*

Some definition of these terms seems appropriate in the present context. Before speaking of extremes, the nature of 'normal' events should be determined. These should be regarded as (i) occurring each year, (ii) being usually of no great intensity (although this will vary with the climatic zone), (iii) being adapted to the climate–soil–vegetation climax system and (iv) not generally disturbing the equilibrium of slopes, i.e. the ratio of debris removed to that derived from weathering. It is not sufficient to accept as an index of normal processes on slopes the mean annual value of denudation as this provides only a first approximation to the process intensity scale for various climates (Corbel, 1964; Leopold, Wolman and Miller, 1964). The rate of denudation is usually calculated on the basis of suspended load removed from large catchments. This tells us little about actual denudation of slopes within the whole basin because (i) mean (multi-annual) values give no indication of the component values diverging from the means; (ii) slope material deposited at the slope-foot does not always reach river-beds (Rapp, 1960; Gerlach, 1966); (iii) drainage basins display denudation rates which vary with factors such as geological structure and degree of afforestation; and (iv) torrential falls of rain seldom cover the whole catchment and their local and annual incidence is such that even the years with rainfall minima reveal the effects of local downpours. It follows that data on 'normal' processes may be limited to static observations of the mechanism and effectiveness of slope processes within an appropriately determined dynamic equilibrium.

On the other hand, extreme events are the result of precipitation of a quantity or intensity rarely experienced, of rapid snowmelt or of strong winds. The extremity can be considered either in terms of meteorological cause or from the viewpoint of their geomorphological effects.

Morphogenetically-effective meteorological phenomena may be characterized and compared by analysis of their intensity and frequency. The indices of rainfall intensity usually employed (apart from approximate number of days with precipitation above 10 mm and above 50 mm) are the amount of hourly precipitation and the intensity of precipitation expressed in millimetres per minute. The duration of rain which is essential for notable activity in the processes due to gravitation may be as much as several days of continuous fall. Typically, the intensity of local convectional downpours is spatially very variable.

The frequency of the so-called extreme events has been estimated in various ways. Downpours of high intensity which control the magnitude of the annual surface runoff peak (Aristachova and Fedorovitch, 1971) or of the annual amount of precipitation have been singled out in this way. More often, phenomena not occurring each year are regarded as extreme (Leopold, Emmett and Myrick, 1966; Temple and Rapp, 1972). Sometimes events of previously unrecorded magnitude have been described (Rapp, 1963). Years with extremely high total precipitation favouring, for example, the formation of landslides may also be regarded as extreme events (Jakubowski, 1964; Grant 1966).

The morphological results of extreme meteorological events can be analysed both quantitatively (volume of mass displaced, index of slope lowering, etc.) and as relief-forming effects, i.e. the creation of new forms. An extreme phenomenon can be spoken of when a single downpour dictates the mean annual value (Verstappen, 1955) or when, as a result of a downpour, the average amount of slope-wash is exceeded (Skrodzki, 1972), or when the quantity and volume of freshly-formed flows or rock-falls greatly exceed average annual values (Temple and Rapp, 1972). In morphogenetic terms, an extreme event is a set of processes and forms different from those determined by the slope system in a mean state of dynamic equilibrium. Gullies form new elements within many slopes and their rapid creation alters the trend of slope evolution. Such forms may come into being at certain time-intervals (Curry, 1966). Alternatively, they may not recur.

Newly-created forms undergo rejuvenation at intervals, so that, in transforming slopes inherited from earlier regimes (glacial, periglacial, etc.) they increasingly dominate the contemporary morphogeny. Such a development 'by jumps' is defined as 'dynamic metastable equilibrium' by Chorley and Kennedy (1971). Most slope processes of extreme type are limited by the principle of self-regulation (Woskresienski, 1971) such that the intensity of the process ceases if the whole of the waste-mantle is removed. Extreme processes may accelerate or retard relief maturation. Deforestation and cultivation of slopes have destroyed the natural circulation of water and of solids. The intensity of processes has increased by a factor of a thousand or more. The processes of extreme intensity described above usually apply to deforested areas. Hence, the concepts of accelerated erosion (Bennett, 1939) and catastrophic erosion (Tricart and coworkers, 1962) or the paroxysmal phase (Tricart and Cailleux, 1972) have emerged.

To sum up, what follows is concerned with extreme meteorological events of importance for slope evolution and especially with short-term events which occur less frequently than once a year and disturb the dynamic equilibrium of slopes created in climax conditions of a given morphoclimatic region or inherited from an earlier period.

*7.1.2 Working method*

Extreme meteorological phenomena are considered against the background of more 'normal' ones. For particular morphoclimatic regions the following values are taken into account: mean annual precipitation and its deviations, mean annual and observed maxima of daily precipitation, number of days with a precipitation above 10 mm and above 50 mm, maximum observed precipitation per hour and per minute. Unfortunately, exact analyses of the morphological effectiveness of rainfall such as those performed by de Ploey (1972) for Kinshasa and Antwerp are generally absent for large regions. Special attention has been paid to the mountains where the greatest extremes have been recorded.

The data on the morphological effects of rainfall on slopes display a very wide range (Table 7.1 and Figure 7.4). While the indices of mean denudation, based mainly on suspended load in large rivers, provide rough values (see above), more can be learned from measurements of load removed from catchments of a few square kilometres. Further information can be gleaned only by measurements on slopes, including water circulation, over a period of years. However, the values of measurements from plots of limited size may vary considerably from the general denudation rates over entire slopes. Unfortunately, extreme values have seldom been recorded at research stations. Consequently, the basic data consist of the effects (more rarely the cause) of rapid downpours, prolonged rainfall events and sudden snowmelts. These usually involve surface changes which can be recorded on maps at various scales, so facilitating estimation of the scale of the morphogenic event and a reconstruction of the processes at work. They are by no means always supplemented by measurements of precipitation or discharge of water in the streams but, in spite of the incompleteness of much of the data, they are usually gathered by specialists and form a reliable record. The plotting of observed extreme events

Table 7.1 Geomorphological effects of selected extreme meteorological events

| No. | Locality region | Climatic zone | Relief lithology | Annual rainfall (mm) | Date of event | Land use |
|---|---|---|---|---|---|---|
| D 1 | Ulvadal, Norway | boreal alpine | steep slopes with debris | 1000 | 26.6.60 | alpine vegetation |
| D 2 | Cairngorms, Scotland | temperate alpine | steep slopes with debris | 1500 | 13.8.56 | alpine vegetation |
| D 3 | Posorty, N. Poland | temperate | morainic hills, till, sand; plots | 600 | May 67 | cultivated |
| D 4 | Piaski Szlacheckie, Poland | temperate | loess upland | 600 | 23.6.56 | cultivated |
| D 5 | Postołów S. Poland | temperate | flysch foothills | 650 | 20.7.52 | fields, meadows |
| D 6 | Szymbark, S. Poland | temperate | flysch foothills, plots | 680 | 20.6.69 | potatoes |
| D 7 | Huta Szklana, S. Poland | temperate | quartzites, shales, low mountains | 750 | 17.7.65 | forests, cultivated |
| D 8 | S. E. Ukraine | temperate semi-arid | gentle slopes 6, plots | 400 | 17.6.57 | cultivated |
| D 9 | Moldavian SSR | temperate | loess upland, plots | 500 | 6.6.59 | cultivated |
| D10 | Little Creek, Virginia | temperate | low mountains, sandstones, shales | >1000 | 17–18.6.49 | forest |
| D11 | Suchumi, W. Caucasus | mediterranean | foothills, plots | 1400 | 6.9.67 | tea garden, orchards |
| D12 | Kishtsai, E. Caucasus | semi-arid | middle–high mountains, partly flysch | >1000 | 28.7.36 | alpine vegetation |
| D13 | White Mts. California | arid | mountains, subalpine belt | 312 | | scarce vegetation |
| D14 | Arroyo de los Frijoles, New Mexico | arid | low hills | 350 | | scarce vegetation |
| D15 | Nawa, Rajasthan | semi-arid | low granitic hills sandy foothills | 400 | 11.7.68 | source vegetation |
| D16 | Nishnyi Creek, Tian-shan | arid | high granitic mountains with debris | 633 | 14.7.56 | alipine vegetation |
| D17 | Mgeta, Tanzania | tropical humid | middle–high mountains meta-ligneous rocks | >1000 | 23.2.70 | fields, forest |
| C 1 | Kärkevagge, N. Sweden | boreal alpine | steep slopes with debris | 850 | 5–7.10.59 | alpine vegetation |
| C 2 | Sludianka, Trans-Baikal | boreal | middle mountains | 1000 | 17–20.6.60 | forests (mainly) |
| C 3 | Tenmile Range, Colorado | temperate alpine | steep sides of glacial corrie | 2500 | 17–18.8.61 | scarce alpine vegetation |
| C 4 | Jaworki, S. Poland | temperate | flysch low mountains | >1000 | 29.6.58 | fields, forests |
| C 5 | Szymbark, S. Poland | temperate | flysch foothills, plots | 680 | 16–19.8.69 | potatoes |
| C 6 | Eastern Hutt, New Zealand | subtropical | sandstone hills with loess | | 25–26.4.66 | cultivated |
| C 7 | Italian Alps | mediterranean, alpine | middle–high mountains | >1000 | 3–4.11.66 | various |
| C 8 | Meshushim, Antilibanon | semi-arid | basaltic hills | 650 | 21–23.1.69 | scarce |
| C 9a | Nagri Farm, Sikkim Himalayas | tropical monsoon | middle mountains, metamorphic rocks | 3100 | 2–4.10.68 | tea gardens, forest, fields |
| C 9b | Kurseong, Sikkim Himalayas | tropical monsoon | middle mountains metamorphic rock's | 4050 | 2–4.10.68 | tea gardens, forest, fields |
| C10 | Darjeeling, Sikkim Himalayas | tropical monsoon | middle mountains, metamorphic rocks | 3100 | 10–14.6.50 | tea gardens, forest, fields |
| M 1 | Mt. Rainier, Washington | temperate alpine | high mountains, volcanic rocks | 3000 | 23.10.47 | nival zone |
| M 2 | Guil Valley, French Alps | temperate mediterranean | high mountains with glacial forms | >2000 | 12–14.6.57 | alpine vegetation, forests |
| M 3 | Almatinka Zailisky Ala-Tau | arid | high mountains with thick debris | 600 | 8.7.21 | scarce alpine vegetation glaciers |

*Rainfall in 24-hours

| Recorded rain | Intensity mean (mm/h) | Intensity max. (mm/h) | max. (mm/min) | Main geomorphological process | % of transformed relief | Calculated erosion (m³/t per km²) | denudation rate (mm) | mean denudation (mm/year) | Author |
|---|---|---|---|---|---|---|---|---|---|
| > 19 | | | | debris-slides and flows | up to 50 | >100,000 | 200 | < 1 | Rapp (1963) |
| > 150 | | | | torrent trunks | | | | | Baird and Lewis (1957) |
| 60·0 22·5 | | > 3·0 > 4·5 | | slope-wash | | 5000 | 3·5 | < 0·2 | Skrodzki (1972) |
| > 81 | | > 40 | | slope-wash, gullying | up to 1 | 3150 | 1·4 | < 0·4 | Maruszczak and Trembaczowski (1959) |
| 111 | > 30 | | | bowl-slides, mud-flows | 1–2 | 2200 | 2·2 | | Starkel (1960) |
| 42 | | | 2·3 | slope-wash, topsoil flows | | | | | Gil (in press); Słupik (1973) |
| 125 | | 250 | 4·2 | suffosion, wash, gullying | | | | | Starkel [after Kaszowski and co-workers (1966)] |
| 43 | | | | slope-wash | | 20,000–50,000 | 20–50 | | Skorodumov (1973) |
| 14·6 | | | 5·8 | slope-wash | | 2150 | 2·15 | | Zaslawski [after Mirchulava (1970)] |
| 183 | | | | chutes, water blow-outs | 1–2 | | | | Hack and Goodlett (1960) |
| 91 | | | | slope-wash | | 6800 | 6·8 | 4 | Darseliya and Gvazava (1970) |
| 42·7 | | 37 | > 1·0 | Sjels | | 50,000 | 50 | > 1·6 | Gagoshidze (1970) |
| 20 | > 100 | | | debris-avalanches | | | | 0·1–0·5 | Lamarche (1968) |
| 38 | | | 1·5 | slope-wash, gullying | | | | | Leopold and co-workers (1966) |
| 488 | 50 | | | Gullying, suffosion, slope-wash | | | | | Starkel (1972) |
| 34·4 | 30 | | 5·0 | debris-flows, sjels | | 2400 | 2·4 | 0·02 | Iveronova (1962) |
| 100·7 | 30 | | | mud-flows, sheet-slides, bottle-slides | up to 10 | 13,500 | 14 | 0·26 | Temple and Rapp (1972) |
| 170 (107)[a] | | | | sheet-slides bowl-slides, debris-flows | | | | | Rapp (1960) |
| 326 (152)[a] | | | 0·43 | sjels | | | | | Joganson (1962); Sołonienko (1962) |
| 295 | 12 | 50 | | debris-flows | | | | | Curry (1966) |
| 130 | 17 | | | slumps, mud-flows | | > 300 | 0·3 | | Gerlach (1966) |
| 153 | | | 0·6 | slope-wash (outside mud-flows) | | 50 | 0·05 | | Słupik and Gil (1974) |
| 132 (110)[a] | 29 | | 0·67 | slips | | | | | Jakcson (1966) |
| 567 | 42 | | | debris-flows, mud-flows, gullying, slides | up to 30 | locally > 100,000 | | | Pellegrini (1969) |
| 120–360 | 10 | | | slope-wash, linear erosion | | | 0·24 | 0·04 | Inbar (1972) |
| 697 (369)[a] | 715 | 40 | > 1·0 | mud-flows, slumps, slides, suffosion | gardens up to 25 forests | 200,000 | 200 | 1·5 | Starkel (1972a) |
| 1091 (638)[a] | 760 | | | mud-flows, slumps, slides, suffosion | up to 2 | 10,000 | 10 | | Starkel (1972a) |
| 1000 (450)[a] | | | | mud-flows, etc. | | similar in size to 1968 | | | Starkel (1972a) |
| 148 | | | | mud-flows (lahars) with ice-, melting snowmelt | | | | | Crandell (1971) |
| 300 (202)[a] | | | | slush-avalanches, mud-flows, gullying | | | | | Tricart and coworker (1962) |
| 72 | | | | snow-melt, ice-melting, sjels | | | | | Gagoshidze (1970); Dujsenov (1971) |

on diagrams to bring out, for example, the relationship between amount and intensity of precipitation (Figure 7.4) as well as their estimated frequency, may suggest sequences of changes in various climatic regions.

### 7.1.3 Regional differentiation of extreme meteorological events

The distribution of extreme phenomena in space and time coincides clearly with climatic zones although the mountain zones are outstanding in this respect. The works of Jennings (1950), Pardé (1961), Common (1966) and Marx (1969) provide data on world extremes, of which the following are examples:

maximum precipitation per minute—16·5 mm (Opid's Camp, California)
maximum precipitation per hour—335 mm (Jamaica)
maximum precipitation per 24 hours—1870 mm (Cilaos, Réunion)
maximum precipitation per 8 days—3430 mm (Cherrapunji).

According to Poggi (1959) snowmelt rates may reach as much as 80 mm per 24 hours.

The size and frequency of extreme values coincide with regional climates and should be considered against the background of annual downpours (Reich, 1963; see Figures 7.2 and 7.3). The review presented below is largely restricted to Eurasia and Africa.

In arctic climates (periglacial) both amount and intensity of precipitation are low; only during rapid warming and snowmelt is more intense runoff of water possible (Washburn, 1965).

In boreal continental climates with some permafrost, 'catastrophic' precipitation has been recorded very rarely (daily precipitation above 50 mm). Only in the mountains, on the passage of cyclones, are downpours totalling 100 mm encountered (Rapp, 1967) and these should be considered as exceptional. The precipitation of 546 mm recorded at Hamar-Daban in the Trans-Baikal Mountains (Solonienko, 1962) must also be regarded as exceptional. However, rapid snowmelts occur in these regions.

In temperate climates with generally high annual precipitation (500–2000 mm) arising from regular cyclonic activity, both frontal and convectional downpours occur, as well as prolonged continuous rainfall, especially in mountainous areas. One or two rainfall events of more than 50 mm a day occur every year, while downpours with a maximum intensity of 2 mm per minute or continuous falls of 200–500 mm in amount (with an intensity usually below 10 mm per hour) occur every few years. Daily precipitation maxima of 200–300 mm or more and downpours of 3–10 mm per minute have been recorded locally. The intensity of downpours increases with continentality of climate. In mountains within the temperate zone rapid snowmelt is common.

Apart from the concentration of precipitation in winter and the occurrence of intense downpours (especially well investigated in the Alps, Apennines and in Georgia, USSR) the mediterranean type of climate is characterized by the

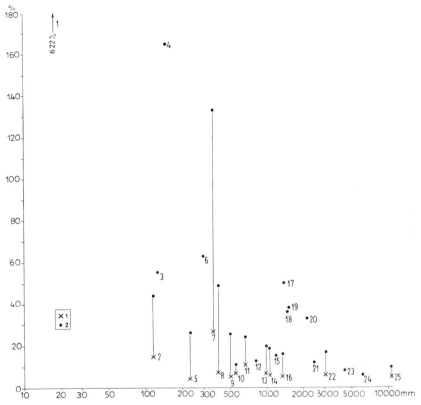

Figure 7.2. Relationship between mean annual (1) and absolute (2) rainfall maximum in 24 hours to the mean annual precipitation (in per cent). Stations: 1—Faya (Tchad), 2—Kazalinsk (central Asia), 3—Khartoum, 4—Naukszot (Mauretania), 5—Baku, 6—White Mountains (California), 7—Jodhpur, 7a—Nawa (Rajasthan), 8—Southern Ukraine, 9—Tbilisi, 10—Alma Ata, 11—Poona (India), 12—Polish Carpathians, 13—Riksgränsen (Lapland), 14—Mgeta (Tanzania), 15—Cairngorms (Scotland), 16—Suchumi (Caucasus), 17—Nelson (Virginia), 18—Hamar-Daban (Trans-Baikal), 19—Italian Alps, 20—Hong Kong, 21—Ten-mile Range (Rocky Mountains), 22—Darjeeling, 23—Conakry, 24—Mahabaleshwar (India), 25—Cherrapunji.

occurrence of continuous rains and periods of rapid warming in the mountains. Every year periods of 5–10 days occur in which precipitation exceeds 50 mm. The wetter (western) mediterranean type is characterized by continuous rains totalling 500–1500 mm which occur every few years or every few tens of years. Maximum intensities often exceed 20 mm per hour and some torrential downpours may amount to 200 mm per hour (Marx, 1969; Zanian, 1961).

Apart from their low annual totals (100–1000 mm), precipitation in the arid climates of central Asia is typically irregular and intense. Every year as many as ten falls of more than 10 mm per 24 hours may occur. Precipitation above 50 mm is very rare (perhaps once in 10 years) but rainfall rates are high

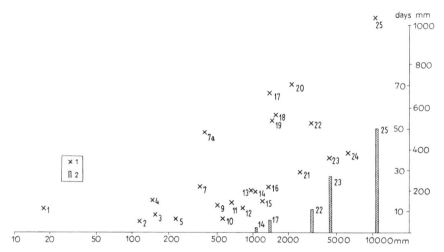

Figure 7.3. Comparison of observed 24 hours maximum rainfall and number of days with rain greater than 50 mm with total annual precipitation. Key: 1—Observed maximum in mm, 2—Number of days with more than 50 mm of rain. Station numbers as in Figure 7.2.

rising to 10 mm per minute (Lenkoran, 16 mm per minute). This zone also displays considerable and rapid temperature changes during invasions of arctic or tropical air-masses (*Jahreszeitenklima*: Troll, 1962). The latter often produces rapid melting of snow and ice (Joganson, 1962). Central Asia is also an area of strong winds. Fort Shevtschenko records up to 84 days with winds exceeding 15 mm per second (Tshelpanova, 1963).

The zone of arid tropical climate displays the greatest irregularity of precipitation with some falls exceeding 40% of the mean annual value (Biel, 1929). Many stations do not record precipitation at all in some years (Kufra Oasis). Nevertheless, and especially on the margins of that zone, precipitation in excess of 50 mm and daily maxima considerably greater than the mean annual values may be recorded every few years, e.g. 250 mm rainfall has been recorded at Naukshot (Mauretania), 488 mm at Nawa (Rajasthan) and also in Tunisia (Mensching, Giessner and Stuckmann, 1970). On the other hand, sudden temperature oscillations approaching 0 °C have not been observed (*Tageszeitenklima*: Troll, 1962).

The zone of humid tropical climates is characterized, on the whole, by distinct rainfall seasons with a broadly constant annual precipitation total (coefficient of irregularity 0–15%) although the range of annual totals is considerable (500–11000 mm). The highest precipitation has been recorded in the equatorial regions and in areas with monsoonal circulations where cyclone tracks converge on mountain barriers, as well as in regions of frequent typhoon routes (oceanic islands). Daily precipitation totals reaching 1000 mm are known from Japan, Taiwan, the Philippines, Assam, Reunion, Angola and Jamaica (Jennings, 1950). Sometimes these are 'typhoon-eye' downpours with a considerable intensity (300 mm per hour); more often, they consist of continuous rains of 500–2000 mm lasting 2–3 days and with intensities of up to 50 mm per hour. While such record rainfalls occur only every few tens of years, annual daily maxima range from 100 mm to 500 mm (Krishnan and coworkers, 1959) and the number of days with a precipitation above 50 mm ranges from 1 to 5 in the monsoon regions (often in a precipitation shadow) and perhaps 20 or more in equatorial and mountain regions.

As can be seen from Figures 7·2 and 7·3 not only the absolute values of precipitation extremes but also their value expressed as a ratio of the 'normal' precipitation are different in particular climatic zones. Moreover, their morphogenetic potential is strongly influenced by the conditions of dynamic equilibrium on slopes in these zones. Other extreme phenomena with morphogenetic potential such as sharp rises in temperature or strong winds are limited to some zones only.

In all the climatic zones the highest extreme values and the highest erosional intensities have been observed in mountain areas. Of course, mountains are typified by steep slopes, recent tectonic movements giving rise to irregular slope and river-bed profiles (Mieshtcheriakov, 1970; Hewitt, 1972; Krivolutskiy, 1971). The altitude of mountains produces the well known phenomenon of the vertical zonation of climatic belts each characterized by its distinctive suite of geomorphic processes. Mountain ranges are a barrier for air-masses and influence even the course of tropospheric air streams. Accordingly, they commonly show a marked climatic asymmetry. Weischet (1969) pointed out that it is marginal zones in compact mountain massifs that receive the heaviest precipitation rather than areas situated in the interior of mountains or in their foreland. At the same time, marginal zones are the most active tectonically and so most rapidly uplifted and dissected (Kostienko, 1960; Starkel, 1972b). The zone of highest precipitation in the tropics lies at about 1500 m above sea level and in the temperate zone in a range up to 3500 m (Weischet, 1965), above which precipitation declines. This precipitation inversion defines the range of the occurrence of extremes, on a global scale. In fact, most of the 'catastrophic' erosional events have been recorded in such areas. Nevertheless, mountains do not necessarily constitute a separate morphogenetic region. The extreme processes acting in the mountains preserve certain zonal features which are dictated by thermal regime and the type of air circulation as well as climax sets of climate–soil–vegetation systems developed in zonal form.

## 7.2 Types of extreme meteorological phenomena and the associated processes

Extreme precipitation may vary in volume, intensity and duration. This, in turn, influences among other things the intensity and type of the slope processes (Figure 7·4). Variation in these processes, such as the relative importance of surface runoff and infiltration, as well as their intensity is also dependent on the state of the ground and the intensity of precipitation (cf. Figure 7·1). On a slope of a given inclination, soil and plant cover, rainfall intensity dictates

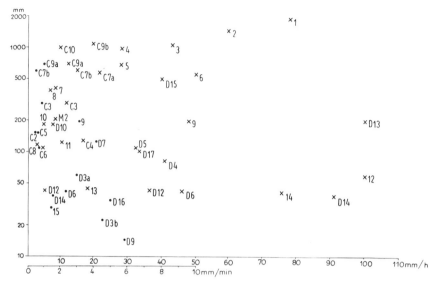

Figure 7.4. Relationship between the heavy rainfall totals and their mean intensity per hour (crosses) and maximum intensity per minute (points). Most events included are those described in the course of geomorphic work. A key to stations identified by a letter and a number can be found in Table 7.1. Other stations are as follows: 1—Cilaos (Réunion), 2—Cangamba (Angola), 3—Cherrapunji, 4—Kurseong (22 September 1952), 5—Nelson (20 August 1969), 6—Hamar-Daban (Solonienko, 1962), 7—Hong Kong (11–12 June 1966 (So, 1968), 8—Leskowiec (Polish Carpathians) (July 1970), 9—Nilolayev (Ukraine) (30 July 1955), 10—Terek (Caucasus), 11—Kazbegi (Caucasus) (16–17 August 1953), 12—Kinshasa (Congo), 13—Moldavian SSSR (Mirchulava, 1970), 14—Harquahala Mountains (15 August 1963) (Rahn, 1967), 15—Gruszowiec (Polish Carpathians) 5 July 1957).

whether surface runoff or infiltration will prevail. Rainfall duration, on the other hand, will determine whether the saturation limit will be exceeded in deeper horizons or whether subsurface flow or liquefaction will occur. In the light of this, precipitation will be considered under two headings: heavy downpours and continuous rain.

### 7.2.1 Heavy downpours

Downpours are rapid rainfalls of storm character due to convection currents or the passage of fronts of intensity usually exceeding 1 mm per minute and lasting from a few minutes to several hours. The total fall may exceed 500 mm while intensities may reach 16 mm per minute. Hack and Goodlett (1960) noted a fall of 762 mm in 270 minutes in the Pennsylvanian Appalachians.

Downpours usually begin quite voilently (Figure 7·5) and this often gives rise to splash erosion of the soil caused by drops of bombarding rain up to 5 mm in diameter (Wischmeier and Smith, 1958). On dusty soils compaction of the

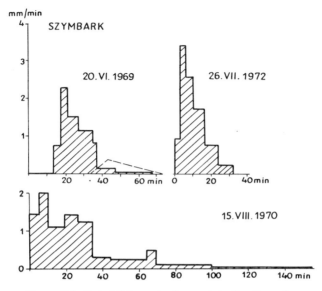

Figure 7.5. Rainfall intensity in mm/min for three downpours at Szymbark physico-geographical station (Polish Carpathians). Broken line shows the measured intensity of surface runoff (Słupik, 1973).

top soil may follow so that surface permeability is rapidly reduced (Duley, 1939). Air becomes imprisoned in the soil which also makes infiltration difficult (de Ploey, 1969; Słupik, 1973). When the rainfall intensity exceeds the infiltration rate and the water surface overcomes the local friction (detention), overland flow begins (Figure 7.5). This phenomenon is most often seen in semi-arid areas devoid of soil and on arable soils especially in regions underlain by shales and loess. The greatest effects are due to downflow on plant roots: this may reach 20 tonnes per hectare during a downpour of 40 mm. There is a close relationship between the amount of water flowing downhill and the quantity of the material removed (Carson and Kirkby, 1972). The coefficient of surface runoff is usually 20–50%. Słupik and Gil (1974) point out that in the Flysch Carpathians the volume of slopewash depends directly on the maximum intensity of surface runoff. During one of seven downpours investigated, this reached as much as 6 m$^2$/s/km$^2$. In the Ukraine the denudation during single downpours amounts to 100–500 tonnes/ha (Skorodumov, 1973) which is equivalent to a lowering of slopes in the area by up to 40 mm.

Certain values of rainfall and runoff intensity may prove critical in relation to events. Zaslavski's investigations in Soviet Moldavia (Mirchulava, 1970) show that mean precipitation intensity differences as small as 0·05–0·30 mm/min may control sediment removal rates from minima of about 0·2 to maxima of 142 tonnes/ha. In the same way, the level of soil moisture prior to a downpour can be critical. For example, during two downpours of 90 mm in western Georgia (USSR) sediment removal varied from 8 to 102 tonnes/ha according

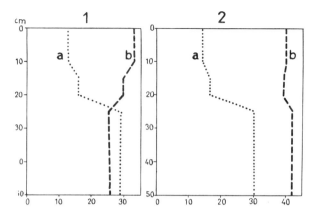

Figure 7.6. Soil moisture on the experimental slope at Szymbark before (a) and after (b) rainfall (after Słupik and Gil (1974)). 1—short-lasting shower 15 June 1969; 2—continuous rain 15–19 August 1969.

to soil moisture variations (Darselija and Gvazava, 1970). Frequently, single downpours effect the total annual denudation and greatly exceed the long-term denudation rate.

The surface runoff in areas under cultivation is usually related to the degree of saturation of the arable layer which may be up to 20 cm thick (Figure 7·6). The base of the arable layer is poorly permeable on the whole and saturation may be followed by liquefaction: such material can be seen to flow on slopes of some 10°. Such phenomena have been observed also in the Carpathians (Figula 1960; Gil and Słupik, 1972). On measurements carried out on standard plots such flow is included in total slopewash.

More detailed records of slopewash over extensive surfaces suggest that rill networks appear initially to be followed by ravines above both erosional and depositional plains which vary in form according to the lithology (Richter and coworkers, 1965). The depth of rills is greater in areas devoid of vegetation as runoff is turbulent on slopes with vegetation densities below 18% (Shiff, 1951). The surface runoff described by Leopold and coworkers (1966), Schumm (1956) and others involves the development of pediments in semi-arid areas. Moreover, the erosional relief of the mountains within the Sahara suggests that rainstorms have played an essential role. Detailed mapping in loess regions of southeast Poland of the effects of heavy rains with a return period of 50 years (Figure 7.7) has shown that the highest runoff and dissection by gullies occurred in the lower parts of convexo–concave slopes and the mean denudation of the catchment reached 1·4 mm. Single downpours of similar type in the Dnieper valley have led to the formation of ravines more than 10 m deep (Drozd, 1962). For the loess-covered uplands of central Germany Hempel (1957) gives a cadastral plan from the eighteenth century marking newly-formed gullies, the result of a local downpour (Figure 7.8).

217

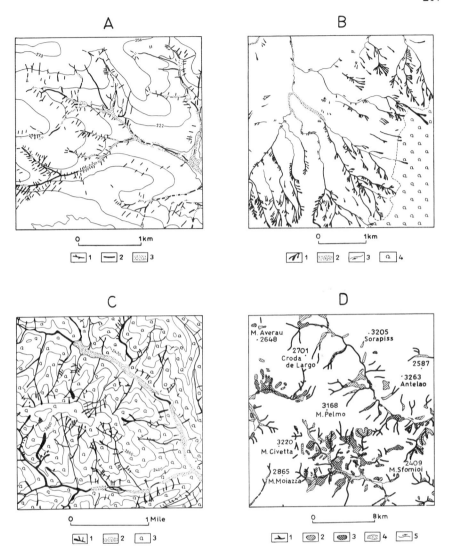

Figure 7.7. Details from maps expressing geomorphological changes after downpours (A, B, C) or continuous rains (D). A—Effects of downpour on 23 June 1956 in Piaski Szlacheckie, loess plateau, eastern Poland (Maruszczak and Trembaczowski, 1959). 1—erosional rills; 2—gullies and new channels; 3—debris accumulation. B—Landslides near Mgeta (Tanzania) after the rainstorm of 23 February 1970 (Temple and Rapp, 1972). 1—slides and mud-flows; 2—channel sedimentation and debris fans; 3—streams; 4—woodland and forest. C—Effects of the 1949 flood in the Little River basin, Virginia (Hack and Goodlett, 1960). 1—chutes and channels created after flood; 2—valley bottoms with erosion and debris accumulation (ornament added); 3—forest. Altitudes in feet. D—Effects of the November 1966 flood in the Italian Alps (after Cons. Nat. d. Res., Inst. d. Geog. Univ. Padova, (1972)). 1—forms of linear erosion; 2—rejuvenated older landslides; 3—newly formed landslides; 4—fluvial accumulation; 5—rivers.

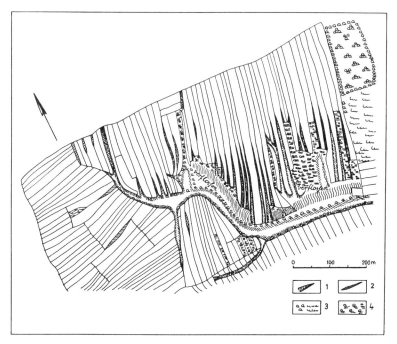

Figure 7.8. Gullies and deep rills on a map of Heyliedsgraben in Eichsfield, central Germany, dated 1768 A.D. (after Hempel (1957)). 1—gullies and rills; 2—edges of terracettes; 3—forest; 4—areas with degraded soils.

If the water capacity of soils is high enough and their infiltration potential exceeds the rainfall intensity (as may often occur in mixed or sandy regolith) or when splash-erosion does not destroy the uppermost soil layer (as when plant cover is present) then a heavy fall of rain will soak into the ground. In such a case, soil displacement depends on the existence of a deeper boundary horizon (the base of the regolith) as well as on the character of the soil mantle. The latter either allows throughflow of water or the liquid limit of the material is exceeded, in which case mud-flows and debris-flows occur. The stability of the whole slope may be reduced and landslides and slumps triggered. Phenomena of this kind are encountered on steeper slopes (say above 15°) in various climatic zones, not only in treeless areas but also in forests which, although they generally reduce the rate of runoff, retain considerable stored water in the soil.

The effects of brief downpours in the mountains of the tropical zone with thick lateritic soils have also been described from Hong Kong (Berry and Ruxton, 1960; So, 1968) and Tanzania (Temple and Rapp, 1972). The latter authors describe the effects of a rainfall lasting 3 hours when 101·7 mm of precipitation preceded by a smaller rainfall brought about widespread sheet-slides and mudflows on slopes of 24°–25° (Figure 7.7). Simultaneous circulation of water under seepage pressure in channels within the waste-mantle and the liquefaction of the soil are suggested. No rills were observed. On more deeply weathered surfaces 'bottle slides' occurred with blow-outs of water circulating in channels. Rejuvenation and dissection of deforested hillsides was effected by a number of shallow (2 m) narrow (5–20 m) and long troughs. Expressed in the form of the denudation index the downpour amounted to 14 mm.

In the mountains of semi-arid regions torrential rains of only a few millimetres in intensity, are the basic impetus of slope change (Joganson, 1962; Rustamov, 1962; Dumitrashko and coworkers, 1970). In the higher parts of the mountains they are represented by snowmelt. From the Caucasian republics of the USSR and from areas in central Asia, as well as from the southern Rocky Mountains, there have been reported tens of such downpours (Table 7.1). In the valley-heads of the High Caucasus, composed of shales and sandstones, the 40° slopes are being actively undercut and are far from being in equilibrium. Each torrential rain sets masses of material in motion, and gullying, mud-flows, debris-flows and even giant landslides occur. Along the slopes and then down the steeply inclined valleys flow 'sjels', the frequency of which ranges up to a few per year in particular valleys.

The term 'sjel' is applied in the Caucasian countries to 'debris-muddy streams' carrying great quantities of material. For example, in the Tsheri valley one sjel amounted to 180,000 m$^3$ of material derived from 1 km$^2$ of the catchment basin: this is equal to a single lowering of the basin by 180 mm (Dumitrashko and coworkers, 1970). Two types of sjel are commonly distinguished: bound (structural) and free (turbulent) sjels. The structural sjel (mud-flow equivalent) is a flowing colloidal muddy or debris mass with a specific gravity in the range 1·7–2·3 tonnes/m$^3$. According to Kerkheulidze (1969) these move with a velocity 1·4–22 m/s on inclinations of 3·8–26%. Gagoshidze (1970) found that on increasing the amount of water to more than 56%, deposition of material occurs, water is released and structural sjels are transformed into turbulent ones. It behaves like an overloaded river with a specific gravity of 1·1–1·3 tonnes/m$^3$, capable of both transportation and erosion.

Traces of sjel-type processes have been described by Lamarche (1968) from the White Mountains, California where, with an annual precipitation of about 300 m close to the treeline, annual downpours wear down the slopes by about 0·1–0·5 mm. In referring to debris avalanches formed in a neighbouring valley after a 200 mm downpour of two hours' duration, la Marche also describes chutes on the slopes and a chaotic accumulation of debris below them which has been dated by dendrochronology to the seventeenth century.

Processes take a different course when there is infiltration into permeable rocks on gentle slopes. In the semi-arid Rajasthan near Sambhar Lake, the author has observed the effects of a downpour lasting less than 12 hours (Starkel, 1972a). On sandy covers mantling the 15° footslopes of inselbergs, some 488 mm of rain fell, exceeding the expected maximum over 250 years. Because of suffosion and liquefaction of sands, dissection of the middle portion of slopes was effected by ravines up to 10–15 m deep, the removed material forming fans at the foot. Gullying also occurred at the contact of rocky hills with waste-mantles resulting in partial exhumation of the rock slopes beneath.

Similar processes of runoff and suffosion are also known from regions of the temperate zone, although they occur on a much smaller scale. In the valley of the San (Flysch Carpathians) the author has observed mudflows, both channel-like and lobe-like, on slopes of less than 25° that were formed in a rainfall event of 110·7 mm (Starkel, 1960). They covered about 50% of the surface of the steeper slopes and led to aggradation of incised sections of the valley. In the twenty years since that event, the headwalls of these features are still not vegetated and small gravitational movements continue.

The effects of a rapid downpour in the Appalachians were worked out in detail by Hack and Goodlett (1960). In spite of the complete afforestation of the basin studied, accelerated subsurface flow occurred in the waste-mantles on Devonian sandstones and shales and this caused bursts at the surface. Debris avalanches cut numerous chutes, generally along the axes of older depressions (Figure 7.7), and water bursts occurred because of the liquefaction of the waste-mantle under pressure. Surface runoff occurred locally as indicated by accumulations of layers of organic matter on some slopes. Runoff estimated at 0·5 m$^3$/s/ha brought about rapid erosion and transport in river-beds. The occurrence of old 'overgrown' forms in the forests points to the recurrence of similar phenomena, though with a frequency of less than once every 100 years. It is evident that the afforested slopes inherited from the Pleistocene are not alway stable in the temperate climate as was considered by Rapp (1967). Similar traces of local disturbance of the slope equilibrium under a quartzite debris cover have been observed by the author in the Holy Cross Mountains (Kaszowski and coworkers, 1966). Following precipitation of 125 mm in 30 minutes, water flowing within the waste-mantle brought about the formation

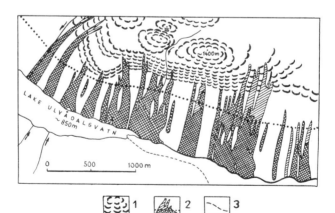

Figure 7.9. Section of a sketch map of debris slides at Ulvadal, western Norway (after Rapp (1963)). 1—form lines on bare bedrock; 2—slide tracks (in upper part eroded down to bedrock); 3—approximate position of 1000 m contour and upper forest limit.

of suffosional holes and the dissection of small gullies from which coarse debris was carried away. Above the treeline the equilibrium of slopes of 30°–40°, covered by a mixed debris mantle, was disturbed. Given good underground drainage these phenomena do not occur frequently. Rapp (1963) described effects of a local downpour at Ulvadal on the fjord side where about 50% of the surface of a till- and debris-mantled concave hillside was involved in channel debris-slips partly transformed into mudflows (Figure 7.9). The amount of precipitation was not measured but the damage caused to a fresh shoreline of early Flandrian age shows that the event was exceptional on the Flandrian scale.

To sum up, it must be stressed that waste-mantle permeability and limiting slope angles determine the effective processes during downpours. Short-lived heavy rains are the main motive power in the transformation of relief in arid areas of the subtropical zone and in less elevated regions of the tropical and temperate zones. The extreme character of these processes consists, however, either in exceeding the intensity of the 'normal' processes (sjels in the mountains of the semi-arid zone, slopewash in farming regions) or in the change of process type which overcomes slope stability.

*7.2.2 Continuous rainfalls*

These are generally local in range and are connected with advection of humid air-masses, often impeded by mountain chains. Therefore, continuous rainfalls are especially characteristic of the mountains of the tropical-monsoon and temperate zones. Precipitation of continuous type (lasting 1–5 days) with only brief pauses generally begins at low intensity and increases to 20–60 mm/h in the middle or toward the end of the event. Light precipitation together with high atmospheric humidity favours infiltration and saturation of the soil and often of the whole waste-mantle (Figure 7.6). Given saturation of the deeper layers, any increase in intensity of precipitation enhances subsurface flow and the stability of the steeper slopes becomes disturbed. Eventually surface runoff begins and all the changes in the intensity of precipitation are reflected in the rate of slopewash according to the work of Ellison (cf. Carson and Kirkby, 1972). A rapid increase in rainfall intensity causes a rise in water pressure and liquefaction of the debris mantle as happens during short downpours on more permeable deposits. Forest areas favour subsurface drainage during continuous rain, although the root-systems of trees impede mass-movements to some extent.

The northeastern slopes of the western Carpathians provide an example of the effect of continuous rainfall. Here, summer rainfalls of 2–5 days' duration with 200–400 mm of precipitation (daily maxima 300 mm) occur every few years and sometimes two or three years in succession. With saturation of the ground, an increase in intensity above 10–20 mm/h initiates only slight washing (Table 7.1) but, at the same time, slumps occur on steep slopes (often on terracette edges) causing mudflows of saturated soils lower down the slope. They

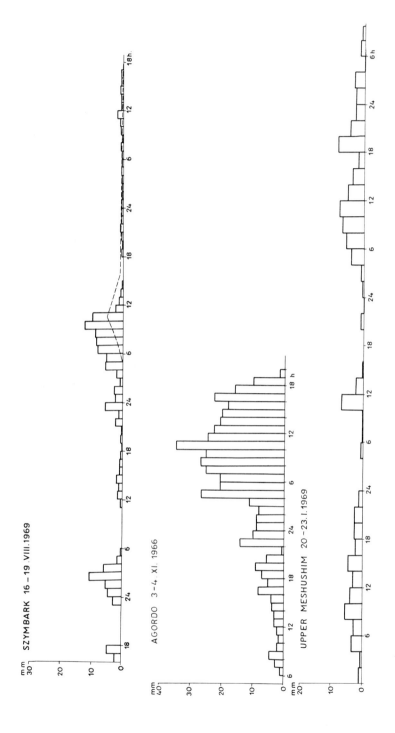

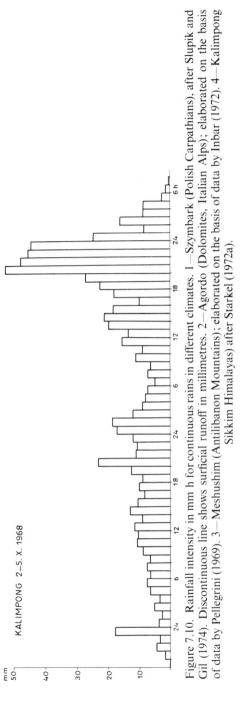

Figure 7.10. Rainfall intensity in mm h for continuous rains in different climates. 1—Szymbark (Polish Carpathians), after Slupik and Gil (1974). Discontinuous line shows surficial runoff in millimetres. 2—Agordo (Dolomites, Italian Alps); elaborated on the basis of data by Pellegrini (1969). 3—Meshushim (Antilibanon Mountains); elaborated on the basis of data by Inbar (1972). 4—Kalimpong (Sikkim Himalayas) after Starkel (1972a).

were recorded in 1934 (Klimaszewski, 1935; Stecki, 1935), in 1958 (Jakubowski, 1964; Gerlach, 1966) and in 1969 (Figure 7.10). Gerlach has calculated that there were 16 slumps per square kilometre displacing some 300 m$^3$ of the soil. In forests the accelerated subsurface circulation led to the formation of many new springs and to the rejuvenation of landslides (Brykowicz and coworkers, 1973). Simultaneous intense gully erosion undermines the slopes and accelerates mass-movements over many years. Above the treeline in the Tatra Mountains, the daily precipitation in 1973 amounted to 300 m with intensities exceeding 20 mm/h for 2 hours. According to Kotarba (personal communication) this stimulated debris-flows in certain areas.

The Flysch Carpathians also display deep structural landslides arising from periodic continuous rains. These were particularly abundant in 1913 (Sawicki, 1917) and 1958 (Zietara, 1968). Many older forms have also been rejuvenated.

Working in the Kärkevagge massif (Scandinavia), Rapp (1960) describes the effects of a three-day precipitation event the last day producing as much as 107 mm. Above the forests there were produced numerous debris-flows furrowing talus-cones as well as bowl-slides and sheet-slides, leading in consequence to the exhumation of bedrock.

Curry (1966) describes the effects of continuous rainfall in the Tenmile Range (Rocky Mountains) between 4200 and 4600 m above sea level. He observed the results of a once-in-a-century precipitation event, 320 mm falling in 49 hours. On the steep (35°–41°) sides of a glacial cirque, tens of debris-flows (inaccurately described by the author as mudflows) developed as, toward the end of the storm, precipitation increased to 50 mm/h. Movement was rapid, amounting to 15·5 m/s, water constituting as little as 9·1% of the mass. Study of the *Rhizocarpon* lichen proved that 5 similar flows have occurred over the past 1500 years. The occurrence of sjels following a four-day continuous rainfall event is also reported by Joganson (1962) and others from the valley of Sludianka in the Trans-Baikal Mountains.

In mediterranean climates continuous falls (with a frequency of 2–5 times in 10 years) are common in autumn and winter. Probably the best known of these is the extremely heavy fall of 3–4 November, 1966, when 711·5 mm fell in the Alps above Venice and up to 437·2 mm in the Apennines of Tuscany (Marx, 1969). Specific runoff reached 6·6 m$^3$/s, exceeding all previous records. Toward the end of the period of rainfall, intensities reached 42 mm/h (Figure 7.10) which brought about a sudden increase in porewater pressures, the liquefaction of the debris cover and the development of channel-like debris-flows such as those described by Pellegrini (1969) from the Dolomites. Most of them came into being on slopes of 15°–20° on poorly permeable but water-soaked till deposits where many episodic springs were formed. Detailed mapping has shown that a number of flows and slumps were located on valleysides whose balance was disturbed by floodwaters. A map of flood damage in the Alps produced in 1962 by the Geographical Institute at Padua indicates changes at various scales due to the twin factors of lithology and intensity of rainfall. In some catchments with areas of some tens of square kilometres, the percentage of the surface transformed by gravitational and erosional processes reached as much as 30% (Figure 7·7). Modification has been at least as great in the deforested Apennines.

Inbar's (1972) work provides data on the effect of continuous rainfall in a semi-arid region within reach of mediterranean winter rains (Figure 7·10). Here some 120–360 mm of rain fell in 4 days, an amount equivalent to one third of the mean annual precipitation. In spite of the modest mean intensities (2–3 mm/h, rarely exceeding 8 mm/h), this precipitation constituted a 100-year event. It caused intense downwash in the basalt catchment of the Meshushim. The mean denudation rate calculated for the basin (0·04 mm/year) was exceeded some sixfold. This example demonstrates that continuous rainfall may be of considerable significance even in semi-arid regions, if the natural vegetation cover had been destroyed.

Southeast Asia is the region of the world with the highest continuous rainfalls. Heavy rains are most often associated with cyclone track centres as the monsoon breaks at the beginning and end of the rainy season (Ananthakrishnan and Krishnan, 1962). The effect of continuous falls in the Darjeeling Hills has also been investigated (Dutta, 1966; Starkel, 1972b). These mountains,

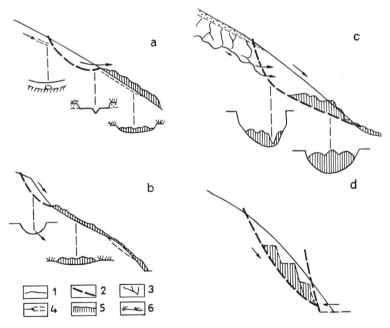

Figure 7.11. Types of mass movements formed during heavy rain in the Darjeeling Hills, Sikkim Himalaya, October 1968 (Starkel, 1972a). a—mudflow associated with suffosion channel; b—slump/mudflow; c—debris/rock-slide associated with deep percolation; d—landslide or slump created by lateral erosion. 1—slope surface before movement; 2—new slope profile; 3—systems of joints; 4—suffosion channel; 5—accumulated material; 6—soil with tea bushes.

with a relative relief of up to 2000 m and steep slopes (20°–40°), suffer heavy continuous rains (12 days with precipitation greater than 50 mm) every year. Sandy-silty mantles derived from schist favour ready soil drainage. Rainfall events of 800–1500 mm in 2–5 days (with a maximum at 1500 m above sea level) have occurred 3–5 times in 100 years (notably in 1899, 1950 and 1968). Typically, intensities rise toward the end of the rainfall event. In the latest example (1968) a rate of 40–60 mm/h was maintained for 4 hours (Figure 7·10). Seepage pressures rose rapidly and the debris cover moved by flow, frequently associated with slumps on the steeper slopes (Figure 7.11). Thousands of slope channels came into being, the flows generally following valley bottoms and changing slowly from structural into turbulent sjels (Figure 7·12). Where infiltration of water

Figure 7.12. Earth and debris mass of a structural sjel transformed into a turbulent sjel in its upper section. Poobong, Darjeeling Hills.

was greater, deep-seated landslides occurred even on wooded slopes (Figure 7·11). The impedance of runoff both on slopes and in river channels by landslipped masses produced wide pulsation in discharge: these have been estimated at 50 m$^3$/s/km$^2$. Flood stages varied by as much as 18 m. The rapid erosion and accumulation in some stretches, producing changes in channel levels of up to 10 m, were accompanied by the straightening of channels, undercutting of slopes and the release of large landslides. Up to 25% of the surface was disturbed in this way in the tea plantations compared with only 1–2% in the forests. This is equivalent to a mean lowering of slopes by about 200 mm. Given that the mean denudation rate is 1–2 mm it is clear that extreme events may accelerate erosion 100-fold. The action of mudflows on continuously undermined hillsides leads to parallel retreat rather than to their dissection. Shallow incisions control the removal of material from slopes (Starkel, 1972b).

Rainfall of continuous character is also well known from the Assam Upland (Khasi-Jaintia Hills) where 2000–5000 mm may fall in a single month. This tectonically uplifted block has been dissected and consists of an upland surface with valleys up to 1000 m deep. Deforestation and the persistence of a robber economy in agriculture (potato culture on slopes up to 30°) have combined with the generally steep slopes to produce a rocky region devoid of soils (Starkel, 1972b). On the plateau slopewash and gullying prevail and only on thick waste-mantles or clay-shales are mass-movements notable. On precipitous slopes produced by gully incision rock-falls occur. The progress of linear erosion is inhibited by the fact that the plateau, on which chemical weathering dominates, is not providing any notable bed load. In fact, the rocky parts of the upland area are an example of an area in which extreme processes are relatively inefficient as the former dynamic equilibrium of the slopes has been destroyed.

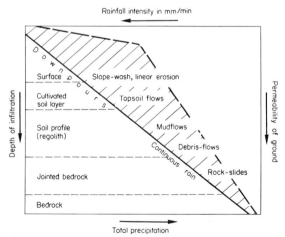

Figure 7.13. Model of relationships between rainfall parameters, permeability of weathered rock horizons and type of slope processes during extreme events.

Continuous rainfall events may be more important than short-lived, severe downpours in the modelling of mountain slopes. This may well be due to the fact that, after the soaking of the waste-mantle by continuous rains, an increase in rainfall intensity of up to 1 mm/min commonly occurs. In comparing the effects of downpours and continuous rains it is evident that, depending on the ratio of the duration and amount of precipitation to its intensity, a different role is played by particular boundary surfaces, these being determined by substratum permeability and various slope processes (Figure 7.13). Slopewash, linear erosion and soil-flows are associated with short-lived heavy falls while mudflows, debris-flows, landslides and sjels in the valleys are associated with continuous rainfall events. Of course, degree of slope and the nature and distribution of surface debris (notably their permeability and liability to liquefaction) are also major factors in determining the kind of displacements which occur on slopes, e.g. debris-flows on talus-cones.

*7.2.3 Rapid warming and snowmelt*

The processes described above are not brought about only by extreme precipitation amounts. An alternative source of abundant water is that derived from the melting of snow or ice which may often be enhanced by rainfall and the thawing of frozen ground. Such conditions may well occur during the following situations: (i) spring snowmelt and midwinter thaws in temperate, boreal and arctic climates; (ii) intense snowfalls in late spring, melting rapidly during periods of rising temperature and associated rainfalls (mountains of the temperate zone); (iii) sudden warming mainly in summer in high-mountain belts above the snowline (frequent in the mountains of Central Asia).

Spring snowmelt or midsummer thaws are efficient when the ground freezes and snow cover is thick. The surficial runoff may reach 95% in such cases (Grin, 1970; Skorodumov, 1973). This process is common in continental climates but is unknown in those of oceanic type. In the transitional zone where deeper freezing occurs once in a few years, runoff on less resistant debris mantles may achieve extreme rates in comparison with mean annual values. Zięmnicki and Orlik (1971) report that in five of six winters investigated (1961–1966), runoff from a loess-covered catchment with an area of 6·2 km², ranged from 0·1 mm to 9·4 mm and slopewash from 0·01 tonnes/km² to 1·90 tonnes/km². In the winter of 1963–64, runoff amounted to 73·8 mm and soil-wash to 67·22 tonnes/km² (Figure 7.14). In comparison with the effects of downpours considered above, these are inconsiderable values but the changes on slopes are, in fact, much greater than the values suggest. Raniger's studies (personal communication) have shown that on slopes of 20° snowmelt accompanied by rainfall results in water amounts some 150% greater than the capillary volume of the arable layer. Soil-flows, 10–20 cm thick, develop in the form of typical mudflows. Similar events were observed by the author in the spring thaw of March 1956 on the loess upland near Cracow (Figures 7.15 and 7.16). The influx of warm air and rainfall produced rill-wash, slush-avalanches and soil-

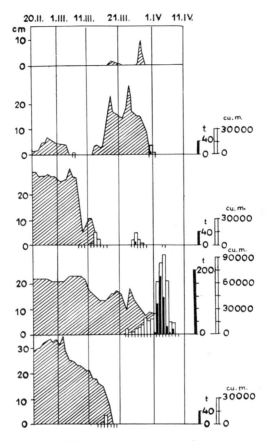

Figure 7.14. Types of snow-melt seasons in eastern Poland from 1960 to 1965 (after Zięmnicki and Orlik (1971)). On the left, thickness of snow cover is shown in centimetres and, on the right, the amount of soil washed down is shown (in black) in tonnes and the amount of runoff in cubic metres (from a catchment area of 6·22 km$^2$).

flows and, in the bottoms of small valleys, so-called slush-flows which are included by Rapp (1960) and Washburn (1965) with the most rapid processes encountered in the Arctic. The displaced snow lumps containing remarkable amounts of rocky fragments and soil aggregates were deposited on the valley floors. During a rapid snowmelt in the eastern Carpathians in December 1957 a rainfall of 135 mm was accompanied by a violent rise of temperature ($c.$ 14 °C): heavy floods and rapid slope modification followed (Figula, 1960; Joganson, 1962).

In the higher mountains, temperature oscillations cause an increase in the density of snow and the frequency of avalanche processes: these are 'avalanche

Figure 7.15. Rapid snowmelt with rainfall and accumulation of washed soil on the snow. Loess upland near Cracow, 2 March 1956.

Figure 7.16. Valley floor just after the passage of a slush-flood. Near Cracow, 2 March 1956.

years'. There is considerable local variation, however. For example, in the Caucasus, chutes modelled by avalanches up to 5 times a year are found adjacent to those known to have been active only once in 16 years (Akifieva, 1970). At the scale of a single gully, the infrequent avalanches might be regarded as an extreme process.

Rain falling at the same time as the rapid melting of late spring snow produces rapid saturation and gravitational processes like those associated with continuous rains. Such events were recorded in 1957 in the French Alps and in 1970 in the Roumanian Carpathians.

Catastrophically heavy rain (202·4 mm) following a snowfall in the Guil valley on 13 July 1957 (Tricart and coworkers, 1962) fell on the snowless southern slopes below 2300 m above sea level and at 2000 m on shaded slopes. Higher up, the snow melted rapidly and above 2600 m above sea level it moved in avalanches. According to Tricart the descent of snow-debris avalanches was preceded by rapid runoff and gravitational processes at lower altitudes. Some of the avalanches turned into mudflows at lower levels. Numerous incisions appeared on the slopes and the overloaded rivers undercut their sides. Some 40 mm of water per day was derived from the melting snow (Poggi, 1959), quite apart from the simultaneous rainfall. The 'independence' of slope and river-channel processes observed by Tricart is related to the width and maturity of Alpine valley floors. Scree volumes calculated suggest that this was an exceptional event on the Flandrian time-scale. However, while the events of 1957 have accelerated slope processes, it seems likely that events of similar intensity have been relatively numerous in the Alps. This is well evidenced in Tricart and coworkers' Figure 6 (1962, p. 289) showing the dissection of a slope by ravines after the 1948 rainfall. In the works of Pardé (1961) or Jäckli (1957) other events connected with rapid snow melting in the Alps are reported.

Rapid summer warming causing snow and ice melting is also common in the high mountains of Central Asia. Sudden weather changes occur in summer particularly associated with occluded fronts. Gontscharov (1962) has shown that, out of 129 sjels observed in the Pamir Mountains, all were due to the advection of cool air-masses in summer. Dujsenov (1971) reports that in the Tianshan Mountains, at a height of 3000 m above sea level, temperatures of 22 °C occurred, followed by rapid melting of snow and ice, the descent of avalanches, the saturation with water of glacial, glacifluvial and solifluction debris covers as well as the formation of structural sjels. Gordon and Trestman (1962) describe a sjel that formed on 9 May 1952 in the western Tianshan Mountains. A warm pulse, giving above-freezing temperatures up to a height of 4500 m above sea level was accompanied by precipitation totalling 72 mm: num-numerous sjels developed and stream discharges exceeded the expected thousand-year maximum. Gagoshidze (1970) states that sjels are sometimes observed in the higher parts of the mountains during sunny warm weather without precipitation. Similar events have also occurred in the Rocky Mountains. On Mt. Rainier, Crandell (1971) observed the occurrence of 'lahars' (debris/mudflows of volcanic deposits) after rainfall totalling 148 mm which melted a great deal of ice.

The melting processes in high-mountain regions (avalanches) differ in their mechanism from slopewash and gravitation due to downpours and continuous rains. Frozen ground often acts as an impervious layer. In the arctic and in temperate-continental climates they are normal processes but in regions of rare ground freezing, such as Poland, they may be an extreme phenomenon. In the middle latitude mountains of Europe and Asia, periods of rapid warming which disturb the slopes involve high precipitation intensities. However, while such occurrences are only one of several extreme events in oceanic climates, in extreme continental climates they are, aside from earthquakes, the unique extreme event.

A comparison of processes acting during downpours as well as those typical of continuous rains and rapid snowmelts makes it possible to establish similarities in their morphogenetic effects. The mutual relationship between runoff and infiltration dictates the thickness of the layer involved in the disturbance (Figure 7.1). Permeability of the weathered mantle and surface slope angle are also important. Accordingly, excess water leads to washing, linear erosion and to valley-broadening on gently-inclined surfaces while on steep slopes instability is determined by the nature of the mantle, itself determined by the balance between chemical and physical weathering.

### 7.3 Comparison of the role of extreme processes in various zones

It is evident from the foregoing review that events of various kinds may act in an extreme fashion in one region or another (Figure 7.17). Diagrams illustrating climate–process systems (Peltier, 1950; Leopold and coworkers, 1964; Wilson,

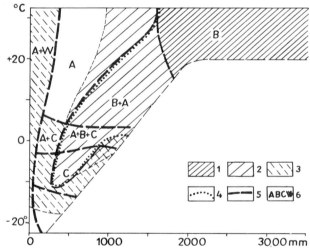

Figure 7.17. Differentiation of extreme events in morphoclimatic regions, with the climate–process systems as the background. 1—very intensive chemical weathering; 2—intensive chemical weathering; 3—intensive physical weathering; 4—forest limit; 5—limits of areas fashioned by different extreme events; 6—major extreme events (A—downpours; B—continuous rains; C—rapid snowmelts; W—strong winds).

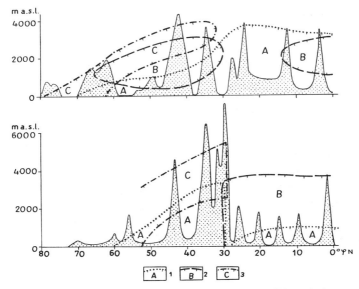

Figure 7.18. Probable latitudinal and vertical differentiation of leading extreme events. Upper profile is drawn across Europe and Africa: lower profile extends across Siberia and southeast Asia. A—short downpours; B—continuous rains; C—rapid snowmelts.

1968 and 1973) reveal that short-lived downpours play an essential role in both arid and semi-arid climates. In tropical and temperate humid climates even in boreal humid ones continuous rainfalls may be more important. In cool climates an essential role is played by rapid oscillations of temperature. On the sketch sections (Figure 7.18) it can be noticed that the parallel mountain barriers of Asia constitute a distinct boundary for various extreme phenomena which also serves to emphasize their vertical zonation. In the section through Europe and Africa the arid zone does not display vertical differentiation, while in the temperate zone extreme phenomena become superimposed, at the same time retaining their altitudinal zonation.

Deforestation caused by man has not only increased the amount and intensity of runoff during extreme events but has also increased the significance of annual downpours in relation to the mean denudation rate. In temperate climates, cultivation has brought about a stepwise increase in denudation from 0·0001 mm/year to 1 mm/year and an increase in the relative importance of extreme phenomena. In addition to the processes of soil-flow (which, in natural conditions, occurs only after once-in-a-century downpours, cf. Hack and Goodlett (1960)), surface wash, gully erosion and shallow gravitation processes are of major importance. In tropical monsoon and mediterranean climates, as much as 25 % of slope surfaces may be in a mobile state at any one time. Sometimes, soil formation is retarded to such an extent that deserts are created. The simultaneity of slope processes and of floods causes rejuvenation of the whole slope landscape.

On uplifted piedmonts erosion of loose and unconsolidated rock masses proceeds apace. In the Apennines, for example, the overall denudation rate, including extreme events, amounts to 8–10 mm/year (Vittorini, 1965).

The role of the wind is also increased by human disturbance. The greatest intensity of aeolian processes is associated with steep pressure-gradients on the margins of anticyclones (Nalivkin, 1969; Fedorovitch, 1970). They are exacerbated by frost processes in winter and the destruction of vegetation in semideserts, steppes and in forests. Losses by blow-out may locally exceed 100 mm/year even in Poland (Gerlach and Koszarski, 1968) and accumulation on slopes in the Pamir Mountains may reach 25 mm. Violent storms of long duration, with wind velocities reaching 55m/s, such as those recorded in central Asia, may cause great local differences in the volume of wind-borne material. For example, the dust-storm of 26–28 April 1928 in southeast Europe raised about 15 million tons from a surface area of about a million square kilometres (Doskath and Gajel, 1970).

Animal husbandry can be a major cause of destruction of the vegetation cover, effectively increasing slope-wash (Krishnan and coworkers, 1963).

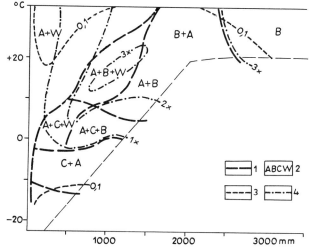

Figure 7.19. Changes in the dominant extreme events in morphoclimatic regions following deforestation (cf. Figure 7.17). 1—limits of areas fashioned by different extreme events; 2—leading extreme events (A—downpours; B—continuous rains; C—rapid snowmelts; W—strong winds); 3—isoline of denudation rate 0·1 mm/year in quasi-natural conditions (after Carson and Kirkby, 1972); 4—isolines of the probable rate of denudation in deforested areas during catastrophic events (1×, 2×, 3×—relative scale of intensity).

Table 7.2

| Locality area | New forms | Frequency of extreme events in 100 years | Denudation during one extreme event | Total of denudation during extreme | Total of mean denudation in 100 years | Relation of mean to extremes in 100 years |
|---|---|---|---|---|---|---|
| Lublin Upland, E. Poland | + | 2 | 1·4 | 2·8 | 40 | 0·07 |
| Szymbark, S-Poland | − | 100 | 2·0 | 200 | 300 | 0·66 |
| W-Carpathians, S-Poland | + | 10 | 30 | 300 | 70 | 4·30 |
| Suchumi, W-Caucasus | − | 33 | 7·0 | 231 | 400 | 0·58 |
| Kishtsai, E-Caucasus | + | 2 | 86 | 172 | 160 | 1·07 |
| Mgeta, Tanzania | + | 10 | 14 | 140 | 26 | 5·38 |
| Darjeeling, India | + | 5 | 100 | 500 | 150 | 3·33 |

Continuous rainfall of only moderate intensity may be effective in semi-arid farming terrains (Inbar, 1972).

The interplay of downpours and other extreme events such as snowmelt and strong winds complicates the picture of the causes of slope disequilibrium in the various morphoclimatic regions (Figure 7.19).

In order to compare the role of extreme events in particular zones precise data are necessary. Unfortunately, observations of extreme phenomena are usually limited to mountain areas. Besides, there is a general dearth of calculations of horizontal and vertical transport of material such as those made by Jäckli (1957) and Rapp (1960). Nevertheless, the author has collected some approximate data which are set out in Figure 7.20 and Table 7.2. The data suggest that, in areas with a mean annual denudation rate of the order from a few thousandths of a millimetre to a few millimetres (such as cultivated areas on relatively weak materials in mountains with a monsoon climate), extreme phenomena display a denudation index ranging from a few centimetres to 200 mm. They exceed by more than 100-fold the mean denudation rate: in loess regions they are the sole determinant of the mean rate. The varying frequency of extreme events with climate has already been mentioned. For example, in humid monsoon climates they may occur a few times each century while in boreal or arid climates events of equivalent magnitude may not occur at all or, at best, only once every few hundred years. A comparison of the denudational efficiency of the number of extreme events per century with the average annual denudation rates (Table 7.2) shows that continuous rains are the major component (moun-

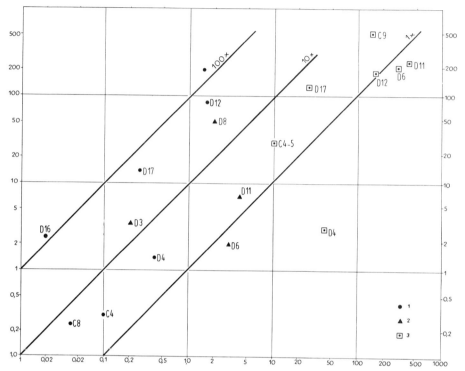

Figure 7.20. Comparison of the normal denudation rate and denudation after extreme events. 1—data representative of large areas or whole catchments; 2—data from plots under cultivation; 3—comparison of normal denudation rate with effects of extreme events in one century. Numbers refer to stations in Table 7.1.

tains in tropical, mediterranean and temperate oceanic climates) while only the extreme downpours of the arid zone appear to equal in denudation effectiveness the more frequent downpours of other regions. Such downpours are rare but they control the morphological evolution (cf. Wolman and Miller (1960)). A comparison of the effects of extreme phenomena and mean denudation rates suggests that the tropical monsoon climate is exceptional (Figure 7.20).

If all the available data is considered, several regions may be distinguished on the basis of magnitude and frequency of extreme events, as follows (Table 7.2).

(a) Regions with a frequency of extreme events of 5–10/century, an efficiency greatly exceeding mean 100-year denudation totals and which are characterized by large-scale landscape modifications (tropical monsoon and mediterranean climates, farming lands of the temperate zone).

(b) Regions in which extreme events are rare: when they occur they do not exceed the mean 100-year denudation total. In such regions the 'normal' processes include the effects of rapid downpours or snowmelts (semi-arid areas).

(c) Regions of very rare extreme events that considerably surpass the 100-year denudation totals due to low mean precipitation (extremely dry areas and some mountains with boreal climates).
(d) 'Stable' regions which show little deviation in denudation rate in particular years. Extreme events which do occur merely cause variations in the mean annual values (arctic, boreal–continental climates and lowlands of the temperate zone).

### 7.4 The role of extremes in the maturation or rejuvenation of slopes

The understanding of accordances or discordances of 'extreme' processes tending to produce slope maturation (by down-wearing, often together with parallel retreat) is possible if tectonic movements and base level oscillations are analysed at the same time (Penck, 1924; Mieshtcheriakov, 1970) and given slope analysis and classification following their genesis and modification (Büdel, 1969; Woskriesienski, 1971).

Slopes in some regions may have a locally falling base-level due to the undercutting or dissection of rejuvenation. This may arise because of uplift in the upper part of a catchment because of changes in discharge and transportation brought about by climatic or anthropogenic factors. In either case, the effect is the same: lowering of the channel, undercutting of slopes by floods or sjels, disturbance of stability, formation of flows, slides or gullies. Depending on the balance between debris transport and slope degradation, concave or straight slope profiles may result. On weak rocks denudation dominates and concave slopes develop, as on shales in the Siwaliks, Roumanian Sub-Carpathians or the Apennines, or a network of badlands may develop as in the Mediterranean region.

Regions with a stable base-level are usually tectonically stable. They have broad-floored valleys and slope processes tend to smooth and elongate the slope partly by accumulation at the base. Processes of extreme type may carry away considerable amounts of material from slope-foot accumulations cutting the scree slopes with debris-flow channels and the colluvial channels with deep gullies. It is evident, therefore, that changes in base-level have a fundamental influence on whether extreme processes accelerate the maturation of slopes or bring about their rejuvenation.

The processes which originally produced the slopes may often differ diametrically from the processes which are currently modifying them. Tectonic escarpments, erosional valley sides and the walls of glacial troughs, for example, may be in process of modification within a single climatic region and by a similar set of processes. Depending on the age of the initial forms, they may preserve many features which do not accord with the current slope 'equilibrium': extreme processes tend to modify such features differentially.

*7.4.1 Accordance of extreme processes with present-day trends in slope evolution*

Currently active slope processes tend toward the degradation of the upper

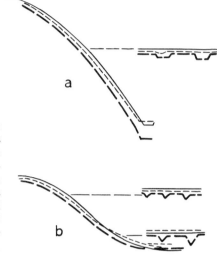

Figure 7.21. Tendency to change in selected slopes. a—mountain areas of uplift, characterized by rejuvenation and slope retreat; b—deforested upland areas of the temperate zone with slope degradation and dissection, plus segments of normal deposition. Dotted line shows the tendency due to normal processes and the discontinuous line the dendency due to extreme processes.

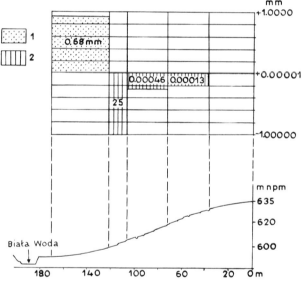

Figure 7.22. Continuation of the evolution of convex–concave periglacial slopes in cultivated areas by 'normal' slope-wash under the present climate (after Gerlach (1966)). Data for winter–spring season 1955–56. 1—accumulation rate in millimetres; 2—denudation rate.

parts of slopes and to accumulation at their bases. Thus, the convexo–concave slope model (Baulig, 1940), partly inherited from Pleistocene periglacial morphogenesis, has been progressively modified under contemporary conditions in agricultural regions of temperate Europe (Gerlach, 1966; Jahn, 1968; Figures 7.21 and 7.22). Catastrophic surface runoff leads to the extension of erosional surfaces to the lower parts of slopes (Figure 7.8; Richter and co-workers, 1965). Because of the lack of vast basins in the temperate zone the colluvial cover accumulates in valley floors which act as local slope base-levels. This, in turn, means that aggradation proceeds upslope. Similarly, in tropical climates, the removal of soil from deforested footslopes exposes solid rock. At other gradients and in other lithologies the gravity-induced movements in water-soaked soils produce similar effects. Avalanches cause dissection and retreat of slopes at Kärkevagge, Sweden, and landslides render the originally steep slopes more gentle, e.g. the Apennines or the Roumanian Sub-Carpathians.

If chemical weathering keeps pace with debris removal in uplifted areas, constant slope retreat with straight profiles and uniform gradients is maintained near the limiting stability value of the slope materials. This value is exceeded in occasional years only, so that a new waste-mantle is produced periodically. This can be spoken of as metastable dynamic equilibrium (Chorley and Kennedy, 1971). When the weathering rate is retarded, mudflow scars persist. In climates with high frequency rainstorms, rock surfaces are swept bare and the prevailing dynamic equilibrium climax ceases to exist as in parts of the mountains of Assam.

*7.4.2 Discordance of extreme processes and present-day trends in slope evolution*

Extreme processes of relatively low frequency disturb the prevailing dynamic equilibrium and result in the development of new forms within the slope or even in changes in its evolution. Excessive subsurface flow at Nawa has brought about the dissection of slopes while saturation has disturbed accumulations on the lower slopes at Ulvadal. In both cases not only were processes accelerated but exhumation of the subjacent rock slope was initiated. Such phenomena are typical of those climates in which extreme meteorological events are rare. Post-glacial climatic changes may result in an increase in valley density due to

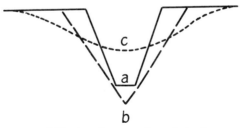

Figure 7.23. Developmental trends in chutes or gullies cutting slopes during heavy rain (a). Dissection can be continued (b) or, in the case of an extremely rare event, can be slowly fossilized (c).

subsurface flow (Starkel, 1960). The destruction of the natural vegetation can often produce similar effects, e.g. dissection of gentle slopes by gullies to produce steep gully-side slopes, as in loess areas.

However, newly-created features may prove very short-lived in the absence of further impulses of the intensity of those which produced them. They may be degraded (like many debris-flow channels above the treeline) or even filled in by wind-borne dust (like some fossil valleys in desert regions). The 'normal' processes bring the slopes back into equilibrium (Figure 7.23). Such forms may also persist untouched in both desert and forested areas if no other process of greater frequency is able to destroy them (Hack and Goodlett, 1960; Lamarche, 1968). In some cases they may involve a negative effect in draining water from the weathered mantle, thus increasing slope stability and weakening the processes of removal or runoff.

## 7.5 Relation of extreme phenomena to climatic changes

Extreme processes of the present day are generally discordant with the trend of slope evolution inherited from past periods (glacial, periglacial, tectonic, due to pediplanation, etc.). While a specific slope may have been in equilibrium during modelling by a glacier or under solifluction, it remains undisturbed by current processes simply because they are not capable of removing the thick debris mantle. It is only when this equilibrium is overcome during rainstorms, as in the mountains of the temperate zone, that gullies, mudflows, debris-flows and deep landslides occur. Moreover, as chemical weathering increases with the time elapsed since formation, these events become more likely as time goes on. Such forms are an element of the post-Pleistocene relief, although they differ from the 'normal' model of chemical weathering and from water circulation in the mantle typical of this zone. Equally, dissection of solifluction mantles in mountain-valleys involves the disturbance of slope-foot equilibrium and the headward retreat of gullies and landslide headwalls (Starkel, 1963). In such a system, the continuous rainfall events and periodic downpours responsible for the formation of these new forms are included in the definition of present-day morphogenesis even if they no longer recur in a given region. In the same way, debris-flows on scree-slopes on cirques and glacial troughs (Rapp, 1960; Curry, 1966) should be regarded as an essential component of contemporary morphogenesis. The exaration-slopes of the Caucasus, produced by extreme events, have undergone the entire evolutionary cycle by way of processes such as rock-fall, avalanche and sjel to produce erosional debris-covered and colluvially-mantled slopes (Shtschebakova, 1971). Any newly-gained equilibrium may be easily disturbed by heavy rains (Hewitt, 1972). In an area of glacial recession in the central Caucasus underlain by shaly-sandstones, the author has observed modification of the steep sides of glacial troughs by modern landslides.

It is evident that any analysis of the processes normal to a given zone of water circulation, weathering and denudation, reveals the effects of extreme events. Accordingly, in interpreting landforms and deposits differing from the average, one should be extremely cautious neither to refer them to long timespells nor to connect them with other climatic conditions simply because they differ from the current morphogenetic model. At the same time, however, one must agree with Rapp (1967) that the slopes of the temperate zone, inherited over wide areas from periglacial morphogenesis, are in most cases stable enough to have avoided, at least before deforestation, disturbance of their equilibrium by extreme rainfall events since the end of the Pleistocene.

## 7.6 Final remarks

Extreme meteorological phenomena, most typically seen in mountain areas, play an essential role in shaping the relief of most morphoclimatic regions. Depending on the particular climate, the most prominent role is played by short-lived downpours, periods of continuous rain, rapid snowmelts or strong winds (Figure 7.17).

Gravity-induced processes or those due to wash are usually associated with the exceeding of critical values of infiltration and water capacity of slope mantles on surfaces varying in structure and gradient: a critical variable in this is the intensity and amount of precipitation (Figure 7.13). Extreme phenomena evoke processes which either exceed the mean denudation rate by at least 100 times or control the annual values of denudation. Their efficiency as expressed in absolute values depends on the 'background', i.e. the intensity of the normal processes.

The processes initiated by extreme meteorological events either accelerate natural slope development, (producing rejuvenation in uplifted areas) or create new forms quite different from the normal ones. The latter activate the processes and change the trend of slope evolution or die out because of the lack of further impulses of similar intensity. They are of essential importance in the transformation of slopes of various origins and those inherited from the past. The destruction of the natural vegetation, together with the various farming practices, has resulted not only in greatly accelerated denudation but also in an increase in the importance of extreme phenomena in relief evolution.

The picture of the influence of extreme events in slope evolution is undoubtedly more complex than that presented in this chapter. This is due in large measure to numerous shortcomings in the available data.

## Acknowledgements

The author wishes to express his sincere gratitude to his colleagues from the Department of Physical Geography, Cracow, namely Dr. A. Kotarba and Dr. J. Słupik. Thanks are also due to Dr. T. Gerlach, E. Gil, R. Soja and to his wife Barbara for valuable discussion and to Dr. A. Rapp for an exchange of views on the role of extreme processes during conference sessions in Montreal and Göttingen. Finally, he wishes to thank the authors of the enclosed figures for their kind permission to reproduce them, Mrs. M. Klimek for drawing the figures, and K. Czekierda for translating the text.

## References

Akifieva, K. W., 1970, Caucasus, in *Avalanche regions of the Soviet Union* (in Russian), Moscow.
Ananthakrishnan, R. and A. Krishnan, 1962, Upper air changes over India and neighbourhood associated with the south-west monsoon, *Current Science*, 31, 133.
Aristarchova, L. B. and B. A. Fedorovitch, 1971, Slope processes in deserts and semideserts (in Russian), *Wopr. geogr.*, Moscow, 85, 25-51.
Baird, P. D. and W. V. Lewis, 1957, The Cairngorm flood 1956, *Scott. Geogr. Mag.*, 73, 91-100.
Baulig, H., 1940, Le profile d'équilibre des versants, *Ann. de Géographie*, 49, 81-97.
Bennett, H. M., 1939, *Soil conservation*, New York.
Berry, L. and B. P. Ruxton, 1960, The evolution of Hong Kong Harbour basin, *Zeitsch. f. Geomorph.*, 4, 97-115.
Biel, E., 1929, Die Veränderlichkeit der Jahressumme der Niederschläge auf der Erde, *Geogr. Jahresber. aus Österriech*, 14/15, 151-180.
Brykowicz, K., A. Rotter and K. Waksmundzki, 1973, Hydrograficzne i morfologiczne skutki katastrofalnego opadu i wezbrania w lipcu 1970 roku w zrodlowej czesci zlewni Wisly, *Folia Geogr., ser. Geogr.-Phys.*, (Cracow), 7, 115-129.
Büdel, J., 1969, Das System der Klima-genetischen Geomorphologie, *Erdkunde*, 23, 165-183.
Cons. Nat. d. Res., Inst. d. Geog. Univ. Padova, 1972, *Carta dell' alluvione del novembre 1966 nel Veneto nel Trentino-Alto Adige, effetti morfologici e allagamenti*, Touring Club Ital., Milano.
Carson, M. A. and M. J. Kirkby, 1972, Hillslope form and process, Cambridge University Press, London.
Common, R., 1966, Slope failure and morphogenetic regions, in *Essays in Geomorphology*, G. H. Dury (Ed.), Heineman, London, 53-81.
Corbel, J., 1964, L'érosion terrestre, étude quantitative (Méthodes-Techniques-Résultats), *Ann. de Géographie*, 78, 386-412.
Chorley, R. J. and B. A. Kennedy, 1971, *Physical geography, a systems approach*, Prentice-Hall, London.
Crandell, D. R., 1971, Postglacial lahars from Mount Rainier Volcano, Washington, *U. S. Geol. Sur. Prof. Pap. 677*.
Curry, R. R., 1966, Observations of alpine mudflow in the Tenmile Range, Central Colorado, *Bull. geol. Soc. Am.*, 77, 771-776.
Darselija, M. K. and S. T. Gvazava, 1970, Soil erosion in subtropic regions of Georgia, *Proc. Int. Water Eros. Symp. Prague*, 3, 111-119.
Doskath, A. G. and A. G. Gajel, 1970, Eolian processes in the steppe zone of the Soviet Union (in Russian), in *Present Exogenic Geomorphic Processes*, Moscow, 138-148.
Drozd, N. J., 1962, Mudflows in gully areas of the Ukraine (in Russian), in *Proc. V Symp. on Sjels*, Baku, 94-98.
Dujsenov, E., 1971, *Sjel streams in the Transilian Alatau Mountains* (in Russian), Alma-Ata.
Duley, F. L., 1939, Surface factors affecting the rate of intake of water by soil, *Soil sci. Soc. Am.*, Proc. 4.
Dumitrashko N. W. and coworkers, 1970, Problem of sjels and their investigation (in Russian), in *Present exogenic geomorphic processes* (Moscow), 131-137.
Dutta, K. K., 1966, Landslips in Darjeeling and neighbouring hillslopes in June 1950, *Bull. Geul. Surv. India*, Ser. B. 15.
Emmett, W. W., 1968, Gully erosion, in *Encyclopedia of Geomorphology*, R. W. Fairlaridge (ed.) Reinhold, New York, 517-159.
Fedorovitch, B. A., 1970, Intensity of present-day eolian processes in the deserts of the Soviet Union (in Russian), in *Present exogenic geomorphic processes* (Moscow), 149-159.

Figula, K., 1960, Erozja w terenach gorskich, *Wiad. IMUZ*, (Warsaw), **1**, 109–147.
Gagoshidze, M. S., 1970, *Sjel phenomena and protection against them* (in Russian), Tbilisi.
Gerlach, T., 1966, Wspolczesny rozwoj stokow w dorzeczu gornego Grajcarka, *Prace Geogr. IG PAN*. Warsaw, **52**, 1–110.
Gerlach, T. and Koszarski, L., 1968, L'activité eolienne en tant qu'un des plus importants processus morphogénetiques actuels dans le climat tempéré humide, *Geogr. Polonica*, **14**, 47–56.
Gierasimov, I. P., 1970, Introduction in *Present exogenic geomorphic processes* (in Russian), Moscow, 7–14.
Gil, E. and J. Slupik, 1972, The influence of the plant cover and land use on the surface run-off and washdown during heavy rain, *Studia Geomorph. Carp.-Balcan.*, **6**, 181–190.
Gontscharov, E. P., 1962, Sjels in Tadshykistan (in Russian), *Proc. Fifth Symp. on Sjels*, Baku, 109–114.
Gordon, S. M. and A. G. Trestman, 1962, Sjel activity in Warzob catchment basin (in Russian), *Proc. of Fifth Symp. on Sjels*, Baku, 119–122.
Grant, P. J., 1966, Variations of rainfall frequency in relation to erosion in Eastern Hawke's Bay, *J. Hydrol. N. Z.*, **5**, 73–86.
Grin, A. M., 1970, Results of stationary studies of run-off and slope wash (in Russian), in *Present exogenic geomorphic processes*, Moscow, 85–95.
Hack, J. T. and J. C. Goodlett, 1960, Geomorphology and forest ecology of a mountain region in the central Appalachians, *U.S. Geol. Surv. Prof. Pap. 347*.
Hempel, L. geb. Tecklenburg, 1957, Das morphologische Landschaftsbild des Unter-Eichsfeldes unter besonderer Berücksichtigung der Bodenerosion und ihrer Kleinformen, *Forsch. z. Deutsch. Landeskunde*, **98**, 1–55.
Hewitt, K., 1972, Mountain environment and geomorphic processes, in *Mountain Geomorphology, University of British Columbia geogr. ser.* **14**, 17–34.
Inbar, M., 1972, A geomorphic analysis of a catastrophic flood in a Mediterranean basaltic watershed, *Univ. Haifa, 22 Intnl. Geogr. Congr. Pub.*
Ivironova, M. I., 1962, Activity of periodic creeks in the high-mountain surrounding of the Issyk-kul Basin (in Russian), *Trudy Inst. Geogr. AN SSSR*, **81**, 30–57.
Jäckli, M., 1957, Gegenwartsgeologie des bunderischen Rheingebietes. Ein Beitrag zur exogenen Dynamik Alpiner Gebirgslandschaften, *Beiträge zur Geol. der Schweiz*, Geotechn. Serie, **36**, 1–126.
Jackson, R. J., 1966, Slips in relation to rainfall and soil characteristics, *J. Hydrol. N. Z.*, **5**, 45–53.
Jahn, A., 1968, Denudational balance of slopes, *Geogr. Polonica*, **13**, 9–29.
Jakubowski, K., 1964, Płytkie osuwiska zwietrzelinowe na Podhalu, *Prace Muz. Ziemi*, Warsaw, **6**, 113–152.
Jennings, A. H., 1950, World's greatest observed rainfalls, *Monthly Weather Rev.*, **78**, 4–5.
Joganson, W. E., 1962, Hydrometeorological conditions of sjel formation in Soviet Union (in Russian), in *Proc. Fifth Symp. on Sjels*, Baku, 21–26.
Kaszowski, L. A. Kotarba, M. Niemirowski and L. Starkel, 1966, Maps of contemporaneous morphogenetic processes in Southern Poland, *Bull. Acad. Pol. Sc., ser. geol.-geogr.*, **14**, 113–118.
Kerkheulidze, J. J., 1969, Estimation of basic characteristics of mud-flows ("sjel"), in *Floods and their computation*, UNESCO, Vol. 2, 940–948.
King, L. C., 1953, Canons of landscape evolution, *Bull. geol. Soc. Am.*, **64**, 721–752.
King, L. C., 1957, The uniformitarian nature of hillslopes, *Trans. Edin. geol. Soc.*, **17**, 81–102.
Klimaszewski, M., 1935, Morfologiczne skutki powodzi w Malopolsce zachodniej w lipcu 1934 r., *Czas. Geogr.*, **13**, 2–4.
Kostienko, N. P., 1960, Slope changes and seismicity with the central Asiatic mountains as an example (in Russian), *Bull. sov. seysmologii*, **8**, 150–156.

Krishnan, A., E. N. Blagoveschensky and P. Rakhecha, 1963, Water balance of *Prosopis spicigera* community, *Indian J. of Meteorol. and Geoph.*, **19**, 181–192.
Krishnan, A., P. K. Raman and A. D. Vernekar, 1959, Probable maximum twenty-four-hour rainfall over India, *Proc. Symposium on Meteorological and Hydrological aspects of floods and droughts in India*, 196–202.
Kirvolutskiy, A. E., 1971, *Life of the earth's surface-geomorphological problems* (in Russian), Moscow.
Leopold, L. B., W. W. Emmett and R. M. Myrick, 1966, Channel and hillslope processes in a semiarid area, New Mexico, *U. S. Geol. Sur. Prof. Paper 352 G*.
Leopold, L. B., M. G. Wolman and J. P. Miller, 1964, *Fluvial processes in Geomorphology*, Freeman, New York.
Lamarche, V. G. 1968, Rates of slope degration as determined from botanical evidence, White Mountains, California, *US Geol. Sur. Prof. Paper 352 I*, 341–377.
Maruszczak, H. and J. Trembaczowski, 1959, Geomorfologiczne skutki gwałtownej ulewy w Piaskach Szlacheckich koło Krasnegostawu, *Ann. UMCS*, B (Lublin), **11**, 129–160.
Marx, S., 1969, Die Unwetter in Norditalien Anfang November 1966 in klimatologischer Sicht und im Vergleich zu mitteleuropäischen Verhältnissen, *Wiss. Z. Pädag. Hochschule Potsdam, Math.–Naturwiss.*, **13**, 1005–1035.
Mensching, H., K. Giessner and G. Stuckmann, 1970, Die Hochwasserkatastrophe in Tunesien im Herbst 1969, *Geogr. Zeitsch.*, **58**, 81–94.
Mieshtcheriakov, J. A., 1970, On the theory of exogenic processes (in Russian) in *Present exogenic geomorphic processes*, Moscow, 15–22.
Mirchulava, C. E., 1970, *Engineering methods of evaluation and prognosis of water erosion* (in Russian), Moscow.
Mortensen, H., 1963, Abtragung und Formung, *Nachr. Ak. Wiss, Göttingen*, **II**, 17–27.
Nalivkin, D. W., 1969, *Hurricanes, storms and whirlwinds* (in Russian), Leningrad.
Pardé, M., 1961, Sur la puissance des crues en diverses parties du monde, *Géographica*, **8**.
Pellegrini, G. B., 1969, Osservazioni geografiche sull' alluvione del November 1966 nella valle del Torrente Mis, *Atti e Mem. Ac. Patavina*, **81**, 277–318.
Peltier, L. C., 1950, The geographic cycle in periglacial regions as it is related to climatic geomorphology, *Ann. Assoc. Am. Geogr.*, **40**, 214–246.
Penck, W., 1924, *Die morphologische Analyse*, Stuttgart.
Ploey, J. de, 1969, L'érosion pluviale: expériences à l'aide de sables traceurs et bilans morphogénétiques, *Acta geogr. Louvain*, **7**, 1–28.
Ploey, J. de, 1972, A quantitative comparison between rainfall erosion capacity in a tropical and a middle-latitude region, *Geogr. Polonica*, **23**, 141–150.
Poggi, A., 1959, La fusion de la neige et les crues de juin 1957 dans les Alpes françaises orientales, *Rev. de Géogr. Alpine*, **47**, 363–373.
Rahn, P. R., 1967, Sheetfloods, stream floods and the formation of pediments, *Bull. geol. Soc. Am.*, **78**, 593–605.
Rajonirovanije SSSR, 1965, (collective work), Regionalisation of USSR territory according to the main erosional factors (in Russian), *Inst. of Geogr.*, Moscow.
Rapp, A., 1960, Recent development of mountain slopes in Kärkevagge and surroundings, northern Scandinavia, *Geogr. Annlr*, **42**, 1–200.
Rapp, A., 1963, The debris slides at Ulvadal, western Norway, *Nachr. Ak. Wiss, Göttingen*, **II**, 13, 197–210.
Rapp, A., 1967, Pleistocene activity and Holocene stability of hillslopes with examples from Scandinavia and Pennsylvania, *Symp. Intnl, de Geomorph. Dynam. Liege-Louvain*, **1**, 229–244.
Reich, B. M., 1963, Short-duration rainfall intensity estimates and other design aids for regions of sparse data, *J. Hydr.*, **1**, 3–28.
Reniger, A., 1955, Spływ gleb na uprawianych zboczach, *Rocz. Nauk Roln.*, **71**, serie F-1, 21–44.

Richter, G. and cowworkers, 1965, Bodenerosioschaden und gefährdete Gebiete in der Bundesrepublik Deutschland, *Forsch. z. D. Landeskunde*, **152**, 1–592.

Rustamov, S. G., 1962, Sjel phenomena in Aserbaijan and their hydrological characteristics (in Russian), *Proc. Fifth Symp. on Sjels*, Baku, 6–12.

Sawicki, L., 1917, Die Szymbarker Erdrutschung und andere westgalizische Rutschungen des Jahres 1913, *Rozpr. Wydz. Mat-Przyr. PAU*, ser. A, Krakow, 56.

Schiff, L., 1951, Surface detention, rate of runoff, land use and erosion relationships on small watersheds, *Trans. Am. Geoph. Union*, **32**, 57–65.

Schumm, S. A., 1956, The role of creep and rainwash in the retreat of badland slopes, *Am. J. Sci.*, **254**, 693–706.

Shtscherbakova, E. M., 1971, Slope processes and types of slopes in the alpine zone of the Great Caucasus in the light of palaeogeography (in Russian), *Woprosy geogr. Moskva*, **85**, 152–163.

Skorodumov, A. S., 1973, *Soil erosion and agricultural productivity* (in Russian), Kiev.

Skrodzki, M., 1972, Present-day water and wind erosion of soils in N. E. Poland, *Geogr. Polonica*, **23**, 77–91.

Słupik, J., 1973, Zroznicowanie splywu powierzchniowego na fliszowych stokach gorskich, *Dok. Geogr. IG PAN*, Warsaw, **2**, 118.

Słupik, J. and E. Gil, (1974 *in press*), The influence of intensity and duration of rain on water circulation and the rate of slope-wash in the Flysch Carpathians, *Nachr. Ak. Wiss. Göttingen*.

So, Chak Lam, 1968, Mass movements associated with the rainstorm of June 1966 in Hong Kong, *Abstracts of Papers Intnl. Geogr. Congr.*, New Delhi, 21.

Solonienko, W. P., 1962, Sjel streams in mountain taiga conditions (in Russian), in *Proc. Fifth Symp. on Sjels*, Baku, 103–109.

Starkel, L., 1960, Rozwoj rzezby Karpat fliszowych w holocenie, *Prace Geogr. IG PAN*, Warsaw, **22**, 1–239.

Starkel, L., 1963, Stand der Forchungen über morphogenetische Prozesse in den Karpathen während des Quartärs, *Nachr. Ak Wiss. Göttingen*, 139–16 ;.

Starkel, L., 1972a, The role of catastrophic rainfall in the shaping of the relief of the Lower Himalaya (Darjeeling Hills), *Geogr. Polonica*, **21**, 103–147.

Starkel, L., 1972b, The modelling of monsoon areas of India as related to catastrophic rainfall, *Geogr. Polonica*, **23**, 151–173.

Stecki, K., 1935, Zerwy ziemne w Beskidzie Zachodnim po ulewach w lipcu 1934 roku, *Kosmos*, Lwow, ser. A, **59**, 391–396.

Temple, P. H., 1972, Measurements of runoff and soil erosion at an erosion plot scale, with particular reference to Tanzania, *Geogr. Annlr*, **54A**, 203–220.

Temple, P. H. and A. Rapp, 1972, Landslides in the Megeta area, Western Uluguru Mountains, Tanzania, *Geogr. Annlr*, **54A**, 157–193.

Tricart., J. and coworkers, 1962. Mécanismes normaux et phénomènes catastrophiques dans l'evolution des versants du bassin du Guil (Hautes Alpes, France), *Zeitsch f. Geomorph.*, **5**, 277–301.

Tricart, J. and A. Cailleux, 1972, *Introduction to climatic geomorphology* (translated de Jong), Longman, London.

Troll, C., 1962, Die dreidimensionale Landschaftsgliederung der Erde, *Wissman-Festschrift*, Tübingen.

Tshelpanova, O. M., 1963, Central Asia, in *Climate of the USSR*, Vol. 3, Leningrad.

Verstappen, H. Th., 1955, Geomorphologische Notizen aus Indonesien, *Erdkunde*, **9**, 134–144.

Vittorini, S., 1965, La valutazione quantitativa dell'erosione nei suoli argilosi pliocenici della Val d'Era, *Atti XIX Congr. Geogr. Ital.*, 83–101.

Washburn, A. L., 1965, Geomorphic and vegetational studies in the Mesters Vig District, NE Greenland, *Meddelelser om Gronland*, **166**, 1.

Weischet, W., 1965, Der tropisch-konvektive und der aussertropisch-advektive Typ des vertikalen Niederschlagsverteilung, *Erdkunde*, **19**, 6–13.

Weischet, W., 1969, Klimatologische Regeln zur Vertikalverteilung der Niederschläge in Tropengebirgen, *Die Erde*, **100**, 287–306.
Wilson, L., 1968, Morphogenetic classification, in *Encyclopedia of Geomorphology*, R. W. Fairbridge (Ed.), Reinhold, New York, 717–728.
Wilson, L., 1973, Relationships between geomorphic processes and modern climates as a method in paleoclimatology, *Climatic Geomorphology*, E. Derbyshire (Ed.), Macmillan, London. 269–284.
Wischmeier, W. H. and D. D. Smith, 1958, Rainfall energy and its relationship to soil loss, *Trans. Am. Geoph. Union*, **39**, 285–291.
Wolman, M. G. and J. P. Miller, 1960, Magnitude and frequency of forces in geomorphic processes, *J. Geol.*, **68**, 54–74.
Woskriesienski, S. S., 1971, Typical profiles of slopes (in Russian), *Woprosy geogr.*, Moscow, **85**, 10–24.
Zanina, A. A., 1961, Caucasus, in *Climate of the USSR*, **2**, Leningrad.
Ziemnicki, S. and T. Orlik, 1971, Charakterystyka okresowych spływow z falistej zlewni lessowej, *Zesz. Probl. Post. Nauk Roln.*, **119**, 7–22.
Ziętara, T., 1968, Rola gwaltownych ulew i powodzi w modelowaniu rzezby Beskidow, *Prace Geogr. IG PAN*, Warsaw, **60**, 1–116.

CHAPTER EIGHT

# Hydrological Slope Models: The Influence of Climate

MICHAEL J. KIRKBY

## 8.1 Introduction

Mathematical models of slope development on landslide-stable hillsides have typically specified rates of sediment removal in terms of topographic variables, principally slope gradient and distance from divide, which are taken as given (Souchez, 1961: Culling, 1963; Young, 1963; Ahnert, 1967). The models used have corresponded to the stippled boxes in Figure 8.1. In attempting here to model the influence of *climate* explicitly, a further crucial component has been added. This component is a sub-model of the hillslope hydrology, the

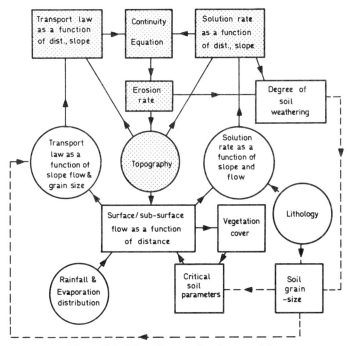

Figure 8.1. Flow diagram for slope models.

importance of which stems from the major role played by water in both wash and solutional slope processes.

Such an additional component is timely because recent work on hillslope hydrology (e.g. Betson, 1964; Dunne and Black 1970; Freeze, 1972; Weyman, 1974) has reached a stage where it is possible to propose alternative models of both water flow and erosion to those in Horton's (1945) classic work.

In this paper the basis of hydrological modelling is the concept of soil water storage capacity, $r_c$. It thus represents one extreme of a range of possible models on a scale between pure-storage and pure-infiltration models. Because daily rainfall data are readily available, the storage capacity is expressed on a daily basis. In the model daily rainfalls in excess of the available storage capacity are assumed to run off as overland flow and potential evaporation is satisfied as far as supply permits. The remaining flow is considered to move downslope beneath the surface, providing an average subsurface flow depth which partly fills the available soil water storage as well as providing a background rate of evaporation between rains. This allows a separation into an average low flow and high flows from individual rain events which are themselves modified by the background low flow. Although this model supresses the detail of individual storm hydrographs and, in its simple form, ignores storm autocorrelation and seasonal effects, it is thought appropriate for budgeting the total volumes of overland flow, subsurface flow and actual evapotranspiration. It has the additional advantage of responding sensitively to the slope profile topography, as can be seen more clearly below.

These predictions of the slope flow components enable estimates to be made of the sediment contributions. Surface wash is empirically related to overland flow volumes (Carson and Kirkby, 1972, Chapter 8) and solution rates to sub surface flow (Carson and Kirkby, 1972, Chapter 9). Creep rates are thought to vary relatively little as climate varies. One remaining link in the model is completed by estimating vegetation cover as the ratio of actual to potential evapotranspiration and relating soil storage capacity to vegetation cover, ranging from 10 mm/day on bare ground to 100 mm/day in densely vegetated areas (Carson and Kirkby, 1972, p. 216). This linkage assumes that vegetation is controlled by hydrology. Where this is not the case, either through human interference connected with cultivation or pollution, or because vegetation cover is limited by the length of the growing season in cold areas, then the soil storage capacity becomes an independently determined factor.

In principle, the model is now as shown by the solid line boxes in Figure 8.1. Lithology remains a major factor which is not incorporated into the model presented here, although its importance is recognized. More weathered material has finer grain sizes, allowing readier transport of material by wash processes. This effect has *not* been included in the present model, although allowance *is* made for chemical composition in determining solution rates.

This framework, in which annual hydrological budgeting forms the basis for sediment estimates, has been used as a model for hillslope development through time. A mass balance equation (Ahnert, 1967) has been used to convert sediment transport rates into rates of surface lowering in both a computer simulation model and an analytical approximation which are described more fully below. A similar mass balance is made for the soil.

$$\frac{\partial y}{\partial t} = -\nabla \cdot (\mathbf{\Delta} + \mathbf{S}) \tag{8.1}$$

$$\frac{\partial w}{\partial t} = \nabla \cdot \left(\mathbf{\Delta} - \mathbf{S}\frac{1-p_s}{p_s}\right) \tag{8.2}$$

where $\mathbf{\Delta}$ = rate of chemical transport,
$\mathbf{S}$ = rate of mechanical transport,
$p_s$ = proportion of original rock substance remaining in near-surface soil layers,
$y$ = elevation of soil surface (rock substance equivalent),
$w$ = total deficit of rock substance in soil,
$t$ = time elapsed.

## 8.2 The hydrological budgeting model

Each day's rainfall, $r$, is assumed to fall on to a soil which has total capacity $r_c$, of which an average depth $h$ is already occupied by subsurface water (Figure 8.2). The following cases may be distinguished:

(1) If $r \geqslant r_c - h$ then
  (a) the excess, $(r + h - r_c)$, will flow off rapidly overland,
  (b) an amount $e_0$ (assumed $< r_c$ always) will be released as evapotranspiration at its potential rate,
  (c) the balance of $(r_c - e_0)$ will contribute to subsurface flow.
(2) If $r_c - h > r \geqslant e_0 - h$ then
  (a) there will be no overland flow,

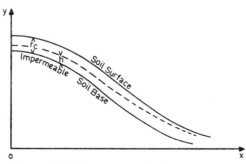

Figure 8.2. Notation used for hydrological sub-model.

(b) an amount $e_0$ will evapotranspire,
(c) $r - e_0$ will contribute to subsurface flow (this may be positive or negative).
(3) If $e_0 - h > r \geq 0$ then
 (a) there will be no overland flow,
 (b) $(h + r)$ will evapotranspire,
 (c) there will be a contribution of $-h$ to subsurface flow.

These contributions may be summed over the year if we assume a distribution of rainfall frequencies. It will be assumed here that the frequency of a rainfall in excess of $r$ is

$$N(r) = N e^{-r/r_0} \tag{8.3}$$

where $N$ = number of raindays per year,
 $r_0$ = mean rain per rainday ($= R/N$),
 $R$ = total annual rainfall.

This distribution provides a good fit for non-seasonal rainfall regimes. Where there is marked seasonality, a series of such terms provides good seasonal fits.

Summing over the appropriate ranges gives:

Total actual evaporation

$$\begin{aligned} E_A &= 365h + R(1 - e^{-(e_0 - h)/r_0}) \text{ if } h \leq e_0 \\ &= E_P \hspace{4.5cm} \text{if } h \geq e_0 \end{aligned} \tag{8.4}$$

where $e_0$ = mean daily potential evapotranspiration,
where $E_P$ = mean annual potential evapotranspiration ($= 365 e_0$).

Overland flow

$$OF = R e^{-(r_c - h)/r_0}$$

where $r_c$ = soil water storage capacity.
Subsurface flow

$$TF = R - OF - E_A \tag{8.6}$$

These equations are linked via the depth of flow and via the vegetation (if hydrologically controlled). Summing the subsurface flow downslope from the divide over the slope drainage area, $a$, gives:

$$\sum_a TF = 365 h K s \tag{8.7}$$

where $K$ is the saturated lateral permeability of the soil in m/day and $s$ is the local ground slope.

Equation 8.7 can be taken as an equation for the depth of flow, $h$, allowing it to be estimated and substituted for in the flow components of Equations 8.4 to 8.6. In the hydrologically simplest case of a parabolic convex slope without contour convergence, Equation 8.7 becomes:

$$TF = 365h\lambda, \text{ where } \lambda = \frac{Ks}{x} \tag{8.8}$$

since on such a slope the gradient is directly proportional to distance from the divide, leading to a constant flow depth, $h$.
Thus:

$$\left.\begin{array}{l}R - Re^{-(r_c-h)/r_0} - \{365h + R(1 - e^{-\{(e_0-h)/r_0\}})\} = 365\lambda h \\ \qquad\qquad\qquad\qquad\qquad\qquad\qquad\qquad \text{if } h \leqslant e_0 \\ R - Re^{-(r_c-h)/r_0} - E_P = 365\lambda h \text{ if } h \geqslant e_0\end{array}\right\} \tag{8.9}$$

The constant $\lambda$ is of order 1.
The vegetation equation is expressed empirically by the relationship:

$$r_c = 100 (E_A/E_P)^{1/2} \tag{8.10}$$

This equation expresses the calculated values of $r_c$ according to vegetation type (Carson and Kirkby, 1972, Chapter 8), on the assumption that $E_A/E_P$ is a relevant predictor of vegetation cover. This simple relationship tends to suppress any influence of seasonality, which would tend to introduce a lag effect, whereby water flow and erosion respond to a vegetation watered by earlier flows.

The hydrological model proposed requires as inputs the slope profile and the climate (specified by $R$, $r_0$, and $e_0$). It gives as outputs the average flow depth in the soil, the actual evapotranspiration, overland and subsurface flow components, the soil water storage capacity and the vegetation cover. The realism or otherwise of these estimates can best be judged by the examples in Table 8.1, all of which are calculated for a convex slope profile on which the constant $\lambda$ in Equation 8.9 is taken as 1·0.

Table 8.1. Predicted annual contributions to flow for a convex hillslope profile in different climates

| Locality | Inputs | | | | | Outputs | | | |
|---|---|---|---|---|---|---|---|---|---|
| | Annual rainfall (mm) | No. of raindays per year | Potential annual evapotranspiration (mm) | Mean flow depth in soil (mm) | Vegetation cover (%) | $r_c$ (mm) | $E_A$ (mm) | OF (mm) | TF (mm) |
| Lowland Britain | 700 | 140 | 450 | 0·9 | 83 | 91 | 373 | $10^{-5}$ | 337 |
| Highland Britain | 1500 | 260 | 250 | 3·4 | 100 | 100 | 250 | $10^{-4}$ | 1250 |
| British extreme (Styhead?) | 5000 | 300 | 200 | 13·1 | 100 | 100 | 200 | 27 | 4773 |
| CNntral Mexico | 700 | 70 | 2000 | 0·6 | 24 | 49 | 483 | 5 | 212 |
| Arizona | 300 | 20 | 3000 | 0·3 | 4 | 20 | 114 | 83 | 103 |

## 8.3 The sediment transport model

Sediment is assumed to be carried by slow processes after the slope has been stabilized with respect to landslides. The most important processes are considered to be soil-creep, wash and solution.

Soil-creep is replaced by rainsplash and unconcentrated wash in arid areas, but the rates of action appear to be comparable (Carson and Kirkby, 1972; Williams, 1973; Kirkby and Kirkby, 1975). The sediment discharge is assumed to be independent of hillslope flow conditions and to be directly proportional to the slope gradient; at a rate

$$S = 10 \tan\beta \text{ cm}^3/\text{cm year} \tag{8.11}$$

Soil wash is strongly dependent on overland flow and empirical estimates suggest that the rate is approximately:

$$S = 170 \, q^2_{OF} \tan\beta \text{ cm}^3/\text{cm year} \tag{8.12}$$

where $q_{OF}$ is the annual overland flow discharge in $m^2$/year. This expression is compatible with Equation 8.10 in Carson and Kirkby (1972, p. 216) and close to experimental estimates of soil erosion from erosion plots (e.g. Musgrave (1974) if it is assumed that overland flow increases linearly with distance.

Transport in solution is calculated for each constituent oxide of the parent material separately. It is assumed that subsurface water is able to pick up additional material until it reaches equilibrium with the proportions of oxides in the parent material, so that these proportions are appropriate throughout. For each oxide, the model assumes that a quantity

$$\Delta_i = 100(q_{TF} + 10^{-3}E_A) k_i p_i \text{ cm}^3/\text{cm year} \tag{8.13}$$

can be dissolved by rainfall which initially penetrates into the soil over a 1 m length of slope profile (even if later transpired), where $q_{TF}$ is the subsurface flow discharge in $m^2$/year, $k_i$ is the saturated mobility of oxide $i$ in mg/l, $p_i$ is the proportion of oxide $i$ in the parent material and $E_A$ is the actual evapotranspiration in mm/year.

After evapotranspiration, a maximum amount

$$\Delta_i = 100 q_{TF} k_i \text{ cm}^3/\text{cm year} \tag{8.14}$$

can be carried away by subsurface runoff. The effective solute transport for each oxide is obtained as the lesser of expressions (8.13) or (8.14) above, which is then summed over all the rock constituents. This approach is developed from the linear model for solution proposed in Carson and Kirkby (1972, Chapter 9).

To solve Equation 8.2 in terms of soil thickness or accumulated substance deficit, $w$, and to obtain an estimate of the degree of soil development from the variable $p_s$, the proportion of rock substance remaining in near-surface soil, an empirical 'rating function' must be used to relate $w$ and $p_s$. In simulation runs so far, this function has been:

$$p_s = 1 - 2w \text{ or } 0.4, \tag{8.15}$$

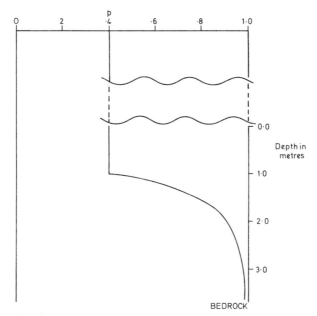

Figure 8.3. Assumed soil profile form, truncated at any level corresponding to the local surface value of $p$.

whichever is the greater. This function corresponds to truncation of the soil profile shown in Figure 8.3 at any point corresponding to the then-surface.

## 8.4 Simulation procedure

The model described above in Equations 8.1 to 8.14 has been programmed on a Wang 600–14 calculator for a 100 m long slope in 5 m units of length, with allowance for convergence or divergence of flow lines to make provision for contour curvature. The hydrological flow components are solved iteratively at successive points downslope and flows are accumulated. The accumulated flows are then used to estimate sediment transport and the mass balance equation is used in its first-difference form to obtain mean rates of slope lowering for both mechanical and chemical removal.

Lowering rates are pro-rated to a suitable period (usually 200 years) as a basis for calculating a new slope profile and the process is repeated for the new profile.

A fixed basal elevation has been assumed, but this can readily be modified.

Input consists of: (1) climatic parameters (annual rainfall, mean rain per rainday and potential evapotranspiration); (2) rock and soil parameters (lateral saturated permeability of the upper horizons and rock constituents in terms of proportions and saturated solubilities for its oxides); and (3) an initial slope profile (elevation and unit width along the contours at 5 m intervals downslope).

254

Output at selected time intervals consists of a downslope sequence of:

(1) Slope profile elevations.
(2) Contributions to subsurface flow, actual evapotranspiration and overland flow (summing to rainfall).
(3) Soil water storage capacity, vegetation cover, and mean depth of subsurface flow in the upper soil horizons.
(4) Rates of lowering due to mechanical and chemical processes of sediment transport.
(5) Soil thickness expressed in terms of total substance deficit, $w$ (Equation 8.2).

The simulation produces a wide range of results which may be compared with published sediment and slope profile material, though no direct field checking has yet taken place. The bases of the model are firmly rooted in short-period experimental results of erosion rates. They lead, by way of the well-established mass-balance or storage equation, to predictions of slope profiles in different climates, and variations of sediment yield in different climates. The results seem to show very clearly that varying proportions of the three processes of creep (or rainsplash), concentrated wash and solution can be controlled through the hillslope hydrology to produce the observed spectrum of variation in both rates and landforms outside periglacial areas, and on slopes which are stable with respect to landslides.

The results of the simulation have also been analysed to show: (1) the extent to which simple approximations, especially that of the 'characteristic form' (Kirkby, 1971), are justifiable as models of hillslope profiles in a more complex system than that for which they were first proposed; (2) the circumstances under which equilibrium soils (Carson and Kirkby 1972, Chapter 9; Kirkby, 1974) are likely to be closely approached during erosional development of hillslope.

## 8.5 Simulation results

### 8.5.1 Development of slope profiles over time

Comparisons have been made between two-dimensional slope profiles, with no flow convergence or divergence, which were 100 m long, initially at a uniform slope of $\tan^{-1}(\frac{1}{2})$, with a fixed divide and with basal removal at a fixed point. Although all of these assumptions are restrictive, they form a valid basis for comparison between climates. In Figure 8.4, the development of a profile is shown for a dry savanna climate with an annual rainfall of 600 mm, a rainfall intensity of 15 mm/rainday and a potential evapotranspiration of 2000 mm. A convexo-concave profile is well established after 100,000 years, while the divide has been lowered from an original 50 m to 38 m elevation. Subsequent development leaves the overall form substantially unchanged, with the maximum slope at about 35 m from the divide.

In Figure 8.5, the profiles obtained after 100,000 years are compared for

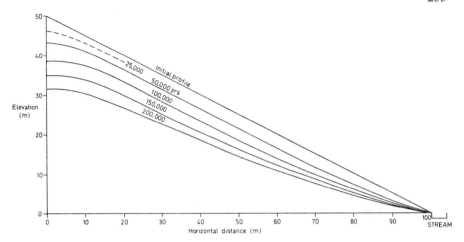

Figure 8.4. Evolution of a simulated slope profile over time. $R = 600$ mm, $r_0 = 15$ mm/rain day, $E_P = 2000$ mm.

different annual rainfalls, keeping intensity and evaporation constant. It can be seen that the profile form becomes more convex as the rainfall increases, the convexity having a width of 12 m at 150 mm rainfall, rising to 85 m at 1800 mm rainfall. The location of downcutting varies correspondingly, being much more concentrated near the divide at high rainfalls.

Some of the components of the erosion are separated out in Figure 8.6 which shows flow and erosion downslope for the 600 mm rainfall profile after 100,000 years. In Figure 8.6(a), the annual average hydrological components are shown for overland flow, subsurface flow and actual evapo-

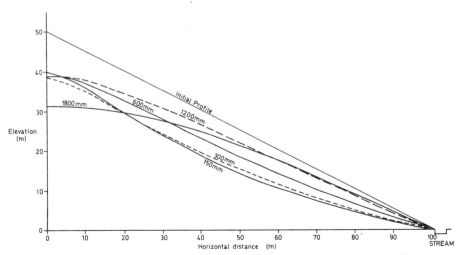

Figure 8.5. Comparison of simulated slope profiles after 100,000 years, for different annual rainfalls. $r_0 = 15$ mm/rain day, $E_P = 2000$ mm for all profiles.

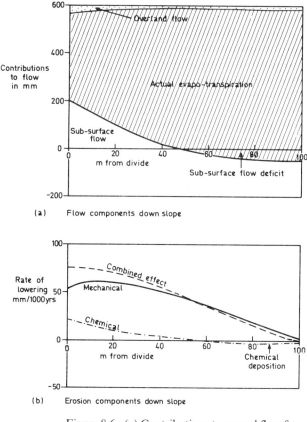

Figure 8.6. (a) Contributions to annual flow for slope shown in Figure 8.4 at 100,000 years. Total flow is accumulations from upslope. (b) Rates of mechanical and chemical lowering for the same example.

transpiration. Total flows at a point can be calculated by accumulating the area under each curve from the divide to the point in question. Thus although downslope points contribute a *net* deficit to the subsurface flow, the *total* subsurface flow remains positive. In the upper parts of the slope, water is able to flow away before it can evaporate, while downslope this additional water is able to boost the local evaporation above 600 mm (the annual rainfall). The increasing evaporation in turn is able to support a denser vegetation cover (16% at divide; 31% at slope base) downslope, leading to increasing soil water storage capacity (40 mm at divide; 56 mm at base) and reducing overland flow.

The chemical removal rate (Figure 8.6(b)) closely follows the curve for subsurface flow, with highest removal rates near the divide and some net deposition near the slope base. Mechanical removal, on the other hand, it influenced by the combination of overland flow and gradient. Accordingly, it is somewhat reduced, relative to the overland flow, near both divide and slope base, where gradients are lowest.

### 8.5.2 Variation of sediment yield with climate

Total rates of sediment yield from a slope depend strongly on total slope length and available relief, but the simulation shows much less sensitivity to slope profile form. The lateral saturated permeability, K, has been held *constant* at 5 m/day, introducing an unknown degree of error. Given low sensitivity to profile form, comparisons have been made for a 1-in-2 slope of 100 m length. Average rates of mechanical and chemical lowering have been calculated for annual rainfalls of 10–3000 mm; intensities of 5, 10 and 15 mm/rainday; and potential evaporation rates of 500, 1000, 2000 and 3000 mm. These evaporation rates correspond roughly to increasing temperature levels.

In Figure 8.7, a set of curves is shown for a constant (high) rainfall intensity of 15 mm/rainday. The curves show the features which ae familiar from the work of Langbein and Schumm (1958), Langbein and Dawdy (1964) and others, with a peak sediment yield in semi-arid climates and a minimum

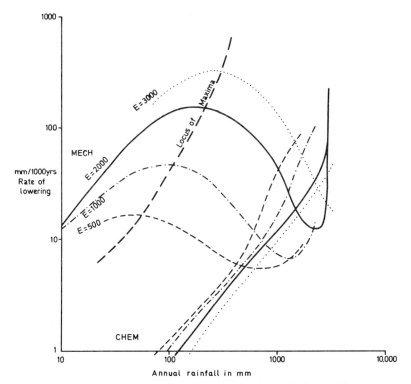

Figure 8.7. Simulated variation of mechanical and chemical lowering rates as rainfall varies, for fixed $r_0 = 15$ mm/rainday. $E =$ potential evaporation values.

mechanical yield in temperate areas. Chemical removal is dominant in temperate and most tropical climates and is negligible in arid and semi-arid areas.

The exact position of these curves, and of the rainfall at peak yield, depends on the particular model parameters used. The curves show, however, a temperature dependence which differs significantly from that proposed by Schum (1965). In the present simulation higher temperatures (or evaporation rates) not only produce peak sediment yields at higher rainfalls, but also produce very much greater yields at all rainfalls. In qualitative terms, two important effects are operating. The first, as suggested by Schumm, is the reduction in runoff with increasing evaporation, so that higher rainfalls are required to produce the same runoff. This effect produces the shift in the peak rainfalls in Figure 8.7. The second effect of increased evaporation is to reduce vegetation cover and so *increase* the *overland* flow even though *total* runoff is reduced. The amount of soil wash erosion is consequently increased to an important extent, raising the overall mechanical yields in Figure 8.7.

In Figure 8.8 the effect of increasing rainfall intensity is shown, for a constant (high) evaporation rate of 3000 mm. Its influence is strictly comparable, and the line joining the peaks of the curves is substantially the same as in Figure 8.7 (Although the behaviour at very high rainfalls is appreciably different). It

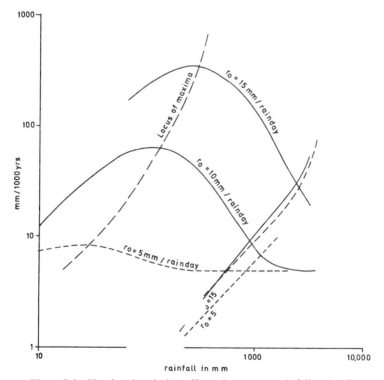

Figure 8.8. Simulated variation of lowering rates as rainfall varies, for fixed $E_P = 3000$ mm.

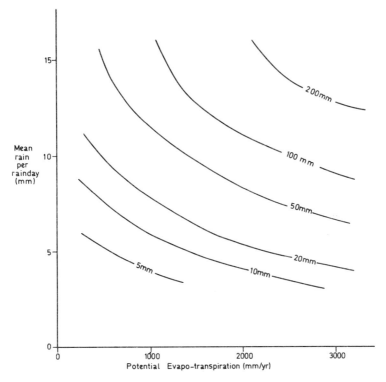

Figure 8.9. Annual rainfall at which erosion rate is a maximum, expressed in terms of intensity and potential evapotranspiration, for model data.

can be seen that the dependence of yield rates on intensity factors is at least as great as the temperature dependence. The two effects are combined in Figure 8.9 which shows the rainfall at peak sediment yield for a range of potential evaporation and daily rainfall rates.

### 8.5.3 'Characteristic forms' as an approximation

It has been suggested that slopes tend to a state in which the rate of lowering is proportional to elevation, over large areas (Ahnert, 1970) as well as on more local scales (Kirkby, 1971). In the more complex set of interactions represented by the present simulation, the characteristic form assumption can no longer be treated as a theoretical concept, but Figure 8.10 shows that it can still act as a useful empirical model. The set of curves shows the relationship between rate of lowering and elevation as it develops over time, for the 600 mm rainfall example used above. A remarkably close convergence on a straight line relationship is found by 200,000 years, with a value of $K = 2 \times 10^{-6}$/year in the relationship:

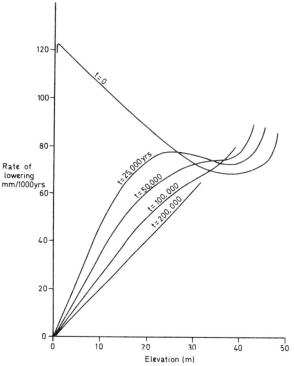

Figure 8.10. Rate of lowering expressed in terms of elevation, for example shown in Figure 8.4. The curves are seen to converge on a straight line, corresponding to a 'characteristic form'.

$$\frac{dy}{dt} = -ky \qquad (8.16)$$

Similar relationships for other climatic conditions show that the value of $K$ can be as low as $10^{-7}$/year and tends to be less at higher (temperate) rainfalls at lower intensities and evaporation rates. These values are somewhat higher than the world average empirical figure of $10^{-7}$/year (Ahnert, 1970).

### 8.5.4 'Equilibrium soils' as an approximation

In Equation 8.2, an equilibrium soil can be defined in which deficit is not accumulating over time so that

$$\Delta = S \frac{1 - p_s}{p_s} \qquad (8.17)$$

or

$$p_s = \frac{S}{S + \Delta} \qquad (8.18)$$

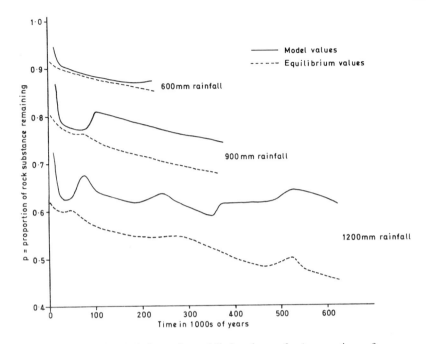

Figure 8.11. Variation of modelled values of $p$ (proportion of bedrock substance remaining in surface soil) and the changing equilibrium over time. Irregularities are thought to reflect properties of the model only.

Whether this equilibrium is reached in practice depends on the rate at which it is approached compared to the rate at which the equilibrium is itself changing, owing to changes in the values of $S$ and $\Delta$. The modelled changes in current and equilibrium values for $p$ are shown in Figure 8.11. An initial rapid approach to equilibrium during the first 25,000 years is evident, followed by a moderate adherence to equilbirium during the next 100,000 years, and then a progressive divergence at longer time periods.

It is suggested that equilibrium soils are characteristic of rather steep slopes where processes are acting rapidly (Kirkby, 1974). On more gentle topography, soils tend to accumulate less rapidly than the rate at which their equilibrium value (of $p_s$) changes, so that imbalances are normal.

## 8.6 Approximate analytical model

An alternative formulation of the slope development equations allows us to obtain an approximation to a mathematical solution for a combination of creep, wash and chemical removal, and to examine the influence of changes in the basal conditions.

Rewriting Equation 8.1 in two-dimensional form:

$$\frac{\partial y}{\partial t} = -\frac{\partial}{\partial x}(S+\Delta) \tag{8.19}$$

If the basal point ($x = x_1$) is undergoing lowering at a constant rate $T_0$, and elevations are measured relative to this changing base-level, then:

$$\frac{\partial y}{\partial t} = T_0 - \frac{\partial}{\partial x}(S+\Delta) \tag{8.20}$$

Let us separate the right-hand side into two components, one of which depends on topography and one of which does not. This separation recognizes that chemical lowering rates may depend less on topography than mechanical rates, but that this dichotomy is far from complete, as is clear from Figure 8.6(b).

In this form

$$\frac{\partial y}{\partial t} = -D' - \frac{\partial S'}{\partial x} \tag{8.21}$$

where $D'$, $S'$ may both contain terms which may be related to lowering of the basal point, chemical lowering and/or mechanical transport rates.

Now let us express $S'$ as a combination of creep and wash-like terms, on the assumption of a uniform rate of overland flow production:

$$S' = (C + \alpha x^2)\left(-\frac{\partial y}{\partial x}\right) \tag{8.22}$$

A particular solution to Equation (8.21) is the solution for $\partial y/\partial t = 0$, which gives the steady state slope profile corresponding to a constant value of the difference between basal lowering and chemical removal rates ($D'$).

We have:

$$\frac{dy}{dx} = \frac{\int D' dx}{C + \alpha x^2} = \frac{2w_b x}{u^2 + x^2} \tag{8.23}$$

where

$$w_b = D'/\alpha \tag{8.24}$$

and

$$u = \sqrt{C/\alpha} \tag{8.25}$$

Writing $z = x/u$ and $z_1 = x_1/u$ for the basal point

$$y_2 = -y_b\left\{1 - \frac{\log_e(1+z^2)}{\log_e(1+z_1^2)}\right\} \tag{8.26}$$

where

$$y_b = w_b \log_e(1+z_1^2) \tag{8.27}$$

It may be noted that this steady state profile is exactly the same as the first approximation to the characteristic form (Kirkby, 1971) for a fixed base level ($D' = 0$). It is concluded that erosional behaviour of the basal point has only a secondary effect on the overall form of the slope profile. The difference between the first and second approximations to the characteristic form is seen to give the magnitude by which profiles differ for the two extremes of (a) equilibrium with a constant downcutting rate and (b) erosion down towards a fixed basal level.

Substitution of $y_3 = y - y_2$ in Equation 8.21 eliminates the $D'$ term, giving:

$$\frac{\partial S}{\partial x} + \frac{\partial y_3}{\partial t} = 0 \tag{8.28}$$

which tends to a characteristic form solution. Following the method described in Wilson and Kirkby (1975), a better approximation to this solution can be obtained by interation of:

$$-\frac{\partial y}{\partial x} \propto \frac{\int y dx}{C + \alpha x^2} \tag{8.29}$$

If we define

$$I(z) = \int_0^x \frac{\int_0^X z dx}{C + \alpha x^2} dx \tag{8.30}$$

Then a sequence of improving solutions

$$y^{(n)} \text{ for } n = 0, 1, 2 \ldots$$

can be formally obtained by substitution in Equation 8.29

$$y^{(n+1)}/y^{(0)} = 1 - k.I\{y^{(n)}/y^{(0)}\} \tag{8.31}$$

or

$$y^{(n)}/y^{(0)} = 1 - kI(1) + k^2 I^2(1) - k^3 I^3(1) \ldots \tag{8.32}$$

where $I^2(z) = I\{I(z)\}$ and so on
and $k$ is chosen to satisfy the boundary condition,

$$y = 0 \text{ at } x = x_1 \tag{8.33}$$

The second approximation (Equation 8.26 was the first approximation) is then found to be:

$$y^{(2)} = y_3 \propto 1 - \frac{\frac{1}{2}\log(1 + z_1^2)\log(1 + z^2) - \frac{1}{4}\log^2(1 + z^2) + \log(1 + z^2) - (\tan^{-1} z)^2}{\frac{1}{4}\log^2(1 + z_1^2) + \log(1 - z_1^2) - (\tan^{-1} z_1)^2}$$

$$= 1 - \phi(z, z_1), \text{ say} \tag{8.34}$$

Then the overall approximate solution for $y$ is:

$$y = y_3 + y_2 = y_0\{1 - \phi(z, z_1)\} - y_b\left\{\frac{\log(1 + z^2)}{\log(1 + z_1^2)} - \phi(z, z_1)\right\} \tag{8.35}$$

Where $y_0$ is the elevation of the divide.
The first term in this expression is a convexo-concave slope profile which is constant in form but reduces in absolute elevation as the divide is lowered.

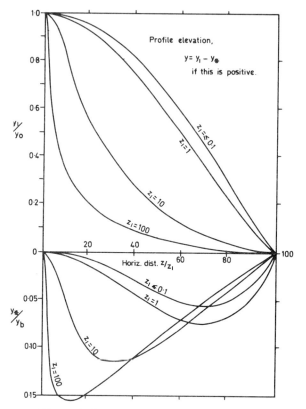

Figure 8.12. Dimensionless curves for profile development according to analytical approximation to slope model.

The second term is zero at $z = 0$ and $z = z_1$, and does not change in value over time. As the divide is lowered, the relative importance of the second term therefore increases, leading to the growth of a basal concavity ($D' > 0$) upslope as the slope profile as a whole is lowered (or growth of a basal convexity if $D'$ is negative owing to downcutting at the basal point). These solutions, which are still approximate, are illustrated in Figure 8.12, and are meaningful only for positive values of $y$.

Where the effect of solution is not too great, as appears usual, the overall scale of the slope profiles obtained can be characterized by the width of the convexity. Using the first approximation of Equation 8.26, this distance is approximately equal to $u$. The value of $u$ can be calculated from climatic figures using Equations 8.25 and 8.5, to express the width of convexity in terms of rainfall intensity ($r_0$), soil storage capacity ($r_c$) and annual rainfall ($R$). This gives

$$u = \frac{243}{R} e^{r_c/r_0} \tag{8.36}$$

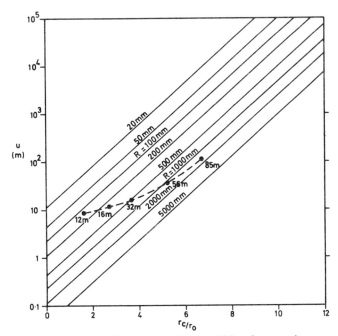

Figure 8.13. Estimates of the width of convexity, $u$, in terms of total rainfall, $R$, and the ratio of soil storage capacity, $r_c$, to mean rain per rain day, $r_0$. The broken line shows the values obtained by simulation, plotted in the positions appropriate to their values of $R$ and $r_c/r_0$. The values written beside each point are the values of $u$ obtained from the simulation programme.

where $R$ is measured in millimetres of rain per year and $u$ is the width of convexity in metres.

This relationship is shown in Figure 8.13. For comparison, the broken line is a plot of modelled values of $R, r_c/r_0$. The values for $u$ obtained from Equation 8.36 may be obtained from the graph and may be compared with the values obtained from the model which are written in Figure 8.13. Model values are seen to be comparable, though slightly larger.

The analytical model, though less precise, has the advantage of rapid applicability. It also helps to bridge the gap between the simulation model described above and previous formulations in terms of characteristic forms (Kirkby, 1971). It may, for example, be more immediately seen that slope profiles form a continuous set of convexo-concave forms in all climates.

In examining the assumptions of the analytical model, serendipity is on the side of the model since its principal simplifying assumption is of constant solution lowering rates downslope. Where solution is a major item in the sediment budget (for humid areas) this assumption is approximately true:

where the assumption breaks down under arid conditions, solution plays such a minor role in slope profile development that the errors are insignificant. A second assumption of the analytical model is that overland flow is produced uniformly over the surface, and this assumption is similarly most nearly true for the arid areas where overland flow is most significant.

Significant differences between analytical and simulation models arise only for humid areas, in hollows and on very long slopes, for which saturation overland flow becomes appreciable in the lower parts of the slope.

## 8.7 Conclusions

(1) The major influences of climate on vegetation, slow slope processes and hillslope development may be effectively modelled at the present state of the art. Several of the interactions are, however, poorly understood and require considerable further investigation, notably the feedback from actual evapotranspiration, *via* vegetation cover, to available soil water storage.

(2) The resulting profiles form a viable basis for designing low angle artificial slopes, at less than landsliding angles, which will minimize erosional losses. It is clear that no design can eliminate erosion, and that the normal engineering criterion of a stable basal level is most appropriately met by a characteristic form slope profile, in which lowering is directly proportional to elevation above the base. The concept of characteristic forms is considered to provide an effective working model of slope development during the complex process interactions simulated.

(3) Simulation results confirm Langbein and Schumm's (1958) sediment yields curves in some detail from a different viewpoint, which is derived indirectly from slope process measurement data. However, it suggests a stronger temperature dependence than has been previously proposed.

## References

Ahnert, F., 1967, The role of the equilibrium concept in the interpretation of landforms of fluvial erosion and deposition, *IGU, Slopes Commission Report 6*, 71–84.

Ahnert, F., 1970, Functional relationships between denudation relief and uplift in large mid-latitude drainage basins, *Am. J. Sci.*, **268**, 3, 132–163.

Betson, R. P., 1964, What is watershed runoff? *J. Geoph. Res.*, **69**, 1541–1552.

Carson, M. A. and Kirkby, M. J., 1972, *Hillslope form and process*, CUP.

Culling, W. E. H., 1963, Soil creep and the development of hillside slopes, *J. Geol.*, **71**, 127–161.

Dunne, T. and Black, R. D., 1970, Partial area contributions to storm runoff in a small New England watershed, *Water Resources Res.*, **6**, 1296–1311.

Freeze, R. Alan, 1972, Role of subsurface flow in generating surface runoff. Base flow contributions to channel flow, *Water Resources Res.*, **8**, 3, 609–623.

Horton, R. E., 1945, Erosional development of streams and their drainage basins: hydrophysical approach to quantitative morphology, *Bull. geol. Soc. Am.*, **56**, 275–370.

Kirkby, M. J., 1971, Hillslope process–response models based on the continuity equation, *Institute of British Geographers, Special Publn. No 3*, 15–30.

Kirkby, M. J., 1974, Landslides and weathering rates, *Dept of Geog, Univ of Leeds, Working Paper 34.*

Kirkby, A. V. and Kirkby, M. J., 1975, Geomorphic processes and the surface survey of archeological sites in semi-arid areas, in *Geoarchaeology: Earth Science and the Past,* Duckworths (in press).

Langbein, W. B. and Dawdy, D. R., 1964, Occurence of dissolved solids in surface waters in the US, *USGS Professional Paper 501-D.*

Langbein, W. B. and Schumm, S. A., 1958, Yield of sediment in relation to mean annual precipitation, *Transactions American Geophysical Union,* **39**, 1076–84.

Musgrave, G. W., 1947, Quantitative evaluation of factors in water erosion—a first approximation, *J. Soil and Water Cons.,* **2**, 133–138.

Schumm, S. A., 1965, Quaternary paleohydrology, in *The Quaternary of the United States,* H. E. Wright and D. G. Frey (Eds.), Princeton University.

Souchez, R., 1961, Théorie d' une évolution des versants, *Bulletin de la societe royale belge de geographie,* **85**, 7–18.

Weyman, D. R., 1974, Runoff processes, contributing area and streamflow in a small upland catchment, *Institute of British Geographers, Special Publication No. 6,* 33–43.

Williams, M. A. J., 1973, The efficacy of creep and slopewash in tropical and temperate Australia, *Aust. Geogr. Studies,* **11**, 62–78.

Wilson, A. G. and Kirkby, M. J., 1975, Mathematical methods for geographers and planners, OUP, Ch. 6.

Young, A., 1963, Some field observations of slope form and regolith and their relation to slope development, *Trans. Inst. Brit. Geog.,* **32**, 1–29.

CHAPTER NINE

# Erosion Rates and Climate: Geomorphological Implications

IAN DOUGLAS

## 9.1 Introduction

> The rate of erosion is the quantity of material actually removed from the soil surface per unit of time and area, and this may be governed by either the transporting power of overland flow or the actual rate of erosion, which ever is smaller. (Horton, 1945, p. 277)

> The main erosional factor is the run-off, which is most active on vegetation-free slopes. The rate of erosion depends on: 1. the climate, 2. the configuration of the relief, 3. the presence of vegetation, and 4. the degree of cultivation of the area. (Kukal, 1971, p. 25)

These two quotations illustrate the concept of the rate of erosion and its relationship to climate. However, removal of soil particles is but part of the supply of material to stream channels. Material is carried in solution in subsurface waters, while in many environments a large part of the material carried by rivers is derived from bank erosion or scour of the channel bed. Kazarinov and coworkers (1966) have proposed that chemical denudation should be expressed in terms of loss per unit volume of rocks drained, the volume of rocks drained being computed from hypsometric maps. Such a suggestion would express the major contribution of baseflow to solute denudation and might be particularly appropriate in studies of karst landforms where the actual land surface may not alter during long periods of subsurface erosion (Chapter 13).

Erosion rates are usually calculated by determining the amount and character of material carried by streams out of their catchment areas or by measuring the rate of sedimentation in lakes or rivers (Kukal, 1971). In the last decade several assessements based on potassium–argon dating of original land surfaces have been made (Ruxton and McDougall, 1967; Ollier, 1969), while layers of volcanic ash and buried soils have been used to date rates of infilling of the Gisborne Plains, New Zealand (Pullar, 1967). Apart from the potassium–argon dating method, all the above techniques measure the overall rate of erosion of a catchment area and fail to reveal which parts of the catchment are being eroded, whether any sedimentation is taking place further upstream, or what the relative contributions of the various tributaries are to the total flow, chemical and solid load of the river. If the basin being studied is developed entirely on a single rock type and under uniform ecological conditions, the water and material evacuated from the catchment accurately define denudation conditions within within that catchment. In larger basins with more complex geological and ecological patterns, studies of material being evacuated by rivers or deposited in reservoirs can reflect only a general regional rate of erosion, not necessarily related to any single system of weathering and transport processes. In these more complex catchments, conditions are often such that material carried by the river is derived in different proportions from various parts of the basin according to the season of the year (cf. Gregory, Chapter 10).

Direct measurements such as those described in Chapter 3 may be more satisfactory, but microclimatic, mesoclimatic, structural and stratigraphic variations may cause difficulties in applying the results of individual plot or small-scale experiments to larger landform assemblages. As the partial source area concept of streamflow initiation and studies of karst denudation both illustrate, there are marked spatial variations in the intensity of erosion. In forested areas, a stream head hollow, an undercut river bank, a cliff-foot debris accumulation or even a termite mound may be eroded many times more rapidly than adjacent areas. Spatial variations of this nature and temporal variations due to changing hydrometeorological conditions, both seasonally and in terms of individual storm events, combine to produce the actual landforms studied by geomorphologists, be they karren on bare limestone, point bars in river channels of all sizes, or gullies and hollows on hillslopes.

## 9.2 Aspects of climate of geomorphic significance

Erosion depends on the activity of external agencies, be they fire, water, wind, waves, ice or a man-made bulldozer. In trying to assess how erosion operates, empirical, experimental and theoretical approaches have been adopted. Empirically derived equations include factors for rainfall, slope length, slope steepness and soil erobibility (Musgrave, 1947; Miller, 1965). Although they were developed for uniform slopes, the equations have been used to predict trends of erosion along curved slopes (Meyer and Kramer, 1969). Experimental plots using rainfall simulators (Boughton, 1967; Hayward, 1967) have improved our understanding of the process of soil particle detachment and the effects of different rainfall magnitudes and raindrop sizes. Data from these experiments is now being incorporated in conceptual models of the erosion process which will allow the influence of irregular land profiles and interaction effects to be analysed (Meyer and Wischmeier, 1969; Rowlinson and Martin, 1971). From such developments erosion equations derived using basic principles of hydraulics, sediment transport and erosion mechanics become possible. Foster and Meyer (1972) have described erosion by water on upland areas by the continuity-of-mass transport equation and an equation expressing an interrelationship between detachment by runoff and sediment load. The latter equation was:

$$\frac{\text{detachment rate by flow}}{\text{detachment capacity by flow}} + \frac{\text{sediment load of flow}}{\text{transport capacity of flow}} = 1$$

The detachment and transport capacities were each taken as the 3/2 power of the flow's shear stress. Using the Chezy uniform flow equation, the closed-form erosion equation for a uniform slope under steady conditions became:

$$G_* = X_* - (1-\theta)(1 - e^{-\alpha}X_*)/\alpha$$

where $G_*$ is the sediment load relative to the transport capacity at the end of the slope, $X_*$ is distance from the top of the slope, $\theta$ is a rainfall detachment parameter, and $\alpha$ is a runoff detachment parameter. The equation recognizes the dependence of water erosion on two climatic variables, namely rainfall intensity (which controls the rainfall detachment parameter) and the effectiveness of rainfall in producing an excess to runoff and cause flow detachment. The erosion process is thus seen in terms of detachment of particles by raindrop impact and their transport by thin flow to rills where sediment transport and flow detachment by runoff occurs.

These theoretical considerations bring out the fundamental climatic controls of erosion: rainfall and those aspects of temperature which affect the rainfall–runoff relationship. The rainfall factor used in soil conservation practice on cultivated areas often takes the form of a rainfall erosion index derived from the rainfall energy expressed in metric-tonne metres per hectare times the maximum 30-minute rainfall intensity (Wischmeier, 1973). There is no particular reason why the maximum 30-minute rainfall intensity is appropriate. In a Sahelian representative basin study (of the Kountkouzout, Niger), Vuillaume (1969) argues that the rainfall during the *first* 20 minutes of the storm is the most appropriate factor. He points out that the initial twenty minutes takes account of: (1) the period at the beginning of the storm when rain is totally retained by the soil whatever the infiltration rate; (2) the inclusion, in most cases, of the time of most intense rain in the first 20 minutes of rain whereas a period of 15 minutes does not; (3) the possibility that a 30-minute period could produce erroneous results as many storms do not last that long. Thus Vuillaume's argument hangs on the water deficit of the uppermost soil horizons being made up more rapidly the more intense the rain is during the first 20 minutes of the storm. To account for water deficits at the onset of rain, Vuillaume uses the time in hours since the last storm as a measure of catchment dryness. In wetter parts of Africa where erosion, especially splash erosion, is almost entirely caused by rain falling at intensities above 25 mm per hour (Hudson, 1971) a measure, $KE > 25$, of the total kinetic energy of the rain falling at intensities of more than 25 mm per hour has been found to be more appropriate than the rainfall erosion index (Rapp and coworkers, 1972). A wide variety of rainfall intensity and storm energy parameters relating to erosion rates has thus been used. As differing rainfall characteristics arise from the general circulation of the atmosphere and local energy exchanges, a parameter which is highly reliable in one area may not be sufficiently sensitive in another. It follows that generalization of these parameters for the quantitative evaluation of differences in erosion rates between major climatic zones at the global scale is difficult.

A spatial problem of generalization of rainfall intensity data also exists. The equations designed for plot experiments are not necessarily suitable for application to large river basins and their reliability for catchments of more than a few square kilometres has seldom been tested (Vice and Heidel, 1972). Storm rainfall intensity is extermely variable over short distances. In individual storms in the Kuala Lumpur district of Peninsular Malaysia, rainfall intensities have varied from 60 mm in 30 minutes to less than 10 mm in 30 minutes over distances of 2 km. Unless a dense rain-gauge network is available, reliable estimates of catchment area rainfall intensity cannot be obtained. Admittedly, over several storms, mean 30-minute intensities over a catchment area of up to 100 km$^2$ tend to be more uniform. However, considerable spatial variations in mean daily rainfall intensities, unrelated to total rainfall or number of raindays, occur over areas of 100,000 km$^2$ or more (Jackson, 1972).

To unravel the aspects of climate of geomorphic significance, the short term temporal variations of hydrometeorological variables, such as storm rainfalls, must be separated from seasonal and longer term regimes which are associated with spatial variations in the general circulation. The hydrometeorological variables are those considered when examining fluctuations in the intensity and magnitude of erosion at individual sites, while the seasonal regime factors are those considered when differentiating between rates of erosion under particular climatic conditions.

Hydrometeorological variables used in studying at a station variations in erosion rates (Table 9.1) include discharge, temperature, precipitation, hydrograph shape and antecedent condition factors. While Guy (1964) indicates that stream sediment transport is more closely related to discharge than any other variable, Imeson (1971) finds that variables expressing catchment wetness and flood intensity improve the accuracy of multiple regressions predicting sediment concentration. Nevertheless the important contrast in sediment concentrations at similar discharges on the rising and falling stages of the hydrograph (Temple and Sundborg, 1972; Loughran, 1974) implies considerable variations during a storm in erosion intensity, most fluvial sediment being supplied to the stream channel while rain is still falling and stream discharge is still rising.

Climatic variables used in comparisons of catchment areas include thermal, precipitation, seasonality and runoff characteristics (Table 9.2). While accurate prediction of erosion rates requires the use of multicomponent models which include many different climatic and non-climatic factors (Wilson, 1973), water availability and the type of cover protecting the ground are the basic causes of differences in sediment yields between catchment areas. Basin relief and lithology cause further variations, the highest sediment yields of major

Table 9.1. Variables used in studying at a station fluctuations of erosion rates

| Discharge variables | Source |
|---|---|
| Total discharge (from the rise point to the point of inflection on the recession limb) | Loughran, 1974 |
| Peak discharge | Loughran, 1974 |
| Spring flood | Brobovitskaya, 1969 |
| Summer-Autumn high water | Brobovitskaya, 1969 |
| Flood intensity index | Imeson, 1971 |
| **Temperature variables** | |
| Water temperature | Imeson, 1971 |
| Mean monthly air temperature in month of flood | Negev, 1972 |
| Mean daily air temperature in month of flood | Negev, 1972 |
| **Precipitation variables** | |
| Maximum observed hourly rainfall in the day | Negev, 1972 |
| Rainfall duration × maximum 30-minute intensity | Imeson, 1970 |
| Rainfall duration in past 48 hours | Imeson, 1970 |
| Kinetic energy value | Dragoun, 1962 |
| Maximum 30-minute intensity | Dragoun, 1962 |
| Maximum 5-minute intensity | Vuillaume, 1969 |
| Rainfall intensity index | Brobovitskaya, 1967 |
| **Hydrograph shape variables** | |
| Peakedness index (peak discharge/total discharge) | Guy, 1964 |
| Rate of hydrograph fall | Imeson, 1970 |
| Time since, and size of, hydrograph peak | Imeson, 1970 |
| **Antecedent condition variables** | |
| Precipitation in previous $n$ days | Dragoun, 1962 |
| Volume of runoff for previous $n$ days | Dragoun, 1962 |
| Discharge before the rise | Guy, 1964 |
| Total rainfall during wet season prior to the storm event under consideration | Vuillaume, 1969 |
| Time in hours from previous storm to beginning of the storm under consideration | Vuillaume, 1969 |
| Number of previous consecutively wet days | Douglas, 1970 |
| Antecedent precipitation index | Negev, 1972 |
| Seasonal cumulative flow volume at a gauging station antecedent to any flood event | Negev, 1972 |

rivers being those of the Chinese streams draining easily eroded loess soils cleared of vegetation. In present-day terms, ground cover implies land use and cannot be related directly to climatic variables. However, under natural conditions, the broad character of the ground cover is related to precipitation and

Table 9.2. Climatic variables considered likely to affect erosion rates

| Temperate variables | Source |
| --- | --- |
| Mean annual temperature | Schumm, 1965 |
| Number of days with frost | Clauzon and Vaudour, 1971 |
| Average actual temperatures below 32 °F all the year | Common, 1966 |
| Average actual temperatures above 70 °F all the year | Common, 1966 |
| Poleward limit of average actual 70 °F isotherm | Common, 1966 |
| Equatorial limit of actual 30 °F isotherm | Common, 1966 |
| Extent of general permafrost | Common, 1966 |
| Area where diurnal range of temperature greater than annual range | Common, 1966 |
| **Precipitation variables** | |
| Areas essentially outside temperate and tropical storm tracks | Common, 1966 |
| Thunderstorm zones of importance | Common, 1966 |
| Areas with monsoon type of climate | Common, 1966 |
| Areas with at least 50 mm of precipitation in January, April, July and October | Common, 1966 |
| Areas with less than 125 mm precipitation per year | Common, 1966 |
| General equatorward limit of snowfall | Common, 1966 |
| Mean annual rainfall | Langbein and Schumm, 1958. |
| Ratio of square of the maximum monthly rainfall to the mean annual rainfall (mm) | Fournier, 1960a |
| Effective precipitation | Langbein and Schumm, 1958 |
| **Discharge variables** | |
| Mean annual runoff | Schumm, 1965 |
| Discharge peakedness | Anderson and Wallis, 1965 |
| Maximum yearly peak discharge | Anderson and Wallis, 1965 |

temperature, edaphic conditions producing local variations of the broad zonal relationships.

Tests of the amount of the spatial variation of erosion rates which may be explained by climatic variables have indicated the importance of effective precipitation, runoff and seasonality of precipitation. Hence Common (1966) emphasizes monsoonal climates, Fournier (1960a) both seasonality and absolute quantity of precipitation and Wilson (1973) the recognition of continental, Mediterranean and tropical wet–dry climates as areas of high sediment yields. In the tropical southwest Pacific both Baltzer (1972) and Douglas (1969a) have stressed the importance of tropical cyclones in erosion in New Caledonia and North Queensland respectively, while Guy and Clayton (1973) have demonstrated the magnitude of erosion caused by a tropical storm in the eastern

United States. In arid and semi-arid regions, the pattern of seasonal occurrence and spacing of storm events plays a major role in determining the total annual sediment yield (Negev, 1969 and 1972). Climatic seasonality is clearly of great geomorphic significance, but one often neglected aspect of seasonality is the type of precipitation and the role of seasonal snowmelt (cf. Starkel, Chapter 7). The highest sediment loads moved by streams in the Snowy Mountains of Australia occur as a result of heavy rainfalls from summer storms, or in late autumn when winter storms may produce more rain than snow, or in October when rainstorms may accelerate snowmelt and produce large volumes of surface runoff (Douglas, 1973). Depending on the total volume of winter snow accumulation, the bulk of erosion in temperate mountain environments may be accomplished either by runoff and floods accompanying intense summer storms (Marchand, 1971) or during spring peak flows in response to snowmelt and to rainstorms on the snow pack (McPherson, 1971; Nanson, 1974). One of the greatest single erosion events in recent decades, the catastrophic Guil Valley flood in the French Alps, was caused by large quantities of warm rain falling on a snow pack which had persisted until June (Tricart, 1959). The equatorward limit of snowfall (Common, 1966) is thus a realistic factor in differentiating the operation of denudation systems.

Climatic variables also affect the dynamics of ecosystems and the type of fauna and flora in a given locality. Biological activity and the passive role of vegetation as ground cover affect both the supply of material for detachment by raindrop impact and the movement of water and material downslope. Differences in the surface aggregation ratio, an indicator of soil erodibility, occur in some combinations of rock types and cover types in Northern California (Anderson and Wallis, 1965). Under brush, for example, high erodibility is associated with both marine and non-marine soft Cainozoic sediment, while under grass, granitic rocks and quartz sericitic schists display high erodibility. On the margin of rainforests in eastern Australia, the denser ground cover beneath the main tree layer of wet sclerophyll forest protects the soil more effectively than adjacent rainforest where the forest floor is subject to surface runoff and soil loss (Plummer, 1963). The close relationship of biological activity to climatic fluctuations may be illustrated by many examples of the response of flora and fauna to weather patterns. Rainforest responds to excessively dry periods by flowering and growth rings in such species as *Agathis borneensis* Warb. may be related to the irregular occurrence of unusually dry spells (Brunig, 1971). In north Queensland the stability of rainforest on exposed slopes is affected by cyclones, especially the high winds. Certain types of tree which normally appear as emergents elsewhere never reach maturity. After reaching a certain hight, they collapse owing to previous weakening by gales, weight of vines or the accelerated erosion of thin soils on steep slopes (Webb, 1958). The collapse of these rainforest trees exposes the soil to raindrop impact and admits light to the forest floor, provoking opportunist insect and vegetative activity.

Insects show both seasonal and irregular responses to climatic fluctuations.

In Australia where the upsurge of biological activity following the flooding or soaking of parts of the interior of the continent is dramatic. Control of insect populations by the weather is well documented (Colinvaux, 1973). The grasshopper *Austroicetes cruciata* relies on sufficient moisture during the crucial 90 days of early spring and summer to survive from the egg to the adult egg-laying stage. If rains fall, massive grasshopper deaths may occur before the eggs essential for the survival of the species are laid (Birch and Andrewartha, 1941). Good conditions of both feed and moisture on the other hand lead to excessive concentrations of grasshoppers which eat large areas of vegetation and leave the soil poorly protected against wind and water erosion. In recent years plagues of spur-throated locusts have caused damage of this type in northern New South Wales and southern Queensland.

Breakdown of organic matter, uptake of chemical elements from the soil, liberation of substances from rock surfaces by chelation and translocation of debris are all achieved by biological activity. The movement of debris may be on the large scale, such as the burrowing activities of mammals like rabbits and wombats, or on the small scale, such as the activity of termites and ants (Tricart, 1957). In high latitudes, seasonal contrasts of this type are readily apparent, insects and microbes being more profoundly affected than higher forms. When this is coupled with changes in the density of the vegetation cover and autumnal leaf fall, a pronounced biological rhythm affecting the dynamics of sediment production is apparent. The marked seasonal contrasts in sediment production and individual solute concentrations in British catchment areas (Edwards, 1973; Imeson, 1973; Imeson and Ward, 1972) must be seen, in part at least, as a response to this seasonal contrast in biological activity.

Climatic data which emphasize seasonality are thus likely to be of great value in the study of erosion rates. Fournier's use (1960a) of the $p^2:P$ parameter and Douglas' demonstration (1969b) of contrasts in sediment yields between Malaysia and Australia stress the role of seasonal concentration of precipitation and runoff. The map of seasonal-climates by Troll and Paffen (1964) makes the kind of distinctions needed to interpret hydrologic and geomorphic process data. West Malaysia, for example, is in the tropical rainy climates (class V1) while the rainforests of northeast Queensland are in the tropical humid summer climates (V2). Class V1 climates may or may not have short interruptions of the rainy season, but they support tropical evergreen rain forest. V2 climates have between 2 and 5 dry months and a vegetation ranging from rainforest to savanna woodland. The definition of a dry month follows that of Lauer (1952). The map is based on three climatic elements:

(1) The seasonal course of illumination and solar radiation between the equator and the poles which corresponds to astronomical conditions varying with latitude.
(2) The seasonal course of temperature which in addition depends markedly on the distribution of water and land and on altitude.
(3) The seasonal distribution of precipitation or the duration of humid periods and humid seasons as mainly conditioned by the circulation of the atmosphere.

The interaction of these elements controls river regimes and the dynamics of ecosystems (Troll, 1964) producing the seasonal fluctuations of erosive activity described earlier. Some of these seasonal climatic factors are incorporated in the classification developed by Oliver (1970) and used by Wilson (1973) in studies of variations in sediment yield.

Of course, the seasonal climates approach could be taken back a stage further and related to energy budgeting and solar radiation receipt. However, as the energy of a given storm depends on both vertical and horizontal transfer of momentum, it is difficult to relate energy inputs to sediment outputs for individual storms on anything larger than the experimental plot scale. Gross annual energy inputs summarized as part of catchment heat budgets could be related to mean annual sediment yields, but their effect is probably well expressed by climatic classifications which incorporate temperature data.

Several types of climate have been alleged to be geomorphically distinct in terms of sediment yield (Table 9.3), but few of these types relate to seasonality. Fournier (1960a) has a different relationship between $p^2:P$, a measure of seasonality and quantity of precipitation, for each of the climatic zones he recognizes. Jansen and Painter (1974), also developing separate multiple regressions for sediment yield prediction for 4 climatic zones, do not emphasize seasonality. Wilson (1972) stresses seasonality by using a sedihydrogram to look at the distribution of sediment yield through the year. The seasonal distribution of Wischmeier's erosion-index value in the eastern United States suggests spatial variations in seasonal sediment yields (Guy, 1970). Overall seasonality of water inputs to, and biological activity within, catchment areas are sufficiently marked over a large enough area of the earth's land surface to require the use of climatic parameters expressing seasonality.

Table 9.3. Geomorphically distinct climatic types affecting sediment yields

| Author | Climatic categories recognized |
| --- | --- |
| Fournier (1960a) | Humid, semi-arid, temperate and subtropical climates. |
| Wilson (1969) | Glacial, desert, selva, mediterranean, tropical wet–dry and continental climates. |
| Fleming (1969) | Desert and scrubland, short grassland and shrub, coniferous and tall grass, mixed broadleaf-coniferous. |
| Jansen and Painter (1974) | A. Tropical, rainy, coldest month 18 °C +<br>B. Dry, where $R$ is less than 2, $R$ being defined as $R = r/t + C$, where $r$ is average rainfall (cm), $t$ is average annual temperature (°C), $C$ is 0 if rainfall is mainly in winter, 7 if rainfall uniform throughout the year, 14 if rainfall mainly in the summer.<br>C. Humid, mesothermal, coldest month average exceeds 0 °C but less than 18 °C, warmest month 10 °C +<br>D. Humid, microthermal, coldest month average less than 0 °C, warmest above 10 °C. |

Even so, many climates have such irregular precipitation events that seasonality measures (such as $p^2:P$, based on long-term average monthly data) are inadequate to express the significance of infrequent storm events. Over much of southeastern Australia, average monthly precipitation may vary little, but individual monthly totals vary greatly from year to year. For example, at Canberra, where the ratio of summer to winter rainfall is 1:1·5 and where there is no pronounced wet season, the average August monthly rainfall of 46 mm was exceeded in a single day in August 1974. Under this type of climate, which is by no means a semi-arid climate (mean annual precipitation 600 mm), individual storm events cause pulses of sediment to move out of drainage basins, a phenomenon normally associated with semi-arid areas (Schumm, 1968). Indeed, as more becomes known of the world's wettest climates, the more it becomes apparent that the bulk of the sediment yield of even humid forested catchment areas is carried during a few days of the year. For example, over 50% of the suspended load of streams in the rainforests of northeast Queensland is carried in seven days of the year, and of the Sungai Gombak in west Malaysia in 24 days of the year (Douglas, 1969b). Perhaps a closer look at the timing of erosive events is required, using a hydrometeorological rather than a seasonal approach.

## 9.3 Hydrometeorological approaches to erosion

Runoff following a rainstorm develops into a flood wave which passes down the river channel as a pulse or surge of water discharge. The flood wave comprises direct overland flow, subsurface throughflow (or interflow), some groundwater and a small amount of direct channel precipitation. As the flood wave carries sediment and solutes with it, the flood hydrograph may be complemented with graphs of sediment and solute load distribution demonstrating how the wave of water is accompanied by a pulse of solid or dissolved material. Apart from the distribution graph of Johnson (1943) and recent work by Loughran (1974) and Walling (1974), the detailed form of sediment or solute hydrographs has seldom been investigated (Gregory and Walling, 1973). The detailed shape of the sediment load graph for individual storm events appears to reflect storm rainfall intensity and duration (Johnson, 1943). By redrawing the sediment load graph with dimensionless load ordinates expressed as the percentage of the total load transported within a specified duration (for example one hour), graphs for a particular period of runoff rise may be superimposed and a general master graph derived. Thus, ultimately, for a given gauging station, a unit sediment or solute load hydrograph could be established, at least for certain seasons if not for the whole year. The unit sediment graph could then be used for extrapolation in a similar manner to the unit hydrograph. Preliminary investigations on Dumaresq Creek at Armidale, New South Wales, show that the timing and shape of increases in sediment concentration with discharge are similar for runoff events of differing magnitude. Characteristic relationships of this type could form the basis of comparison between basins.

The relationship of sediment concentration to the hydrograph depends on the size of catchment and spatial distribution of precipitation. If the distance of travel from sediment sources on slopes to stream channels is short, or if there is little water in the streams before storm runoff begins, the peak concentration in the main channel may be determined by the entry of this unrepresentative flow. Indeed, the peak of the concentration of fine material may lag far behind the peak of the flow if the fine material originated far upstream or if, just before the storm runoff, the stream channel contained large volumes of water having low sediment concentrations. Peak sediment concentrations normally precede water discharge peaks in small basins, but in larger allogenic rivers the sediment peak may lag well behind the water discharge peak (Guy, 1970).

The character of the rising stage of the sediment concentration graph is a direct reflection of storm characteristics and antecedent conditions. The rise may be extremely rapid, varying from less than 10 hours during cyclones in a 210 km² basin in New Caledonia (Baltzer and Trescases. 1971) to less than 10 minutes in a 0·26 km² catchment in southwest England (Walling, 1974). Once the peak sediment concentration is reached, the decrease in wash load transport is rapid, falling off much more quickly than the water discharge. At the current stage of fluvial denudation research, the storm-period hydrometeorological approach to erosion rates is in its infancy. Empirical and statistical relationships have been established for a few isolated catchments, but as sediment data collection is improved through the use of automatic samplers and recorders, analysis of variation of storm-period sediment flows on regional and larger scales should become possible.

## 9.4 Hydrometeorological variables and sediment yield

As large proportions of annual sediment loads are carried by a few flood flows (Table 9.4) analysis of flood events and their probability may help establish sedimentation hazards and, by extrapolation, long term erosion rates. Negev (1972) argues that a flood-oriented model of suspended sediment discharge will serve as a convenient tool for quick estimates of annual loads. Such a model overcomes the problems of unequal sediment loads carried at the same discharges on the rising and falling stages of the hydrograph, but has the disadvantage of a reduction in the data input for regression analysis. Negev (1972) was led to conclude that the flood-load model is an effective tool for studying the factors influencing soil erosion and sediment transport.

Establishment of sediment loads carried by individual floods provides a set of data for evaluating the frequency of occurrence of sediment transport events of different magnitudes. Probabilities of sediment yields per unit area of given magnitudes may be established and frequencies of given yields could be mapped. Average annual sediment yields could be treated in a similar way. Plotting combined average annual sediment yields from Deadhorse and Lexer watersheds, Fraser Experimental Forest, Colorado (Leaf, 1966) as a cumulative

Table 9.4. Number of days per year required to transport 50% of annual sediment yield

| Stream | Catchment area (km²) | Days | Source |
| --- | --- | --- | --- |
| Experimental basin near Tiassale, Ivory Coast | 0·02 | 0·5 | Mathieu, 1971 |
| 5 small catchments, east Devon, England | 0·2 to 6·4 | approx. 2 | Walling, 1971 |
| Murrumbidge, above Tantangara, N.S.W. | 102 | 2 | Douglas, 1966 |
| Rio Puerco, Rio Puerco, New Mexico | 13,364 | 4 | Leopold and coworkers 1964 |
| Cheyenne, Hot Springs, S. Dakota | 22,559 | 4 | Leopold and coworkers, 1964 |
| Experimental basin, Kountkozout, Niger | 0·04 | 4 | Vuillaume, 1969 |
| Wild, northeast Queensland | 585 | 5 | Douglas, 1966 |
| Amitioro, near Tiassale, Ivory Coast | 170 | approx. 5 | Mathieu, 1971 |
| Freshwater Ck., northeast Queensland | 44 | 6 | Douglas, 1966 |
| Snowy, above Guthega, N.S.W. | 76 | 7 | Douglas, 1966 |
| Barron, northeast Queensland | 225 | 11 | Douglas, 1966 |
| Strike-a-light, N.S.W. | 216 | 11 | Douglas, 1966 |
| Mogoro, Mogoro Town, Tanzania | 19 | 11 | Rapp and coworkers, 1972 |
| Millstream, northeast Queensland | 92 | 15 | Douglas, 1966 |
| Marmot Ck., near Calgary, Alberta | 9·4 | 19 | Inland Waters Branch, 1968, 1969, 1970, 1972. |
| Happy Jack's Ck., Snowy Mtns., N.S.W. | 110 | 20 | Douglas, 1966 |
| Gombak, Kuala Lumpur, Malaysia | 140 | 24 | Douglas, 1969b |
| South Saskatchewan at Saskatoon | 139,601 | 29 | Inland Waters Branch 1968, 1969, 1970, 1972 |
| Tumut, above Happy Jack's Pondage, N.S.W. | 133 | 31 | Douglas, 1966 |
| Colorado, Grand Canyon, Arizona | 356,902 | 31 | Leopold and coworkers, 1964 |
| Tooma, above Pondage N.S.W. | 116 | 33 | Douglas, 1966 |
| Niobara, Cody, Nebraska | 7700 | 95 | Leopold and coworkers, 1964 |

frequency curve showed that while during a two-year recurrence interval mean annual sediment yield is likely to be just over 1 m³/km², during a ten-year recurrence interval it will be of the order of 6 m³/km². However, there is an appreciable probability of experiencing a larger event in that ten-year period, with an even chance that the highest sediment yield will exceed 7 m³/km², and a 30% chance that it will exceed 8·5 m³/km².

It must be borne in mind that the nature of flood events is such that occasional major floods will change catchment conditions, scouring tributary channels, initiating landslides, and displacing large quantities of material in major channels. Such changes cause sediment yields to increase for several years after the major flood event. In northern California, for example, sediment yields in poorly managed, logged catchments increased in the first year after the flood by 30% to 200% depending on the time since logging (Anderson, 1970). Second and third year increases declined to 99% and 66%, respectively, of the first year increases. Similar effects are produced by other natural hazards, for example the increase in sediment yield following a bushfire in the Yarrangobilly catchment in New South Wales (Brown, 1972).

Unfortunately, the analysis of flood events and sediment yields has not been taken far enough for world-scale comparisons to be effected. Despite greater understanding of sediment transport processes and erosion rates, prediction equations still rely on broad climatic parameters, few of which are any more satisfactory than Fournier's general equation embodying $p^2:P$ and relief (1960b):

$$\log D.S. = 2\cdot 65 \log \frac{p^2}{P} + 0\cdot 46 \log \frac{H^2}{S} - 1\cdot 56$$

where  $D.S.$ is sediment yield in tonnes/km²/year,
  $H$ is mean catchment height (difference between mean altitude calculated from a hypsographic curve and minimum altitude),
  $H^2$ is the coefficient of massivity,
  $S$ is the surface of the basin in km².

Jansen and Painter (1974) developed equations for different climates which show that sediment yield increases with increasing runoff, altitude, relief, precipitation, temperature and rock softness but that it decreases with increasing area and with increase in protective vegetation. However, they note an anomaly in arid climates where a negative correlation between sediment yield and discharge was thought to be due to the few high intensity floods which carry vast sediment loads in dry climates. Such an observation is really a confirmation of the 'Langbein–Schumm Rule' (Wilson, 1973), but does at least point out the value of the hydrometeorological approach.

In the present state of knowledge, climatic geomorphologists, soil conservationists and design engineers are relying on prediction equations which imply simple relationships between mean annual climatic factors and mean annual sediment yield. Detailed analyses have shown that the role and importance of climatic variables often undergo a seasonal change and that mean annual values are useless for predicting sediment yield (Wilson, 1973). Sediment yield variations will be explained best by multivariate models with several variables expressing climatic influences (Anderson and Wallis, 1965).

## 9.5 Scale and erosion rates

While sediment yield per unit area generally decreases with catchment size, save when the lower tracts of rivers are greatly disturbed by man, or when a river passes from a densely vegetated to a less well-vegetated zone (Douglas, 1967), the influence of catchment variables on the yields of basins of differing size has seldom been discussed.

Douglas (1966) compared sediment yields from two areas of $G$-scale 4·2 and 4·0 (Haggett, Chorley and Stoddart, 1965) in eastern Australia. These two areas were considered representative of larger units of climatic significance with $G$ scale values of approximately 2·0. At such a scale broad relationships between erosion rates and mean annual precipitation and runoff clearly exist and broad climatic variables are probably the most significant factors affecting denudation rates. However, the catchments studied in the field vary in size from $G$ scale 8·7 to 7·0. At this scale differences between catchments are expressed in terms of such factors as vegetation, soils, land use and relief.

These observations in tropical and temperate areas have since been complemented by additional data from cool temperate and subarctic environments in Wales and Canada (Slaymaker, 1972a and 1972b). By studying sampling site ($G$ scale 15), sample plot ($G$ scale 12), small catchment ($G$ scale 8·7–7·1), medium catchment ($G$ scale 5·6–4·8) and large catchment data ($G$ scale 4·3–3·2), the scale-linked changes in importance of factors affecting erosion rates may be evaluated (Table 9.5). Although variation at a given scale is often as great as variation between scales, at G scale 3–4 the major rivers of western Canada have similar sediment yields all falling within the range 70 to 220 $m^3/km^2/year$. However, at the $G$ scale 4·8–5·6 in Rocky Mountain Streams, variability becomes marked due to differences in lithology, degree of glacier development and relief. At G scales 7·1–8·7, the catchment condition variables become important, particularly vegetation and land use. At higher scale values, the actual properties of erodible materials, such as grain size, have greater significance.

Table 9.5. Influence of scale on factors affecting erosion rates (based on observations by Douglas (1966) and Slaymaker (1972a and 1972b))

| G scale value | Factors considered important |
|---|---|
| 2 | Mean climatic variables, such as precipitation and runoff related to the general circulation of the atmosphere. |
| 3–4 | Climatic variables such as seasonal climates. |
| 4·8–5·6 | Lithology, relief, presence of glaciers. |
| 7·0–8·7 | Vegetation, land use, bank erosion, landslides. |
| 12 | Nature of slope, presence/absence of undercutting stream, soil grain size. |
| 15 | Soil properties, slope angle, (depending on type of erosion process). |

If yields over time-spans of less than a year are considered, various hydrometeorological variables become important at the plot and small catchment scales. Variables of this type are listed in Table 9.1 and 9.2.

## 9.6 Problems in assessing erosion rates for geomorphological studies

The study of present-day processes in geomorphology is at least in part based on the assumptions of uniformity, in that analogies of past climates should exist somewhere at the present time. Yet, when attempts are made to argue from both present-day process and stratigraphic information, wide discrepancies in estimates of erosion and sedimentation rates arise. The evolution of vegetation through geological time (Schumm, 1968) and the response of vegetation to Quaternary climatic change has greatly affected past erosion rates. Inevitably there will be errors in using even the most accurately measured present-day erosion rates to estimate denudation through past periods of landform evolution (Meade, 1969).

The use of dating techniques to determine the age of original land surfaces and the measurement of the volume of material removed provides an alternative method of computing denudation rates. Three types of original landscape which have relatively smooth, even surfaces and which can sometimes be dated are erosion surfaces, newly emerged or upraised surfaces, and volcanic landforms (Ruxton and McDougall, 1967). Such techniques are more likely to yield reliable estimates of long term denudation rates, but may neglect climatic fluctuations during the period of dissection. In recent volcanic areas and regions of little climatic change, this technique is of great value.

The sediment yield data problem is not merely one of interpretation, it is also one of data quality. Reliable sediment yield data are scarce. Many mean annual sediment yield data are derived from a few measurements at different discharges and the calculation of the annual yield by the flow-duration/sediment method. At high discharges, single stage sediment sampler data are often used to supplement depth-integrated samples, the logistics of sediment sampling often resulting in samples for the rising stage of storm runoff being single-stage, point samples, while those for equivalent discharges on the falling stage are depth-integrated, manual samples. Fortunately, the growing use of double-stage samplers and automated pumping and photoelectric samplers may help to overcome some of these problems. Nevertheless, even when the samples reach the laboratory, there are difficulties of standardization of analytical techniques (Loughran, 1971; Douglas, 1971).

The measurement of rates of reservoir sedimentation poses similar problems because of the variability of the packing and settling of sediment on a reservoir floor. Even if the original topography of the reservoir floor is known, the weight of water and sediment in the reservoir may cause some compression and settling of the underlying materials.

The climatic variables correlated with sediment yields are also subject to error. The difficulty of interpreting results from catchment experiments has led many to criticize the use of representative and experimental basins for the understanding of fundamental hydrological relationships (Edwards, 1970). A combination of several scales of field experiment together with plot and laboratory studies may be needed to understand fully the nature of fluvial denudation (Piest and Heinemann, 1965). Perhaps the recent phase of emphasis on process measurement in geomorphology is helping to overcome the data problem. While the process-oriented geomorphologists might be accused of forgetting the fundamental objective of explaining landform distributions and evolution, their work will eventually lead to an understanding of how regolith and rock materials are eroded and redeposited and the application of this understanding to the interpretation of correlative deposits, palaeosols and erosional landforms at a variety of scales (Birot, 1970).

## References

Anderson, H. W., 1970, Principal components analysis of watershed variables affecting sediment discharge after a major flood, *Publs. Int. Assoc. Sci. Hydrol.*, **96**, 404–416.

Anderson, H. W. and J. R. Wallis, 1965, Some interpretations of sediment sources and causes, Pacific Coast basins in Oregon and California, *Proc. Federal Interagency Sedimentation Conf. 1963, U. S. Dept. Agric. Misc. Publ.*, **970**, 22–30.

Baltzer, F., 1972, Quelques effects sédimentologiques du cyclone Brenda dans la plaine alluviale de la Dumbéa (Côte Ouest de la Nouvelle-Calédonie), *Rev. Geom. Dyn.*, **21**, 97–114.

Baltzer, F. and J. J. Trescases, 1971, Erosion, transport et sédimentation liés aux cyclones tropicaux dans les massifs d'ultrabasites de Nouvelle-Calédonie, *Cahiers ORSTOM Sér. Géol.*, **3**, (2), 221–224.

Birch, L. C. and H. G. Andrewartha, 1941, The influence of weather on grasshopper plagues in South Australia, *J. Dept. Agric. S. Aust.*, **45**, 95–100.

Birot, P., 1970, *L'influence du climat sur la sédimentation continentale*, Paris: Centre de Documentation Universitaire, 219 pp.

Boughton, W. C., 1967, Plots for evaluating the catchment characteristics affecting soil loss, 1.—Design of experiments, *J. Hydrol. (N. Z.)*, **6**, 113–119.

Brobovitskaya, N. N., 1967, Discharge of suspended sediment as a function of hydrologic characteristics, *Soviet Hydrol.*, **2**, 173–183.

Brobovitskaya, N. N., 1969, Determination of the normal annual discharge of suspended sediments and its cyclic fluctuations, *Soviet Hydrol.*, **4**, 447–462.

Brown, J. A. M., 1972, Hydrologic effects of a bushfire in a catchment in south-eastern New South Wales, *J. Hydrol.*, **15**, 77–96.

Brunig, E. F., 1971, On the ecological significance of drought in the equatorial wet evergreens (rain) forest of Sarawak (Borneo), *University of Hull Department of Geography Miscellaneous Series*, **11**, 66–88.

Clauzon, G. and J. Vaudour, 1971, Ruissellement, transports solides et transports en solution sur un versant aux environs d'Aix-en-Provence, *Rev. Geog. Phys. Geol. Dyn.*, **13**, 489–503.

Colinvaux, P. A., 1973, *Introduction to ecology*, Wiley, New York, 621 pp.

Common, R. 1966, Slope failure and morphogenetic regions, in *Essays in Geomorphology*, G. H. Dury (Ed.), Heinemann, London, 53–81.

Douglas, I., 1966, *Denudation rates and water chemistry of selected catchments in eastern Australia and their significance for tropical geomorphology*, Unpublished Ph.D. thesis, Australian National University, 472 pp.

Douglas, I., 1967, Man, vegetation and the sediment yields of rivers, *Nature*, **215**, 925–928.

Douglas, I., 1969a, Sediment sources and causes in the humid tropics of north east Queensland, Australia, *Brit. Geom. Res. Group Occ. Pap.*, *5*, 24–33.
Douglas, I., 1969b, The efficiency of humid tropical denudation systems, Trans. Inst. *Brit. Geog.*, **46**, 1–16.
Douglas, I., 1970, Sediment yields from forested and agricultural lands, in *The role of water in agriculture*, J. A. Taylor (Ed.), Pergamon, Oxford, 57–58.
Douglas, I., 1971, Comments on the determination of fluvial sediment discharge, *Australian Geographical Studies*, **9**, 172–176.
Douglas, I., 1973, Rates of denudation in selected small catchments in Eastern Australia, *University of Hull Occasional Papers in Geography*, *21*, 128 pp.
Dragoun, F. J., 1962, Rainfall energy as related to sediment yield, *J. Geoph. Res.*, **67**, 1495–1501.
Edwards, A. M. C., 1973, The variation of dissolved constituents with discharge in some Norfolk rivers, *J. Hydrol.*, **18**, 219–242.
Edwards, K. A., 1970, Sources of error in agricultural water budgets, in *The Role of Water in Agriculture*, J. A. Taylor (Ed.), Pergamon, Oxford, 11–23.
Fleming, G., 1969, Design curves for sediment load estimation, *Proc. Inst. Civ. Engnrs.*, **43**, 1–9.
Foster, G. R. and L. D. Meyer, 1972, A closed-form soil erosion equation for upland areas, in *Sedimentation*, H. W. Shen (Ed.), The Editor, Fort Collins, 12-1–12-19.
Fournier, F., 1960a, *Climat et Erosion*, Presses Universitaires de France, Paris, 210 pp.
Fournier, F., 1960b, Debit solide des cours d'eau enterre subie par l'ensemble du globe terrestre, *Publs. Int. Assoc. Sci. Hydrol.*, **53**, 19–22.
Gregory R. J. and D. E. Walling 1973, *Drainage Basin Form and Process*, Edward Arnold, London, 456 pp.
Guy, H. P., 1964, An analysis of some storm period variables affecting stream sediment transport, *U.S. geol. Surv. Prof. Pap.*, *462B*.
Guy, H. P., 1970, Fluvial sediment concepts, *Techniques of Water Resources Inventory U.S. geol. Surv.*, Book 3, Chapter C1, 55 pp.
Guy, H. P. and T. L. Clayton, 1973, Some sediment aspects of tropical storm Agnes, *J. Hydraulics Div. Proc. Am. Soc. Civ. Engnrs*, **99** (HY9), 1653–1658.
Haggett, P., R. J. Chorley and D. R. Stoddart, 1965, Scale standards in geographical research, *Nature*, **205**, 844–847.
Hayward, J. A., 1967, Plots for evaluating the catchment characteristics affecting soil loss, 2. Review of plot studies, *J. Hydrol. (N. Z.)*, **6**, 120–137.
Horton, R. E., 1945, Erosional development of streams and their drainage basins; hydrophysical approach to quantitative morphology, *Bull. geol. Soc. Am.*, **56**, 275–370.
Hudson, N., 1971, *Soil Conservation*, Batsford, London, 320 pp.
Imeson, A. C., 1970, Variations in sediment production from three East Yorkshire catchments, in *The role of water in agriculture*, J. A. Taylor (Ed.), Pergamon, Oxford, 39–56.
Imeson, A. C., 1971, Hydrological factors influencing sediment concentration fluctuations in small drainage basins, *Earth Science Journal*, **5**, 71–78.
Imeson, A. C., 1973, Solute variation in small catchment streams. *Trans. Inst. Brit. Geog.*, **60**, 87–99.
Imeson, A. C. and R. C. Ward, 1972, The output of a lowland catchment, *J. Hydrol.*, **17**, 145–159.
Inland Waters Branch 1968, *Sediment Data for Selected Canadian Rivers 1965*, Dept. of Energy, Mines and Resources, Ottawa, 102 pp.
Inland Waters Branch, 1969, *Sediment Data for Selected Canadian Rivers 1966*, Dept. of Energy, Mines and Resources, Ottawa, 156 pp.
Inland Waters Branch, 1970, *Sediment Data for selected Canadian Rivers, 1967*, Dept. of Energy, Mines and Resources, Ottawa, 210 pp.
Inland Waters Branch, 1972, *Sediment Data for Selected Canadian Rivers, 1968*, Dept. of Energy, Mines and Resources, Ottawa, 256 pp.

Jackson, I. J., 1972, Mean daily rainfall intensity and number of rain days over Tanzania, *Geog. Annlr*, **54A**, 369–375.
Jansen, J. M. L. and R. B. Painter, 1974, Predicting sediment yield from climate and topography, *J. Hydrol.*, **21**, 371–380.
Johnson, J. W., 1943, Distribution graphs of suspended matter concentration, *Trans. Am. Soc. Civ. Engnrs.*, **69**, 941–956.
Kazarinov, V. P., A. E. Kontorovich and L. M. Gerasimova, 1966, Mechanical and chemical denudation of drainage areas, *Internat. Geol. Rev.*, **8**, 1199–1207.
Kukal, Z., 1971, *Geology of Recent Sediments*, Academia, Prague, 490 pp.
Langbein, W. B. and S. A. Schumm, 1958, Yield of sediment in relation to mean annual precipitation, *Trans. Am. Geophys. Un.*, **39**, 1076–1084.
Lauer, W., 1952, Humide and aride Jahreszeiten in Afrika and Südamerika und ihre Beziehung zu den Vegetationsgürteln, *Bonner Geogr. Abhandl.*, **9**, 15–98.
Leaf, C. F., 1966, Sediment yields from high mountain watersheds, Central Colorado, *U.S. Forest Service Research Paper, RM-23*, 15 pp.
Leopold, L. B., M. G. Wolman and J. P. Miller, 1964, *Fluvial processes in geomorphology*, Freeman, San Francisco, 522 pp.
Loughran, R. J., 1971, Some observations in the determination of fluvial sediment discharge, *Australian Geographical Studies*, **9**, 54–60.
Loughran R. J., 1974, Suspended sediment and total solute transport in relation to the hydrograph, *Search*, **5**, 156–158.
McPherson, H. J., 1971, Dissolved, suspended and bed-load movement patterns in Two O'Clock Creek, Rocky Mountains, Canada, *Am. J. Sci.*, **274**, 471–486.
Marchand, D. E., 1971, Rates and modes of denudation, White Mountains, Eastern California, *Am. J. Sci.*, **270**, 109–135.
Mathieu, Ph., 971, Erosion et transport solide sur un bassin versant forestier tropical (Bassin de l'Amitioro, Côte d'Ivoire), *Cahiers ORSTOM Ser Geol.*, **III**, 2, 115–144.
Meade, R. H. 1969, Errors in using modern stream-load data to estimate natural rates of denudation, *Bull. geol. Soc. Am.*, **80**, 1265–1274.
Meyer, L. D. and Kramer, L. A., 1969, Relation betweeen land-slope shape and soil erosion, *Paper Pres. 1968 Winter meeting Am. Soc. Agric. Engrs.*, Paper No. 68/749.
Meyer, L. D. and W. H. Wischmeier, 1969, Mathematical simulation of the process of soil erosion by water, *Trans. Am. Soc. Agric. Engnrs.*, **12**, 754–758.
Miller, C. R., 1965, Advances in sedimentation relevant to watershed problems, *Trans. Am. Soc. Agric. Engnrs.*, **8**, 146–152.
Musgrave, G. W., 1947, The quantitative evaluation of factors in water erosion—a first approximation, *J. Soil and Water Conservation*, **21**, 133–138.
Nanson, G. C., 1974, Bedload and suspended-load transport in a small steep, mountain stream, *Am. J. Sci.*, **274**, 471–486.
Negev, M., 1969, Analysis of data on suspended sediment discharge in several streams in Israel, *State of Israel Hydrological Service, Hydrological Paper*, **12**, 41 pp.
Negev, M., 1972, Suspended sediment discharge in western watersheds of Israel, *State of Israel Hydrological Service, Hydrological Paper*, **14**, 73 pp.
Oliver, J., 1970, A genetic approach to climatic classification, *Ass. Am. Geog. Ann.*, 615–637.
Ollier, C. D., 1969, *Volcanoes*, ANU press, Canberra, 179 pp.
Piest, R. F. and H. G. Heinemann, 1965, Experimental watersheds contribute useful sedimentation facts, *Publs. Int. Assoc. Scient. Hydrol.*, **66**, 372–379.
Plummer, B. A. G., 1963, Preliminary field investagation into soil slope and vegetation relationships in the New England National Park, in *New England Essays*, R. F. Warner (Ed.), University of New England, Armidale, 53–60.
Pullar, W. A., 1967, Uses of volcanic ash beds in geomorphology, *Earth Science Journal*, **1**, 164–177.
Rapp, A., V. Axelsson, L. Berry and D. H. Murray-Rust, 1972, Soil erosion and sediment

transport in the Morgoro River Catchment, Tanzania, *Geog. Annlr*, **54A**, 125–155.
Rowlinson, D. L. and G. L. Martin, 1971, Rational model describing slope erosion, *J. Irr. Drainage Div. Am. Soc. Civ. Engnrs.*, **97** (IR1), 39–50.
Ruxton, B. P., 1967, Slope wash under mature primary rain forest in northern Papua, in *Landform studies from Australia and New Guinea*, J. N. Jennings and J. A. Mabbutt (Eds.), ANU Press, Conberra, 85–94.
Ruxton, B. P. and I. McDougall, 1967, Denudation rates in northeast Papua from potassium-argon dating of laves, *Am. J. Sci.*, **265**, 545–561.
Schumm, S. A., 1965, Quaternary paleohydrology, in *The Quaternary of the United States*, H. E. Wright and D. G. Frey (Eds.), Princeton University Press, 783–794.
Schumm, S. A., 1968, Speculations concerning paleohydrologic controls of terrestrial sedimentation, *Bull. geol. Soc. Am.*, **79**, 1573–1588.
Slaymaker, H. O., 1972a, Sediment yield and sediment control in the Canadian Cordillera, *Mountain Geomorphology*, B. C. Geographical Series, **14**, 235–245.
Slaymaker, H O., 1972b, Patterns of present sub-aerial erosion and landforms in mid-Wales, *Trans. Inst. Brit. Geog.*, **55**, 47–68.
Temple, P. H., and Å. Sundborg, 1972, The Rufiji river, Tanzania: hydrology and sediment transport, *Geog. Annlr*, **54A**, 345–368.
Tricart, J., 1957, Observations sur le role ameublisseur des termites, *Rev. Géom. dyn.*, **8**, 170–172, 179.
Tricart, J., 1959, Evolution du lit du Guil au cours de lacrue de Juin 1957 en aval de Ristolas, *Bull. Sect. Géogr. Com. Trav. hist. sci.*, **72**, 169–408.
Troll, C., 1964, Karte der Jahreszeiten-Klimate der Erde, *Erdkunde*, **18**, 5–28.
Troll, C. and K. H. Paffen, 1964, Karte der Jahreszeiten-Klimate der Erde, *Erdkunde*, **18**, Supplement of map and index.
Vice, R. B. and S. G. Heidel, 1972, Sedimentation forecasting, *Status and trends of research in hydrology 1965-74, UNESCO Studies and Reports in Hydrology*, **10**, 106–112.
Vuillaume, G., 1969, Analyse quantitative du rôle du milieu physico-climatique sur le ruissellement et l'érosion a l'issue de bassins de quelques hectares en zone sahélienne, (Bassin du Kount kouzout, Niger), *Cah. ORSTOM Ser. Hydrol.*, **6**, (4), 87–132.
Walling, D. E., 1971, Sediment dynamics of small instrumented catchments in southeast Devon, *Trans. Devonshire Assoc.*, **103**, 147–165.
Walling, D. E., 1974, Suspended sediment and solute yields from a small catchment prior to urbanization, *Inst. Brit. Geog. Spec. Publ.*, **6**, 169–192.
Webb, L. J., 1958, Cyclones as an ecological factor in tropical lowland rain forest, north Queensland, *Aust. J. Bot.*, **6**, 220–228.
Wilson, L., 1969, Les relations entre les processus geomorphologiques et le climat modern comme methode de paleoclimatologie. *Rev. géog. phys. geol. dyn.*, **11**, 303–314. Also available as, Relationships between geomorphic processes and modern climates as a method in paleoclimatology, in *Climatic Geomorphology*, E. Derbyshire (Ed.), Macmillan, London, 269–284.
Wilson, L., 1972, Seasonal sediment yield patterns of United States rivers, *Water Resources Res.*, **8**, 1470–1479.
Wilson, L., 1973, Variations in mean annual sediment yield as a function of mean annual precipitation, *Am. J. Sci.*, **273**, 335–349.
Wischmeier, W. H., 1973, Upslope erosion analysis, in *Environmental Impact on Rivers*, H. W. Shen (Ed.), The Editor, Fort Collins, 15-1–15-26.

CHAPTER TEN

# Drainage Networks and Climate

KENNETH J. GREGORY

**10.1 Introduction**

That drainage networks possess a distinctive character in particular morphoclimatic zones is demonstrated by long-accepted regional names. In West Africa the term *marigot* is used to describe seasonal streams that dry up during the dry season and are reduced to a chain of ponds (Tricart, 1972). In Zambia the U-shaped vales at the head of perennial streams are referred to as *dambos*: they are moist during the rainy season although they lack a definite streambed and evidence of concentrated flow (Ackermann, 1936). In the eastern steppe zones of the USSR the *balki* are shallow valleys which are dry for much of the year but which may experience water flow for a short period during spring peak runoff (Tricart, 1953). The *washes* of the USA refer to infrequently used elements in the drainage networks of semi-arid areas; and in semi-arid Australia a watercourse which is often dry is denoted a *creek* (Heathcote, 1965). Experience of the seasonal expression of drainage networks has thus provided a number of appropriate terms but more recently the quantitative definition of the drainage network has often presented difficulties. In Romania, for instance, Zavoianu (1964) noted that many first order elements of the drainage network are wide, open features modelled by slope processes where mass-movement prevails and where the flow of water is not organized.

The fluvial geomorphologist is concerned with the way in which landforms are fashioned by running water and therefore with the way in which precipitation is transformed into streamflow. One expression of this transformation is provided by the drainage network and the fact that drainage networks vary with climate was appreciated by G. K. Gilbert (1880) and by W. M. Davis (Chorley, Beckinsale and Dunn, p. 169, 1973). More recently the means for the quantitative expression of the drainage network was provided by the definition by R. E. Horton (1932, 1945) of drainage density as the total length of stream channels per unit area. Despite proposals for climatic geomorphology during the last thirty years, there have been few attempts to establish the worldwide variation of drainage density and of other fluvial features of drainage basins in relation to climate. Although it has been claimed that consideration of the worldwide variation of drainage density is premature (Stoddart, 1968) knowledge of the magnitude of world variation is necessary if present spatial variations are to be understood and temporal changes interpreted. Budel (1972) has recently distinguished types of valley according to climato-geomorphological zones and he has contended that the definition of valley in geomorphology should include the idea of river transport and linear erosion. To proceed towards an understanding of the way in which river transport and linear erosion vary according to climate it is necessary to focus upon the dynamic function of the drainage network, and thus necessarily requires cognizance of recent developments in theories of runoff formation (e.g. Black, (1970) and Jamieson and Amerman (1969)).

Although drainage networks have long been considered to embrace several types of component, such as perennial, intermittent and ephemeral streams, the dynamic nature of drainage networks, and therefore of drainage density (Gregory and Walling, 1968), has only recently figured extensively in drainage basin analyses. Although the dynamic nature of drainage density is fundamental to future analyses of the variation of this drainage basin parameter, the dynamic nature has hitherto been responsible for difficulties which have surrounded the measurement of drainage density. The debate concerning derivation of the drainage network derives primarily from the fact that different map agencies portray the dynamic drainage network in a variety of ways although portrayal is also affected by map scale. Thus whereas in many parts of the USA it has been demonstrated that the most reliable networks are obtained by including elements indicated by contour crenulations as well as those indicated by blue lines (e.g. Morisawa (1957)), in Britain the presence of extensive dry valley networks and of glacial drainage channels renders the mapped stream network the best approximation to the present functioning stream network (Gregory, 1966; Werritty, 1972). Substantial differences between stream networks derived from map sources and those apparent on aerial photographs (e.g. Selby (1968) and McCoy (1971)) have been identified and differences have also been established between these values and measurements of field situations. As several types of water flow contribute to the drainage network, including base flow, overland flow and throughflow, and also because the contribution of these elements varies from one area to another, it is inevitable that some manifestations of the drainage network are neither recorded on topographical map series nor detectable from aerial photographs. A further and related problem arises from the fact that subsurface flow networks have now been identified and networks of pipes (Jones, 1971; Institute of Hydrology, 1972) may provide a further manifestation of the scale problem in areas with a particular rock type, soil cover or vegetation history.

Regional reactions to drainage networks, illustrated by terms such as those cited above, indicate that world variations in drainage networks, and therefore in drainage density, should exist. Differences in runoff characteristics may not be expressed exclusively in the character and density of the drainage network because seasonal variations in precipitation and runoff can be accommodated by adjustments of channel pattern or of channel geometry instead of, or as well as, adjustments of the drainage net. Thus a markedly seasonal climate may be associated with an annual variation from a meandering to a braided stream channel pattern. However, it is expectable that on the world scale there should be some expression of climatic character in the drainage network, and knowledge of such expression is vital to the utilization of drainage networks for predicting present and past spatial variations in drainage basin mechanics. Indications of the existence of spatial variation of drainage networks are afforded by comparing the drainage map and the mean annual rainfall maps in the *Atlas of Tasmania* (Davies, 1965); a consequence of spatial variation is found in the way in which the hydrological map of Poland distinguishes permanent, periodic and episodic streams (Galon, 1964); and general views have contended that consistently high or low drainage density throughout a region depends upon rainfall characteristics (Cotton, 1964). Although a map indicating the relative number of permanent streams in the United States was published in 1954 by Visher, little further mapping has been accomplished until recently.

To review the magnitude of variations of drainage networks it is necessary to consider the nature of variations, in relation to climate at the world scale in order to establish the variation between morphoclimatic regions; at the scale within a particular morphoclimatic zone including variations over short time-periods; and at the temporal scale to indicate the extent of changes which have been identified in relation to changes of climate.

## 10.2 World variation

Sufficient results have now been obtained from a range of world areas to sketch the extent of drainage density variation. Data from 44 areas is collected in Table 10.1 to indicate the magnitude of world variation although there is a dominance of data from middle latitude areas, a paucity of data for the tropics, and an absence of data for high latitudes. Where feasible the rock type and vegetation character for the areas studied have been included and the values are tabulated approximately according to latitudinal distribution. Despite the difficulties of drainage density determination outlined above, and the fact that the values collected (Table 10.1) were obtained using a number of methods, the range of world values is very considerable. The values of up to 821 reported for a Perth Amboy industrial dump (Schumm, 1956) are exceptional in view of the unconsolidated material and lack of a protective vegetation mantle, but values above 100 have been quoted for areas in South Dakota and Arizona (Table 10.1). Values greater than 20 are reported from 5 of the 42 areas and values greater than 15 from a further 8 studies. This range of drainage density values therefore tends to be substantially greater than the range of values normally encountered within a single area or climatic type. The scope of the values indicates how values in humid temperate landscapes often tend to be less than 5·0 except where annual precipitation totals are high or where seasonal

Table 10.1 Reported drainage densities

| No. | Area | | Drainage density (km/km²) | Mean annual precipitation(mm) | Predominant rock type | Vegetation | Climato-Genetic zone (Büdel, 1963) | Source |
|---|---|---|---|---|---|---|---|---|
| 1 | Great Britain | | 0·54– 7·14 | 584–3250 | — | — | Extratropical zone of former valley formation (3) | Figure 10.5 |
| 2 | England | : Southwest | 0·62– 2·30 | 760–2030 | — | Farmland, heath, woodland | 3 | Gregory, 1971a |
| 3 | England | : Otter Basin | 0·40– 5·5 | 1005 | — | Farmland, heath, woodland | 3 | Gregory, 1971a |
| 4 | England | : Dartmoor | 2·2 | 1725 | Granite | Heath, grass, woodland | 3 | Chorley Morgan, 1962 |
| 5 | Wales | | 1·3 – 2·7 | 760–2540 | — | Heath, woodland, farmland | 3 | Howe, Slaymaker and Harding, 1967 |
| 6 | USA | : Michigan | 2·1 – 4·4 | 813 (Snowfall 2920–3300) | Glacial drift over sandstone shale, lava series | — | 3 | Hack, 1965 |
| 7 | USA | : Appalachian Plateau | 1·7 – 3·6 | 1264 | Sandstone | Coniferous and deciduous forest | 3 | Morisawa, 1962 |
| 8 | USA | : Pennsylvania | 1·9 – 3·9 | 1099 | Sandstone | Coniferous and deciduous forest | 3 | Smith, 1950 |
| 9 | USA | : Appalachians | 1·9 – 5·0 | 950–1575 | Granite, gneiss, shale | Coniferous and deciduous forest | 3 | Carlston, 1963 |
| 10 | USA | : Connecticut | 3·63– 5·46 | 1220 | Glacial drift over crystalline rocks | Forest, abandoned fields | 3 | Wilson, 1970 |
| 11 | Europe | : Little Mill Creek | 1·03– 8·00 | — | Sandstone, shale | Woodland, pasture, cultivated | 3 | Pinchemel, 1957 |
| 12 | USA | | 4·20– 4·40 | 940 | | | 3 | Morisawa, 1957 |
| 13 | USA | : New Jersey | 341·9 –820·6 | 1075 | Clay and sand fill | Woodland, some heath | 3 | Schumm, 1956 |
| 14 | USA | : Unaka Mountains, Ca., Tenn. | 7·0 | 2095 | Gneiss, meta-morphoted sedimentaries | | 3 | Chorley and Morgan, 1962 |
| 15 | USA | : Clinch Mountain Area, Va., Tenn. | 7·34– 9·89 | 1143–1270 | Sanestone, quartzite, argillites | Forest | 3 | Miller, 1953 |
| 16 | USA | : Ozark Plateau | 8·70 | 1001 | Cherts and limestone | Deciduous forest and pasture grass | 3 | Strahler, 1952 |
| 17 | Romania | | 2·2 – 4·1 | 600 | — | — | 3 | Zăvoianu, 1964 |
| 18 | Tasmania | | 4·8 – 8·2 | 1220–2540 | — | Shrubs, grassland | 3 | Abrahams, 1972 |
| 19 | Australia | : Victoria | 3·9 – 5·4 | 533–1524 | — | Woodland, shrubs, grassland | 3 | Abrahams, 1972 |

293

| # | Country | Location | Range 1 | Range 2 | Geology | Vegetation | Zone | Reference |
|---|---|---|---|---|---|---|---|---|
| 20 | USA | Gulf Coastal Plain | 2.9 | — | — | — | Subtropical zone (4) | Schumm, 1956 |
| 21 | Italy | Apennines | 4.5 – 8.00 | 711 | — | Farmland | — | Pinchemel, 1957 |
| 22 | USA | Nebraska | 5.5 – 11.8 | 555 | — | Grassland | 3 | Brice, 1966 |
| 23 | New Zealand | Volcanic Plateau | 5.4 | 1397 | Pumice and ignimbrite | Scrub and grass | 3 | Selby, 1968 |
| 24 | New Zealand | Auckland Experimental Basins | 15.7 / 4.15 – 20.51 | 1247 / 1050–1550 | Graywacke / Graywacke and others | Pasture grass / Native grassland | 3 / 3 | Selby, 1968; Leamy, 1972 |
| 25 | New Zealand | Central Otago | 8.12 – 16.16 | 325 | Greywacke gravels (16.16) and schist (8.12) | Native grassland | 3 | Leamy, 1972 |
| 26 | USA | California | 11.2 – 19.9 | 500–1010 | Igneous | Chaparral | 4 | Maxwell, 1960 |
| 27 | USA | San Gabriel Mountains | 9.7 | 790 | — | Chaparral | 4 | Strahler, 1952 |
| 28 | USA | Verdugo Hills | 16.3 | 723 | — | Chaparral | 4 | Strahler, 1952 |
| 29 | USA | California | 9.4 – 22.7 | 723 | Igneous | Sparse chaparral | 4 | Smith, 1950 |
| 30 | USA | Colorado | 2.9 – 33.4 | 305 | Granite, gneiss, schist | Montane forest | 4 | Melton, 1957 |
| 31 | USA | South West | 2.1 – 20.8 | 178–727 | — | Montane forest, grass, none | 4 | Melton, 1957 |
| 32 | USA | Colorado | 5.1 – 9.4 | 300 | Igneous | Montane forest | 4 | McCoy, 1971 |
| 33 | USA | South Dakota | 49.7 – 161.6 | 375 | Clays and shale | Sparse grass or none | 4 | Smith, 1958 |
| 34 | Australia | | 7.6 – 16.6 | 252 | — | Sparse trees | Arid zone (5) | Woodyer and Brookfield, 1966 |
| 35 | USA | Arizona | 157.2 – 167.8 | 432–635 | Shale | None | 5 | Smith, 1950 |
| 36 | Australia | New South Wales | 1.8 – 4.9 | | | Woodland, shrubs, grassland | 4 | Abrahams, 1972 |
| 37 | India, Luni Basin | | 0.8 – 1.3 | 356 | Alluvium over granite and volcanics | Sparse | 5 | Ghose and coworkers, 1967 |
| 38 | Australia | Queensland N.S.W. | 3.3 – 14.0 | 890–3988 | — | Woodland, savanna | Circumtropical zone (6) | Abrahams, 1972 |
| 39 | Uganada | | 0.7 – 5.7 | 1000–1250 | — | Savanna | 6 | Doornkamp and King, 1971 |
| 40 | Malaysia | | 0.93 – 7.2 | 2480 | — | Tropical rain forest | Intertropical zone (7) | Eyles, 1968 |
| 41 | Fiji | | 15.6 – 18.8 | 2540 | — | Grass and scrub | 6 | Wright, 1973 |
| 42 | Sri Lanka | | 6 – 15 | 1270–2100 | Granite | Tea plantations | 6 | Madduma Bandara, 1974 |
| 43 | New Zealand | | 15.5 | 1143–1270 | Deep weathered greywacke | — | 3 | Cotton, 1964 |
| 44 | Japan | | 28.3 – 32.9 | 1120 | Granite, sedimentary | — | 4 | Cotton, 1964 |

precipitation distribution is more pronounced. Higher values are notable from mediterranean areas, from temperate continental, from semi-arid, and from mountainous areas. The broad picture revealed tends to indicate that markedly seasonal climates are associated with the highest drainage density values and to substantiate the suggestion made by Peltier (1962) that drainage density is lowest in desert, attains maximum values in semi-arid, and is lower in temperate areas. Peltier made sample measurements of drainage density in relation to mean slope and concluded that values in the tropics may be greater than those of temperate latitudes.

In addition to the several methods available for drainage density determination, many of which have been employed to obtain the values cited in Table 10.1, a further complicating factor is introduced by the problem of scale. The values (Table 10.1) were calculated for a variety of sizes of drainage basins ranging from the large basins up to 167 km$^2$ studied by Ghose and coworkers (1967) to the small experimental catchments for which values are cited by Leamy (1972) or the small basins of 0·002 km$^2$ analysed by Melton (1957). It is known that drainage density is not independent of basin area and that drainage density values tend to be greatest in the smallest basins (Gregory and Walling, 1973, Figure 2.1). Therefore, a method is required which will allow comparison of drainage density values between areas independently of scale and, ideally, the method should also separate the general character of drainage density in a particular area from the local 'noise' which occurs arising from the characteristics of particular basins. A simple method meeting these requirements is provided by the relation between basin area ($A$) and the total channel length ($\Sigma L$) which the basin supports. The law of contributing areas (Schumm, 1956) is in some ways analogous but that relation utilizes mean basin values according to stream order. This relation ($\Sigma L \propto A$) is similar to that used to relate meander dimensions and basin area (Dury, 1964). Although sinuosity variation may be a significant complicating factor in the relation between basin area and mainstream length (Smart and Surkan, 1967) this should be less significant when total channel length is related to drainage area. The logarithmic relation between total channel length (log $\Sigma L$) and basin area (log $A$) should have a constant exponent of unity, unless drainage density ($\Sigma L/A$) varies with scale ($A$). If a sufficient sample size is available to characterize each area, the exponent should approach a constant value between areas. Deviation from the general relation (log $\Sigma L \propto$ log $A$) in a particular area could reflect the nature of individual basin characteristics. Although the exponent should maintain a standard value from one area to another, the constant of the relation should provide a general method for the comparison of drainage density between areas.

Data already published for thirteen areas allows this relationship to be used to indicate the extent of drainage density variation more precisely than in Table 10.1. Although no data were available for high latitude situations, analysis of map data at a scale of 1:62,500 for sixteen basins in Alaska illustrates the manner in which the relation may be employed. Despite the small map

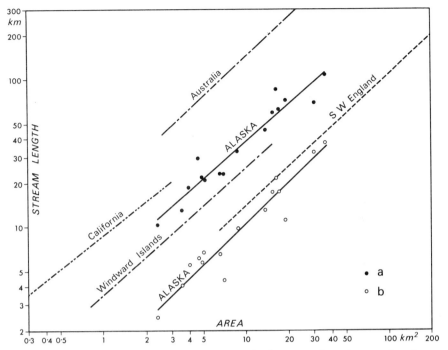

Figure 10.1. Logarithmic relationship between basin area and total stream length. The solid lines refer to data from northwest Alaska (upper, based upon contour crenulations; lower, according to watercourses mapped). These relations are compared with similar relationships for data from semi-arid Australia (Woodyer and Brookfield, 1966); from California (Smith, 1950); from southwest England (Gregory, 1971); and from the Windward Islands.

scale a clear discrepancy exists between the perennial stream network recorded by blue lines and the pattern of valleys testified by contour crenulations. Therefore, two relations can be plotted for the Alaska data, one relating the length of streams ($\Sigma L_s$) to area ($A$) and one relating the length of streams plus valleys as indicated by contour crenulations ($\Sigma L_v$). These two relations are illustrated in Figure 10.1 and are contrasted with the relations obtained for data from 30 basins of the Windward Islands, based upon 1:25,000 map series, and for data from semi-arid Australia (Woodyer and Brookfield, 1966), for California (Smith, 1950) and for southwest England (Gregory, 1971a). Data published for ten other areas allows analysis in terms of the same relation between area and stream length and provides the results included in Figure 10.2. The exponent of the 15 relationships ranges from 0·52 to 1·1235 but 12 of the 15 lie between 0·73 and 0·89 and the three extreme values are for small samples or for studies based upon the smallest basin areas. In general the exponent of the relation between basin area and channel length appears to approximate to 0·8 from a variety of world areas.

In an analysis of some of this data Gregory and Gardiner (1975) suggested

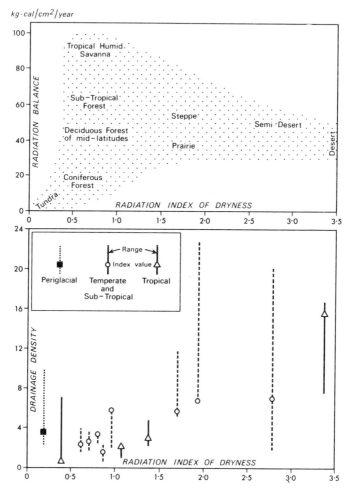

Figure 10.2. Drainage density values compared with a graph of geobotanical zonality. The upper graph of geobotanical zonality was proposed by Budyko (1958) and in the lower diagram drainage density values are illustrated in relation to the radiation index of dryness.

that the constants of the regression equations may be employed to indicate the general character of drainage density and that the values indicated can be related approximately to mean annual precipitation. This confirms that the highest values of drainage density are obtained for semi-arid areas or for areas with a markedly seasonal precipitation regime. Thus the consistently high values analysed from central Australia by Woodyer and Brookfield (1966) in an area with 256 mm annual precipitation confirm the supremacy of areas with low annual rainfall totals, including high intensity rains and sparse vegetation cover. In this study it was noted that there was more variation within than between land systems and that relief and slope parameters exercised an influence upon local variations. The values reported by Melton (1957) from the southwest United States also demonstrate that the highest densities occur in semi-arid areas. Subsequently it appears that values for seasonal climatic regimes are

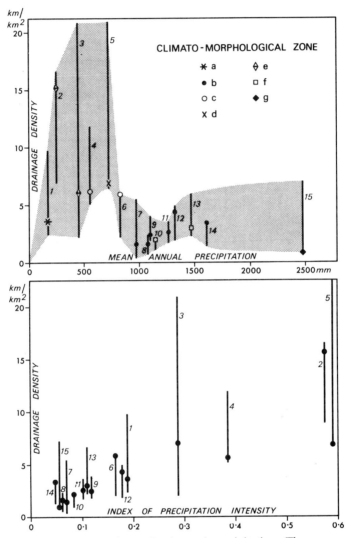

Figure 10.3. Drainage density and precipitation. The upper diagram shows the pattern of drainage density in relation to mean annual precipitation amount. Key to climato-morphological zones: a, extratropical zone of pronounced valley formation; b, (humid) and c, (continental) extratropical zone of former valley formation; d, subtropical; e, arid; f, circumtropical; g, intertropical. The lower graph shows drainage density values plotted in relation to an index of precipitation intensity (maximum recorded 24-hour rainfall/average annual precipitation).

characterized by high drainage densities as shown by the data from California and for Nebraska. The values for humid mid-latitude areas demonstrate that drainage densities tend to be lower in these areas and that there is a general tendency for values to increase with increase in mean annual precipitation (Figure 10.3). Although comparatively few data are as yet available for humid tropical zones it is notable that values of drainage density may exceed those of humid temperate landscapes and thus the values from the Windward Islands exceed those quoted for many parts of Britain and the eastern United States. In addition to the average values of drainage density deduced for the 15 areas relating broadly to mean annual precipitation amounts, it is notable that the range of values reported varies also. The range of drainage density values is greatest in semi-arid areas and in areas of pronounced seasonal precipitation including California and Nebraska; the range is much less in humid temperate areas, but increases again in humid tropical areas. The range of values from Alaska is also higher than those of middle latitudes. The variation in range of drainage density values may partly reflect the fact that seasonal variation in network density is greatest in semi-arid areas and in those with a marked seasonal precipitation regime. Thus in the case of the data from southeast United States, from Australia, from California, from Nebraska and from Alaska the range of values may indicate the magnitude of seasonal variation.

Whereas the world pattern of drainage density values therefore indicates a pattern which conforms broadly to morphoclimatic zones (Figure 10.2), the reason for this correspondence is not solely the result of climatic character but rather of climate in combination with other factors, notably vegetation. For this reason Melton (1957) showed that 92% of the variability of drainage density could be explained by the Thornthwaite (P–E) index, the infiltration and the intensity of rainfall defined as the amount of rain in one hour every five years. Similarly Chorley (1958) had related the drainage density of areas of Exmoor, Pennsylvania and Alabama to a vegetation index incorporating precipitation effectiveness index, precipitation amount and precipitation intensity. More recently the world pattern of landscape zones has been expressed in relation to the radiation index of dryness and radiation amount (Budyko, 1958; Ye Grishankov, 1973). Although, as Hare (1973) has noted, the energy budget approach may not be a very fruitful one in geomorphology, it is possible to compare (Figure 10.2) the world pattern of drainage density values reported with the pattern of landscape zones proposed (Budyko, 1958).

The magnitude and range of drainage density values for particular climatomorphological zones can be illustrated in relation to mean annual precipitation amount (Figure 10.3). This illustrates the way in which drainage density values and ranges tend to be greatest in semi-arid areas, to decrease in humid extratropical areas but perhaps to increase again in the seasonally humid and humid tropics. The broad world pattern of drainage density may therefore be analogous to the pattern recently proposed for sediment yield in relation to mean annual precipitation (Wilson, 1973). Precipitation intensity is one possible reason for the existence of high drainage density values in markedly seasonal climatic regimes. An indication of intensity may be afforded by the ratio between average maximum monthly precipitation and average annual precipitation. According to this index drainage density values are low in areas with index values up to 12–13% but they are substantially higher where the index approaches or exceeds 20%. This index is not satisfactory because it may disguise the magnitude of actual storm rainfall intensities. For example, in Malaya (Eyles, 1968) there were 16 storms in 1966 with intensities greater than 76 mm/hour although the ratio of maximum month to annual precipitation of 11·9% does not readily reflect this. An improved index of precipitation intensity may be afforded by the ratio of the maximum reported 24-hour rainfall to the average annual precipitation amount. In Figure 10.3 drainage density values are related to this index and this shows how world drainage density values increase as precipitation intensity increases. Bearing in mind the difficulties of drainage network definition, it therefore appears that there are substantial differences in drainage network density according to the pattern of world climates. The highest values hitherto reported are found in areas with a markedly seasonal climate and network density appears to be broadly related to an index of precipitation intensity.

The drainage network appears to be related to high intensity storms and this may be analogous to conclusions reached concerning channel geometry and river channel pattern. It has been demonstrated that the magnitude of river channel corss-section is related to an event of a particular frequency. Thus Leopold, Wolman and Miller (1964) suggested that, for many rivers, the channel-forming discharge appears to possess a recurrence interval of approximately 1·5 years. More recently a number of parameters of peak streamflow have been related to channel capacity as for example by Brown (1971) who found a direct positive relationship between channel capacity and the five-year flood discharge. Studies of river channel pattern have also provided relationships between indices of meander geometry and flood discharge and the relationship between meander wavelength and bankfull discharge was employed by Dury (1964). More recent work has pointed to the fact that the bankfull discharge does not necessarily have a recurrence interval of 1·5 years (e.g. Kennedy (1972)), and that the channel-forming discharge is necessarily represented by a range of flows rather than by a single flow value. Thus Carlston (1965) concluded that meander wavelength is related to a 'dominant discharge' which is a range of flows, possibly falling stage flows, between the mean discharge of the month of maximum discharge and the mean annual discharge and that this range of flows is equalled or exceeded for 10–40% of the year in the humid regions which he studied in the United States. In a study of meandering rivers in Indiana, Daniel (1971) showed that meander change was influenced by a range of higher flows commencing just above mean flow. The most recent studies have exposed the complexity of the relations between discharge and channel geometry and river channel pattern and they have indicated that the significance of other factors, including bed and bank material, vegetation and slope, must be included. However, it has emerged that the geometry of a river channel and of the channel pattern may be related to a range of discharges greater than the average discharge. The channel cross section and the pattern of a particular reach may thus be in quasi-equilibrium with this range of channel-forming flows. These flows, above the average discharge, must usually be occasioned by quickflow or storm runoff and therefore related to storm precipitation amount. Thus in just the same way that a general relation has been indicated between channel and pattern geometry with storm discharges, the drainage network appears to be related to high intensity storms. The drainage network and the channel geometry are both in quasi-equilibrium with an event or range of events of a particular magnitude. Although the precise significance of the variation of the drainage network is not yet fully appreciated in relation to the range of variation in river channel geometry, it is apparent that a closer understanding of the covariation of the two is necessary.

## 10.3 Variation within morphoclimatic zones

The significance of climatic characteristics and particularly of precipitation intensity has already been established by a number of studies (Melton, 1957; Chorley and Morgan, 1962). A study of drainage densities in eastern Australia allowed Abrahams (1972) to conclude that the density of valleys is related to climate indexed by mean annual precipitation. In this study a polynomial relationship indicated that maximum densities characterize areas with less than 180 mm annual precipitation, that values fall to a minimum at about 460 mm annual precipitation, and that subsequently density values increase again to a secondary maximum corresponding approximately to a mean annual rainfall between 1270 mm and 1520 mm. Analysis of drainage composition in relation to the broad climatic pattern of Iowa (Karsten and Tuttle, 1971) indicated that drainage densities are highest in the most southerly locations, and this was interpreted to signify that drainage composition (as indicated by drainage density and stream frequency) is more advanced or better organized in the areas of Iowa with more rainfall and with more frequent freezing and thawing. The study by Melton (1957) suggested an inverse relation between drainage density and the Thornthwaite (P–E) index but Madduma Bandara (1974) has demonstrated a significantly positive relation between the two variables for 24 basins of the central hills of Sri Lanka. Comparison of results from the southeastern parts of the United States with those from Sri Lanka led to the conclusion that above a certain critical level of precipitation the relationship between drainage density and the P–E index becomes positive (Madduma Bandara, 1974).

Whereas several analyses have concluded that some attribute of climate, possibly combined with vegetation character, exercises an important control over drainage networks, other studies have demonstrated that surface mantle or rock type are of paramount influence. Twenty-five basins in Connecticut were analysed by Wilson (1971) and indicated the way in which drainage density was adjusted to lithology. Similarly Donahue (1972) attributed the presence of three drainage intensity regions of the Allegheny Plateau to differences in glaciation and to variations in surface mantle characteristics. An analysis of the estimated drainage density of Dartmoor (Gardiner, 1971) underlined the significance of climate in combination with land use, vegetation and soil type. An extension of this work led to the conclusion that the pattern of estimated drainage density in southwest England must be interpreted according to the nature of rock types (Gregory and Gardiner, 1974). Although the influence of climate, expressed as mean annual precipitation, could be discerned in the pattern of estimated drainage density of southwest England, allowance had to be made for the fact that networks behave characteristically on particular rock types.

The consensus of opinion, from the limited number of studies available to date, is that within morphoclimatic zones values of drainage density reflect climate, but that rock type is responsible for details of the spatial pattern. It may be that areas with sparse vegetation cover and with higher values of drainage density (Figure 10.2) are more susceptible to the effects of precipitation intensity than is the case in humid landscapes where precipitation intensities are lower, drainage densities are also lower, and where network variation is contributed substantially by contrasts in rock type.

The way in which drainage densities vary within a single morphoclimatic zone is illustrated by the pattern of densities for fifteen areas in Britain. The areas (Figure 10.4) were selected to represent a variety of mean annual rainfall totals, and drainage density measurements from ten sample basins in each area were based upon the stream networks shown on Provisional Edition 1:25,000 maps. Although the stream networks on this series consistently underestimate actual drainage densities, the map series is the only one available with wide coverage. Although the sample basins in each area were selected ranging in area from $0.11$ km$^2$ to $2.86$ km$^2$, the effect of basin area upon drainage density could still be significant and so the regression technique was employed to provide a comparable value of drainage density for each of the fifteen areas. When the average densities from each area are considered in relation to mean annual precipitation totals it is notable that there is a general association of increase in drainage density with increased precipitation amount. However, the range of density values in each area is also appreciable and there appear to be two groups of areas. By grouping the fifteen areas according to whether permeable or impermeable rock type prevails, it was possible to distinguish two relationships (Figure 10.5). In each case there is a significant relationship between drainage density and mean annual precipitation and the association of increase of drainage density with increasing mean annual precipitation is probably a reflection of increasing precipitation intensities. Rodda (1967) has indicated that precipitation intensity is broadly related to mean annual precipitation in the UK and in Table 10.2 results from this analysis are referred to mean annual precipitation totals and also to precipitation intensity characteristics. Whereas precipitation intensity may account for the way in which drainage density increases with rainfall over impermeable rock types, on

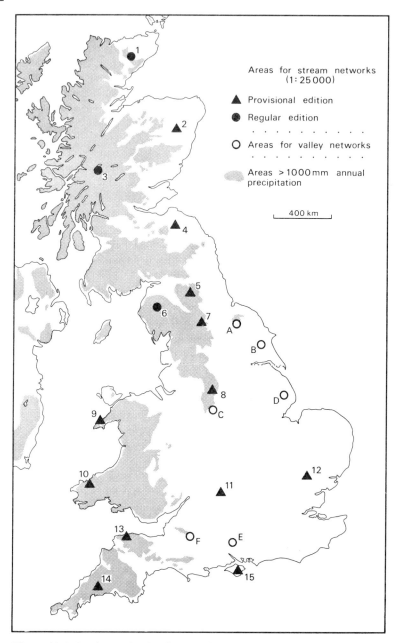

Figure 10.4. Drainage density for sample areas in relation to mean annual precipitation. Areas 1–15 are explained in Table 10.2 and valley networks analysed for areas A to F are explained in Table 10.3.

permeable outcrops the amount of water available to recharge groundwater flow may be a significant determinant so that the relation between drainage

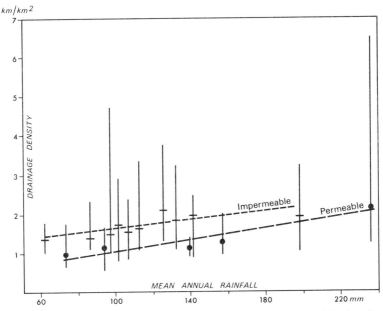

Figure 10.5. Drainage density in relation to mean annual precipitation for 15 areas in Britain. The 15 areas are located on Figure 10.4, the value of drainage density for each area is obtained from the regression relating total stream length and basin area, and the 15 areas are subdivided according to the dominance of permeable or impermeable rock types.

density and mean annual precipitation may then partly reflect the higher base flow discharge facilitated by a greater rainfall amount. It is notable therefore that the range of drainage density values within a specific area (Figure 10.5) is less on permeable than on impermeable rock types.

Distinction between permeable and impermeable rock types, reflecting the different contributions to the stream network, may have its counterpart in other areas. An analysis of the data from basins in Alaska (Figure 10.1) shows that the mapped streams are significantly less dense than the network of valleys indicated by contour crenulations. The significance of seasonal runoff due to thaw and melting snow in high latitude areas (e.g. Rudberg (1961) and Pissart (1967)) is known to produce a network of valleys which experience flow for a short portion of the year and these may be represented by the difference between the stream and valley networks (Figure 10.1). Analysis of the Alaska data indicates that whereas the stream densities and the densities of total streams and valleys are not significantly related to relief ratio of the basins or to average slope, the density of the valleys alone is significantly related to both indices. This may be indicative of the fact that seasonal expansion of the network is controlled by relief through its influence upon snowmelt, whereas the permanent network is largely independent of the relief factor.

Studies within particular zones therefore prompt a consideration of the

Table 10.2. Drainage Densities for 15 areas in Britain

| Area (Figure 10.4) | Drainage density (km/km²) | Average relief ratio (m/km) | Predominant rock type | Mean annual precipitation (mm) | Estimated daily rainfall of mean frequency once in 10 years[a] (mm) |
|---|---|---|---|---|---|
| 1  | 1·70 | 56  | I (Impermeable) | 1016 | 53  |
| 2  | 1·16 | 56  | P (Permeable)   | 940  | 53  |
| 3  | 2·28 | 183 | P | 2362 | 104 |
| 4  | 1·44 | 279 | I | 940  | 61  |
| 5  | 1·84 | 89  | I | 1321 | 53  |
| 6  | 1·89 | 64  | I | 1981 | 89  |
| 7  | 1·78 | 96  | I | 1118 | 48  |
| 8  | 2·06 | 190 | I | 1244 | 52  |
| 9  | 1·50 | 48  | I | 1067 | 56  |
| 10 | 1·81 | 99  | I | 1397 | 56  |
| 11 | 0·95 | 49  | P | 737  | 48  |
| 12 | 1·35 | 18  | I | 610  | 36  |
| 13 | 1·30 | 95  | P | 1575 | 81  |
| 14 | 1·12 | 92  | P | 1397 | 64  |
| 15 | 1·41 | 51  | I | 864  | 43  |

[a]Estimated daily rainfall of mean frequency once in 10 years is derived from a map in J. C. Rodda, A study of magnitude, frequency and distribution of intense rainfall in the United Kingdom, *British Rainfall 1966*, 204–215.

components of drainage density and of the way in which particular drainage densities may be the result of specific events. Understanding of this problem may be achieved in three ways: by study of the behaviour of small individual watersheds; by the analysis of particular and often extreme events; and by the interpretation of controlled experiments. Thus in a humid temperate area in southeast Devon Gregory and Walling (1968) showed that drainage density values in a single catchment can range from 0·6 to 3·1 and from 1·2 to 6·2 according to the input of individual storms transformed into expanded drainage densities according to antecedent conditions, temperature conditions and basin characteristics of soil and vegetation. In a study of the fluctuation of the source of a first order stream in a tropical area Morgan (1971b) showed how the source points of first order streams were related to individual climatic events and to changes in the importance of throughflow and overland flow above stream heads. He showed that the amount of precipitation in the two hours previous was a major determinant of stream source position.

Individual events can also indicate the potential extent of drainage networks and a catchment in southeast Devon which normally has a drainage density ranging between 2·7 and 4·1 (Gregory, 1971b) gave rise to a density of 4·4 as a result of a storm on 18–19 July 1972 when 74·9 mm rain fell during two hours with a maximum intensity of 40 mm/hour. That drainage density did not

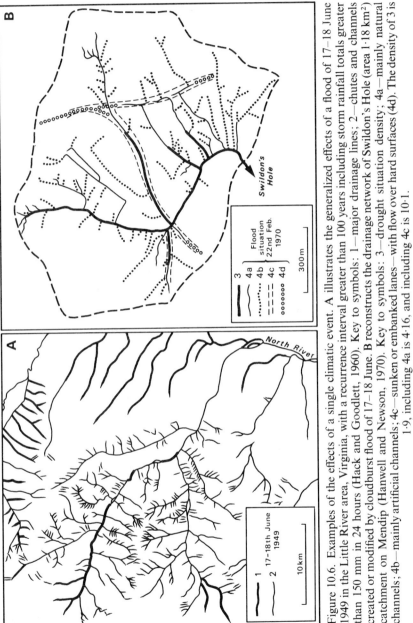

Figure 10.6. Examples of the effects of a single climatic event. A illustrates the generalized effects of a flood of 17–18 June 1949 in the Little River area, Virginia, with a recurrence interval greater than 100 years including storm rainfall totals greater than 150 mm in 24 hours (Hack and Goodlett, 1960). Key to symbols: 1—major drainage lines; 2—chutes and channels created or modified by cloudburst flood of 17–18 June. B reconstructs the drainage network of Swildon's Hole (area 1·18 km$^2$) catchment on Mendip (Hanwell and Newson, 1970). Key to symbols: 3—drought situation density; 4a—mainly natural channels; 4b—mainly artificial channels; 4c—sunken or embanked lanes—with flow over hard surfaces (4d). The density of 3 is 1·9, including 4a is 4·16, and including 4c is 10·1.

increase further was due to the fact that catchment characteristics, particularly vegetation and soil type, dictated that the watershed could contain the further excess by overland flow and by adjustment of the character of existing channels. More extreme events such as the Lynmouth flood of August 1952 included instances of extension of the drainage network (e.g. Green (1955)) but it is notable that this increase was limited and that substantial changes of pattern and channel geometry were occasioned by the extreme rainfall. The effect of an intense rainstorm may lead to an expansion of the drainage network and one example studied by Hack and Goodlett (1960) is illustrated in Figure 10.6A. In this case it was notable that the effect of the storm was manifested in some network change but was accompanied by mass movements and by channel changes. An intense storm over a permeable catchment was instanced by the Mendip flood of 10 July 1968 when up to 160 mm rain fell during 24 hours. Hanwell and Newson (1970) reconstructed the drainage network for the catchment during a lower flood situation and showed how drainage density varied (Figure 10.6B). According to the relationships which usually obtain between drainage density and discharge it was argued that maximum drainage density during the Mendip flood may have been as much as 4·73 km/km$^2$ but this necessarily supposes that channel character was unchanged.

A final approach to the problem may be provided by results from simulation models. Although these have not yet been used extensively some measurements under controlled conditions have been undertaken. Williams and Fowler (1969) indicated the way in which drainage density may vary with precipitation input by comparing the number of segments of fourth order basins in two areas. One area received 30% more precipitation than the other and more segments per unit area occurred in this area. More specifically Blyth and Rodda (1973) measured the length of flowing channel in a small clay catchment at weekly intervals for one year. The variation in network density was equivalent to a range from 0·55 to 2·70 and specific values of the network state were shown to be related to effective rainfall, preceding rainfall and soil moisture condition.

Comparatively few theoretical approaches have hitherto been directed towards the nature of drainage density in relation to climatic elements and in relation to processes more generally. Thus Wooding (1966), in a study of the catchment stream problem, concluded that a better geometrical description of the stream network is desirable. In general the availability of moisture for the drainage net should be directly related to the density of that net and precipitation intensity is significant because it overcomes the 'resistance' afforded by basin characteristics, by storage changes and by evapotranspiration. However, it appears, provisionally at least, that drainage density in a particular area, although dynamic in nature, will not continue to increase indefinitely and at the same rate, as precipitation intensity increases. This was implicit in the variable system proposed by Melton (1958) who suggested that negative feedback probably operates to keep the channel length constant if basin area, slope runoff rate, slope erosion rate and perhaps soil creep rate are constant.

On a small scale an increase in channel discharge at a valley head would lead to an increase in channel length which would reduce the area drained by the channel head and so tend, by negative feedback, to maintain channel length, which is basic to drainage density, at a constant mean value. However if P–E index and precipitation intensity are changed the resulting morphological changes could be permanent. More generally, King (1970) has suggested that negative feedback appears to be dominant in fluvial geomorphology and therefore the most dramatic changes of drainage density may require a change in climate or a change in basin character.

## 10.4 Drainage network changes

In view of the sensitive nature of the drainage network it is expectable that changes of climate should be reflected in drainage network changes. River metamorphosis (Schumm, 1969) has been identified as a response to climatic change and it is likely that accompanying changes of drainage networks should also have occurred. Such network metamorphosis can occur either by expansion of the existing network or by contraction.

Expansion of the drainage network has been identified most extensively in areas where the seasonality of precipitation is pronounced and where vegetation cover is sparse. Although it has prove difficult to distinguish between the effects of climatic change and the impact of man-induced change, as expressed through a change in vegetation cover for example, gullies indicative of network expansion have been recorded from the Mediterranean basin, from New Zealand and from the southeastern part of the United States. The significance of climatic change in relation to gullying is not easily resolved because, whereas an increased wet season with more intense storms could lead to an extended drainage network, it can alternatively be argued that a period of more pronounced drought may be more instrumental in leading to an extended drainage network because of the concomitant reduction in the effectiveness of the vegetation cover (Yi Fu Tan, 1966).

Whereas extension by gullying is a much-studied feature of semi-arid areas, in humid temperate landscapes the dry valley, indicative of network contraction, is more apparent. Classically the dry valley was visualized as a landform of permeable limestone outcrops but it has been appreciated that dry valleys occur on other rock types. It has also been realized that a variety of dry valleys exist and that they include: steep-sided features often with a steep head in scarp situations; more subdued but definite valley forms on dip slopes; a variety of shallow cradle-like valleys (*vallons en berceau*) and *dellen* forms; and suffosion valleys. Thus the problems of definition of the drainage network are compounded when one attempts to identify the valley network precisely. A useful guide may be to include those valleys and depressions which appear, from map evidence or from field examination, to continue the lines of the present stream network. Morphological evidence of changes in the valley network is provided by the study of the Roztocze of Goraj in Poland by Buraczynski (1968). He identified, on this siliceous limestone plateau area, *vallons en berceau* as small features with a convexo-concave cross profile which were developed under the action of processes of solifluction and ablation. Dry valleys with a mean density of 1·33 were developed during cold conditions by processes of solifluction and melting. These two types of fossil elements in the valley network were compared with the network of recent erosion with densities up to 8·52. The forms of this area illustrate the way in which climatic history may be retained in the network and accordingly Buraczynski (1968) proposed a sequence of stages of development from the Pliocene to the present.

Evidence of network metamorphosis can also be provided by sections and by studies of sediments. In the Manuherikia valley of South Island New Zealand, Leamy (1972) identified ridge line notches and he assembled pedologic evidence to indicate that periodic dissection and soil formation alternated with periods during which gullies were being filled under regimes which were characterized by mass transport. He concluded that the fine dissection of semi-arid central Otago is not a contemporary feature but that the valley densities of as much as 16 are a legacy from the waning phase of a glacial period when surface runoff from frozen ground produced runoff volumes higher than those of the present.

Some studies of network metamorphosis have endeavoured to illuminate the significance of specific climatic parameters. Thus in a study of a limestone area of Barbados, Fermor (1972) compared the valley densities of 26 sample areas, representing a range of valley densities from 0·13 to 5·86, with the present frequency of annual extreme 24-hour rainfall and showed that there was a significant correlation between present precipitation intensity and dry valley density. This was interpreted to suggest that surface runoff was responsible for the development of the now-dry valleys, although this could have taken place during former climates which had a spatial pattern comparable to that of the present.

These three studies indicate that network metamorphosis by contraction is a feature of a variety of areas and that it may be studied initially be reference to the density of the valley network. The general relationships (Figure 10.2, 10.3) may indicate the way in which changes of climate involving changes in precipitation amount and intensity could be expressed in network change. Evidence of former greater network densities is available from areas of limestone outcrop and, although dry valleys of these areas have been studied in Britain for more than a century, the application of network analysis to the problem has been a fairly recent development.

In Britain the dry valley densities of the Chalk of the Yorkshire Wolds were sampled by Lewin (1969) from randomly located quadrats of different size and found to range from 0·9 to 1·1. The Chalk basins of the Ivinghoe area of the Chilterns were shown by Brown (1969) to have valley densities of 1·65 and 2·39 and, in a study of 54 basins of the Chalk landscapes of southern England, Morgan (1971a) found a mean density of 2·3. Although Morgan (1971a) suggested the presence of equilibrium and non-equilibrium segments in the Chalk dry valley networks it has been shown that there is no sound morphometric basis for this distinction (Ellis Gryffydd, 1973). The values of dry valley density quoted from these several areas thus range from 0·9 to 2·4 and it is necessary to establish the extent to which variation occurs in other areas.

This objective can be achieved by sampling valley densities for basins in six areas (Figure 10.4) including outcrops of Chalk, Jurassic Limestone and Carboniferous Limestone. For each area a sample of basins ranging in size from 0·8 km² to 75 km² was selected to indicate the magnitude of valley densities. Analysis of this data revealed a consistent density of dry valleys in each of the six areas and the average density of dry valleys in all areas is 1·86. Using the regression between valley lengths ($\Sigma L_v$) and basin area ($A$) the average densities range from 1·14 to 2·32. It is apparent that the highest average densities occur in the areas with the greatest average slope indexed by relief/basin area. This relation to relief and to slope can be further illuminated because in all six areas there is a significant correlation of dry valley density with relief ratio and also with average slope, whereas the relief ratio or average slope do not relate significantly in any area to the present stream density. Therefore although dry valley densities are related to relief of the basins, the present stream network is not related to a parameter of relief in this way. This may indicate that, whereas the present stream networks are the result of perennial groundwater flow, the valley densities are an expression of networks produced by intermittent or ephemeral flow in relation to relief. Furthermore, when the average valley densities of limestone outcrops (Table 10.3) are compared with the present densities on other permeable outcrops in Britain (Figure 10.5) it is apparent that these valley densities are broadly similar to the stream densities which at present occur on permeable outcrops.

Two principal groups of reasons have been offered to explain the presence of dry valleys as remnants of former drainage networks. One group visualizes the occasional flow of water in ephemeral or intermittent courses as a possible but infrequent occurrence. Such explanations are supported by the occasional

Table 10.3. Network densities for limestone areas of Britain

| Area (Figure 104) | Rock type | Stream density | Valley density Average | Range | Relief per unit area (m/km²) | Present mean annual precipitation |
|---|---|---|---|---|---|---|
| A | Jurassic Limestone | 0.47 | 2·32 | 1·45–3·73 | 97 | |
| B | Chalk | 0.35 | 1·14 | 0·94–1·58 | 13 | 719 |
| C | Carboniferous Limestone | — | 1·72 | 0·75–2·94 | 59 | 1029 |
| D | Chalk | 0.71 | 1·51 | 1·25–1·79 | 16 | 635 |
| E | Chalk | 0.38 | 1·89 | 1·71–2·02 | 28 | 912 |
| F | Jurassic Limestone | — | 1·74 | 1·36–2·13 | 23 | |

flow recorded in dry valleys, by the dependence of the valley density upon relief which should directly influence precipitation and runoff, and by the similarity in form between dry valleys and valleys at present drained by a stream. The subsequent abandonment of the dry valleys (Gregory, 1971a) may have been occasioned by water table lowering through reduced precipitation, scarp recession, sea level lowering or stream incision which led to a lowering of the saturation level in the interfluves. Manley (1964) has suggested that the persistence of seasons slightly wetter than those of today could have led to spring lines substantially higher than present. A second group of explanations requires recourse to climatic conditions substantially different from those of the present time and these suggestions have two implications. First, is the proposal that the dry valley networks may have been fashioned during periglacial conditions when snowmelt produced higher runoff rates which would have been greatly accentuated over permafrost: a striking valley of the Chalk of the North Downs has been ascribed to the effects of niveofluvial processes between 10,800 B.C. and 10,300 B.C. (Kerney, Brown and Chandler, 1964). Secondly, it has been recognized that past valley development under a contrasted periglacial morphogenetic system could have seen a much less pronounced distinction between mass-movement by solifluction and water-movement by streamflow. Thus water may have flowed seasonally, but at other times of the year solifluction movement may have been directed along the line of the valley. This is supported by the fact that in some areas small convexo-concave *dellen* occur tributary to larger dry valleys recording the nature of network and valley development during cold climatic conditions. On sandstones in southeast Devon, *dellen* have been identified (Gregory, 1971a) but in the same area, and on some southern Chalk outcrops, some large dry valleys are infilled by solifluval deposits and have slopes mantled by these deposits so that it has been argued that these larger forms were not substantially modified by morphogenesis under cold conditions.

It appears likely that in Britain the remnants of former valley networks cannot be explained by a single model of development. Although extensive areas of dry valleys may represent a former seasonal expansion of the drainage net, it seems that at least two types of element occur in the present valley network. The larger valley elements appear to be a phenomenon not produced under climatic conditions substantially different from those of the present, whereas shallow dry valleys and *dellen* are manifestations of a different climate when snowfall was considerable and mean annual temperature was perhaps 9 °C lower than present. These climates saw some valleys created by seasonal expansion of the drainage net and others created by a combination of water flow and solifluval processes. The present valley network may thus be the cumulative result of the effect of a number of contrasted climates.

## 10.5 Conclusion

Drainage networks vary substantially with climate and particularly in relation to precipitation minus evapotranspiration (P–E) ratio. Acceptance of the drainage network as a dynamic characteristic of drainage basins necessarily directs attention towards two aspects of the network. The *basic network* which flows for much of the year is governed by groundwater flow or by throughflow and therefore depends upon mean annual precipitation as modified by basin characteristics and particularly by rock type. The *expanded network* which occurs once or more each year is a response to individual climatic events and is related particularly to precipitation intensity. This duality of the drainage network, which incidentally has been responsible for variations in approaches to drainage network definition, is apparent in several ways at different scales. Worldwide values in drainage density tend to be highest in semi-arid areas, to be lowest in humid temperate areas, and perhaps to increase again in the tropics. However, this picture is essentially a comparison of maximum network densities. Whereas the density of the perennial network may increase simply with mean annual rainfall amount, the maximum densities in several areas directly reflect the effect of precipitation intensity. Thus the greatest range of drainage density values, both spatially and temporally during the year, may feature in semi-arid areas. Within morphoclimatic zones the significance of precipitation intensity is also apparent. Within a single basin it is evident that the network expands and contracts annually over a substantial range but, according to the studies made of specific events, this range is not infinite because negative feedback operates to reduce the rate of expansion of the network under a specific climatic regime or to eliminate some of the effects of a particular event. Stoddart (1968) isolated two alternative hypotheses regarding drainage density. The first states that drainage density increases with rainfall as a greater amount of rainfall per unit area would require a greater amount of channel per unit area. The second states that drainage density decreases with increasing rainfall as vegetation density, transpiration, moisture retention and other factors increase. Both hypotheses appear to apply: the first to the basic network and the second to the expanded network of a specific area.

Appreciation of the components of the drainage network and their spatial pattern (Morgan, Chapter 11) is fundamental in interpreting adjustments of networks in relation to climatic change. The dry valleys of many areas may represent the expanded element of former networks which did not survive when the network contracted to a largely perennial basic one.

**References**

Abrahams, A. D., 1972, Drainge densities and sediment yields in eastern Australia, *Australian Geographical Studies*, **10**, 19–41.
Ackermann, E., 1936, Dambos in Nordrhodesian, *Wiss. Veroffentl. Dtschen Mus. Landerkunde zu Leipzig*, n.s. **4**, 149–157.
Black, P. E., 1970, Runoff from watershed models, *Water Resources. Res.*, **6**, 465–477.
Blyth, K. and J. A. Rodda, 1973, A stream length study, *Water Resources. Res.*, **9**, 1454–1461.
Brice, J. C., 1966, Erosion and deposition in the loess-mantled Great Plains, Medicine Creek drainage basin, Nebraska, *U.S. Geol. Surv. Prof. Paper*, *352H*, 255–339.

Brown, D. A., 1971, Stream channels and flow relations, *Water Resources. Res.*, **7**, 304–310.
Brown, E. H., 1969, Jointing, aspect and orientation of scarp-face dry valleys, near Ivinghoe, Bucks., *Trans. Inst. Brit. Geog.*, **48**, 61–73.
Büdel, J., 1963, Klima-genetische geomorphologie, *Geog. Rundschau*, **7**, 269–286.
Büdel, J., 1972, Typen der Talbildung in verschiedenen klimamorphologischen Zonen, *Zeitschrift für Geomorphologie*, Suppl. 14, 1–20.
Budyko, M. I., 1958, *The heat balance of the earth's surface*, US Dept. Commerce.
Buraczynski, J., 1968, Type dolin Roztocza Zachodniego, *Annales Universitatis Mari Curi-Sklodowska Lublin-Polonia*, **23**, 47–86.
Carlston, C. W., 1963, Drainage density and streamflow, *U.S. Geol. Surv. Prof. Paper*, 422C.
Carlston, C. W., 1965, The relation of free meander geometry to stream discharge and its geomorphic implications, *Am. J. Sci.*, **263**, 864–885.
Chorley, R. J., 1958, Climate and morphometry, *J. Geol.*, **65**, 628–638.
Chorley, R. J. and M. A. Morgan, 1962, Comparison of morphometric features, Unaka mountains Tennessee and North Carolina, and Dartmoor, England, *Bull. geol. Soc. Am.*, **73**, 17–34.
Chorley, R. J., R. P. Beckinsale and A. J. Dunn, 1973, *The history of the study of landforms: the life and work of W. M. Davis*, Methuen, London.
Contton, C. A., 1964, The control of drainage density, *N. Z. J. Geol, Geophys.*, **7**, 348–52.
Daniel, J. F., 1971, Channel movement of meandering Indiana streams, *U.S. Geol. Surv. Prof. Paper*, 732A.
Davies, J. L. (Ed.), 1965, *Atlas of Tasmania*, Lands and Surveys Department, Hobart.
Donahue, J. J., 1972, Drainage intensity in western New York, *Ann. Ass. Am. Geog.*, **62**, 23–36.
Doornkamp, J. C. and C. A. M. King, 1971, *Numerical analysis in Geomorphology*, Edward Arnold, London.
Dury, G. H., 1964, Principles of underfit streams, *U.S. Geol. Surv. Prof. Paper*, 452A.
Ellis Gryffydd, I. D., 1973, Morphometric studies and their relevance to a study of the denudational history of the English Chalklands, *Trans. Inst. Brit. Geog.*, **58**, 131–132.
Engelen, G. B., 1973, Runoff processes and slope development in Badlands National Monument, South Dakota, *J. Hydrol.*, **18**, 55–80.
Eyles, R. J., 1968, Stream net ratios in west Malaysia, *Bull. geol. Soc. Amer.*, **79**, 101–112.
Fermor, J., 1972, The dry valleys of Barbados: a critical review of their pattern and origin, *Trans. Inst. Brit. Geog.*, **57**, 153–165.
Galon, R., 1964, Hydrological research for the needs of the regional economy, *Problems of Applied Geography II, Geographica Polonica.* **3**, 239–50.
Gardiner, V., 1971, A drainage density map of Dartmoor, *Trans. Devon Assoc.*, **103**, 167–180.
Gilbert, G. K., 1880, *Geology of the Henry Mountains*, US Geog. and Geol. Survey of the Rocky mountain region, Washington.
Ghose, B., S. Pandey, S. Singh and G. Lal, 1967, Quantitative geomorphology of the drainage basins in the central Luni basin in western Rajasthan, *Zeitschrift für Geomorph.*, **11**, 146–160.
Green, G. W., 1955, North Exmoor floods, August 1952, *Bull. Geol. Surv. G.B.*, **7**, 68–84.
Gregory, K. J., 1966, Dry valleys and the composition of the drainage net, *J. Hydrol.*, **4**, 327–340.
Gregory, K. J., 1971a, Drainage density changes in south west England, *Exeter Essays in Geography*, K. J. Gregory and W. L. D. Ravenhill (Eds.), Exeter.
Gregory, K. J., 1971b River networks in Devon, *Transactions and Proceedings Torquay Natural History Society*, **16**, 4–11.
Gregory, K. J. and D. E. Walling, 1968, The variation of drainage density within a catchment, *Bull. Int. Assoc. Sci. Hyd.*, **13**, 61–68.

Gregory, K. J. and D. E. Walling, 1973, *Drainage basin form and process*, Edward Arnold, London.
Gregory, K. J. and V. Gardiner, 1975, Drainage density and climate, *Zeits. für Geom.*, **19**, 287–298.
Hack, J. T., 1965, Postglacial drainage evolution and stream geometry in the Ontonagon area, Michigan, *U. S. Geol. Surv. Prof. Paper, 504B*, 1–40.
Hack, J. T. and J. C. Goodlett, 1960, Geomorphology and forest ecology of a mountain region in the central Appalachians, *U. S. Geol. Surv. Prof. Paper, 347*.
Hanwell, J. and M. D. Newson, 1970, The great storms and floods of July 1968 on Mendip, *Wessex Cave Club Occasional Publication 1*, 1–72.
Hare, F. K., 1973, Energy-based climatology and its frontier with ecology, *Directions in Geography*, R. J. Chorley (Ed.), 171–192.
Heathcote, R. L., 1965, *Back of Bourke. A study of land appraisal and settlement in semi-arid Australia*, Melbourne University Press.
Horton, R. E., 1932, Drainage basin characteristics, *Trans. Am. Geophys. Union*, **13**, 350–361.
Horton, R. E., 1945, Erosional development of streams and their drainage basins: hydrophysical approach to quantitative morphology, *Bull. geol. Soc. Am.*, **56**, 275–370.
Howe, G. M., H. O. Slaymaker and D. M. Harding, 1967, Some aspects of the flood hydrology of the upper catchments of the Seven and Wye, *Trans. Inst. Brit. Geog.*, **41**, 33–58.
Institute of Hydrology, 1972, Subsurface processes, *Institute of Hydrology Research 1971–2*, 25–6.
Jamieson, D. G. and C. R. Amerman, 1969, Quick return subsurface flow, *J. Hydrol.*, **8**, 122–136.
Jones, A., 1971, Soil piping and stream channel initiation, *Water Resources. Res.*, 7, 602–610.
Karsten, R. A. and S. D. Tuttle, 1971, The distribution of climatic factors in Iowa and their influence on geomorphic processes, *Iowa Acad. Sci. Proc.*, **77**, 266–281.
Kennedy, B. A., 1972, 'Bankful' discharge and meander forms, *Area*, **4**, 209–212.
Kerney, M. J., E. H. Brown and T. J. Chandler, 1964, The late-glacial and post-glacial history of the Chalk escarpment near Brook, Kent, *Phil. Trans. Roy. Soc.*, Series B, **248**, 135–204.
King, C. A. M., 1970, Feedback relationships in Geomorphology, *Geogr. Annlr.*, **52A**, 147–159.
Leamy, M. L., 1972, Fine-textured dissection in semi-arid central Otago, *N. Z. J. Geol. Geophys.*, **15**, 394–405.
Leopold, L. B., M. G. Wolman and J. P. Miller, 1964, *Fluvial Processes in Geomorphology*, Freeman, New York.
Lewin, J., 1969, The Yorkshire Wolds: A study in geomorphology, *University of Hull Department of Geography Occasional Paper*.
Madduma Bandara, C. M., 1974, Drainage density and effective precipitation, *J. Hydrol.*, **21**, 187–190.
Manley, G., 1964, The evolution of the climatic environment in *The British Isles*, J. W. Watson and J. B. Sissons (Eds.), Nelson, London, 152–176.
Maxwell, J. C., 1960, Quantitative geomorphology of the San Dimas experimental forest, California, *Office of Naval Research, Geography Branch, Project NR 389–042, Technical Report 19*.
McCoy, R. M., 1971, Rapid measurement of drainage density, *Bull. geol. Soc. Amer.*, **82**, 757–762.
Melton, M. A., 1957, An analysis of the relations among elements of climate, surface properties and geomorphology, *Office of Naval Research, Geography Branch, Project NR 389–042: Technical Report 11*, Columbia University.
Melton, M. A., 1958, Correlation structure of morphometric properties of drainage systems and their controlling agents, *J. Geol.*, **66**, 442–460.

Miller, V. C., 1953, A quantitative geomorphic study of drainage basin characteristics in the Clinch mountain area: Va and Tenn., *Office of Naval Research, Geography Branch, Project NR 389–042, Technical Report 3*, Columbia University.

Morgan, R. P. C., 1971a, A morphometric study of some valley systems on the English Chalklands, *Trans. Inst. Brit. Geog.*, **54**, 33–44.

Morgan, R. P. C., 1971b, Observations on factors affecting the behaviour of a first-order stream, *Trans. Inst. Brit. Geog.*, **56**, 171–185.

Morisawa, M. E., 1957, Accuracy of determination of stream lengths from topographic maps, *Trans. Amer. Geophys. Union*, **38**, 86–88.

Morisawa, M. E., 1962, Quantitative geomorphology of some watersheds in the Appalachian plateau, *Bull. geol. Soc. Amer.*, **73**, 1025–1046.

Peltier, L. C., 1962, Area sampling for terrain analysis, *Prof. Geogr.*, **14**, 24–28.

Pinchemel, P. H., 1957, Densités de drainage et densités des vallées, *Tijd. Kon. Ned. Aard.-Kun. Genoots.*, 373–376.

Pissart, A., 1967, Les modalites de l'écoulement de l'eau sur l'Ile Prince Patrick, *Biul. Peryglacjalny*, **16**, 217–224.

Rodda, J. C., 1967, A countrywide study of intense rainfalls for the United Kingdom, *J. Hydrol.*, **5**, 58–69.

Rudberg, S., 1961, Geomorphological processes in a cold semi-arid region, *Axel Heiberg Island Research Reports*, McGill University, 139–150.

Schumm, S. A., 1956, The evolution of drainage systems and slopes in badlands at Perth Amboy, New Jersey, *Bull. geol. Soc. Amer.*, **67**, 597–646.

Schumm, S. A., 1969, Rivert Metamorphosis, *Proc. A. S. C. E. J. Hyd. Div.*, **HYL 6352**, 255–273.

Selby, M. J., 1968, Morphometry of drainage basins in areas of pumice lithology, *Proc. fifth New Zealand Geog. Conference. New Zealand Geog. Soc.*, 169–74.

Smart, J. S. and A. J. Surkan, 1967, The relation between mainstream length and area in drainage basins, *Water Resources. Res.*, **3**, 963–974.

Smith, K. G., 1950, Standards for grading texture of erosional topography, *Am. J. Sci.*, **248**, 655–668.

Smith, K. G., 1958, Erosional processes and landforms in Badlands National Monument, South Dakota, *Bull. geol. Soc. Amer.*, **69**, 975–1008.

Stoddart, D. R., 1968, Climatic geomorphology: review and reassessment, *Progress in Geography* **1**, 160–222.

Strahler, A. N., 1952, Hypsometric (area–altitude) analysis of erosional topography, *Bull. geol. Soc. Amer.*, **63**, 1117–1142.

Tricart, J., 1953, Géomorphologie dynamique de la Steppe Russe, *Rev. de Géom. Dyn.*, **4**, 1–32.

Tricart, J., 1972, *The Landforms of the Humid Tropics, Forests and Savannas*, Longmans, London.

Visher, S. S., 1954, *Climatic Atlas of the United States*, Harvard University Press.

Werritty, A., 1972, Accuracy of a stream link lengths derived from maps, *Water Resources. Res.*, **8**, 1255–1271.

Williams R. E. and P. M. Fowler, 1969, A preliminary report on an empirical analysis of drainage network adjustment to precipitation input, *J. Hydrol.*, **8**, 227–238.

Wilson, L., 1971, Drainage density, length ratios and lithology in a glaciated area of south Connecticut, *Bull. geol. Soc. Amer.*, **82**, 2955–2956.

Wilson, L., 1973, Relationships between geomorphic processes and modern climates as a method in paleoclimatology: in *Climatic Geomorphology*, E. Derbyshire (Ed.), Macmillan, London, 269–284.

Wooding, R. A., 1966, A hydraulic model for the catchment-stream problem. III Comparison with runoff observations, *J. Hydrol.*, **4**, 21–37.

Woodyer, K. D. and M. Brookfield, 1966, The land system and its stream net, *C. S. I. R. O. Div. of Land Resources Tech. Mem. 66/5*.

Wright, L. W., 1973, Landforms of the Yaruna granite area: Viti Levu, Fiji: A morphometric study, *J. Trop. Geog.*, **37**, 74–80.

Ye Grishankov, G., 1973, The landscape levels of continents and geographic zonality, *Soviet Geography*, **14**, 61–78.

Yi Fu Tan, 1966, New Mexican Gullies: A critical review and some recent observations, *Ann. Ass. Am. Geog.*, **56**, 573–97.

Zavoianu, I., 1964, Determination of the drainage density of the hydrographic network based on Horton's Laws *Rev. Roum. Geol. Geophys. et Geogr., Serie de Geographie*, **13**, 171–179.

CHAPTER ELEVEN

# The Role of Climate in the Denudation System: a Case Study from West Malaysia

ROYSTON P. C. MORGAN

## 11.1 Introduction

The validity of climatic geomorphology depends on being able to establish a relationship between climate and process and on being able to show that regional variations in critical climatic factors cause spatial differences in landform. The importance of climate in relation to other landforming factors must be evaluated. In this way climatic geomorphology can progress beyond the descriptive morphogenetic studies of Büdel (1963) and Tricart and Cailleux (1965, 1972) and improve upon the models of climate–process systems of Peltier (1950), Leopold, Wolman and Miller (1964), Common (1966) and Wilson (1973) which are constrained by their dependence on mean climatic data. As an example of this approach a case study is presented of a climatic geomorphological analysis of West Malaysia. A model of the denudation system is applied to a study of spatial variations in two gross landform parameters, drainage density and texture. By analysing the interaction between the various components of the system, the relative importance of climate is evaluated. Some indication of the value of the study to resource evaluation in the context of national development is also presented.

## 11.2 Framework of Study

### 11.2.1 The denudation system

The denudation system provides the basic model for studies in climatic geomorphology. It is a functional system in which inputs of energy, both potential and kinetic, carry out work and its varied components control the form of the landscape which is, therefore, a surface expression of the system. The basic components of the system, according to the work on landscape dynamics (Kalesnik, 1961 and 1968; Isachenko, 1968; Rikhter, 1970; Zonneveld, 1972), are rocks, structure, relief, solar radiation, air masses, water, soils, vegetation, man and animal life. The components can be described in more detail by a number of carefully chosen parameters (see below).

*11.2.2 Drainage density and texture*

The selection of drainage density and texture as gross landform parameters, in preference to other aspects of landform, has several advantages. First, both are indices of the drainage system. Second, the drainage system is the surface expression of a sub-system of the denudation system, being influenced by inputs of precipitation and solar energy and outputs of discharge, evaporation and re-rediation (Chorley, 1969). Third, previous studies have shown that drainage density is one of the parameters most sensitive to environmental conditions, varying with rock type (Pinchemel, 1957; Strahler, 1964), climate (Melton, 1957; Chorley and Morgan, 1962), relief (Slaymaker, 1968) and time (Ruhe, 1952).

Drainage density, a statement of the length of drainage lines per unit area, is an index of gross landform, being associated with the horizontal distance between stream channels (Horton, 1945) and the degree of ruggedness (Strahler, 1950). It is also an index of the hydrological characteristics of an area and appears to be both a response to the factors controlling runoff (Carlston, 1966; Orsborn, 1970) and a control of the runoff itself (Kirkby and Chorley, 1967). The importance of this central position of drainage density to studies of fluvial systems has been emphasized by Gregory (Chapter 10). Drainage density may also have considerable value as an index of water yield (Schumm, 1968).

Although drainage density is commonly used as a measure of landscape texture it is not particularly valuable as such because, as shown by Leopold, Wolman and Miller (1964), two basins may have the same density but different stream spacing owing to differences in the number of streams and their component lengths. To take account of this, a very simple index of drainage texture is adopted in this paper, namely the number of stream source points (taken as equal to the number of first-order streams) per unit area. This index, concentrating as it does on the headwater areas is thought to have considerable potential as an index of erosion, being very closely allied to the indices of gully density which form an important part of studies evaluating the severity of soil erosion (Keech, 1968; Stocking, 1972).

Kirkby and Chorley (1967) have indicated that drainage density is partly controlled by events in streamhead hollows but there is some evidence to suggest that this is only partially true. In some instances, the behaviour of first-order streams is a poor indicator of the behaviour of larger drainage basins to which the index of drainage density is more closely related. Morgan (1972) quotes a case during the January 1971 floods in West Malaysia when a major and widespread rise in river levels of the large drainage basins had no parallel in the behaviour of first-order streams where only a limited rise in water was recorded. Rapid rises in first-order streams appear to be closely controlled by meteorological events influencing the occurrence of subsurface and overland flow on adjacent hillslopes (Morgan, 1972) whereas, in larger watersheds, part of the runoff, and possibly a high percentage in West Malaysia (Douglas, 1971), is contributed by groundwater flow. The different behaviour of small and large drainage basins has also been observed in studies of sediment yield. McGuinness, Harrold and Edwards (1971) have shown that in Ohio, USA, sediment yield from basins of less than 1 ha is closely correlated with rainfall intensity and energy whereas that from larger basins is correlated with runoff.

Drainage density is thus employed in this paper as an index of the response of fluvial systems to the denudation system on a medium and large scale, and texure is used as an index of response on a smaller scale. That the two indices express different properties of the drainage system may explain the poor correlation obtained between them for West Malaysia (Morgan, 1971a) and a similarly poor correlation between drainage density and gully density found in Rhodesia (Stocking, 1972).

### 11.2.3 West Malaysia

Although the use of West Malaysia as the area of study was largely conditioned by the author's affiliation at the time the work was carried out (Morgan, 1971a), there are other advantages in choosing this area. First, there is no evidence for climatic change during the Tertiary and Quaternary periods and it is therefore realistic to assume that the present landforms are closely adjusted to the present denudation system. Second, there exists adequate documentation of most aspects of the environment.

West Malaysia is dominated by two landform associations: strongly dissected highlands and low-lying coastal and riverine plains (Figure 11.1). Intrusive granitic rocks underlie much of the area and differential erosion has removed most of the overlying sediments, exposing the granites which form the north–south trending Banjaran Titiwangsa and associated ranges. Sedimentary rocks form low hills and scarplands in the west and centre but in the east they remain as the high Banjaran Tahan (Gunung Tahan, 2190 m) and the Trengganu Plateau. Extensive lowlands, formed in part by alluviation following a post-Pleistocene rise in sea level, occupy almost all the west coast and the lower portions of the main river valleys in the east. Hillslopes rise sharply from the lowlands and the change from low to high relief is rapid. Lowland granitic areas may be weathered to depths of 20 m or more but thinner regoliths characterize sandstone areas and most mountain regions. Rock outcrops are rare, being largely restricted to tors and corestones on granitic and sandstone terrain, areas of tower karst, and razorback ridges of vein quartz.

The climate is an equatorial monsoon type with two main seasons: a south-west monsoon from April to September and a northeast monsoon from October to March. The mean daily maximum temperature is 32°C and the mean annual precipitation is between 2290 mm and 2510 mm. In spite of an apparent uniformity of climate over the whole country there is an important regional distinction between the east coast and the west. The east coast has a strongly seasonal precipitation regime, $p^2/P$ values (Tables 11.1) being over 100, and

Figure 11.1 West Malaysia.

receives rains of high intensity, the daily total with a 10-year return period being over 250 mm compared with less than 150 mm over much of the west where the regime is more equable (Morgan, 1971b).

## 11.3 Procedure

### 11.3.1 Selection of parameters

The selection of parameters to describe the various components of the denudation system is governed largely by the availability of suitable information and

Table 11.1. Parameters selected for landscape description

| Component | Parameter | Comments |
|---|---|---|
| Rock type | Porosity<br>Permeability | In the absence of sufficient information on mineralogy and jointing, only a general description is given. The dominant rock type is determined from the 1/500,000 Geological Map of Malaya (Alexander, 1965), and the appropriate values of porosity and permeability given in Leopold, Wolman and Miller (1964) are used. |
| Soil type | Percentage clay<br>Percentage sand<br>Silt/clay ratio | The dominant soil type is obtained from the 1/500,000 Provisional Soil Map of West Malaysia (Law and Selvadurai, 1968). The clay and sand fractions of the soil influence its erodibility (Ellison, 1944) and the silt/clay ratio is an index of weathering (van Wambeke, 1962). |
| Landform | Average slope<br>Hypsometric integral | The dominant landform type, according to the classification of Eyles (1968b) is obtained from his map. Average slope and hypsometric integral are two of the parameters on which his classification is based. |
| Land cover | Runoff ratio | The dominant land cover is obtained from the 1/63,360 topographical maps and the parameter is expressed by the following runoff ratios: 25% for tropical rainforest (Tan, 1969; Kenworthy, 1969); 35% for tree-crop cultivation (Tan, 1969); 30% for padi cultivation on wet, alluvial lowlands (an estimate based on an application of Cook's Method, US Soil Conservation Service, 1953); and 50% for urban areas (also based on Cook's Method). More recent research, although not affecting the relative values, has indicated that the ratios may be too low: values of 45–50% for rainforest in upland areas (Low, 1971) and 60–65% for urban areas (Low and Goh, 1972) have been quoted. |

| Component | Parameter | Comments |
|---|---|---|
| Precipitation | Mean annual precipitation<br>Seasonality of precipitation<br>Number of months with at least a 1% probability of 400 mm<br>Return period of 50 mm daily total<br>Return period of 100 mm daily total<br>Daily total with 10-year return period | Although precipitation is the most variable element of the Malaysian climate, the number of parameters expressing this component is limited to six. A previous study (Morgan, 1971b) has shown that as many as twelve parameters expressing characteristics of precipitation in West Malaysia can be represented by four principal components. The six parameters used here cover these components and include some of the frequency–magnitude parameters commonly adopted in hydrological studies. The seasonality of the precipitation regime is expressed by $p^2/P$ which Fournier (1960) has successfully correlated with sediment yield ($p$ = highest mean monthly precipitation; $P$ = mean annual precipitation; values in millimetres). Data are taken from Wycherley (1967) and frequencies and magnitudes are estimated by semi-logarithmic extrapolation of data in Lockwood (1967). |
| Evaporation | Mean annual evaporation<br>Difference between mean annual precipitation and evaporation | Mean annual evaporation is estimated according to the procedure of Nieuwolt (1965). |
| Temperature | Mean annual temperature | Data from Dale (1963). |
| Sunshine | Mean annual sunshine hours/day<br>Seasonality of sunshine | Data from Dale (1964). The seasonality index is $sh^2/SH$, where $sh$ = highest mean monthly sunshine hours/day and $SH$ is mean annual sunshine hours/day. Sunshine is used as a guide to solar radiation for which insufficient data are available. |
| Altitude | Altitude of the meteorological station | Altitude influences the geomorphology of drainage basins in two ways: (1) climatic factors change with altitude; (2) basins at different altitudes differ in their degree of dissection (Eyles, 1969). |
| Map error | Date of publication of map sheets | The accuracy of information on drainage shown on the maps varies with the date of publication, the more recent maps being more accurate (Eyles, 1966). |

Note 1. Selection of terrain parameters, describing the non-climatic aspects of the landscape, is based on criteria used in the various systems of land classification and evaluation (Beckett and Webster, 1962; Zonneveld, 1972).

Note 2. Climatic parameters are chosen to represent energy conditions in the denudation system. Whilst mean climatic data are used, an attempt is made to adopt other, more meaningful indices, based on daily, monthly and seasonal variations in climate.

the ease with which it may be numerically expressed. Details of the parameters, together with reasons for their selection and the source material used, are given in Table 11.1. The climatic parameters are chosen to represent energy conditions in the system while the terrain parameters largely describe those aspects of the physical and cultural environment which influence the circulation of energy. The input of potential energy for erosion is also expressed to some extent by the two parameters describing landform type. An additional parameter is included to determine to what extent spatial variations in drainage density and texture reflect variations in the accuracy of the topographical maps from which their values are derived.

*11.3.2 Measurement of drainage density and texture*

The values of drainage density and texture used in this study are derived from the 1/63,360 topographical map sheets of the Jabatanarah Pemetaan Negara, Malaysia (Department of National Mapping, Malaysia). The measurement of density and texture from maps raises certain problems regarding 'standardization'. These are discussed fully elsewhere (Gregory, 1966 and 1968; Giusti and Schneider, 1963) but two aspects may be considered here. First, in a study of climatic geomorphology, it is unrealistic to standardize density and texture values on the permanent drainage network, as indicated by the blue lines on the maps, because most landforming processes are more active during conditions of high runoff when water is flowing in the intermittent first-order channels than during base flow. Clearly, the value of density and texture should be representative of the 'effective' drainage system under these conditions. Second, Eyles (1966) has shown that the drainage network is often portrayed inconsistently on the Malaysian topographical maps. Not only are first-order streams generally omitted but the accuracy of the network shown appears to vary from one map sheet to another, largely in relation to the date of publication. The more recent maps are the more accurate. Unfortunately, aerial photographs do not provide a feasible alternative to maps for stream measurement because of the difficulty of seeing the stream system under a rainforest canopy.

For the purpose of this study, the drainage shown on the maps is taken as commencing at the second Strahler order and a map error factor is included as a parameter in the analysis (see Table 11.1). Correction factors, indicated below, are applied to obtain values for density and texture which, it is believed are representative of the 'effective' drainage system.

Two maps (Figure 11.2 and 11.3) show spatial variations in density and texture. The texture map was based on a stratified random sample of third-order drainage basins. Within each 100 km$^2$ grid square on the maps sheets, a single 1 km$^2$ grid square was selected, using random numbers as coordinates. As third-order basins do not always fall within or across the first square selected, particularly in areas of incoherent drainage and swamp, twelve squares were examined in turn. If no third-order basin was found after twelve tries, that 100 km$^2$ grid square remained unsampled. The number of first-order streams in each

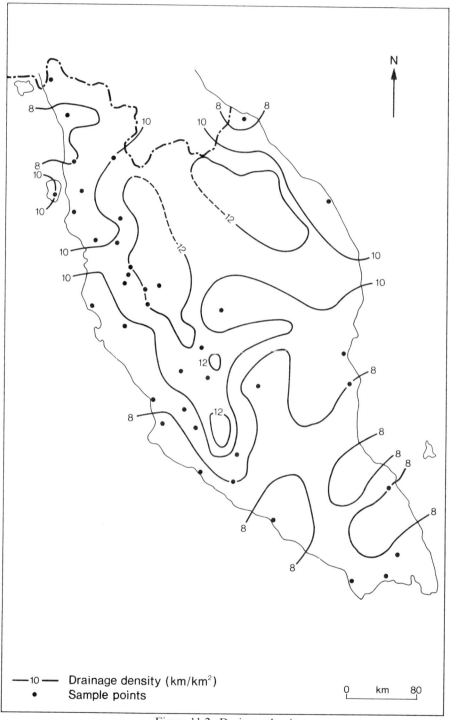

Figure 11.2 Drainage density.

325

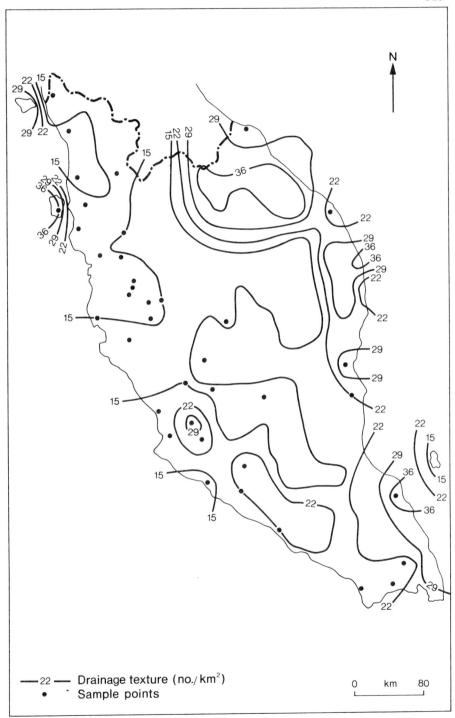

Figure 11.3. Drainage texture.

sample basin was estimated by multiplying the number of second-order streams observed on the map by 3·5 which is the mean bifurcation ratio for West Malaysia (Eyles, 1968a). Average texture values were calculated for each of the 132 map sheets, each sheet generally containing between 9 and 12 sample basins, and these values were used to construct the texture map (Figure 11.3).

To ensure independent sampling of drainage density, a slightly different technique was adopted. The average values for each map sheet are based on the density values for the 410 fourth-order basins sampled by Eyles (1968, unpublished) which gives between four and five sample basins on most map sheets. The values, taken directly from Eyles's work, have already been adjusted by his correction factor (Eyles, 1968, unpublished, p. 42). It should be stressed, however, that the applicability of this factor to mountainous, forested areas remains unproven, as it is based on a sample study of fifty basins which are biased towards lowland areas under tree-crop cultivation.

Although there is a difference of one Strahler order in the magnitude of the basins sampled for estimating texture and density, the area sampled in both cases is comparable, because of the difference in the number of sample basins, and represents about 3% of their respective total basin populations in West Malaysia. Further comments on the validity of this size of sample to show regional trends are given by Eyles (1968b, unpublished).

### 11.3.3 Spatial variations in drainage density and texture

As indicated in Figure 11.2, there is only a small range in drainage density. Density is highest inland and decreases towards the coast and is also highest in the north, decreasing southwards, although the coastal plains of Kedah and Perlis are anomalous in this respect. The highest densities are recorded in the Trengganu Plateau and in much of the Banjaran Titiwangsa.

In contrast, texture values (Figure 11.3) show a greater range, being particularly high on the east coast, although some areas of low texture are found in the lowlands of Kelantan and Trengganu and extending inland along the Pahang valley. The lowest textures occur inland, particularly over much of the Banjaran Titiwangsa and on the west coast where Pulau Pinang and the Kuala Lumpur areas stand out as enclaves of high texture.

### 11.3.4 Assignment of sample areas

Only thirty-nine sample points are used in the regional analysis, their distribution (Figure 11.1) being biased towards the western side of West Malaysia. The lack of sufficient climatic data for most of the eastern and central parts of the country precludes more extensive random or stratified sampling. Such sampling would require estimating climatic data by extrapolation from isoline maps which would introduce considerable sample error. The thirty-nine sample points correspond to major recording stations of the Perkhidmatan Kajichuacha Malaysia (Malaysian Meteorological Service). Where certain climatic information is not available, equivalent values for the nearest station in the same climatic sub-region (Morgan, 1971b) are used. Terrain parameters describe the landscape of an area, 15 km in radius, surrounding the sample points.

*11.3.5 Statistical analysis*

In its simplest form the statistical problem is one of assessing the association between drainage density and texture on the one hand and the twenty-one parameters of the denudation system on the other. Although multiple correlation and regression analyses serve this purpose, they are often difficult to interpret when so many parameters are involved because of the problem of intercorrelation. Before using correlation and regression techniques it is helpful to examine the structure of the denudation system independently of the density and texture values. This structure can be stated in the form of a correlation matrix and analysed by examining the correlations between every pair of parameters comprising the matrix. Such an analysis is simplified, however, if the matrix is re-ordered so as to emphasize the major trends. Further, if the re-ordering reduces the number of parameters required to describe the denudation system, it is probable that the relationship between them and drainage density and texture will be easier to interpret.

The principal components solution is the simplification procedure adopted here. This solution emphasizes the main trends of a correlation matrix by analysing the pattern of intercorrelation between parameters. The correlation matrix is replaced by a component matrix which is derived from the identification of a number of statistically uncorrelated (orthonormal) combinations of the parameters comprising the correlation matrix through an application of the algebra of eigenvalues and eigenvectors (Kendall, 1957). The importance of each component is expressed by its eignevalue. The higher the eigenvalue, the more important is the component because of the larger number of intercorrelated parameters contributing to it. In this study only components with eigenvalues greater than unity are identified.

The strength of association between a parameter and a component is expressed by the loading value of the parameter on that component. The parameters comprising a particular component can thus be recognized by their high loadings and, by setting an arbitrary limit to the loading values, in this study equal to or greater than $\pm$ 0·50, the composition of each component can be determined. The association of several parameters on a single component raises the question of whether causal relationships exist between the parameters. Similarly, the identification of statistically independent components raises the question of whether conceptual independence also exists. Generally, these questions can be answered by interpretation and it is only when a logical interpretation of association or lack of it can be given that the components have meaning in anything other than a statistical sense.

The positions of the orthogonal axes of each component in space can be manipulated or 'rotated' to aid interpretation. A variety of rotation procedures is available but the one most suited to this analysis is the varimax rotation (Kaiser, 1958) which satisfies the condition of 'simple structure'. The rotation is achieved by maximizing some loadings and minimizing others so that fewer parameters contribute to each component. Once individual components have been identified from the rotated component matrix, component scores for each sample area are calculated in respect of each component. The scores show how each area contributes to the variation in each component.

The adoption of a principal components solution for simplifying a correlation matrix has the advantage that it can be used with various types of data (e.g. ordinal, nominal, ratio) and with normal and skewed distributions. The scores derived from the rotated component matrix however, are normally distributed (King, 1969) and therefore in suitable form for use in correlation analysis.

A problem of principal components analysis which is rarely stressed concerns the ratio between the number of parameters and the number of observations. The ratio affects the number of degrees of freedom and the importance of having considerably more observations than parameters has been emphasized (T. Leinbach, personal communication). The ratio used here, consisting of thirty-nine observations to twenty-one parameters, is workable but by no means ideal.

Drainage density and texture are analysed for their association with the scores of the individual components. The association is expressed by linear and multiple correlation coefficients. Regression analysis is used to predict values of drainage density and texture in terms of the score values of the significantly correlated components and the residual values are examined.

It should be emphasized that each component is treated separately at this stage. It is possible to carry out the analysis using average scores for all the components combined but to do so is of little advantage. It is difficult to evaluate the contribution of individual components in a correlation analysis based on average component scores. Further, an average score usually assumes that each component is of equal weight whereas, as is obvious from the eigenvalues, some components are more important than others. Although this importance could be expressed by a weighting system, the value of such a procedure is somewhat dubious because the importance of an individual component can be spurious. This arises where a large number of like parameters are included in the analysis. In this study, the majority of the parameters are climatic and it is to be expected that these form the most important components because of their numerical superiority and not because of any real significance. Any weighting procedure would produce average scores with values biased towards the climatic components. In order to be as objective as possible, the problems of weighting and average scores are avoided. Each component is analysed separately for its association with drainage density and texture. Thus, even though the majority of the components are identifiable as climatic ones, it is theoretically possible for them all to show poor correlations with density and texture and for all the non-climatic components to show high correlations.

## 11.4 Results of the analysis

### 11.4.1 Identification of principal components

The application of a principal components solution reduces the twenty-one parameters listed in Table 11.1 to eight components which, together, explain 83% of the variation in the original data. Using a varimax rotation of the component matrix (Table 11.2) and a cutting-off point of ± 0·50 for parameter loading values, the components are identified in Table 11.3. Although some, particularly those with low eigenvalues, are difficult to identify because of the association of parameters with no obvious causal connection, none can be discarded. This is because all the components are required for computing component scores for correlation with drainage density and texture in the next state of the analysis. The scores are derived from the varimax component matrix for each sample area in respect of each component separately.

### 11.4.2 Correlation analysis

Significant linear correlations exist between drainage density and the scores of components 1, 2, and 3 (Table 11.4) and, when combined, these components explain 36% of the variation in density. Although component 8 is not significantly associated with density in a linear correlation, its addition to the other three components improves the coefficient of determination to 56%. A significant linear correlation exists between texture and the scores of component 3 which explain 12% of the variation in texture. The addition of component 6, itself not significantly associated with texture, improves the coefficient of determination to 17%. The addition of further components does not significantly improve the correlation.

## 11.5 Interpretation

Isoline maps of the scores of components 1, 2, 3 and 8 aid interpretation by enabling regional trends in the score values to be compared visually with the regional variations in drainage density and texture (compare Figure 11.4 with Figures 11.2 and 11.3).

### 11.5.1 Variations in drainage density

The association between density and the four components significantly related to it in a multiple correlation (Table 11.4) is examined for each component in turn.

(a) *Component 1.* This component explains 15% of the variation in density. The contrasting signs of the parameters contributing to this component, the frequency of daily rainfall of 100 mm, average slope and altitude (positive) and mean annual evaporation, mean annual daily sunshine and mean annual

Table 11.2. Varimax component matrix

|  | Components | | | | | | | | |
|---|---|---|---|---|---|---|---|---|---|
|  | 1 | 2 | 3 | 4 | 5 | 6 | 7 | 8 | h |
| Eigenvalue | 4·4 | 3·6 | 2·6 | 2·0 | 1·4 | 1·3 | 1·2 | 1·0 |  |
| Cumulative percentage explanation | 21 | 38 | 50 | 60 | 67 | 73 | 79 | 83 |  |
| Mean annual precipitation | −04 | 93 | 18 | 08 | −12 | −08 | −08 | −11 | 94 |
| Precipitation seasonality | −06 | 26 | 81 | 17 | −17 | 02 | 22 | −15 | 85 |
| Probability of 400 mm/month | −16 | 81 | −04 | −01 | 05 | −04 | −32 | −03 | 80 |
| Return period of 50 mm/day | 32 | −27 | 30 | −38 | 19 | 61 | −11 | −02 | 83 |
| Return period of 100 mm/day | 91 | −06 | −15 | 11 | 04 | −20 | −08 | 03 | 91 |
| Rain/day with 10-year return period | −23 | 17 | 81 | −06 | −23 | 27 | 15 | 02 | 88 |
| Mean annual evaporation | −56 | −40 | 17 | 13 | −05 | −21 | −25 | −29 | 71 |
| Precipitation less evaporation | 13 | 96 | 11 | 03 | −10 | −01 | 01 | −01 | 96 |
| Mean annual temperature | −76 | 06 | −25 | 10 | 13 | −11 | −04 | −35 | 80 |
| Mean annual sunshine hours/day | −58 | 07 | 08 | 16 | 05 | 01 | −09 | −66 | 81 |
| Sunshine seasonality | −10 | 08 | −03 | 10 | 03 | 12 | 03 | −87 | 81 |
| Rock porosity | −12 | −25 | 27 | −01 | 02 | −06 | 85 | 00 | 88 |
| Rock permeability | −11 | −10 | −05 | 18 | 81 | 02 | −02 | −05 | 71 |
| Percentage clay in soil | −10 | −15 | 21 | −80 | −16 | 03 | −21 | 25 | 85 |
| Percentage sand in soil | −04 | −09 | 19 | 90 | −01 | 04 | −18 | −07 | 90 |
| Silt/clay ratio | 28 | 55 | −26 | −26 | 45 | −18 | 22 | −26 | 87 |
| Hypsometric integral | 31 | 04 | 03 | −10 | −02 | −83 | 02 | 08 | 80 |
| Slope | 60 | 25 | −06 | −05 | 28 | −37 | −35 | −21 | 80 |
| Altitude | 93 | 01 | −02 | 05 | −05 | −13 | −08 | 05 | 90 |
| Runoff ratio | −29 | 24 | −68 | 12 | −25 | 34 | 06 | −10 | 81 |
| Date of map publication | 08 | −04 | −20 | −17 | 57 | 17 | 03 | 53 | 71 |

$h$ = communality. For information on the parameters, see Table 11.1.

Table 11.3. Principal components of the denudation system

| Component | Identification |
|---|---|
| 1 | Mainly climatic: return period of 100 mm daily rainfall; mean annual temperature; mean annual evaporation; mean annual sunshine hours/day; average slope; altitude. The association of the last two parameters with this component reflects the change in climatic parameters with height. |
| 2 | Mainly climatic: mean annual precipitation; excess of mean annual precipitation over evaporation; number of months with at least a 1% probability of 400 mm of rain; silt/clay ratio. |
| 3 | Mainly climatic: seasonality of precipitation; daily rainfall with a 10-year return period; runoff ratio. |
| 4 | Soil: percentage clay and percentage sand fractions. |
| 5 | Rock permeability; date of map publication. |
| 6 | Return period of 50 mm daily rainfall; hypsometric integral. |
| 7 | Rock porosity. |
| 8 | Sunshine: mean annual sunshine hours/day; seasonality of sunshine; date of map publication. |

Note. Identification is based on the varimax component matrix (Table 11.2) using a cutting-off point of ± 0·50 for parameter loading values.

Table 11.4. Correlation coefficients

| Variables | | Correlation coefficients | | Significance level |
|---|---|---|---|---|
| Dependent | Independent (Components) | Linear (r) | Multiple (R) | % |
| Drainage density | 1 | 0·39 | | 5 |
| | 2 | 0·34 | | 5 |
| | 3 | −0·40 | | 1 |
| | 8 | −0·19 | | 25 |
| | 1, 3 | | 0·51 | 0·1 |
| | 1, 2, 3 | | 0·60 | 0·1 |
| | 1, 2, 3, 8 | | 0·75 | 0·1 |
| Drainage texture | 3 | 0·34 | | 5 |
| | 5 | −0·33 | | 5 |
| | 6 | 0·29 | | 10 |
| | 3, 5 | | 0·39 | 5 |
| | 3, 6 | | 0·41 | 1 |
| | 3, 5, 6 | | 0·45 | 1 |

Significance levels at 0·1, 1, 5, 10 and 25% are based on Students' *t*-Test, assuming normal distributions.

temperature (negative), indicate that in areas of high precipitation, cloud conditions are sufficient to decrease temperature and evaporation. Such conditions are common at high altitudes and it is these areas which have high drainage densities. The isoline map (Figure 11.4(a)) shows that high scores occur in the mountain areas such as the Cameron Highlands and in the interior of West Malaysia, for example Kuala Lipis. The scores decrease in value towards the coast and southwards showing the same regional pattern as drainage density.

(b) *Component 2.* This component explains 12% of the variation in density. High densities are associated with areas of high mean annual precipitation, more months with a probability of 400 mm of rain and a high excess of precipitation over evaporation. A comparison of Figures 11.2 and 11.4(b)) shows that the high densities of the Kinta Valley and the foothills of the Banjaran Titiwangsa and the low densities of the Selangor and Perak coasts, the Muar Valley and the Perlis–Kedah lowlands are strongly influenced by high and low scores respectively on this component. Insufficient information exists to assess the significance of this component on the high densities of the Banjaran Timor.

(c) *Component 3.* This component accounts for 15% of the variation in density to which it is inversely related, indicating that a high density is associated with low values of daily rainfall with a 10-year return period and a low $p^2/P$ value. This implies that a greater length of stream per unit area is maintained where rainfall is equally distributed throughout the year and where

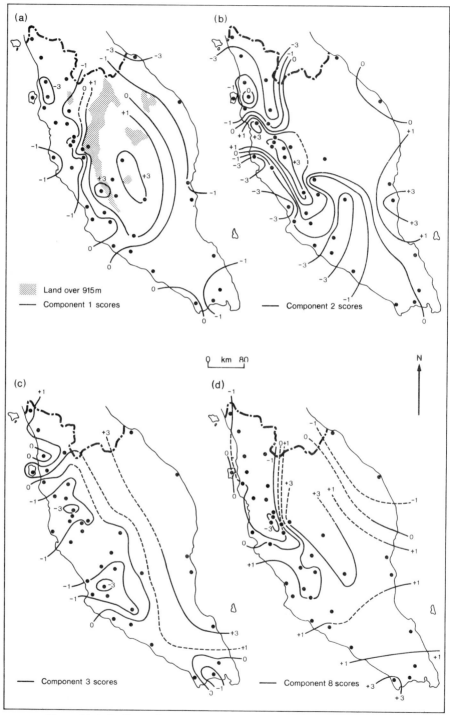

Figure 11.4. Isolines of component scores (a) component 1; (b) component 2; (c) component 3; (d) component 8.

storms of only moderate magnitude are experienced. Strongly seasonal regimes, such as those of the east coast, do not result in great channel lengths. The third parameter contributing to this component is the runoff ratio which loads negatively, indicating that a high density is associated with a high runoff. The influence of this parameter on drainage density is not strong, however, for, with the exception of the Ipoh area (Kinta Valley), few of the urban and mining areas record high densities.

(d) *Component 8.* Although this component, which significantly improves the multiple correlation coefficient when added to components 1, 2 and 3, is made up of the two parameters expressing sunshine and the date of map publication, it seems probable that its association with drainage density largely reflects sunshine conditions. This is indicated by the low linear correlation ($r = 0.03$) between the date of map publication and drainage density. Further, the low correlation between density and mean annual daily sunshine ($r = 0.05$) compared to the high correlation between density and the seasonality of sunshine ($r = 0.52$) indicates that the seasonal regime of solar energy is the most important parameter. The negative association between this component and density, and the negative loading of the parameter for sunshine seasonality on the component, means that high densities are associated with strong seasonality of sunshine. The reason for such an association is not at all clear although it has been suggested elsewhere (Budyko, 1958; Douglas, 1969) that higher radiation in the dry seasons of a tropical climate might lead to the greater availability of energy for geomorphological work and that this might be expressed in greater degrees of denudation. The relationship is, however, extremely complex. Density values, for example, are relatively low on the east coast where sunshine seasonality is fairly high. Possibly, in this seasonal climate, evaporation rates during the dry season are also high and much solar energy must be utilized in this process. In contrast, in the Kinta Valley and in the interior of West Malaysia, strong seasonality of sunshine is accompanied by a more equable rainfall regime; evaporation rates are lower and so more energy may be available for other work. It is in these areas that high drainage densities are recorded. It is, however, premature to draw conclusions too readily. Much more research is required into energy balances at the ground surface before relationships between solar energy and drainage density can be explored. Even then, the proportion of solar energy which becomes available for geomorphological work is probably so small that it is obscured by observational errors in the measurement of radiation (I. Douglas, personal communication). A further possibility which needs to be eliminated before the drainage density–solar energy relationship is pursued is that the association is not seriously affected by a 'map error factor' in the measurement of drainage density. Although map error does not seem to affect this study adversely in general terms it is possible that the high densities recorded, for example in the Kinta Valley area, could reflect local aberrations in the accuracy of the maps.

(e) *Residual analysis.* Regression formulae for predicting drainage density in terms of combinations of components 1, 2 and 3 and 1,2,3 and 8 respectively

Table 11.5. Regression equations

| Variables | Equation | $F$-value | Significance level % |
|---|---|---|---|
| $Dd \times 3$ | $Dd = 14 \cdot 73 - 0 \cdot 45A$ | 6·91 | $2\frac{1}{2}$ |
| $Dd \times 1, 3$ | $Dd = 14 \cdot 73 - 0 \cdot 39A + 0 \cdot 23B$ | 6·45 | 1 |
| $Dd \times 1, 2, 3$ | $Dd = 14 \cdot 73 - 0 \cdot 39A + 0 \cdot 21B + 0 \cdot 25C$ | 6·49 | 1 |
| $Dd \times 1, 2, 3, 8$ | $Dd = 14 \cdot 83 - 0 \cdot 35A + 0 \cdot 46B + 0 \cdot 16C - 0 \cdot 72S$ | 10·81 | 0·1 |
| $T \times 3$ | $T = 50 \cdot 91 + 2 \cdot 38A$ | 4·95 | 5 |
| $T \times 3, 6$ | $T = 50 \cdot 53 + 2 \cdot 05A + 2 \cdot 15H$ | 3·61 | 5 |

$n = 39$.
$Dd$ = drainage density; $T$ = drainage texture.
1, 2, 3, 6 and 8 as variables refer to components identified in Table 11.3
$A, B, C, S$ and $H$ refer to scores on components 3, 1, 2, 8 and 6 respectively.

are given in Table 11.5 and the residual values have been mapped (Figure 11.5). Three areas have high residuals. First, the high density of the Kinta Valley is underestimated unless all four components are considered, thus emphasizing the importance of component 8 (solar energy or local map error) in explaining the density values of this area. Secondly, several areas of the west coast are overestimated. Thirdly, high residuals occur for Mentekab. In these second and third cases, it is probable that sampling errors resulting from either the measurement of drainage density or the assignment of climatic and terrain characteristics to the sample areas are the main cause of the residuals. A further possibility could be the omission of important parameters from the analysis (see below).

*11.5.2 Variations in drainage texture*

The association of texture with component 3 indicates that conditions of strong seasonality of rainfall with high daily totals with a 10-year return period give rise to a large number of stream source points per unit area. A comparison of Figures 11.3 and 11.4(c) shows that the high texture values of the east coast strongly reflect the influence of this component. Additional factors associated with high values of texture, as indicated by its correlation with component 6, are high frequencies of daily rainfalls of 50 mm and low hypsometric integrals. Evidence has been presented elsewhere (Morgan, 1972) to indicate, under West Malaysian conditions, a close association between the positioning of stream sources on a hillslope and the return period of 50 mm of rain per day. The nature of the association between texture and the hypsometric integral is less easy to determine partly because of the poor linear correlation between the parameters ($r = -0 \cdot 32$). Strahler (1952) has suggested that the hypsometric integral may be interpreted as representing a 'stage' of landform

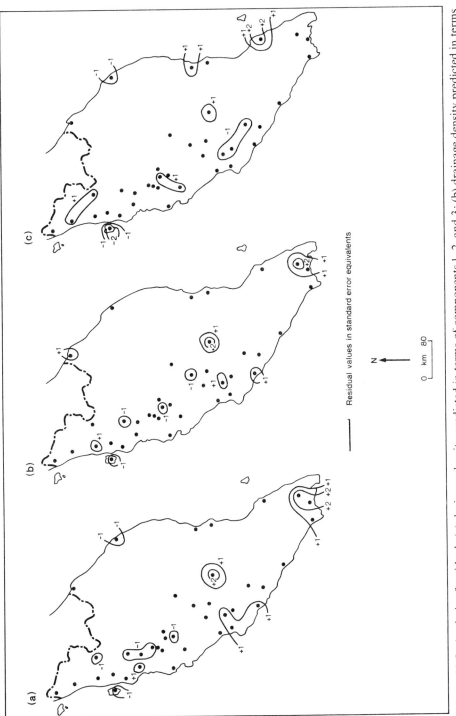

Figure 11.5. Analysis of residuals (a) drainage density predicted in terms of components 1, 2, and 3; (b) drainage density predicted in terms of components 1, 2, 3 and 8; (c) drainage texture predicted in terms of components 3 and 6.

evolution. Eyles (1969), however, in a study of hypsometric integrals in West Malaysia, has shown that much of the country is in a 'transitory monadnock phase' and what this means in terms of drainage texture and drainage development is not clear.

Residual values of texture predictions using a regression combining components 3 and 6 (Table 11.5) have been mapped (Figure 11.5(c)). Most of the overestimates of texture probably reflect sample errors similar to those indicated in the residual analysis of drainage density. The underestimates for Bayan Lepas (Pulau Pinang), Kuala Lumpur and Kajang are more interesting. It may be that the high textures of these areas reflect accelerated erosion because they are more densely populated or that the maps of the areas show more detail in their drainage information. There are no obvious reasons for the underestimates for Kuala Pilah and Kuala Trengganu.

*11.5.3 Limitations of the study*

Before summarizing the main findings, three limitations of this study should be mentioned. First, it is difficult to assess adequately the importance of map error in obtaining the values of density and texture. The inclusion of the date of map publication as a map error factor in the analysis proved inconclusive because it was unable to identify local variations in map accuracy. Second, the study is based on only 39 sample areas because sufficient climatic data was not available. These areas are biased towards the western side of the country and, although the east coast is reasonably well-sampled, large sections of the country, such as the Banjaran Timor and the Trengganu Plateau, are excluded. Third, the sample areas are biased by including few areas of either very high or very low land so that the importance of relief and altitude as controlling factors is not rigorously tested. This may account for the overall low degree of explanation of the variation in texture and for the consistently high residuals obtained for much of Johor and Melaka which are very low-lying.

*11.5.4 Summary*

For both drainage density and texture, the most significant parameters accounting for regional differences are the seasonality of the rainfall regime and the magnitude of daily rainfall with a 10-year return period. Where rainfall is evenly distributed throughout the year and of low intensity, high densities occur. On the other hand, strong seasonality and high intensity of rain appear to favour high textures. Thus a high density seems to be a necessary adjustment of the stream system to transport runoff from moderate, regular falls of rain, whilst more intense rains lead to more numerous source points and to runoff in a large number of short streams, giving rise to high textures but low densities. An exception to this pattern is found in the Banjaran Timor where both high densities and high textures occur.

## 11.6 Morphogenetic Regions

Gregory and Gardiner (1975) state that, because of its dependence on climate and basin form, drainage density provides a suitable basis for delimiting geomorphological regions in relation to morphoclimatic systems. Further, the regions so delimited may have considerable value for the evaluation of resources and the planning of land development. Regions of high density are associated with larger flood flows and low proportion of groundwater contribution to the discharge (Carlston, 1966; Orsborn, 1970). Regions of high texture are associated with greater intensity of streamhead extension and probably with greater sediment yield. Thus density and texture can be used as indices of water yield and conditions of erosion.

An extremely simple and arbitrary regionalization procedure is adopted here for West Malaysia. Three density regions are distinguished, based on isolines at 8 and 11 km/km$^2$, with subdivisions on the basis of texture, using the isoline of 23 source points/km$^2$. The regions themselves are essentially 'functional' in that their features express a certain relationship between gross landform, in so far as this is reflected in indices of density and texture, and the denudation system. The regions thus differ in concept from those recognized by Ho (1964), M.E.X.E. (1964), Swan (1970), Eyles (1971) and Lawrance (1972) which describe the 'form' of the landscape. The two concepts of regions need not be incompatible, however, for it is possible for regions based on form criteria to grade into those based on function by grouping at higher levels of the regional hierarchy. Indeed, such a gradation is implicit in many works on regional hierarchical systems (Vinogradov and coworkers, 1962; Mabbutt, 1968).

### 11.6.1 High density regions

These regions are centred on the highlands of the Banjaran Titiwangsa, Banjaran Gunung Bintang, Banjaran Timor, the Trengganu Plateau and the Banjaran Benom (Figure 11.6) but extend into the lowlands of the Kinta and Perak Valleys and the Dindings area. Particularly on the western side of the country, these areas are characterized by high annual rainfall totals and equable rainfall distribution throughout the year, the bulk of their rain falling in storms of moderate to low intensities. All these factors contribute to a high drainage density. Towards the east, drainage texture increases within this system as the seasonality and intensity of the rainfall become greater.

Much of the land in these regions is mountainous, with steep slopes, and a rainforest vegetation. The importance of the forest vegetation as a regulator of runoff in West Malaysia has been indicated by the drainage basin studies of Low and Goh (1972) who stress that the forests of the headwater regions must be protected. The high drainage densities indicate that runoff from these areas is naturally high. Thus any clearance of the forest cover and its replacement by other forms of land use is likely to increase considerably the risk of flood. A further hazard affecting the development of these areas is erosion.

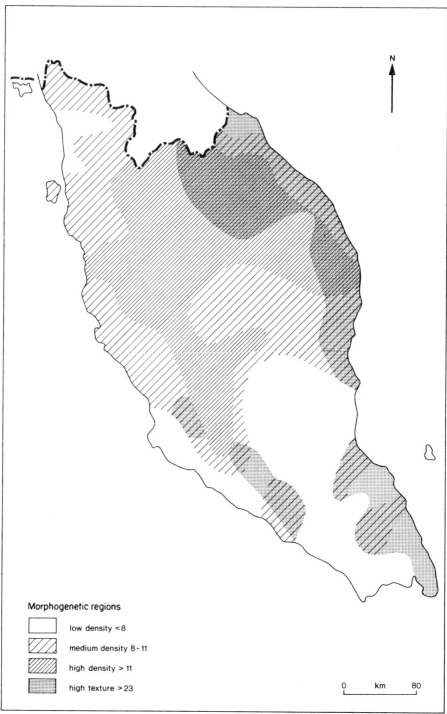

Figure 11.6. Morphogenetic regions.

Soil loss in a natural rainforest ecosystem is relatively low (Douglas, 1967) but increases rapidly where the forest is cleared for agriculture (Shallow, 1956) or worked for timber (Burgess, 1971). The erosion hazard is naturally high throughout the country once the delicate equilibrium of the ecosystem is disturbed, but is all the greater where drainage texture is high, as evidenced by the very rapid erosion associated with road construction for timber extraction in Kelantan (Berry, 1956). Great care is required in managing the development of the land in these regions otherwise the dangers of flooding and erosion will be much increased with consequent detrimental effects both locally and in areas downstream.

*11.6.2 Moderate density regions*

These regions form a belt encircling the regimes of high drainage density. Almost all the areas with any relief in the northern part of the country are included, together with a number of isolated hill masses in Johor. The regions are characterized by average conditions of the factors controlling density. Several areas of high texture occur, however, particularly on the east coast where the rainfall regime is more seasonal.

Much of the land already used for tree-crop cultivation lies within these regions as do those areas scheduled for agricultural development under the Second Malaysia Plan. Soil erosion has long been recognized as a problem and increases in sediment concentrations in rivers have been measured in the agricultural, mining and urban areas of this system (Ow Yang, 1965; Douglas, 1967, 1968, 1970). Few studies have been made, however, of soil loss from hillslopes or of the critical factors influencing that loss and no national study of the hazard of soil erosion has been made even though the implications of inadequately managed development in these areas have now been documented (Low and Leigh, 1972). Moderate to high texture values indicate those parts of the country where the erosion risk is moderate to high and in the Kuala Lumpur (Douglas, 1968) and Pulau Pinang areas (Soper, 1938) reflect the high degree of accelerated erosion which has already taken place. Further research into the dynamics of erosion in man-influenced ecosystems is required in West Malaysia together with very careful management of the land during its development.

*11.6.3 Low density regions*

These regions include most low-lying land such as the Perlis–Kedah plains, the Kelantan lowlands and the lowlands of Johor. The high texture of the east coast and parts of western Johor parallel the high textures found in similar areas with moderate drainage densities. Because a high proportion of the low density regions is made up of coastal and riverine plains, few problems of erosion and flooding are generated within them, but these plains lie downstream of the areas of moderate and high density systems and are therefore considerably affected by flooding and erosion generated within those areas. Environmental stability in the areas of low drainage density can only be maintained by good management of those areas with moderate and high drainage densities.

## 11.7 Evaluation

All four components contributing to the explanation of drainage density in West Malaysia contain climatic parameters, as do both of the components associated with texture. It must be concluded, therefore, that climatic factors strongly outweigh other aspects of the denudation system in controlling density and texture. Thus, climatic geomorphology appears to be a reasonable approach to the study of gross landforms in West Malaysia in so far as these forms can be represented by indices of drainage density and texture. In addition to providing a basis for a regional analysis of landforms in relation to climate, the study focuses attention on the relationship between climate and process in so far as drainage density and texture appear to be controlled by different sets of climatic parameters. The dependence of previous climatic geomorphological studies on mean climatic values has tended to obscure the dynamic relationship between climate and process. The status of climatic geomorphology will be improved if the frequency–intensity–duration patterns of climatic events are used as parameters in future studies. The scope of the subject should also be expanded to develop studies in water resources, soil erosion, conservation and resource management. It has only been possible in this study to hint at the enormous potential of the subject in applied fields and to detail some items of research, urgently required for the planning of resource development in West Malaysia, which would benefit from studies relating form and process to the denudation system.

It is concluded that climatic geomorphology is best developed as an 'energy system' approach to geomorphology, using the framework of the denudation system. Further work with this approach is required for studies of drainage density and texture in other areas and for other aspects of landform such as hillslopes. Such an approach has much to offer geomorphologists, pedologists, engineers, resource planners and many other workers requiring a multivariate but integrated approach to environmental studies.

## Acknowledgements

The author acknowledges the assistance of Messrs. J. H. Slaugh and K. Y. Choong of the Computer Centre, University of Malaya, for advice on the principal components analysis; Dr. J. G. Lockwood (University of Leeds), the Drainage and Irrigation Department, West Malaysia and the Meteorological Services, Malaysia, for supplying climatic data; and Professor I. Douglas (University of New England) for advice at various stages of the study.

# References

Alexander, J. B., 1965, *Geological map of Malaya*, 6th Edition, 1963, 1:500,000, (2 sheets).
Beckett, P. H. T. and R. Webster, 1962, *The storage and collation of information on terrain*, Interim Report, Military Engineering Experimental Establishment, Christchurch.
Berry, M. J., 1956, Erosion control on Bukit Bakar, Kelantan, *Malayan Forester*, **19**, 3–11.
Büdel, J., 1963, Klima-genetische Geomorphologie, *Geogr, Rundschau*, **15**, 269–285.
Budyko, M., 1958, *The heat balance of the earth's surface*, (transl. N. A. Styepanova) Washington.
Burgess, P. F., 1971, The effect of logging on hill dipterocarp forest, *Malay. Nature J.*, **24**, 231–237.
Carlston, C. W., 1966, The effect of climate on drainage density and streamflow, *Bulletin, Inter. Ass. scient. Hydrol.*, **11**, 62–69.
Chorley, R. J., 1969, The drainage basin as the fundamental geomorphic unit, in *Water, earth and man*, R. J. Chorley (Ed.), Methuen, London, 77–99.
Chorley, R. J. and M. A. Morgan, 1962, Comparison of morphometric features, Unaka Mountains, Tennessee and North Carolina, and Dartmoor, England, *Bull. Geol. Soc. Am.*, **73**, 17–34.
Common, R., 1966, Slope failure and morphogenetic regions, in *Essays in geomorphology*, G. H. Dury (Ed.), Heinemann, London, 53–81.
Dale, W. L., 1963, Surface temperatures in Malaya, *J. trop. Geog.*, **17**, 57–71.
Dale, W. L., 1964, Sunshine in Malaya, *J. trop. Geog.*, **19**, 20–26.
Douglas, I., 1967, Natural and man-made erosion in the humid tropics of Australia, Malaysia and Singapore, *Inter. Assoc. scient. Hydrol., Publication 75*, 17–30.
Douglas, I., 1968, Erosion in the Sungai Gombak catchment, Selangor, Malaysia, *J. trop. Geog.*, **26**, 1–16.
Douglas, I., 1969, The efficiency of humid tropical denudation systems, *Trans. Inst. Brit. Geog.*, **46**, 1–16.
Douglas, I., 1970, Measurements of river erosion in West Malaysia, *Malay. Nature J.*, **23**, 78–83.
Douglas, I., 1971, Aspects of the water balance of catchments in the Main Range near Kuala Lumpur, in *The water relations of Malesian forests*, J. R. Flenley (Ed.), Department of Geography, University of Hull, 23–48.
Ellison, W. D., 1944, Studies of raindrop erosion, *Agric. Engineering*, **25**, 131–136, 181–182.
Eyles, R. J., 1966, Stream representation on Malayan maps, *J. trop. Geog.*, **22**, 1–9.
Eyles, R. J., 1968a, Stream net ratios in West Malaysia, *Bull. Geol. Soc. Am.*, **79**, 701–712.
Eyles, R. J., 1968b, *A morphometric analysis of West Malaysia*, unpublished Ph.D. thesis, University of Malaya.
Eyles, R. J., 1969, Depth of dissection of the West Malaysian landscape, *J. trop. Geog.*, **28**, 23–31.
Eyles, R. J., 1971, A classification of West Malaysian drainage basins, *Ann. Assoc. Am. Geogr.*, **61**, 460–467.
Fournier, F., 1960, *Climat et érosion: la relation entre l'érosion du sol par l'eau et les précipitations atmosphériques*, Presses Univ. de France, Paris, 201 pp.
Giusti, E. V. and W. J. Schneider, 1963, Comparison of drainage on topographical maps of the Piedmont Province, *U.S. geol. Surv. Prof. Paper 450-E*, 118–120.
Gregory, K. J., 1966, Dry valleys and the composition of the drainage net, *J. Hydrol.*, **4**, 327–340.
Gregory, K. J., 1968, The composition of the drainage net: morphometric analysis of maps, *Brit. Geomorph. Res. Group, Occ. Paper 4*, 9–11.
Gregory, K. J. and V. Gardiner, 1975, Drainage density and climate. *Zeits. für Geom.*, **19**, 287–298.

Ho, R., 1964, The environment, in *Malaysia*, W. Gungwu (Ed.), Kuala Lumpur, 24–43.
Horton, R. E., 1945, Erosional development of streams and their drainage basins: a hydrophysical approach to quantitative morphology, *Bull. Geol. Soc. Am.*, **56**, 275–370.
Isachenko, A. G., 1968, Fifty years of Soviet landscape science, *Soviet Geog.*, **9**, 402–407.
Kaiser, H. F., 1958, The varimax criterion for analytic rotation in factor analysis, *Psychometrika*, **23**, 187–200.
Kalesnik, S. V., 1961, The present state of landscape studies, *Soviet Geog.*, **2**, 24–33.
Kalesnik, S. V., 1968, The development of general earth science in the U.S.S.R. during the Soviet period, *Soviet Geog.*, **9**, 393–402.
Keech, M. A., 1968, Soil erosion survey techniques, *Proc. & Trans. Rhod. scient. Assoc.*, **53**, 13–16.
Kendall, M. G., 1957, *A course in multivariate analysis*, Charles Griffin, London.
Kenworthy, J. B., 1969, Water balance in the tropical rain forest: a preliminary study in the Ulu Gombak Forest Reserve, *Malay. Nature J.*, **22**, 129–135.
King, L. J., 1969, *Statistical analysis in geography*, Prentice-Hall, New York.
Kirkby, M. J. and R. J. Chorley, 1967, Throughflow, overland flow and erosion, *Bull. Intnl. Assoc. scient. Hydrol.*, **12**, 5–21.
Law, W. M. and K. Selvadurai, 1968, *The 1968 reconnaissance soil map of Malaya*, Third Malaysian Soil Conference, Kuching, unpublished.
Lawrance, C. J., 1972, Terrain evaluation in West Malaysia: Part 1—Terrain classification and survey methods, *Transport and Road Research Laboratory Report, LR 506*.
Leopold, L. B., M. G. Wolman and J. P. Miller, 1964, *Fluvial processes in geomorphology*, Freeman, New York.
Lockwood, J. G., 1967, Probable maximum 24-hour precipitation over Malaya by statistical methods, *Met. Mag.*, **96**, 11–19.
Low, K. S., 1971, The relationship between storm rainfall and runoff, *Geographica*, **7**, 17–26.
Low, K. S. and K. C. Goh, 1972, The water balance of five catchments in Selangor, West Malaysia, *J. trop. Geog.*, **35**, 60–66.
Low, K. S. and C. H. Leigh, 1972, Floods, soil erosion and water quality in West Malaysia—adjustments to disruption of natural systems, *J. Instn. Engrs. Malaysia*, **14**, 14–19.
Mabbutt, J. A., 1968, Review of concepts of land classification, in *Land evaluation*, G. A. Stewart (Ed.), Macmillan, London, 11–28.
McGuinness, J. L., L. L. Harrold and W. M. Edwards, 1971, Relation of rainfall energy and streamflow to sediment yield from small and large watersheds, *J. Soil & Water Cons.*, **26**, 233–235.
Melton, M. A., 1957, An analysis of the relations among elements of climate, surface properties and geomorphology, *Office of Naval Research, Project NR 389–042, Technical Report 11*, Department of Geology, Columbia University, New York.
M. E. X. E., 1964, Classification of terrain intelligence, Malaya, *4th Combined Pool Report*, Christchurch.
Morgan, R. P. C., 1971a, *A test of climatic geomorphology by morphometric analysis in West Malaysia*, unpublished Ph.D. thesis, University of Malaya.
Morgan, R. P. C., 1971b, Rainfall of West Malaysia—a preliminary regionalization using principal components analysis, *Area*, **3**, 222–227.
Morgan, R. P. C., 1972, Observations on factors affecting the behaviour of a first-order stream, *Trans. Inst. Brit. Geogr.*, **56**, 171–185.
Nieuwolt, S., 1965, Evaporation and water balances in Malaya, *J. trop. Geog.*, **20**, 34–53.
Orsborn, J. F., 1970, Drainage density in drift-covered basins, *J. Hydr. Div., Amer. Soc. Civ. Engnrs.*, **96** (HY1), 183–192.
Ow Yang, H. C., 1965, Rivers in Malaya—their deterioration and remedial measures, *Malay. Agric. J.*, **45**, 17–20.
Peltier, L. C., 1950, The geographic cycle in periglacial regions as it is related to climatic geomorphology, *Ann. Assoc. Am. Geogr.*, **40**, 214–236.

Pinchemel, Ph., 1957, Densités de drainage et densités de vallées, *Tijds. v. h. Konin. Neder, Aardr. Genoot.*, **74**, 373–376.
Rikhter, G. D., 1970, A system of natural areal complexes of the earth, *Soviet Geog.*, **11**, 252–255.
Ruhe, R. V., 1952, Topographic discontinuities of the Des Moines lobe, *Amer. J. Sci.*, **250**, 46–56.
Schumm, S. A., 1968, Aerial photographs and water resources, in *Aerial surveys in integrated studies*, Proceedings of the Toulouse Conference, UNESCO, Paris, 70–79.
Shallow, P. G., 1956, River flow in the Cameron Highlands, *Hydro-electric Techn. Memo.*, *3*, Lembaga Letrik Negara, Kuala Lumpur.
Slaymaker, H. O., 1968, Morphometric analysis of the West Wales region—a practical class project, in *Morphometric analysis of maps*, Brit. geom. Research Group, Occasional Paper 4, 5–9.
Soper, J. R. P., 1938, Soil erosion on Penang Hill, *Malay. Agric. J.*, **26**, 407–413.
Stocking, M. A., 1972, Relief analysis and soil erosion in Rhodesia using multivariate techniques, *Zeits. f. Geomorphologie*, **16**, 432–443.
Strahler, A. N., 1950, Equilibrium theory of erosional slopes approached by frequency distribution analysis, *Am. J. Sci.*, **248**, 673–696, 800–814.
Strahler, A. N., 1952, Hypsometric (area–altitude) analysis of erosional topography, *Bull. Geol. Soc. Am.*, **63**, 1117–1142.
Strahler, A. N., 1964, Quantitative geomorphology of drainage basins and channel networks, in *Handbook of applied hydrology*, V. T. Chow (ed.), McGraw-Hill, New York, 4.39–4.76.
Swan, S. B. St. C., 1970, Land surface mapping, Johor, West Malaysia, *J. trop. Geog.*, **31**, 91–103.
Tan, H. T., 1969, Standard catchments for the derivation of rainfall-runoff relation required for flood estimation of small catchments in Malaysia, in *Natural resources in Malaysia and Singapore*, B. C. Stone (Ed.), Kuala Lumpur, 185–193.
Tricart, J. and A. Cailleux, 1965, *Introduction à la géomorphologie climatique*, S.E.D.E.S., Paris.
Tricart, J. and A. Cailleux, 1972, *Introduction to climatic geomorphology*, transl. C. J. K. de Jonge, Longmans, London.
US Soil Conservation Service, 1953, *Engineering handbook for farm planners. Upper Mississippi Valley Region III*, (Agr. Handbook 57) US Govt. Printing Office.
van Wambeke, A. R., 1962, Criteria for classifying tropical soils by age, *J. Soil Sci.*, **13**, 124–132.
Vinogradov, B. V., K. I. Gerenchuk, A. G. Isachenko, K. G. Raman and Yu. N. Tsesel'chuk, 1962, Basic principles of landscape mapping, *Soviet Geog*, **13**, 15–20.
Wilson, L., 1973, Relationships between geomorphic processes and modern climates as a method in Paleoclimatology, in *Climatic Geomorphology*, E. Derbyshire (Ed.), Macmillan, London, 269–284.
Wycherley, P. R., 1967, *Rainfall in Malaysia*, (Planting Manual No. 12), Pusat Penyelidekan Getah, Malaysia, Kuala Lumpur.
Zonneveld, I. S., 1972, *Land evaluation and land (scape) science*, I. T. C. Publications, Enschedé.

CHAPTER TWELVE

# Lithology, Landforms and Climate

Ian Douglas

## 12.1 Introduction

The dependence of both erosional landforms and weathering products on the rock materials from which they are derived and the type of energy which acts upon them is well expressed by the single equation:

Source materials + energy = sedimentary rocks (Keller, 1954)

The nature of the material on which denudation processes act is a fundamental factor in landform evolution. The structure or texture of the rock, whether macrocrystalline, microcrystalline or non-crystalline, uniform or heterogeneous, determines the pattern of mechanical and chemical weathering.

In considering the significance of the relationships between lithology, climate and landforms, not only does the stability of particular minerals require attention (Chapter 2), but the whole character of rock formations must be examined. The resistance of a sedimentary, igneous or metamorphic rock sequence is only as great as that of its weakest component. While a rock may contain some highly stable minerals, such minerals may be cemented together by a soluble cement which is easily eroded. The decay of the cement determines the rate of rock decomposition and thus the pattern of landform evolution. Petrology and stratigraphy are thus of geomorphic significance and probably deserve more attention than they have received hitherto.

Even if relatively common rocks that are frequently used for comparisons in climatic geomorphology are considered, petrological contrasts can be striking. Granitic rocks may have a wide range of grain sizes: this, together with slight differences in mineral composition, may be sufficient to produce differences in rates of weathering when subjected to surface environment conditions (Chapter 14). Variations in composition of both basic and acidic extrusive rocks produce variations in stability from one rock type to another in any given environment. Sedimentary and metamorphic rocks exhibit even greater variety of texture and composition. As Sweeting and Sweeting (1969) have shown, petrological contrasts in calcareous rocks are responsible for considerable contrasts in karst topography. Sandstones with gypsum cements are less resistant to weathering than sandstones with calcareous cements, which, in turn, are less resistant than those with siliceous cements.

It follows that any attempt to make broad generalizations about the stability of different rocks in different major climatic zones on either a global or a continental scale runs the risk of gross oversimplification. Such an examination of climatic controls must be made on several different scales.

## 12.2 Theoretical aspects of the stability of rock materials in relation to climate

The stability of minerals is related to certain thermodynamic principles (Chapter 2). The greater the amount of energy evolved by an ion during its passage into the crystalline state, the more stable is the crystal obtained, the more difficult it is to reduce to a dispersed state, to dissolve and to melt, or to divide again the atoms of the lattice into free ions. This relationship of mineral stability to crystal structure and energy bonds has led some Soviet writers to

Table 12.1. Geochemical types of rock weathering related to climatic conditions (after Lukashev (1970))

| Type of residual weathering product | Geochemical nature of the process | Weathering environment and solute transfer conditions |
| --- | --- | --- |
| Skeletal, clastic | Formation of mixture of debris; slight removal of solutes | Low temperature, slight chemical and biological breakdown of rocks |
| Siallitic, argillaceous (iron-pan type) | $SiO_2$ and $Al_2O_3$ hydrate mixtures formed with accumulation of $SiO_2$ in podzol horizons, and removal of $Al_2O_3$ and $Fe_2O_3$ to underlying horizons, leaching of such elements as Cl, Na, Ca, Mg and K | Moderate humidity and temperature; active organic and humic acids; downward migration of solutes |
| Siallitic–carbonatic (calcrete-type) | Silica, iron and aluminium hydrates formed, together with accumulation mainly of calcium carbonate, but also Mg, K and Na carbonates | Mediterranean and related semi-arid, seasonal climates; organic and humic activity; both upward and downward migration of solutes |
| Siallitic–chloride–sulphate type (gypsum-type) | Formation of hydrated weathering products (siallites). High mobility of $SiO_2$, accumulation of chloride and sodium, calcium and magnesium sulphates | Warm, arid conditions; upward migration of solutes dominant; greatly reduced organic activity |
| Siallitic–ferritic and allitic (ferricrete–bauxite type) | Accumulation of iron and aluminium with general loss of silica and more soluble elements | Hot, wet climates; widespread leaching and migration of solutes |

look at weathering in terms of the degradation of the mineral lattice structure. Mineral disintegration proceeds more rapidly along those directions in the lattice which require maximum energy for their maintenance. Breakdown of the lattice involves a series of stages, with substitution of ions as new lattices are formed. The principal reactions involved in this stagewise transformation is hydrolysis, oxidation, hydration and substitution. Stagewise transformation produces selective changes in the mineral lattice, the nature of the changes depending largely on the pH of the weathering environment. The local pH is determined both by lithology and by local climatic conditions (Lukashev, 1970; cf. Trudgill, Chapter 3).

Lukashev (1970) argues that the nature and pattern of chemical weathering are mainly controlled by climatic conditions, such weathering being insignificant in polar regions and of highest intensity in the humid tropics and subtropics. He recognizes five geochemical types of rock weathering, as shown in Table 12.1. He points out that organisms of all types play a vital role in weathering and that the intensity of biological activity is closely related to the intensity of weathering. Lukashev goes on to argue that weathering and the migration of substances cause 'alterations in relief, so that the landscape often assumes characteristic forms. In polar regions this is manifested in the formation of the polygonal landscape. In humid areas with warm climate and abundant atmospheric precipitation it manifests itself in changes of dimensions and the smothening [sic] of surfaces' (Lukashev, 1970, p. 44). However, he further points out that 'Differences in lithology also produce different landscapes. Igneous rocks (granites, diabases, etc.) mostly give rise to rounded weathering forms such as domes, spherical and pillow-like masses, etc., whereas stratified sediments and metamorphic rocks are weathered to step-like, flat relief forms, cornices, overhangs, niches, etc.' (Lukashev, 1970, p. 44; cf. Thomas, Chapter 14).

While the principle of general climatic control of chemical weathering is readily acceptable, it is difficult to establish the scale at which climatically controlled weathering processes obscure the lithologically determined forms of rock outcrops. Sedimentologists argue that relief plays an important role in the intensity of erosion. 'High relief promotes a high rate of erosion, whereas low relief is associated with a retarded rate of erosion' (Pettijohn, 1957, p. 510), a principle enshrined in Davis's (1909) cycle of landscape denudation. Tectonic history and structure are clearly part of the landform-producing system and have to be considered at the appropriate scales.

Yatsu (1966) has stressed the ways in which rock properties influence the formation of landforms. Placing emphasis on the mechanical properties of rocks and their weathering products, Yatsu demonstrates that landform evolution depends to a great extent on both the physico-chemical stability of minerals and on the mechanical properties of rock outcrops. The latter are dependent on the origin and history of the rocks and may be independent of climate. Clay deposits of varied types may occur in the same area, and the contrasts in clay mineralogy will produce differing slope stability on each clay formation, such as the well known cracking characteristic of clay soils with a high montmorillonite content. Slopes on shales will have clay minerals inherited from the underlying rock, while those on other parent materials may have only the clay minerals derived from the present phase of weathering.

Within the constraints and frameworks provided by rock structure, mineralogy, stratification, texture, porosity and degree of fissuring, climatic influences can determine the nature and rate of weathering. Decomposition and disintegration of a rock may depend on its most soluble or structurally least stable components. Relatively minor components may be the elements most easily leached out of a rock. A carbonate-cemented sandstone might be largely composed of silica in the form of quartz, but water draining from such a rock will remove far more calcium and bicarbonate ions than silica ions.

The most elegant way of examining the overall stability of minerals is in terms of Eh–pH diagrams. Oxidation potentials (Eh) and hydrogen ion concentration (pH) are the basic controls which determine the nature of many sedimentary processes. Krumbein and Garrels (1952) have prepared a chemical classification of non-clastic sediments (evaporites excepted) based on pH and Eh. While most sub–aerial weathering environments are oxidizing environments, parts of the landscape, especially the water-filled depressions and areas of waterlogged soils are reducing environments. Where climatic conditions allow peat and organic matter to accumulate, weathering proceeds under conditions quite different from those in areas devoid of such accumulations (Chapter 3). Only surface waters with good circulation are oxidizing, whereas confined waters rapidly lose their oxygen content, whether confinement results from fixation in rock or soil pores, or by prevention of overturn of open waters. The significance of Eh and pH is probably best illustrated by the solubility of iron.

Iron occurs in two oxidation states, the divalent or ferrous form and the trivalent or ferric form. Iron in aqueous solution is subject to hydrolysis. The iron hydroxides formed in these reactions, especially the ferric form have very low solubility. Over most of the pH–Eh range ferric iron is only slightly soluble. Figure 12.1 shows that ferric iron is stable over the bulk of the area of natural environments, especially the area bounded by pH 5·0 and 8·0 and Eh 0·3 and 0·5 in which most natural waters exposed to the atmosphere lie. Ferrous iron dominates the dissolved iron in natural waters, increasing in solubility as pH and Eh decrease. The dividing line between the ferrous and ferric state is indicated in terms of activity, indicating that at a given Eh, a small decrease in pH produces a great increase in the activity of dissolved iron, and thus in the solubility of iron compounds in the environment. Relatively small shifts in Eh and pH thus cause great changes in solubility However, the picture is complicated by the formation of complexes of ferric and ferrous iron in natural water with inorganic anions and organic chelates.

Chemical considerations of the type outlined above suggest that at least two scales of environmental variation have to be considered when the relationship between climate, lithology and landform is explored. One is the global climatic zonation, probably best exemplified by Troll's (1964) map of the seasonal

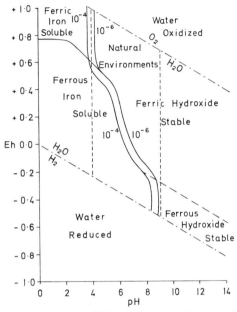

Figure 12.1. Solubility of iron as a function of Eh and pH. Labelled contour lines refer to activities of dissolved iron species of the indicated values, $T = 25$ °C, Total pressure = 1 atm.

climates of the earth, the other the topographical sequence of hydrological conditions associated with position on a catena or landsurface where differences in drainage conditions can cause significant variations in Eh and pH over distances of a few metres or tens of metres (Chapter 5). The significance of such short distances is exemplified by Drever's explanation (1971) of why a higher proportion of montmorillonite is found in the suspended sediments of the Rio Ameca in Mexico than in the weathering profiles of the catchment area. Rainwater falling on high ground leaches the rock extensively and kaolin minerals are formed. When the same water re-emerges at lower levels, it contains high concentrations of dissolved solids and may increase in concentration as a result of evaporation. Reaction between this water and primary rock minerals (or previously formed kaolinite) will characteristically produce montmorillonite. When the water finally reaches the river, its chemistry will be influenced by the montmorillonite-forming stage and, since low-lying soils close to the river will be rich in montmorillonite, montmorillonite will tend to be the dominant mineral in the river sediment.

Into this geochemical picture must be incorporated the physical characteristics of energy distribution, especially seasonal variation in the receipt of solar radiation, the depth, are, duration and frequency aspects of various forms of precipitation, seasonal changes in plant cover and resistance of regolith materials to erosion. Just as spatial differences in drainage conditions cause variations in mineral solubility, so do seasonal variations in moisture conditions. Recent studies of weathering have emphasized the role of bacteria and micro-organisms in the weathering process, indicating further repercussions of seasonal thermal and moisture controls on organic activity. On the global scale, the tropics have little annual thermal variation, but often a seasonal moisture contrast, and always a diurnal variation of temperature. Outside the tropics seasonal thermal contrasts are paramount. Thus similar moisture and thermal conditions may occur at certain seasons in both tropical and temperate latitudes. Similar weathering environments may exist for part of the year in widely separated parts of the world. Thus conditions leading to the formation of gibbsite exist in the Massif Central of France, even though gibbsite formation is largely restricted to wet low latitude environments. A more general example of similar weathering environments through a wide range of latitudes is provided by the forest zones, which extend from equatorial to temperate latitudes as in eastern China and eastern Australia. Such areas are well suited to the investigation of the manner in which thermal and seasonal contrasts influence landforms developed on similar parent material with the same geological history.

## 12.3 Concept of ultimate landforms in response to denudation

Albrecht Penck (1910, 1914) distinguished a number of climatically-determined landform assemblages, but believed that 'pure' climatic forms could be found in the centres of climatic zones, separated from each other by areas of polygenetic landforms. From these ideas developed the notion that different climates, by affecting processes, develop unique assemblages of landforms. Yet it has long been recognized that landforms of similar geometry may have differeing origins (Rich, 1938). Domed inselbergs, long considered characteristic of tropical climates, develop under a wide variety of conditions as a result of differing processes and are described by Bartels (1973) as convergent forms. Hettner (1972) illustrates this concept of convergent development by suggesting that the loess of northern Germany constitutes, in itself, no reason for asserting that Germany experienced a desert climate in the recent geological past. The peculiar rock forms of the Quadersandstein, and in part the Butner Sandstone too, are not the result of desert conditions but of the porosity of the rock. Hettner emphasizes that the particular characteristics of internal build can have so marked an effect that differences of climate become secondary. Their effect can even be so great as to produce landforms which would in other circumstances be seen as belonging to another kind of climate. Such conditions produce convergent phenomena.

The concept of convergence in climatic geomorphology is an application of the principle of equifinality—the recognition that different causes may produce the same effect. Birot (1955) sees the possibility of similar landforms having differing origins as a prime reason for the use of evidence from correlative deposits in geomorphological analysis. For an evaluation of the types of convergent tendencies present in landform evolution, one must turn to Wilhelmy's study (1958) of the climatic geomorphology of massively jointed rocks. Wilhelmy differentiates between macroclimatic and microclimatic–edaphic influences on landform development. Divergence of landform evaluation may occur through the weathering of similar rocks under different macroclimates, and also through the influence of microclimate on the weathering of similar rocks within a single macroclimatic zone. The former cause of variety is the fundamental postulate of climatic geomorphology—the normal case, in Wilhelmy's words. The latter type of influence is the result of aspect and may be illustrated by contrasts in the weathering of different sides of ancient Egyptian monuments (Wilhelmy, 1958), contrasts in soils on slopes of different aspect (Chapter 5), or by contrasts in valley-side slopes in humid temperate areas (Hack and Goodlett, 1960, Chapter 6), even though Carson and Kirkby (1972) emphasize that it is probably only under rather rare circumstances that valley asymmetry is a product solely of microclimatic contrasts between opposite slopes.

Wilhelmy sees convergence of landform evolution possibly occurring under the following sets of circumstances: (a) the weathering of similar rocks in different macroclimatic zones as a result of microclimatic influences; (b) the weathering of similar rocks in different climates without pronounced microclimatic effects; (c) the weathering of similar rocks in analogous climates; (d) the weathering of different rocks in a single macroclimatic zone with (e) the weathering of coarse conglomerates under favourable climatic circumstances producing the most perfect example of convergence of landform evolution.

Citing cavernous weathering (*Hohlblockbildung*) as an example of type (a) circumstances, Wilhelmy (1964) has described *tafonis* in East Africa, Hong Kong, Iceland, Elba, Aruba (W. Indies), coastal Peru, Uruguay, the arid southwest of North America, the rocky Peshan desert in Central Asia and the Aravalli Mountains. He ascribes them to crumbling on shaded rock surfaces, while surfaces exposed to direct insolation are protected from *tafoni* development by crust formation due to evaporation. Evans (1970) extends the range of evidence to include polar regions, emphasizing the relationships of salt weathering to aridity. Similar evidence of convergence of landforms from polar and hot desert areas due to salt weathering has been derived from comparative studies in the Sahara and Westspitsbergen (Meckelein, 1965).

Wilhelmy (1958) suggests that the granular disintegration of massive rocks is a worldwide phenomenon, occurring from the humid tropics to the polar regions, and so does not require a specific microclimatic cause. Granular disintegration in humid regions is produced by chemical weathering while in arid climates it is produced by a combination of insolation, hydration and salt weathering. Although blockstreams are well known periglacial landforms, they occur as weathering residuals in warmer, humid climates, such as Hong Kong (Wilhelmy, 1958) and in the heads of humid tropical valleys as, for example, in Penang, Malaysia (Douglas, 1967) where they are formed of a chaos of granite corestones overgrown with rainforest. Core boulders themselves develop under a variety of climatic conditions. Mainguet (1966), for example, describes corestone formation in an orthoclase-rich pink porphyritic granite in the coastal semi-desert of Peru, 7°south of the equator. Her photographs show boulder-strewn slopes which look similar to slopes cleared of rainforest in the foothills of the Main range of Peninsular Malaysia. However, the process paths leading to these similar landforms are somewhat different. In the humid tropical case chemical weathering, initially along joints and then microfissures, isolates boulders of unweathered rock amid a matrix of decomposed material. In the Peruvian coastal desert, the boulders evolve through onion-skin weathering and granular disintegration. However, the dominant factors affecting the pattern of boulder development in this case are the macroporosity and dense fissure network found in these Peruvian granites.

The weathering of similar rocks in analogous climates (type (c) convergent landform evolution) is especially applicable to domed inselbergs, the particular case described by Bartels (1973). Although Hurault (1963) argues that the domed inselberg is a characteristic landform of the humid tropics, similar forms occur under a variety of climates, particularly on the Gondwanaland shield areas of South America, Africa and Australia. In the latter country, similar forms are found on Mesozoic granites from the humid tropics of northeast Queensland to the humid temperate montane climate of the Snowy Mountains of New South Wales. While some of these domed inselbergs may be relict landforms developed under different past climates, others are similar landforms developed by different processes due to particular structural or lithological conditions (Kesel, 1973). Evolution under seasonal climates, be they winter wet season or summer wet season, is a response to periodic water availability.

The development of similar landforms on different rocks due to the efficiency of particular processes within a given climatic zone (type (d) landform evolution) is perhaps best illustrated either by the effects of chemical deep weathering in the humid tropics (which tends to make slope morphology a response to regolith stability) or by frost shattering in polar or high mountain climates (which produces angular fragments from diverse rock types). Wilhelmy's claim that coarse conglomerates weather to produce the most perfect example of convergent landforms in a variety of climates is equivalent to saying that, given suitable material of heterogeneous calibre and composition, it will break down into its component parts whatever agency is involved. Comments made earlier about the nature of cements in sedimentary rocks apply here. The examples cited by Wilhelmy from Catalonia, Ischia and Korea have parallels in the conglomerates of the Catombal Range of central New South Wales.

The evidence of convergence brought forward here to illustrate Wilhelmy's ideas indicates the importance of attempting to specify thresholds between the influences of climate and rock characteristics, including not only lithology, but structure, granulometry and porosity. In establishig such a set of threshold values, an attempt must also be made to specify the *spatial and temporal scales* for which such thresholds are valid.

## 12.4 Implications of structure

Tricart and Cailleux (1965) attempt to summarize the scales on which different climatic and structural factors affect landforms, suggesting that on the local scale of a few kilometres, lithological influences predominate. On a much wider scale, tectonic factors are dominant: the regional topography of shield areas is in sharp contrast to that of folded mountain chains, while that of sedimentary basins contrasts with that of ancient massifs.

The influence of structure is readily apparent in topography derived from folded sedimentary rocks, be they merely slightly downwarped as in the London Basin, or folded, faulted and refaulted and compressed several times as in the Precambrian and Cambrian sediments of the Flinders Ranges of South Australia (Twidale, 1967). Such tectonic movements produce the basic relief pattern of an area, but the detailed topography is the product of the work of external processes through time. Mainguet (1972) makes this important distinction between relief, which reflects structure, and topography (*modelé*), which is the expression of external dynamics, essentially climatically-determined processes. As de Martonne (1951) emphasizes, structural characteristics intervene at different stages of landform evolution. Using the cyclic terminology characteristic of his time, de Martonne indicated that the majority of structurally determined landforms only emerge at the stage of maturity and when old age is reached they become obliterated. Some features, such as discordant structures appear only at later stages of evolution while others, such as volcanoes, dominate the landscape from the beginning of erosive activity.

The clearly-demonstrated importance of relief and tectonic activity in determining rates of denudation (Schumm and Hadley, 1961; Schumm, 1963; Simonett, 1967; Pain, 1972) suggests that not only are landform assemblages likely to differ from tectonically-active to tectonically-inactive areas, but that areas at different stages of geological evolution within the same macroclimatic zone are likely to have contrasted landform assemblages (topography or *modele*). The geomorphology of the equatorial rainforest-covered areas of New Guinea, of the Tertiary sediments of Malaysian Borneo, the Palaeozoic sediments and Mesozoic intrusives of Peninsular Malaysia and of the Gondwanaland sediments of Sri Lanka differs as a result of structural contrasts.

On the local scale of a single hillslope, less than a kilometre in length, the significance of structural influences again emerges, for Savigear (1952) has shown that slopes formed by cliff retreat in South Wales owe their form and inclination to the structure and lithology of the rock. The nature and inclination of a caprock exerts a major influence on slope form. Where thick sandstone bands or dolerite sills are absent, hillsides in interior Natal, South Africa, are made up of a series of small free faces with intervening rocky slopes, and the slopes as a whole are governed by the occurrence of these thin outcrops (Fair, 1948). Slopes under these conditions are not as steep as those which occur in presence of a thick caprock. Clearly, slope steepness in interior Natal is strongly controlled by this single structural characteristic: the thickness of the caprock.

Between these two scales represented by the major tectonic units on the one hand and the outcrops on a single slope on the other, lie the folded and faulted sedimentary rock structures. The classic case of the breached dome, exemplified by the Pays de Bray (de Martonne, 1951), illustrates this scale of structural influence. On a map, the Pays de Bray appears as an elongated *boutonniere* ('buttonhole'), with a continous inward-facing chalk escarpment, below which lies the varied topography of the lower Cretaceous and Jurassic clays and sands. The whole form of the *boutonnière* is determined by the structure. The rectilinear northern flank is in sharp contrast to the curved southern edge, the former being determined by a fault which throws the lower chalk against the downwarped Portland limestone, the latter by the manner in which the strata dip away from the centre of the dome. However, the centre of the Pays de Bray is not lower than the surrounding chalk in its entirety for the Portland limestone forms ridges as high as the northern chalk escarpment. While the Pays de Bray has been truncated in the past by an erosion surface, present processes are continuing the work of Quaternary climatic changes in dissecting a pattern of landforms in which lithological and structural controls are clearly defined.

## 12.5 Experimental designs to test the theories of climatic geomorphology

An experiment to test the relative influence of climate and lithology on the development of landforms requires a two-way analysis of variance approach with a set of climatic zones and a set of lithologies. Parameters describing landform and process for each site have to be measured and means calculated for every combination in the climate–lithology matrix. Stoddart (1969) uses both rates of denudation and morphometry as tests of the fundamental postulates of climatic geomorphology. Like other workers, he has insufficient data to isolate lithological influences in a wide range of climatic types and his analysis of erosion rates takes no account of lithology. However, Stoddart does indicate that both climate and lithology have to be taken into account in explaining drainage density.

In an experiment to test the impact of climate on fluvial processes, measurements were made of fluvial sediment and solute loads from different lithologies in northeast Queensland and southeastern New South Wales (Douglas, 1973). Even though water quality is subject to seasonal fluctuations, ratios of mean ionic concentrations may be used to test relationships between geological conditions and the elements supplied to streams. The clear chemical differences between black water and white water streams in the Amazon are well known (Sioli, 1968), while Hack (1960) has indicated more localized relationships in the Schenandoah valley. Test of the $Ca + Mg : Na + K$ ratio for the eastern Australian rivers showed that contrasts due to lithological influences were greater than those due to climatic factors (Douglas, 1969). However, if single lithologies are considered, climatic contrasts appear: on granite catchment areas, values of the ratio for New South Wales were approximately six times greater than those for Queensland, the means (derived from measurements in

Table 12.2. Levels of significance between rates of denudation of catchments of differing lithology in Queensland and New South Wales (after Douglas, 1973)

| Lithology compared | Total load (%) | Dissolved load (%) | Suspended load (%) |
|---|---|---|---|
| Basalt | n/s | 5 | 5 |
| Metamorphic rocks | 10 | n/s | 20 |
| Granite | 10 | 5 | 20 |

Levels of significance are determined by analysis of variance using variance ratio tables in Fisher and Yeates (1963). n/s denotes not significant at the 20% or higher levels.

equivalents per million) being 1·83 and 0·30 respectively. While some of the difference is undoubtedly due to some of the New South Wales samples being collected under drought conditions when ionic concentrations in river water had been increased by evaporation, the trend in the data is clearly climatically-determined. When the fluvial sediment and solute loads of the streams in the two eastern Australian study areas are compared on the basis of lithology (Table 12.2) differences in denudation rates from similar lithologies but different climates are significant only at low levels, if at all (Douglas, 1973). Such differences that do exist are more probably due to contrasts in precipitation quantity and distribution than to any thermal contrasts. The demonstration of such regional contrasts may be sufficient to indicate real differences in the rate of operation of processes, but the convergence argument suggests that the factor of time may intervene to create similar landforms even if processes operate at different rates. Morphometry, hitherto little used in climatic geomorphology (Stoddart, 1969), must be the real test.

Morgan (1973) has demonstrated a correlation between precipitation intensity, quantity and seasonality, and drainage density, arguing that at a $G$-scale of 3·5, climate appears to outweigh the influence of other denudation system components over drainage density. Similar regional diversity in drainage density on the Allegheny Plateau is, however, a response to conditions imposed by glacial and periglacial processes (Donahue, 1972). Post-glacial stream concentrations are highest in the north and decrease southward as the degree of glacial modification diminishes, permitting the survival of relict landforms better adapted to normal erosional processes. Drainage density is a sensitive parameter, as channel networks may enlarge in response to changes in runoff regimes, changes which may simply stem from a change in vegetation or land use. In Peninsular Malaysia, drainage densities on granite under rainforest are of the order of 3 to 9 miles per square mile, but in rubber plantations short first-order streams have developed by gully erosion from the stream head hollows of the original rainforest drainage lines. Drainage density thus reflects the relationship between infiltration and surface runoff.

A clear indication of the sensitivity of drainage density to climate and lithology is provided by Selby's comparison (1967) or New Zealand greywacke feral relief with the data for Dartmoor and the Unaka Mountains (Chorley and Morgan, 1962). Whereas Dartmoor and the Unakas are developed on crystalline igneous and metamorphic rocks which may be more resistant to erosion than the greywackes, the mean drainage densities are 3·5, 6·5 and 23·5 respectively. The difference between the Unakas and Dartmoor is related to the intensity of precipitation, the maximum expected 24-hour falls being 375 and 75 mm respectively. However, the much finer texture of the New Zealand area is a result of mass-movement which itself is a response to hydrological conditions, depth of regolith, slope angle, vegetation cover and physical properties of the soil.

These few examples indicate the difficulty of using drainage density for comparisons in a matrix of lithological and climatic units. No really satisfactory analysis of the postulates of climatic geomorphology on the basis of morphometry has yet been carried out. Nor does any analysis on the basis of slope form seem likely to yield any satisfactory solution for, even within a given region of uniform structure, lithology and climate, no unique manner of slope development is likely to exist (Chorley, 1964). Perhaps more morphometric analyses of the type developed by Williams (1969) for karst landforms could be applied to a wider range of landforms, but for the present it is necessary to rely on qualitative, or at best semi-quantitative, analyses of the climate–lithology landform relationship.

## 12.6 Examples of sandstone, schist and basaltic landforms in different climates

As the influence of climate on landforms developed on limestone and granite is considered in later chapters, the discussion here will deal with three other types of rock which may each show some types of convergence in landform evolution in different climates.

### *12.6.1 Sandstone*

As sandstone is a sedimentary rock composed of material which has already been through at least one phase of erosion, transportation and deposition, it tends to be made up of the less readily weathered, the least easily decomposed and the hardest to dissolve particles of the original parent rock. As suggested earlier, during deposition and diagenesis a sandstone may acquire a cement which is less resistant to weathering than the coarse particles which form the bulk of the rock, Variations in cements thus create variations in resistance of sandstone to weathering.

Tricart and Cailleux (1965) use sandstones to test the idea that, in areas of similar rocks which have undergone a long evolution and have similar tectonic structure but different climates, any variation in landform will be due to climatic influences. Three areas chosen were Fouta Djalon in Guinea, the Adrar in Mauretania and the Vosges in France. Under the seasonally wet climate of

Fouta Djalon, a fluvially-eroded landscape of intense dissection with steep slopes has evolved, even isolated sandstone residuals retaining steep slopes. River profiles are closely related to structure, every resistant layer in the sandstone supporting rapids or waterfalls.

In the dry climate of the Adrar, structural influences on landforms are even more apparent. The role of wind action is subordinate to that of running water, the latter washing debris downslope for it to be reworked by the wind later. Hard sandstone layers stand out as massive, vertical-sided caprocks, while softer clayey sandstones have been undercut by water action and furrowed into little channels which carry away the debris supplied from the block falls and granular weathering of the hard layers above.

The sandstones of the Vosges show regional contrasts due to the tectonic history of the area, the eastern flank of the Vosges Horst being truncated by the Rhine Rift Fault and the sandstones dipping to the west beneath the younger rocks of the Paris Basin. Here the abrupt exposed sandstone edges are missing, save for a few sharp ridge crests formed of resistant conglomerates. The topography (*modelé*) of the sandstone Vosges is the result of the Quaternary climatic succession of periods of fluvial erosion and periods of periglacial activity. Although the rivers have profiles approaching the profile of equilibrium, valley-side slopes are steep and straight.

These three cases illustrate the general impact of climate on sandstone landforms and topography. Mainguet (1972) has considerably widened the scope of sandstone landform studies and, after examining sandstones in a variety of African and European climates, concludes that three different systems of sandstone morphogenesis may be recognized: (a) those expressing a response to the action of running water; (b) those expressing a response to predominant wind action; and (c) those arising from the action of water in the form of ice. While this may seem to be simply a restatement of the Davisian scale of climatic variations on the geographical cycle, any particular set of sandstone landforms is a response to a particular combination, intensity and frequency of morphogenetic processes. Within this broad process categorization, Mainguet (1972) discusses morphogenetic systems, not in the ecosystem manner of Tricart and Cailleux (1965), but in terms of landform phenomena which can be recognized, much in the way that land systems are recognized, from aerial photographs. In the humid tropical zone, sandstone landforms evolve under one of three different systems. Gradual retreat of slopes and the burying of rock under a thick weathering mantle is characteristic of the 'polyconvex sandstone morphogenetic system'. Structural characteristics of the sandstone emerge more sharply in the 'system of thin weathering and duricrust mantles'. Convex summits give way to concave valley heads or step-like valley sides with marked structural benches. Bare sandstone forms the third humid tropical sandstone landform type, in which steep-sided residuals with rounded summits stand out from erosion surfaces whose character is determined by fractures and structural planes. The 'bare rock system' often occurs in close proximity to the 'thin weathering mantle system'.

In Mali and Mauretania, Mainguet found a juxtaposition of worn and ruin-like sandstones. The drier the climate, the more the structural features of the sandstone became dominant so that slopes develop staircase-like topography. However, this system is out of phase with the present climate and must be seen as the product of an undetermined number of climatic oscillations with either Mediterranean or tropical affinities. The juxtaposition of the worn and ruin-like forms is probably the result of legacies of various types of process, for ruin-like sandstone landforms are azonal, being found wherever there is sufficient water to enlarge structural planes and where bare rock is exposed, to be broken up into prisms with a quadrangular base. The worn forms are azonal only at the scale of a few metres or tens of metres, larger units requiring an environment in which large thermal variations promote flaking and granular disintegration.

Around Tibesti, the sharp crests and sculpted forms show striations developed by wind-blown sand abrasion (Mainguet, 1972), but the broad features of this morphogenetic system are due to past climates. Aeolian abrasion works easily on well-jointed sandstone, eventually developing long corridors whose width is a function of the difference between the angle of points and that of the mean annual resultant wind. However, many of the escarpments and gorges of Nubian sandstone have the same hillslope forms as those of the seasonally-wet tropics.

In the Vosges, tors and related forms occur, some being analogous to the gritstone tors of the Pennines (Palmer and Radley, 1961). Many slopes are boulder-strewn like the English Millstone Grit Edges of Derbyshire, Staffordshire and South Yorkshire. Mainguet (1972) considers such slopes to be Late-Glacial features, although glacial and periglacial processes probably acted on materials which had been prepared by Tertiary weathering under warmer, humid climates, as in other parts of western Europe, e.g. Northern Ireland (Reffay, 1972).

One type of arenaceous rock, namely conglomerate, produces similar forms in a wide variety of climates. For example, large blocks of conglomerate form upstanding ridges in many parts of Australia. In the Devonian-conglomerates of the Kimberley district of northwestern Australia, weathering of the outer layers of crystalline pebbles produces small depressions between the core of the pebble and the surrounding matrix. Such *Hohlkehlen* have been noted in coastal areas of Tenerife in the Canary Islands and in the limestone boulders of Tasmanian tillites (Schwarzbach, 1966). Conglomerate forms slab-like blocks in many climates, the nature of the material and its jointing and bedding sets being more responsible for the landform than any particular set of climatic conditions. After an extensive review of sandstone morphology, Mainguet (1972) concludes that it would be a mistake to argue *a priori* for a systematic relationship between climate and landforms. On the one hand, when a classification of sandstone landforms as a function of major climatic zones is attempted, the manner in which sandstone creates similar forms in a wide range of climates, such as the great sandstone escarpments from Spitsbergen to the equator, proves a major obstacle. On the other hand, sandstone landforms may differ greatly from those on other rocks in the same climatic region.

*12.6.2 Schist*

Schists vary in mineral composition and therefore in resistance to chemical weathering. However, because of their foliated structure they tend to flake or break off into platy fragments. When shapes of river pebbles of differing lithology are compared, schists are among those with the highest indices of flattening. In tropical Africa, schists often produce landforms similar to those on granite, successions of low hills or monotonous long, gentle slopes (Michel, 1973). However, certain schists create small steep-sided eminences, such as the high points of the Bassari Mountains in Senegal which are composed of amphibolite and epidote schists. In the more humid conditions of Peninsular Malaysia, schists are often deeply weathered and, while not readily distinguishable from adjacent granitic terrain, schist country has less stable slopes owing to the more uniform grain size composition of the schist regolith. The fissile nature of schists makes them particularly susceptible to frost action. To the east of Beauraing in Belgium, the depression of Focant developed through intense periglacial activity in a wide band of pure schists (Seret and Béthune, 1967). In this part of Belgium those slopes developed on schists are longer, gentler and smoother then those on any other rock in the vicinity.

Similar depressions in general erosion surfaces occur in schists in Northern Ireland (Reffay, 1972) in the Calabber Valley and the Crana Basin. However, the black schists of Ards form the southern flank of the Muckish–Errigal chain, while the Killybegs schists form the eastern rampart of the southwestern peninsula. These more resistant schists have a siliceous armour and, although possessing cleavages, do not have any clear stratification. By contrast, the depression-forming schists are coarse grained, with calcareous inclusions and a schistosity parallel to the stratification. Thus, while these Northern Ireland examples reveal the general susceptibility of schist to rapid erosion under periglacial and glacial processes, detailed variations within the region are due to lithological contrasts. However, Tricart (1969) argues that periglacial conditions produce schist slopes which differ little from those on chalk or clay and that there is a marked difference between the relief on schists in periglacial and subtropical areas. In Brittany, periglacial processes have created undulating topography with schist slopes of $5°-10°$, while in Portugal slopes on schist occur in the range $20°-25°$.

One important aspect of schists and indurated shales is that, when deeply weathered, they react dramatically to changes in environmental conditions. The Bungsar Heights housing development on the Kenny Hills Formation in Kuala Lumpur is the scene of rapid and extensive gullying of weathered shales (Douglas, 1972) while deep ravines due to anthropogenic erosion exist in the schists of the valley of the middle Bogoé in West Africa (Vogt, 1962). Such rapid response to change in vegetation indicates that schists react quickly to climatic changes and may be less likely to retain vestiges of past climates than more resistant rocks.

*12.6.3 Basalt*

Classified by many writers as tectonic landforms, basaltic landforms have seldom been discussed in terms of climatic geomorphology. Noting that basalt plains and plateaux displaying similar assemblages of forms are commonplace in the eastern highlands of Australia, from Tasmania to north Queensland, Twidale (1968) notes that 'Wherever they occur, the basalts rise to plains and plateaux with distinct steps and treads, stony rises and soil-filled depressions, caverns and, where the relief amplitude permits, deep, narrow gorges'. As de Martonne (1951) suggests, no relief form has been as widely and minutely studied as volcanoes. Lava flows become dissected in a similar manner in most climates, but Ollier (1967) argues that within basaltic areas several geomorphic provinces exist. Hawaii, Iceland, the Snake River district (USA) and western Victoria, Australia, are all basaltic regions although they differ considerably physiographically. However, most of the differences are due to volcanicity rather than to external factors. In periglacial areas, coarse debris forming blockfields may be produced under some conditions, while in others the product is fine silt. It seems that slight prior weathering favours reduction to silt. Vesicular or scoriaceous basalts are remarkably resistant to freeze–thaw (Tricart, 1969).

The Plateau des Coirons on the eastern flank of the Massif Central of France is formed of a succession of Miocene lava flows which have been affected in places by late Tertiary deep weathering creating depressions in the plateau surface (Tricart and Usselmann, 1969). Despite Quaternary climatic changes, the bulk of the plateau remains a relict Tertiary subtropical landform, although around its margin frost shattered fragments from glacial periods and basalt boulders which were caught up in mud-flows during the Mindel–Riss interglacial are in evidence. Such periodicity of landform evolution suggests a corresponding spatial differentiation associated with climate.

Nevertheless, as Twidale stated, it appears that in eastern Australia similarities of basalt landforms are more striking than their differences. Certainly in country wet enough to support rainforest on basalt, similar topography is found from the Atherton Tableland in North Queensland to Brown Mountain in southern New South Wales. Rounded ridge crests, steep straight-sided valley slopes and dry stream-head hollows are characteristic of the deeply weathered basalts of Australia's humid fringe. But less than 100 kilometres west of these areas, the basalt country takes on quite a different appearance. The westward tongue of the Atherton Basalt along the Millstream, a headwater of the Herbert River, extends from rainforest through wet sclerophyll forest into an open, ironbark woodland where the ground surface is littered with basalt boulders and where the depth of weathering is less than 5 m, compared with the 30 m or more to the east. About 2000 km further south, the same contrast exists between the Dorrigo–Ebor basalts in the east and those of Guyra–Armidale further west in the New England district of New South Wales. The weathering of basalt into boulders has been related to periodicity in the water content of the soil (in other words, to a regular alternation of a dry season with a wet season: Schmid, de la Souchère and Godard, 1951), but may just as well be due to the original structure of the rock (Mohr and Van Baren, 1959). Transformation of columnar basalt is likely to produce boulders.

Specific cases of climatically-determined basaltic landforms are few, but the interaction of lava flows with ice sheets in Iceland has produced a glacial variant of the steps or treads of sub-aerial basalt flows. Where lava flows met icecaps, the lava was halted and cooled, leaving an escarpment when the icefront retreated (Kjartansson, 1966).

Step or terrace-like sequences are produced by successive lava flows such as those of the southern flank of the Mount Warning shield volcano in the Richmond River valley of northern New South Wales or in the Antrim basalts of Northern Ireland (Reffay, 1972). Such distinct steps are still being formed by lava flows from volcanoes in New Guinea and other tectonically-active areas. The initial forms are determined by the nature and quantity of lava and the pre-basalt topography. In all climates the characteristic stony rises, pressure ridges, frilled flow domes and marginal waterfalls occur. Even volcanic cones seem to go through the same sequence of degradation. Throughout areas where fluvial erosion is dominant, stream channels cut into the flanks of the volcanic cone, dividing it into a series of plateaux termed *planèzes* (Tricart, 1974). When the cone is dissected to this *planèze* stage, the original relief has disappeared and structural influences are dominant. Forms of this type may be found in humid tropical climates, for example the Manna Volcanics in Papua (Ruxton, 1967), or in rainforest–sclerophyll forest transitional environments as on the Atherton Tableland of north Queensland: others have been subjected to Quaternary climatic fluctuation prior to the present temperate climate, as at Le Cantal in France (de Martonne, 1951). Thus, it appears that basalts and related volcanic rocks produce landforms which show little variation with climate. Birot (1966) argues that, save in damp tropical climates, basalt pleateau and *planèze* landforms persist over long periods, despite the relative ease of chemical weathering of basalt. The argument may be taken further to suggest that basalt and related volcanic landforms go through a sequence of erosional evolution. In environments which favour rapid decomposition of basalt, the sequence proceeds rapidly. In drier environments it proceeds more slowly.

## 12.7 Climate, time, lithology and landform

The contrasts due to time in the evolution of basalt landforms suggest that many convergent landforms may result through the operation of processes for differing periods of time. The magnitude and frequency of the events responsible for one feature may be very different from the magnitude and frequency of events responsible for another (Wolman and Miller, 1960). Climatic geomorphology must encompass this notion of time and be able to cope not only with temporal relationships of landforms, but also with the time-spans for which particular processes have persisted in particular places. Temporal relationships imply evolution and the existence of inherited landforms created by processes which have ceased to function. Reference has already been made to relict periglacial landforms in this chapter. Much of the present temperate zone has such relict landforms, but with elements of Tertiary tropical deep weathering still surviving in places (Tricart, 1965).

Time also influences the relationship between tectonics and landforms. Using the term 'relief' for tectonically influenced features, Tricart (1965) argues that tectonic features, such as sedimentary basins, fault scarps and *cuestas* influence the morphoclimatic *modelé* (or forms) developed upon them. To clarify and evaluate such assertions, the concept of scale in climatic geomorphology has to be re-examined.

Time and space scales in geomorphology help to distinguish general evolutionary tendencies from minor oscillations and spatial variation in the significance of land forming factors. Distinctions between cause and effect in the moulding of landforms depend on the span of time involved and on the size of the geomorphic system under consideration (Schumm and Lichty, 1965). Scales to cope with this problem range from considerations of orders of magnitude (Cailleux and Tricart, 1956; Haggett, 1972) to the specific areal *G*-scale (Haggett, Chorley and Stoddart, 1965) and the temporal *T*-scale (Clayton, 1968; Sugden and Hamilton, 1971). Nevertheless, despite the availability of such schemes, few of them have been used in climatic geomorphology. Tricart has however indicated that basic mechanisms controlling landforms vary with scale (Tricart, 1965), even though some forms vary with size whereas others may not (Pitty, 1971). River meanders have the same dimensions in plan regardless of scale (Zeller, 1967), while in small-scale sand landforms the coarsest material collects on the crests but the reverse is invariably the case for large-scale dunes.

Adopting Clayton's version of the *T*-scale, in which the fundamental unit is the estimated age of the earth at 4550 million years, the dominant influences at the lower values of the *G*- and *T*-scales are the major global tectonic and crustal characteristics which determine the evolution of the oceans and continents. At *G*- and *T*-scale values of 2 the major shield areas and geosynclines control relief patterns, major climatic zones only affecting the dissection of these structural entities. Even at *G*-scale 3 and *T*-scale 4, basic tectonic structures such as the London Basin or the Pennines have tectonically determined landforms. Climatic and palaeoclimatic influences emerge when the influence of climatic types of *G*-scale 2 on structural units of *G*-scale 3 are consider, for example the morphological contrasts between sandstones of similar age in different climates. At *T*-scales 4 to 7, palaeoclimatic influences are apparent, for example the contrasts in the morphology of the chalklands of France and Britain due to glaciation which are only apparent at *G*-scales 7 and higher, the overall chalk escarpment and dip slope topography being at *G*-scale 3.

Nevertheless, although climatic influences on landforms only become dominant at $G$-scale 10 and $T$-scale 8 in which individual lithologic units acquire variations in form due to mesoclimatic and microclimatic influences, tectonic influences may still intervene at $T$-scales of 8 and higher. Neotectonics, whether in unconsolidated glacial or proglacial sediments, in Quaternary deposits (Blake and Ollier, 1969), or in consolidated materials provoke changes in stream patterns and alter overland and subsurface flows. Age of exposure of Quaternary sediments influences the development of slope form and channel gradient, time-dependent landforms existing on younger glacial materials in the Upper Mississippi Valley (McConnell, 1966). The significance of such recent tectonic movements for rates of landform evolution is demonstrated by Kukal (1971) in the following table:

| | |
|---|---|
| Rate of weathering (soil development) | average 100–300 cm/1000 years |
| Rate of denudation of continents | average 100 |
| Rate of sedimentation | average 50 |
| Rate of recent tectonic deformations | average 500–1000 |

Kukal concludes that tectonic deformations are not usually compensated by either denudation or sedimentation and that tectonics are, and were throughout geological history, the basic factor in the development of the physical environment affecting sedimentary processes. This review of the role of lithology and structure in climatic geomorphology leads to the same conclusion. A glance at the Cambrian palaeogeography of Australia shows the existence of structural lineaments still apparent in the drainage pattern, such as the path of the Darling River across New South Wales. The failure to emphasize structure and lithology in recent geomorphological studies probably stems from the increasing sophistication of geological studies and a legacy from the era when the search for erosion surfaces, or evidence of the power of process over structure, dominated geomorphology. From global tectonics to individual textures and mineral composition, the influence of materials on surface form is marked. Only by comparing like with like can the real effects of climate on particular lithologies be evaluated. Sufficient numbers of such comparisons do not yet exist to permit adequate quantitative evaluation of the scales at which the influence of climate is most marked.

## References

Bertels, G., 1973, Über Glockenberge und verwandte Formen, *Catena*, 1, 57–70.
Birot, P., 1955, *Les Méthodes de la Morphologie*, Presses Univ. France, Paris, 177 pp.
Birot, P., 1966, *General Physical Geography*, Harrap, London, 360 pp.
Blake, D. H. and C. D. Ollier, 1969, Geomorphological evidence of Quaternary tectonics in southwestern Papua, *Rev. géom. dyn.*, 19, 28–32.
Davis, W. M., 1909, *Geographical essays*, Ginn, Boston.
Cailleux, A. and J. Tricart, 1956, Le problème de la classification des faits geomorphologiques, *Annales de Géographie*, 65, 162–186.
Carson, M. A. and M. F. Kirkby, 1972, *Hillslope Form and Process*, Cambridge University Press, 474 pp.

Chorley, R. J., 1964, The nodal position and anomalous character of slope studies in geomorphological research, *Geogr. J.*, **130**, 70–73.

Chorley, R. J. and M. A. Morgan, 1962, Comparison of morphometric features, Unaka Mountains, Tennessee and North Carolina, and Dartmoor, England, *Bull. geol. Soc. Am.*, **73**, 17–34.

Clayton, K. M., 1968, *The Evolution of River Systems*, Introductory note for Informal Discussion, Hydrological Group, Institution of Civil Engineers, London, 25 January 1968 (reported by J. R. Hardy, *Proc. Instn. Civ. Engnrs. Lond.*, **40**, 397–399).

De Martonne, E., 1951, *Traité de géographie physique: Tome II—le relief du sol*, 9th ed. Armand Colin, Paris, pp. 499–1057.

Donahue, J. J., 1972, Drainage intensity in western New York, *Ann. Ass. Am. Geog.*, **62**, 23–36.

Douglas, I., 1967, Erosion of granite terrains under tropical rainforest in Australia, Malaysia and Singapore, *Publs. Assn. Internat. Hydrol Scient.*, **75**, 31–39.

Douglas, I., 1969, Sediment sources and causes in the humid tropics of north east Queensland, australia, *Geomorphology in a Tropical Enrironment*, Brit. Geomorph. Res. Group, *Occ. Paper 5*, 27–39.

Douglas, I., 1972, *The Environment Game*, Inaugural Lecture, the University of New England, Armidale, 28 pp.

Douglas, I., 1973, Rates of denudation in selected small catchments in Eastern Australia, *University of Hull Occasional Papers in Geography*, *21*, 128 pp.

Drever, J. I., 1971, Chemical weathering in a subtropical igneous terrain, Rio Ameca, Mexico, *J. sedim. Petrol.*, **41**, 951–961.

Evans, I. S., 1970, Salt crystallization and rock weathering: a review, *Revue Géom. dyn.*, **19**, 153–177.

Fair, T. J. D., 1948, Slope form and development in the interior of Natal, South Africa, *Trans. Geol. Soc. South Africa*, **50**, 105–119.

Fisher, R. A. and F. Yeates, 1963, *Statistical Tables for Biological Agricultural and Medical Research*, 6th ed., Oliver and Boyal, Edinburgh, 146 pp.

Hack, J. T., 1960, Relation of solution features to chemical character of water in the Shenandoah Valley, Virginia, *U.S. geol. Surv. Prof. Pap.*, *400B*, B387–B390.

Hack, J. T. and J. C. Goodlett, 1960, Geomorphology and forest ecology of a mountain region in the Central Appalachians, *U.S. geol. Surv. Prof. Pap. 347*.

Haggett, P. 1972, *Geography: a modern synthesis*, Harper, New York, 483 pp.

Haggett, P., R. J. Chorley and D. R. Stoddart, 1965, Scale standards in geographical research: a new measure of area magnitude, *Nature*, **205**, 844–847.

Hettner, A., 1972, *The Surface Features of the Land* (translated P. Tilley), Macmillan, London, 193 pp.

Hurault, J., 1963, Recherches sur les inselbergs granitiques nus en Guyane Française, *Revue Géom. dyn.*, **14**, 49–61.

Keller, W. D., 1954, The energy factor in sedimentation, *J. Sedim. Petrol.*, **24**, 62–68.

Kesel, R. H., 1973, Inselberg landform elements: definition and synthesis, *Revue Géom. dyn.*, **22**, 97–108.

Kjartansson, G., 1966, Sur la récession glaciaire et les types vocaniques dans la région du Kjolur sur le plateau central de l'Islande, *Revue Géom. dyn.*, **16**, 23–39.

Krumbein, W. C. and R. M. Garrels, 1952, Origin and classification of chemical sediments in terms of pH and oxidation potentials, *J. Geol.*, **60**, 1–33.

Kukal, Z., 1971, *Geology of Recent Sediments*, Academia, Prague, 490 pp.

Lukashev, K. I., 1970, *Lithology and geochemistry of the weathering crust*, Israel Program for Scientific Translations Jerusalem, 368 pp.

Mainguet, M., 1966, Un exemple de formation de 'boules' dans une roche cristalline du désert littoral Péruvien, *Revue Géom. dyn.*, **16**, 49–53.

Mainguet, M., 1972, *Le modelé des Grés*, Institut Géographique National, Paris, 2 vols,. 657 pp.

McConnell, M., 1966, A statistical analysis of spatial variability of mean topographic slope on stream-dissected glacial materials, *Ann. Ass. Am. Geog.*, **56**, 712–728.

Meckelein, W., 1965, Beobachtungen und Gedanken zu geomorphologischen Konvergenzen in Polar-und Warmewusten, *Erdkunde*, **19**, 31–39.

Michel, P., 1973, Les bassins des fleuves Sénégal et Gambie: Etude géomorphologique, *Mémoires ORSTOM*, **63**, 2 vols., 752 pp.

Morgan, R. P. C., 1973, The influence of scale inclimatic geomorphology: a case study of drainage density in West Malaysia, *Geog. Annlr*, **55A**, 107–115.

Mohr, E. C. J. and F. A. Van Baren, 1959, *Tropical Soils*, Van Hoeve, The Hague, 498 pp.

Ollier, C. D., 1967, Landforms of the Newer Volcanic Province of Victoria, in *Landform Studies from Australia and New Guinea*, J. N. Jennings and J. A. Mabbutt (Eds.), ANU Press, Canberra, 315–339.

Pain, C. F., 1972, Characteristics and geomorphic effects of earthquake-initiated landslides in the Adelbert Range, Papua, New Guinea, *Eng. Geol.*, **6**, 261–274.

Palmer, J. and J. Radley, 1961, Gristone tors of the English Pennines, *Z. Geomorph*, N. F. **5**, 37–52.

Penck, A., 1910, Versuch einer Klimaklassifikation auf physiographischer Grundlage, *Sitzungsbericht. d. Preuss. Akad. d. Wissensch, Math-Phys. Klasse*, **12**, 236–246. Also available as: Attempt at a classification of climate on a physiographic basis, in *Climatic geomorphology*, E. Derbyshire (Ed.), Macmillan, London, 51–60.

Penck, A., 1914, Die Formen der Landoberflache und Verschiebungen Klimagurtel, *Sitzungsbericht. d. Preuss. Akad. d. Wissensch., Math-Phys. Klasse*, **14**, 77–97.

Pettijohn, F. J., 1957, *Sedimentary Rocks*, Harper, New York, 718 pp.

Pitty, A. F., 1971, *Introduction to Geomorphology*, Methuen, London, 526 pp.

Reffay, A., 1972, *Les montagnes de l'Irlande Septentrionale*, Imprimerie Allier, Grenoble 614 pp.

Rich, J. L., 1938, Recognition and significance of multiple erosion surfaces, *Bull. geol. Soc. Am.*, **49**, 1695–1722.

Ruxton, B. P., 1967, Geomorphology of the Safia-Pongani Area, *CSIRO Land Res. Ser.*, **17**, 86–97.

Savigear, R. A. G., 1952, Some observations on slope development in South Wales, *Trans. Inst. Brit. Geog.*, **18**, 31–51.

Schmid, M., P. de la Souchère and D. Godard, 1951, Les sols et la végétation an Darlac et sur le plateau des Trois-Frontières, *Arch. Rech. Agron. Cambodge, Laos, Viêtnam*, **8**, 107 pp.

Schwarzbach, M., 1966, Bemerkenswerte Konglomerat-Verwitterung. *Zeit. für Geom.* N. F., **10**, 169–182.

Schumm, S. A., 1963, The disparity between present rates of denudation and orogeny, *U.S. geol. Surv. prof. Pap.*, 454-H, 13 pp.

Schumm, S. A. and R. F. Hadley, 1961, Progress in the application of landform analysis in studies of semi-arid erosion, *U.S. geol. Surv. Circ.*, 437.

Schumm, S. A. and R. W. Lichty, 1965, Time space, and causality in geomorphology, *Am. J. Sci.*, **263**, 110–119.

Selby, M. J., 1967, Aspects of the geomorphology of the greywacke ranges bordering the lower and middle Waikato basins, *Earth Sci. J.*, **1**, 37–58.

Seret, G. and P. de Béthune, 1967, Compte Rendu de l'Excursion du samedi 11 Juin 1966, la Roche-en-Ardenne-March-Han Sur Lesse-Namur, in *Evolution des Versants*, Université de Liège, 323–349.

Simonett, D. S., 1967, Landslide distribution and earthquakes in the Bewani and Torricelli Mountains, New Guinea, in *Landform Studies from Australia and New Guinea*, J. N. Jennings and J. A. Mabbutt (Eds.), Australian National University Press Camberra, 64–84.

Sioli, H., 1968, Hydrochemistry and Geology in the Brazilian Amazon Region, *Amazoniana*, **1**, 267–277.

Stoddart, D. R., 1969, Climatic geomorphology: review and re-assessment, *Progress in Geography*, **1**, 159–222.
Sugden, D. and P. Hamilton, 1971, Scale, systems and regional geography, *Area*, **3**, 139–144.
Sweeting M. M. and G. S. Sweeting, 1969, Some aspects of the Carboniferous Limestone in relation to its landforms with particular reference to N. W. Yorkshire and Country Clare, *Etudes et Travaux de Mediterranée*, **7**, 201–209.
Tricart, J., 1965, Schéma des mécanismes de causalité en géomorphologie, *Ann. de Géog.*, **74**, 322–326.
Tricart, J., 1969, *Geomorphology of Cold Environments*, Macmillan, London, 320 pp.
Tricart, J. and A. Cailleux 1965, *Introduction à la géomorphologie climatique*, SEDES, Paris, 306 pp.
Tricart, J. and P. Usselmann, 1969, Feuille géomorphologique Privas 7–8, *Revue geom. dyn.*, **19**, 115–127.
Tricart, J., 1974, *Structural geomorphology* (trans. S. H. Beaver and E. Derbyshire), London, Longmans, 301 pp.
Troll, C., 1964, Karte der Jahreszeiten-Klimate der Erde, *Erdkunde*, **18**, 5–28.
Twidale, C. R., 1967, Hillslopes and pediments in the Flinders Ranges, South Australia, in *Landform Studies from Australia and New Guinea*, J. N. Jennings and J. A. Mabbutt (Eds.), Australian National University Press, Canberra, 95–117.
Twidale, C. R., 1968, *Geomorphology with special reference to Australia*, Nelson, Melborne, 406 pp.
Vogt, J., 1962, Une Vallée soudanaise: la moyenne Bagoé, *Revue Géom. dyn.*, **13**, 2–9.
Wilhelmy, H. 1958, *Klimamorphologie der Massengesteine*, Georg Westermann Verlag, Braunschweig, 238 pp.
Wilhelmy, II., 1964, Cavernous rock surfaces (tafoni) in semiarid and arid climates, *Pakistan Geogr. Rev.*, **19** (2), 9–13.
Williams, P. W., 1969, The geomorphic effects of ground water, in *Water, Earth and Man*, R. J. Chorley (Ed.), Methuen, London, 269–284.
Wolman, M. G. and J. P. Miller, 1960, Magnitude and frequency of forces in geomorphic processes, *J. Geol.*, **68**, 54–74.
Yatsu, E., 1966, *Rock control in geomorphology*, Sozoshu, Tokyo, 135 pp.
Zeller, J., 1967, Meandering channels in Switzerland, *Publs. Int. Assoc. Hydrol. Scient.*, **75**, 174–186.

CHAPTER THIRTEEN

# Process, Landforms and Climate in Limestone Regions

DAVID I. SMITH AND TIMOTHY C. ATKINSON

## 13.1 Introduction

This chapter attempts to answer the question whether or not the landforms which typify limestone terrains, and the erosion processes by which they are produced, are different in the various climatic belts of the world. It is certainly true that some limestone landforms such as karst towers and cone karst occur only in the humid tropics. Others such as limestone pavements are found only in areas which suffered glaciation in the Pleistocene. Numerous attempts have have been made in the past to relate less obvious differences in landforms to differences in climate, usually by visually comparing the landscapes of two or more regions with constrasting climates. It has been argued that climatic differences cause such variations in landforms as may exist through their undoubted influence upon vegetation and soil types, and amounts and patterns of runoff. In turn, these factors influence the processes by which limestone is eroded, different processes producing different landforms. Most authors on this topic have asserted, rather than measured, the precise differences in erosion processes. Thus Lehmann, one of the founders of the climatic school of geomorphology, seeks to explain the large proportion of the landscape occupied by closed depressions in tropical cone karst by the assertion that solutional erosion rates are more intense in the tropics. He says that this is a result of the greater runoff produced by higher rainfall, and the greater production of carbon dioxide in the soil at the higher temperatures (Lehmann, 1964, 1970). In fact neither of these assertions is generally true, as will be shown in what follows.

During the last twenty years, various workers have made measurements of of erosion rates in limestone areas all over the world. Much of the initial impetus for these studies came from the work of the French geomorphologist Corbel, who held the view that rate of erosion was determined by climate, the greatest rates occurring in cold, humid climates and the lowest rates in hot, arid areas. In particular, Corbel claimed that the cooler it was the greater would be the hardness of runoff water (i.e. the concentration of dissolved limestone it contains) because carbon dioxide, which is an essential reagent in the solution of limestone, is more soluble at lower temperatures (Corbel, 1957, 1959a;

Corbel and Muxart, 1970). Few other workers now agree with Corbel; but the general belief that erosion rates were at least partially determined by climate was widely held and led to many studies comparing erosion rates under different climates.

In the remainder of this chapter two hypotheses will be critically examined. The first of these proposes that there is a causal relationship between climate and differing landforms and landscapes in limestone regions while the second postulates that there is a similar relationship between climate and erosion processes. In the past, workers in this field have drawn broad distinctions between the Tropical, Temperate and Arctic/Alpine zones' and then considered finer differences between landforms, processes and climate within these boundaries. This same broad classification of climates, namely Tropical, Temperate and Arctic/Alpine, is adopted here.

## 13.2 Erosion processes and their Measurement

Limestones have a fairly simple erosion chemistry because they contain only two major minerals, calcite ($CaCO_3$) and dolomite ($CaMg(CO_3)_2$), both of which are soluble in natural waters containing dilute carbonic acid. The details of the solution process have been the subject of much study (see, for example Picknett (1964, 1972) but the overall chemical reaction may be summarized by the following equations.

$$CaCO_3 + H_2CO_3 \rightleftharpoons Ca^{++} + 2HCO_3^-$$
$$CaMg(CO_3)_2 + 2H_2CO_3 \rightleftharpoons Ca^{++} + Mg^{++} + 4HCO_3^-$$

The ions on the right-hand side are the major species which occur in dilute solutions of limestone or dolomite. The carbonic acid, $H_2CO_3$, is derived by the solution of carbon dioxide from the air and its reaction with water, thus:

$$CO_2 + H_2O \rightleftharpoons H_2CO_3$$

The solubility of carbon dioxide depends upon two factors, its concentration in the air, measured in terms of its partial pressure, and the temperature of the solution. The greater the partial pressure the more carbon dioxide will be dissolved before an equilibrium is established between air and water. At a constant partial pressure, an increase in temperature has the effect of reducing the solubility of carbon dioxide. The mechanisms behind these effects have been studied and explained in detail by Picknett (1964, 1972), Jacobson and Langmuir (1972) and Ek (1969). Smith and Mead (1962) provide a convenient summary in the form of a graph relating the percentage of carbon dioxide in the air to the concentration of pure calcium carbonate which the resulting carbonic acid will dissolve at an overall pressure of 1 bar and a temperature of 10° C. This graph is shown here as Figure 13.1. Note that two curves are shown. The 'equilibrium curve' describes the case in which the air, water and solid calcium carbonate are all in contact together, so that as carbonic acid is used up by reacting with calcite, more carbon dioxide may dissolve

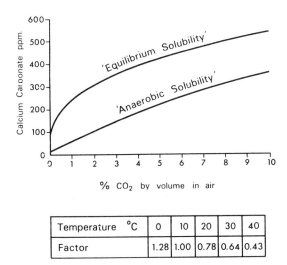

Figure 13.1. Calcium carbonate solubilities at 10 °C.

from the air. Clearly, if the carbonic acid can be replenished the resulting 'equilibrium' solubility of calcite will be higher than in the 'anaerobic' case, in which air and water are first brought to equilibrium with each other and the solution is then removed from contact with the air and allowed to react with the calcite. In this case the supply of carbonic acid is limited to that present initially, and a lower solubility of calcite results. The effect of temperature on solubility is shown in the table in Figure 13.1, which presents factors by which the solubility of calcite at 10° C must be multiplied to give the equivalent solubility at other temperatures.

Limestones are moderately soluble rocks when an adequate supply of carbon dioxide is available. What distinguishes them from silicate rocks, which may also be quite soluble, is that pure limestones leave very little residue whereas the leaching of cations from silicates produces insoluble clays and hydroxides. In addition, limestones tend to respond to folding and other movements of the earth's crust by fracturing and joint formation rather than by plastic deformation. This results in most limestones being criss-crossed by a network of joints and fractures which may be widened by solution, leading to the development of underground drainage, except where special factors such as permafrost or impermeable deposits prevent it. In silicate rocks solutional weathering produces clays which tend to impede water movement along joints and prevent the development of underground drainage. The net erosion from limestone areas with underground drainage is predominantly solutional, partly because in pure limestones there is only a small insoluble fraction of the rock to be mechanically transported and partly because subterranean drainage itself tends to isolate the products of mechanical weathering at the surface. The processes by which particulate debris is produced are separated from the underground streams which are the main means by which the debris may be removed (Newson, 1971).

It is clear from the brief discussion of solutional chemistry given above that the hardness of water leaving a limestone area will depend upon its temperature and the supply of carbon dioxide. A third factor is the length of time which the water has spent in contact with the limestone, because it takes a finite time for the various chemical reactions involved in the solution process to come to equilibrium. If the water has spent less time than this in the system it will not be saturated with calcium bicarbonate. The time taken to reach equilibrium depends upon the concentrations of the different dissolved ions and upon the shape, size and surface area of the reacting volumes of water, air and limestone. It varies between a few minutes or hours for water moving through fine pores (Weyl, 1958) or in thin films over limestone (Bögli, 1960) to about 10 days for water moving through voids the size of natural cave passages (Rocques, 1969). In general, percolation water draining from limestone areas remains within the system for longer periods than seem to be required to achieve equilibrium, whereas swallet water (i.e. fast-flowing turbulent streams in caves or conduits) does not (see, for example, Drew (1968), Smith, Atkinson and Drew (1975) and Atkinson and Drew (1974)). This is the reason why waters from conduit-fed springs show a variation in hardness with discharge and are often undersaturated with calcite, whereas those from springs fed by diffuse percolation have more constant hardness and tend to be saturated (Newson, 1971, 1972; Shuster and White, 1971; Ede, 1972). Rivers in limestone areas show a similar variability of hardness, though not always the same close relationship with discharge (Douglas, 1968; Williams, 1968). In testing the relationship between climate and the hardness of runoff waters it is necessary to compare values of the *mean* hardness at different sites. To obtain reliable estimates of the mean, water samples should be collected and analysed for a range of discharge conditions. In the case of percolation-fed springs, relatively few samples need to be taken although allowance should be made for possible seasonal variations. For rivers and conduit-fed springs, estimates of the mean hardness should be based upon between 30 and 50 water samples taken over the whole range of discharge conditions and seasons.

In principle, the mean hardness of runoff water should depend upon the available carbon dioxide and the water temperature, though some scatter in this relationship can be expected in the case of rivers and conduit-fed springs. The influence of climate, if any, should be exerted through its effect upon temperature, vegetation and soil and the effect that these have upon the carbon dioxide supply.

One further source of variation in mean hardness has not yet been mentioned and that is the variation due to differences in lithology. The hardness of water is measured by titrating the concentration of $Ca^{++}$ and $Mg^{++}$ in solution. Before 1958, the reagent used was generally soap solution which gives results for total hardness (calcium plus magnesium) which are of doubtful precision and accuracy. Since 1958 the complexinetric method using ethylene-diamine-tetra-acetic acid (EDTA) has been widely used (Schwarzenbach, 1957). This method estimates $Ca^{++}$ separately and as a result many field workers have measured only the concentrations of calcium. Magnesium concentrations may often be appreciable, especially in waters draining dolomites, and taking them into account raises the estimate of mean hardness. Also Picknett (1972) has shown that the presence of $Mg^{++}$ has a considerable effect upon the solubility of calcite, while the solubility of dolomite has long been known to differ from calcite, though not very greatly (Yanateva, 1954; Jacobson and Langmuir, 1972). These effects of data quality and solution chemistry both act to increase the scatter in the values of mean hardness which are reported in the literature.

Solutional erosion rates depend upon the product of hardness and rate of runoff. To obtain the most accurate estimate of erosion rate, a long record of discharge from the river or spring under study should be used to construct the flow duration curve. Then the relationship between discharge and solution load (i.e. instantaneous rate of removal of dissolved limestone) should be determined by least squares analysis. This relationship can then be applied to the flow duration curve and the overall erosion rate calculated as the average of the instantaneous rates corresponding to each discharge class, weighted by the proportions of time for which each discharge occurs. However, the method requires a lengthy flow record (30–50 years) which is not often available, and very few rates calculated in this way have appeared in the literature (for an example see Smith and Newson (1974)).

Most other methods of calculating the erosion rate rely upon variations of the relationship,

$$X = \frac{\bar{Q}}{A} \cdot \frac{T}{10^6 \cdot r} \cdot \frac{1}{n}$$

where $X$ is mean erosion rate in $m^3/km^2/year$, $Q$ is mean annual runoff in $m^3/year$, $T$ is mean water hardness in mg/l $CaCO_3$ (or $CaCO_3 + MgCO_3$), $r$ is bulk density of limestone in $g/cm^3$, $A$ is the catchment area in $km^2$ and $n$ is the fraction of the catchment area occupied by limestone.

The annual runoff may sometimes be available from direct measurements at the site under study, but the period of record is often too short to give a reliable estimate of the mean. In addition, the catchment area must be known in order to express the erosion rate in terms which are comparable with other catchments. This is often difficult to determine and, in the case of subterranean drainage, may involve lengthy water tracing experiments. However, some estimates of erosion rate have been prepared in this way by the authors (unpublished data), Pulina (1971) and Corbel (1959a) amongst others. An alternative method is to estimate $\bar{Q}/A$ as a single term, using records of a gauged stream in the same area. This presents problems of extrapolation and involves the assumption that runoff in the ungauged catchment is exactly similar to that in the gauged stream. In many cases it is easier to estimate the mean annual runoff as the difference between mean annual precipitation and mean annual potential evapotranspiration, or else as the sum of the monthly differences between these quantities. Potential evapotranspiration may be known in a few areas from direct measurements but at most localities it must be estimated from other climatic parameters such as temperature, by means of formulae like those of Thornthwaite (1948) and Thornthwaite and Mather (1957). If even the temperature values needed for the Thornthwaite calculation are not available, the runoff may be estimated from the precipitation by means of an empirical runoff coefficient, a factor which is equal to the fraction of precipitation which forms runoff in a gauged stream in the same region. For many countries small-scale contoured maps of the runoff coefficient have been published, and these have been used by Corbel (1965), for example, in many of his estimates of erosion rates. Clearly, these various methods of estimating mean annual runoff decrease progressively in accuracy in the order in which they are described here.

The value of the hardness used ($T$) should be that which corresponds to the mean annual discharge, based upon the relationship established between discharge and hardness from the results of 30 or more representative samples. In practice, the mean value of a similar number of samples is often used, perhaps because discharge was not estimated when the samples were taken, or because there is only a weak relationship between the two variables. In some cases, very few samples are taken and $T$ is estimated as the mean of only two or three. This is likely to give a very poor estimate of erosion rate unless the hardness at the particular site is almost constant. Ideally, the total carbonate hardness in mg/l $CaCO_3 + MgCO_3$ should be used, although many authors have ignored the effect of $MgCO_3$ in their calculations.

Compared with the errors commonly introduced into calculations of erosion rate by the poor quality of hardness or runoff data, the common practice of assuming a value of 2·5 g/cm³ for the bulk density of limestone, $r$, introduces only a minor inaccuracy. The density of calcite is 2·71 and of dolomite 2·86, while the bulk density of massive limestone varies from 2·5 to around 2·7 (see, for example, Muir and coworkers (1956)). Assuming a value of 2·5 introduces an error of only 7·4% if the true density is 2·7. While large, this error is often smaller than those due to other causes.

The most commonly used method of calculating erosion rates is the simplified formula proposed by Corbel (1957), namely

$$X = \frac{4ET}{100} \cdot \frac{1}{n}$$

where $E$ is mean annual runoff in decimetres (equivalent to $\bar{Q}/A$), $T$ is hardness and $n$ is the proportion of limestone in the catchment. Like all the methods discussed so far, this formula makes allowance for the occurrence of non-limestone rocks in the area under study by assuming that *all* of the calcium and magnesium in the runoff is derived from the limestone. This assumption is often untrue and, ideally, extra water samples should be taken to establish the proportions of solutes derived from non-limestone rocks and from rainfall. The value of $T$ should be adjusted so as to include only those solutes derived from the limestone. Uncritical application of the factor, $n$, in the Corbel formula will usually result in a serious overestimate of the rate of solution which will be greater the smaller the proportion of limestone in the catchment.

From this discussion of data quality it is clear that the values of erosion rates which are reported in the literature may vary very widely in their accuracy, depending upon the data used and on the methods employed in combining it to calculate a result.

## 13.3 Erosion Rates, Water Hardness and Climate

Corbel was led by his knowledge of the basic chemistry of the solution process to postulate that erosion rates would be greatest in cool climates (Corbel, 1957, 1959a). He argued that in areas with similar runoff, low temperatures would enhance the solubility of carbon dioxide and so lead to a greater concentration of dissolved limestone. Therefore, the greatest erosion rates at a given value of runoff should be found in the Arctic and high Alpine regions, and the lowest values should occur in the Tropics. This view has been disputed by a large number of writers (for example, Lehmann (1964), Sweeting (1964) and Jakucs (1973) who point out that Corbel ignores the fact that biogenic carbon dioxide occurs in the soil air in concentrations from thirty to one hundred times greater than that found in the atmosphere. They argue that erosion rates will therefore be greater in soil-covered areas than in permafrost climates where there is little soil development. That a soil cover actually has this effect is illustrated by an experiment made by Bauer (1964) who analysed the runoff from plots of a few square metres at altitudes between 1400 m and 2000 m above sea level in the Dachstein Alps, Austria. Plots of bare limestone and of soil-covered areas were both examined with the results shown in Table 13.1. Clearly, both hardness and erosion rate are greater in soil-covered areas than on bare limestone, even where these are present side by side in mountainous terrain.

Table 13.1. Erosion on small plots in Dachstein Alps, Austria (after Bauer (1964)

|  | Bare limestone | | Soil-covered limestone |
| --- | --- | --- | --- |
|  | I | II |  |
| Mean hardness (mg/l $CaCO_3$) | 24·6 | 29·0 | 49·4 |
| Range of hardness (mg/l $CaCO_3$) | 14–63 | 15–76 | 30–96 |
| Erosion rate (m³/km²/year) | 9·0 | 12·5 | 28·0 |

On the other hand Lehmann (1970) has argued that erosion rates should be greater in the tropical soil-covered karst than in temperate areas, because the higher soil temperatures and greater production of tropical vegetation should lead to faster generation of carbon dioxide and higher concentrations in the soil air. Other writers (e.g. Balazs (1971, 1973a) have compared erosion rates in the humid tropics with those from drier temperate climates and have suggested that high rates are typical of tropical areas because of large values of annual runoff.

Thus, the literature contains two opposing hypotheses, each asserting a relationship between erosion rate and climate. Corbel and his adherents (Corbel, 1957, 1959a, 1960, 1965, 1971; Corbel and Muxart, 1970) hold that erosion rates are greater in cold climates, whereas the opposite view is advanced by Lehmann (1970), Balazs (1971, 1973a), Sweeting (1966) and Sweeting and Gerstenhauer (1960) who suggest that the greatest rates occur in the humid tropics. We shall attempt to test both of these hypotheses below by reference to published data.

It follows from this discussion that the principal controls upon erosion rate are the mean hardness and mean annual runoff. Therefore, in order to test the various contentions about the relationship between erosion rate and climate, it would be sensible to examine water hardness first. A search of the literature revealed 134 estimates of the rate of solutional erosion in the catchments of large springs and rivers all over the world, and 231 reports of the mean hardness of spring and river waters. These data are average values for areas between several square kilometers and several hundred square kilometres in extent. They do not reflect local variations within drainage basins. They are very uneven in quality, although most of the erosion rate measurements were made using the Corbel formula, estimating the runoff as the difference between mean annual precipitation and evapotranspiration. A few are based upon records of stream discharge and some upon runoff coefficients. The number of water samples analysed in order to estimate the mean hardness varies from one or two in some cases to several hundred in others. Some reports do not state what the number of samples was, merely stating a mean value! For river sites, at which hardness varies much more than at springs, estimates which were not based upon at least ten seasonally-spaced samples were rejected. In spite of their uneven quality, the hardness data were treated as a single 'grab' sample of comparable values, partly because of the difficulty of assessing the quality of individual mean values and partly because separating them into groups of even quality produces sub-samples which are too small for estimation of variations due to climate. In any case, it is believed that all of the estimates of mean hardness used are reasonably reliable, although they include values based upon calcium alone as well as total hardness.

The two hundred and thirty-one values of mean hardness are grouped in Figure 13.2 into Arctic/Alpine, Temperate and Tropical categories. Under the climatic hypothesis, there should be differences between the distributions of hardness in each climate and between the means of each distribution.

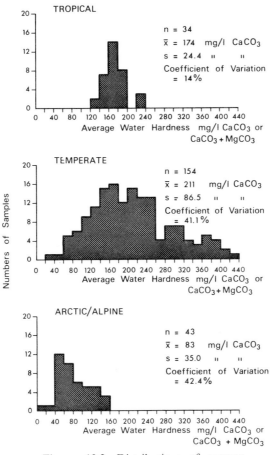

Figure 13.2. Distributions of average water hardness values for rivers and springs under different climates.

Whether these differences exist or not can best be judged by applying statistical tests to the data. A chi-square test of goodness-of-fit (Krumbein and Graybill, 1965, p. 172) showed that both the Arctic/Alpine and Tropical groups have distributions which are significantly different from the normal curve at the 95% confidence level. The Temperate group, for which there is the greatest number of values, shows no significant difference from a normal distribution with the same mean and variance as the sample data. Since not all three samples are normally distributed, it is impossible to apply parametric statistics to test for differences between their means and distributions. No non-parametric tests are suitable for testing differences of means, but the chi-square test may again be used to test for differences in distribution between the samples (Siegel, 1956, p. 175). Using this test it appears that there are statistically signficant differences between the distributions of hardness in each of the three climatic groups at

Table 13.2. Values of chi-square in testing differences of hardness with climate

| Groups tested | $\chi^2$ | df | α | $\chi^2_\alpha(df)$ | Significant difference |
|---|---|---|---|---|---|
| All three | 139·68 | 10 | 0·05 | 18·31 | Yes |
| Arctic/Alpine + Temperate | 92·94 | 7 | 0·05 | 14·07 | Yes |
| Temperate and Tropical | 29·98 | 5 | 0·05 | 11·07 | Yes |
| Arctic/Alpine + Tropical | 59·60 | 4 | 0·05 | 9·49 | Yes |

a 95% level of confidence, both when all three groups are taken together and when they are combined in pairs (Table 13.2). This means that the chance of drawing these three distributions by sampling at random from a homogeneous population of hardness in all climates is less than 5%.

Although no non-parametric test can be used to ascertain the significance of differences between mean hardnesses under different climates, it is quite clear from inspection of Figure 13.2 that there is a great difference between the mean hardness in the Arctic/Alpine group (82·6 mg/l $CaCO_3$) and the other two climates. The latter have the same modal class (161–180 mg/l) but different means, of 210·5 mg/l for the Temperate group and 174·0 mg/l for the Tropical. The absolute scatter of data is greatest in the Temperate group and least in the Tropical group, but the relative dispersion, as indicated by the coefficient of variation is approximately equal in Temperate and Arctic/Alpine groups and least in the Tropical. It should be noted that the coefficient of variation is used here purely as an index of relative dispersion. Its use in statistical inference would imply the assumption that the distributions of hardness were normal, which is not true.

In summary, the hardness data show that there are very clear differences in the distributions of mean annual hardness under different climates. Although it cannot be tested statistically, it would appear that these differences may arise from two underlying trends in the data. There is a trend in the overall means from a minimum value in the Arctic/Alpine group to a maximum in the Temperate region with a slight decline in the Tropical group. Absolute variability shows an opposite trend, being least in the tropics, at a maximum in the Temperate group and intermediate in Arctic/Alpine areas. Relative variability is very small in the Tropical group but about three times greater in both of the other two regions. These trends indicate that climate may well exert an influence or partial control upon water hardness, but they support neither the hypothesis of Corbel (1959a) nor that of his opponents, since the greatest values, as well as the greatest variability, are found in the Temperate group.

We shall now examine possible hypotheses to explain these differences. The most obvious differences is between low values in Arctic/Alpine areas and the high ones in the other two groups. This is almost certainly due to lack of soil cover in the Arctic/Alpine areas. The presence of soil brings percolating water into equilibrium with air of a carbon dioxide content of 0·1 to 2·5% (Russell, 1961; and see below) in the Temperate zone. There is a consequent increase in hardness as the work of Bauer (1964) quoted above shows. Clearly, most Arctic/Alpine waters will also be colder than Temperate waters, giving an increase in carbon dioxide solubility at any given partial pressure. But this effect will increase limestone solubility by only 28% between 10 °C and 0 °C, whereas an increase of carbon dioxide content in the air from 0·03% at 0 °C to 1·0% at 10 °C (at 1 bar pressure) raises the solubility of limestone by 153%, from 95 mg/l to 240 mg/l $CaCO_3$. 10 °C is a typical mean annual water temperature for springs in areas such as southern Britain, and 1·0% is a reasonable estimate for the average carbon dioxide content of soil air in such areas (Russell, 1961). Note the similarity of the last two hardness figures to the means in each climatic group in Figure 13.2.

The difference between hardness of waters in Tropical and Temperate regions may also depend upon temperature and carbon dioxide supply. Lehmann (1970) maintains that the effect of higher carbon dioxide concentrations in tropical soils outweighs the effect of higher temperatures. Russell (1961, p. 366) states, 'under tropical conditions the carbon dioxide content of the soil air may rise much higher [than the 0·1–1·6% normally found in temperate soils]...during the warm rainy seasons, presumably because of the very rapid evolution of carbon dioxide by the soil organisms on the one hand, and the heavily restricted air space in the soil on the other'. There are very few reported figures available with which to test this statement which is based upon the results of Leather (1915) and Vine, Thompson and Hardy (1943). Also it is not clear from Russell's account whether Temperate soils do not sometimes show similar values in summer. Table 13.3 presents all of the data on soil carbon dioxide which is available to the authors. Some are from recent studies in which measurements were made at a variety of depths at different seasons. Earlier studies are often rather incomplete in this respect. In general, carbon dioxide concentrations increases with depth and it is the carbon dioxide concentration at the base of the soil profile which determines how much limestone is dissolved. Certainly, from the few figures available it would appear that carbon dioxide values are higher in the deeply weathered soils of the Tropics. However, when individual measurements made at the same depth are compared it can be seen that there is little difference between Temperate and Tropical soils. Of the latter, only the results from Trinidad support the contention that levels of carbon dioxide are higher in the Tropics, or Russell's view that wet season levels of carbon dioxide will be much higher than summer levels in temperate soils. In particular, soils developed solely by weathering of the underlying limestone are usually thin, in both the Tropical and Temperate zones. Thus it is the carbon dioxide at shallow depths which should be used as a basis for comparison. The average carbon dioxide content for Tropical soils at a depth of around 30 cm can be calculated from the data in Table 13.3 to be about 2·0%, with a range from

Table 13.3. Carbon dioxide concentrations in soil air

| Source | Soil/Vegetation | | Usual percentage $CO_2$ | Summer | Winter | Extreme values |
|---|---|---|---|---|---|---|
| **TROPICAL** | | | | | | |
| Zonn and Li, 1960 | Evergreen forest | 10 cm | 0·5 –1·0 | — | — | — |
| | | 200 cm | 3·4 –6·3 | — | — | — |
| | Bamboo forest | 10 cm | 0·2 –3·5 | — | — | — |
| | | 200 cm | 4·1 –10·8 | — | — | — |
| | | | | Wet season | Dry season | |
| Vine, Thompson and Hardy, 1943 | Cacao plantation, Trinidad | 10 cm | — | 2·8–6·5 | 0·2–0·8 | — |
| | | 25 cm | — | 3·0–8·5 | 0·8–1·7 | — |
| | | 45 cm | — | 4·2–9·7 | 1·4–3·8 | — |
| | | 90 cm | — | 4·5–14·3 | 3·4–7·6 | — |
| | | 120 cm | — | 3·6–17·5 | 3·7–6·8 | — |
| Nicholson and Nicholson, 1969 | Limestone soils, Jamaica | 15 cm | 0·3 –1·6 | — | — | — |
| | | 30 cm | 0·4 –3·0 | — | — | — |
| | | 60 cm | 1·0 | — | — | — |
| **TEMPERATE** | | | | | | |
| Russell, 1961 | Arable | | 0·9 | | | 0·5 –11·5 |
| Russell, 1961 | Pasture | | 0·5 –1·5 | | | 0·05– 3·0 |
| Russell, 1961 | Sandy arable | | 0·16 | | | 0·07– 0·55 |
| Russell, 1961 | Arable loam | | 0·23 | | | 0·28– 1·4 |
| Russell, 1961 | Moorland | | 0·65 | | | 0·01 –1·4 |
| Russell, 1961 | Arable | | 0·1 –0·2 | | | 0·03– 3·2 |
| Russell, 1961 | Manured arable | | 0·4 | | | 0·3 – 3·3 |
| Russell, 1961 | Grassland | | 1·6 | | | |
| Chulakov, 1959 | Dark chestnut | 7 cm | 0·1 | | | |
| | | 300 cm | 1·7 | | | |

| | | | | |
|---|---|---|---|---|
| Matskevitch, 1957 | Steppe: tree coenoses | 2·5 –3·4 | | |
| | herbaceous | 1·2 –2·0 | | |
| Gestenhauer, 1969 | Sandy loam | 30 cm | 2·5 | 0·3 | 0·2 – 3·6 |
| | Sandy loam | 30 cm | 1·5 | 0·1 | 0·1 – 1·9 |
| | Loamy sand | 50 cm | 0·8 | 0·2 | 0·2 – 1·1 |
| | Loamy sand | 20 cm | 0·9 | 0·1 | 0·05– 2·0 |
| Nicholson and Nicholson, 1969 | Brown earth on limestone | | 0·27–0·41 | | 0·08– 0·7 |
| Boynton and Compton, 1944 | Orchard/grass | 30 cm | 1·5– 2·5 | 0·1–1·0 | |
| | | 90 cm | 2– 5 | 1–3 | |
| | | 150 cm | 4– 9 | 2–6 | |
| Sheikh, 1969 | Valley bog | 5 cm | 1– 3·5 | | |
| Boussingault and Levy, 1852 | 'Sandy soil' | | 1·06 | | |
| | 'Manured sandy soil' | | 9·74 | | |
| | 'Black clay' | | 0·66 | | |
| | 'Fertile moist soil' | | 1·79 | | |
| Authors (unpublished) | Grassland + pasture | | 1·09 | | 0·1 – 2·7 |

0·4% to 3·2%. Of course, peak values in the wet season may be much greater than this as the data from Vine, Thompson and Hardy (1943) show.

In temperate areas the average carbon dioxide content of soil air at a depth of 30 cm may be determined from Table 13.3 as 1·1%, this being the average of determinations for which the depth is given. The average of all determinations in temperate soils is 0·91% with a range from 0·1% to 1·8%. As Table 13·3 shows, extreme values as high as 11·5% have been recorded from Temperate soils, but the figures used here are the 'usual' values referred to by Russell (1961, p. 366, Table 82) which it is hoped represent the average annual carbon dioxide supply for limestone solution.

Available data suggest that overall average carbon dioxide levels may be somewhat higher in Tropical soils than in Temperate ones although in individual areas there is a large overlap of values and, in at least one Tropical area, Jamaica, the carbon dioxide concentrations found in soil overlying limestone are indistinguishable from those in the temperate zone. Without more data it is difficult to prove or disprove the contentions of Lehmann (1960, 1970), Gerstenhauer (1960) and others, that Tropical soils have greater carbon dioxide contents than Temperate soils. More fieldwork is clearly required.

It is instructive to use the available carbon dioxide data to predict the hardness of runoff waters in the different climatic zones. In the Temperate zone, at a temperature of 10 °C, the mean carbon dioxide level of 0·91% corresponds to a hardness of 230 mg/l $CaCO_3$ which is in moderate agreement with the actual mean of 210 mg/l in Figure 13·2. The range of hardness extends from 30 to 430 mg/l, whereas the range predicted from 'usual' carbon dioxide values of 0·1% to 1·8% is about 120 to 295 mg/l. This discrepancy may be interpreted as indicating that in some cases at least carbon dioxide values greater than 1·8% must occur in Temperate soils. Values of up to about 5·3% are necessary to account for hardness of 430 mg/l $CaCO_3$. About 17% of the values lie above the range produced by 'usual' carbon dioxide concentrations while 14% lie below it. Soil carbon dioxide values as low as 0·01% have been recorded from acid moorland soils (see 'Extreme values' in Table 13.3), but it is not necessary to suppose that such values are widespread in order to explain these anomalously low figures. It may be that they represent waters which have not dissolved limestone under the 'equilibrium' conditions shown in Figure 13.1, but have acted under 'anaerobic' conditions or in an intermediate state, thus achieving a lower final hardness with the same initial input of carbon dioxide.

The hardness that would result from the average carbon dioxide level of 2·0% in Tropical soils would be about 310 mg/l at 10 °C, but higher temperatures of, say, 25 °C would reduce this to 220 mg/l. This value is in poor agreement with the measured mean for Tropical waters of 174 mg/l $CaCO_3$ and suggests that the data on carbon dioxide levels may not be altogether representative of average conditions for limestone soils. The range of carbon dioxide levels from 0·4% to 3·2% indicates that hardness values should lie between 120 and 260 mg/l $CaCO_3$ at 25 °C. This is in close agreement with the actual range of 120 to 240 mg/l.

There is, therefore, a broad agreement between actual hardness values in the Temperate and Tropical zones and the values and ranges predicted from available data on carbon dioxide levels. The measure of agreement is imperfect, however, and the carbon dioxide data is so inadequate as to make it uncertain whether there is much real difference between soils in different climates. In individual cases the carbon dioxide levels in soils of Tropical and Temperate areas may be identical and differences in hardness may be accounted for by changes of temperature between the two areas. For example, in Jamaica Nicholson and Nicholson (1969) determined the mean carbon dioxide level of soil air to be $1.0\%$ (47 observations). At 10 C this would give a hardness of 240 mg/l $CaCO_3$. The actual temperature of the spring water was 23 °C and if the hardness is corrected for the change in temperature a result of 170 mg/l may be expected. The actual hardness of Dromilly Clear Spring, a percolation-fed spring draining the area is 172 mg/l $CaCO_3 + MgCO_3$ and this value is typical of much of Jamaica (Smith, 1969, 1970). In contrast, the Cheddar spring in the Mendip Hills in southern Britain has an average level of $1.09\%$ carbon dioxide in the soil air of its catchment, a value very similar to that found in Jamaica. The spring water has an average temperature of 10 °C and a mean hardness of 245 mg/l $CaCO_3 + MgCO_3$. The hardness which could be expected on the basis of soil air carbon dioxide is 250 mg/l $CaCO_3$. Clearly, there is no difference between the limestone solution processes in these areas of Jamaica and Britain, save that due to temperatures.

Differences of temperature and carbon dioxide supply do provide a partial explanation of the differences between distributions of hardness shown in Figure 13.2. However, they provide only a partial explanation as the correlation between carbon dioxide concentrations in the soil air and hardness is far from perfect. In any case, a great deal of variation occurs within localities with an almost homogeneous climate. Southern Britain is underlain by several different limestones. Figure 13.3 shows the range of hardness found at springs draining these different limestones. For example, ten springs are shown which rise from the Chalk of the Berkshire Downs draining an area of about 250 km². (Paterson, 1971). The lithology is remarkably constant in the Chalk and there is little variation between catchments. All the springs show a very low variability of hardness indicating a similar hydrology (Schuster and White, 1971). Nevertheless, a wide variation occurs between springs with mean annual hardness in the range 215–357 mg/l $CaCO_3$. As Figure 13.3 shows, similar ranges are found in other areas of uniform lithology and with similar climate to the Berkshire Downs, both on the Jurassic Limestones (mainly oolitic) of the Cotswolds and on the cavernous Carboniferous Limestone of the Mendip Hills. In rivers and springs in northwest Yugoslavia, under a different but nonetheless Temperate climate, a similar range (120 to 340 mg/l) occurs (Gams, 1969, 1972). No one region contains as great a range of hardness as the Temperate zone as a whole and it may be that significant differences exist in hardness values between regions with different subdivisions of Temperate climate. Unfortunately, given the limited data available, this possibility cannot be tested easily. Such

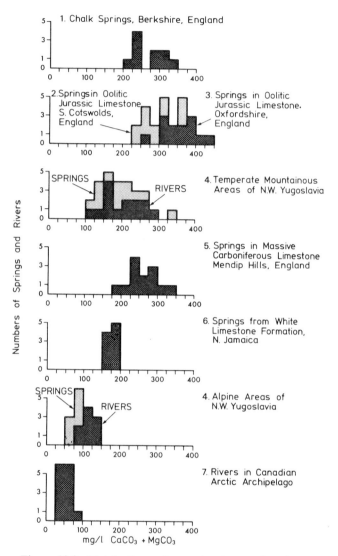

Figure 13.3. Distributions of mean hardness values for springs and rivers in small areas of uniform lithology and climate. Data from: 1—Paterson, 1971; 2—Smith, 1965; 3—Paterson, 1972; 4—Gams, 1969, 1972; 5—Authors; 6—Smith, 1970, 1972; Cogley, 1972.

great variation between sites is obviously not due to climate and is presumably a function of local and complex variations in soil type, soil pH, carbon dioxide content of soil air, and in the drainage pattern, fissuring and hydrology of the bedrock (see Atkinson and Smith, 1975).

Within the Arctic/Alpine zone there is a similar wide spread of mean values among sites in a single region with a homogeneous climate. A range of 36 to 95 mg/l $CaCO_3 + MgCO_3$ occurs among thirteen rivers in the permafrost areas of northern Somerset Island and southern Devon Island, Canada, and of 83 to 141 mg/l among nine rivers in the Julian Alps, northwest Yugoslavia. Springs do not occur in the Arctic permafrost areas, but those in the Julian Alps also show some variation, from 70 to 98 mg/l. Although largely devoid of vegetation, a limited cover of plants and soil is found in Arctic/Alpine regions, being more extensive the less cold they are. Thus, much of the variation within an area may be due to varying amounts and types of vegetation cover. In Alpine areas, in particular, there is often a transition over short distances from almost no vegetation to a full cover. Fully vegetated areas would probably be classified as Temperate, but there is an overlap in the distributions of hardness in waters from Temperate and Arctic/Alpine areas as a result of this transition. The importance of vegetation cover is shown by the results of Ford (1971a) who demonstrates that waters from above the treeline in the Alpine areas of Alberta have a hardness of less than 85 mg/l $CaCO_3$ and are unsaturated, whereas those from below the treeline contain 85 to 140 mg/l and are saturated.

The hardness values of springs and rivers in the tropics show less scatter than those for the other two climates in Figure 13.2, as do the nine springs on the north coast of Jamaica, values for which are shown in Figure 13.3. There is very little seasonal variation at these sites, all of which drain areas of tens of square kilometres of massive limestones with very similar soils and vegetation cover. Their mean annual hardness values show a range of 155 to 188 mg/l $CaCO_3 + MgCO_3$ in spite of the apparent uniformity of their catchments.

It is quite clear from these results that a great deal of the variation between springs and rivers draining sizeable areas is not directly due to climate, but to the operation of other factors which act to vary the supply of carbon dioxide or the conditions under which the solution process takes place. These factors probably include the species composition of the vegetation, the soil type and texture, the organic matter content of the soil and the seasonal regime of soil moisture, all of which will influence the carbon dioxide supply to percolating water. In addition, the nature of the drainage, fissuring of the limestone, and the presence of solutional cavities, caves and conduits, will all influence the conditions under which reaction occurs between water and rock. Not all of these factors are directly influenced by macroclimate, but it would appear that the influence they exert upon the process of limestone solution is just as great as that of macroclimate. These factors are responsible for a very wide variation of average hardness figures within a single area.

We should now examine the effect of climate upon erosion rates. For this purpose, Figure 13.4 shows bar diagrams of the frequency distributions of 134 individual determinations of erosion rate, grouped by climate. There is little apparent difference between the three distributions. The chi-square test of goodness-of-fit shows that none of them conforms to the normal distribution and, more important, that there are no statistically significant differences between them. Table 13.4 shows the results of applying this test.

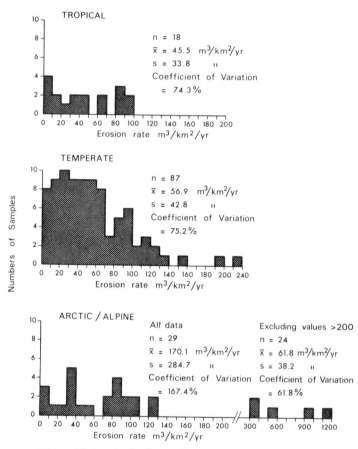

Figure 13.4. Distributions of erosion rates from limestone areas under different climates.

Once again, because the data are not normally distributed it is not possible to apply rigorous statistical tests for differences between means. The means, standard deviations and coefficients of variation in the three samples are shown in Figure 13.4 from which it can be seen that erosion rates tend to be lowest in the Tropics (mean 45·5 m$^3$/km$^2$/year) and rather higher in the Temperate zone (mean 56·9 m$^3$/km$^2$/year). The highest values are found in Arctic/Alpine regions with a mean of 170 m$^3$/km$^2$/year. However, this sample includes five extremely high erosion rates between 320 and 1200 m$^3$/km$^2$/year which are influenced by very high estimates of runoff and the assumption that all of the observed hardness in the rivers sampled is derived from the small proportion of the area underlain by limestone. This is probably untrue, as pointed out above. If these values are excluded, a new mean of 61·8 m$^3$/km$^2$/year is obtained, which remains higher than the value for the Temperate zone, but has a smaller standard deviation and a coefficient of variation which is comparable with the other two zones.

Table 13.4. Values of chi-square in testing differences of erosion rate with climate

| Groups tested | $\chi^2$ | df | α | $\chi^2_\alpha$ (df) | Significant difference |
|---|---|---|---|---|---|
| All three | 8·83 | 4 | 0·05 | 9·49 | No |
| Arctic/Alpine Temperate | 8·68 | 5 | 0·05 | 11·07 | No |
| Temperate + Tropical | 0·88 | 2 | 0·05 | 5·99 | No |
| Arctic/Alpine + Tropical | 2·62 | 2 | 0·05 | 5·99 | No |

These results support Corbel's hypothesis (1959) that the greatest erosion rates occur in Arctic/Alpine regions. However, the effect is much less marked than Corbel claimed, as it involves an increase of only 36% from the Tropics to Arctic/Alpine areas. It should also be borne in mind, that the statistical significance of the differences between the means of the groups is uncertain and that there is no variation in the *distribution* of hardness between climatic zones. Thus the climatic hypothesis (i.e. of climatic control over erosion rate) must be regarded as uncertain, though the effect is small if it occurs at all. It certainly explains nothing to invoke general differences in erosion rate in order to account for differences between Temperate and Tropical karst landforms, for such general differences do not apparently exist.

The variability of erosion rate within a climatic zone is apparently due partly to the variability of hardness discussed above, partly to variations in runoff, and partly to variations in the proportion of limestone in the area under study. The effects of these different factors are apparent in Figure 13.5 in which erosion rate is plotted as a function of runoff. The data are classified by climate and also by the method of calculation used in deriving each value. The method by which the runoff was determined had no effect upon the scatter in the data, but separate graphs are shown for those values for which the percentage of limestone in the area was taken into account. The least squares regression lines of erosion rate on mean annual runoff are shown on the figure, and have the equations shown in Table 13.5.

Unfortunately, the distributions of erosion rates and runoff in the sample data do not conform to a normal curve (at the 95% confidence level). It is therefore impossible to assess the statistical significance of the correlation coefficients, $r$. However, all are fairly high and suggest that variations in runoff account for between about 50% and 77% of the variations in erosion rate within a particular climatic zone. The residuals of the erosion rate about these regression lines (i.e. the differences between the actual erosion rates and those predicted by the equation) were calculated and their frequency distribution tested for normality. In all cases there were significant differences at the 95% confidence level between the actual distributions of residuals and the corresponding normal distribution. In consequence, we cannot assess the confidence limits of

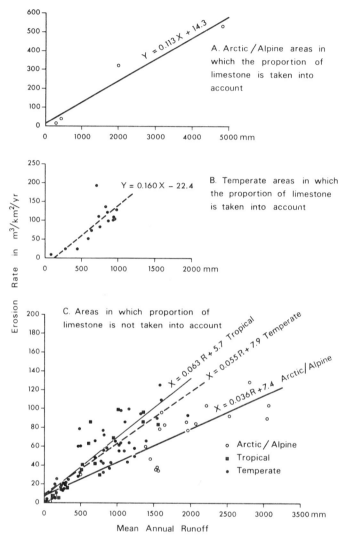

Figure 13.5. Erosion plotted as a function of runoff in different climatic zones.

the regression coefficients and intercepts, or infer whether statistically significant differences exist between them. Instead, they must be examined qualitatively.

From Figure 13.5 it is clear that estimates of erosion rate which take into account the percentage of limestone in the area give values which are roughly twice as great for the same runoff as those which do not take this factor into account. Between the two regression lines which do include this factor, there appears to be little difference, though the gradient of the Temperate zone data is steeper than that of the Arctic/Alpine data. However, the latter line is based

Table 13.5. Regression equations of erosion rate upon mean annual runoff (Erosion rate = $X$ in m³/km²/year; $R$ = runoff in mm/year)

*Arctic/Alpine* areas in which the proportion of limestone was not taken into account:
  $X = 0.0356R + 7.41$   $r = 0.8785$   $N = 18$
*Arctic/Alpine* areas in which proportion of limestone was taken into account:
  $X = 0.1133R + 14.33$   $r = 0.9778$   $N = 4$
*Temperate* areas in which the proportion of limestone was not taken into account:
  $X = 0.0552R + 7.93$   $r = 0.8558$   $N = 51$
*Temperate* areas in which the proportion of limestone was taken into account:
  $X = 0.1604R - 22.44$   $r = 0.7103$   $N = 14$
*Tropical* areas (proportion of limestone not taken into account):
  $X = 0.0633R + 5.69$   $r = 0.8488$   $N = 14$

upon only four samples and probably gives a rather poor estimate of the true relationship.

Among the remaining three lines (i.e. those which do not take the proportion of limestone in the area into account), there is a clear similarity between Temperate and Tropical data. The data cover the same range and the lines both have intercepts of nearly zero and almost identical slopes. These two soil-covered zones are sharply distinguished from the Arctic/Alpine data which show a lower slope, reflecting the generally lower hardness values in such areas.

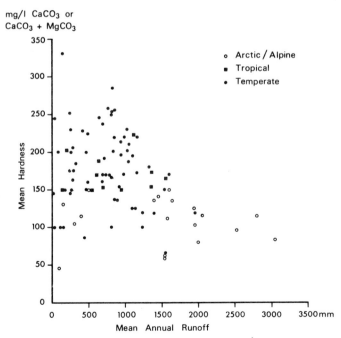

Figure 13.6. Mean hardness of springs and rivers plotted against mean annual runoff.

From these considerations it is clear that erosion rate depends very largely upon runoff and comparatively little upon latitudinal climatic zones. It is worth noting that water hardness appears to vary independently of runoff, as Figure 13.6 shows. Tropical areas do not show any tendency to higher runoff than Temperate areas, at least in the present sample, and in both areas the ranges of runoff and erosion rate are similar. Arctic/Alpine areas, in which precipitation may be very large but evapotranspiration very small because of the low temperatures, do show a greater range of runoff than the soil-covered Temperate and Tropical sites. This leads to some Arctic/Alpine erosion rates being greater than the highest values from soil-covered regions, in spite of lower hardness. The largest erosion rates of all occur in high mountains, where runoff may reach extreme values of 3000 mm/year or more.

## 13.4 The distribution of erosion in the landscape

So far in this essay, we have only considered 'lumped' measurements of erosion rate, that is measurements made at the outflow point of a spring or river catchment and representing the average for that catchment. There is no difference in 'lumped' erosion rates between Temperate and Tropical areas, in spite of the fact that there are differences in landforms. It is possible that the differences in landforms may be due to differences in the *distribution* of solutional erosion in the landscape. For example, the cone karst or cockpit topography of some tropical karst areas may be a response to a greater concentration of solution close to the surface than in temperate areas. It is this type of hypothesis which we now set out to test.

There are very few published results of experiments in which the distribution of erosion was measured, but those that exist are shown in Table 13.6. In all cases, the estimates are based upon analyses of water samples taken at different points in the limestone drainage system, usually in conjunction with a programme of hydrological measurements which determine the amounts of water following various pathways through the system. Smith, Atkinson and Drew (1975) present a review of limestone hydrology from this point of view, and Atkinson and Smith (1975) and Trudgill and Atkinson (1975) describe erosion budgets in detail. A detailed breakdown of the erosion budget of an area in the Mendip Hills, England, is shown in Table 13.6. The total annual erosion is 81 m$^3$/km$^2$/year and the proportions which are removed from the swallet catchments of the area, the soil over the limestone and the uppermost 10 m or so bedrock are all assessed independently. Their total is approximately equal to the rate of surface lowering and amounts to 69% of the overall erosion. Table 13.6 also shows the seasonal distribution of erosion, roughly two thirds of the total occurring in the winter months, when evapotranspiration rates are lower and runoff rates greater.

The proportion of erosion occurring near the surface in the Mendip Hills seems to be typical of the other Temperate areas in Table 13.6, which shows an overall range of 50% to 80% and a mean of 67%. Caves account for very little

Table 13.6. The distribution of erosion in the landscape under different climates

## TEMPERATE AREAS

### 1. Mendip Hills, England[a]

| Location | percentage of annual total | percentage occurring October to March | percentage occurring April to September |
|---|---|---|---|
| Swallet catchments | 1·3 | 69·3 | 30·7 |
| Soil profile | 10·1 | 87·5 | 12·5 |
| Uppermost bedrock | 57·1 | 57·7 | 42·3 |
| Main mass of bedrock | 31·4 | 77·9 | 22·1 |
| Cave passages | < 0·1 | 55·7 | 44·3 |
| Totals | 100 | 67·4 | 32·6 |

Overall erosion rate (81 m$^3$/km$^2$/year)

### 2. Other areas

| | Percentage of annual total | | | |
|---|---|---|---|---|
| | Blue Waterholes,[b] N.S.W. Australia | Northwest[c] Yorkshire | Joux Valley,[d] Jura Mts. | Fergus R.,[e] Ireland |
| Surface and uppermost bedrock | | 50 | 58 | 80 |
| Main mass of bedrock | 78 | | 37 | |
| Cave passages | 22[f] | 50 | 5 | 20 |
| Overall erosion rate, m$^3$/km$^2$/year | 24 | 83 | 98 | 51 |

## TROPICAL AREAS

| | Percentage of annual total | |
|---|---|---|
| | N. Jamaica[g] | Average for[h] Caribbean |
| Surface and soil | 23 | 43 |
| Uppermost bedrock | 59 | 26 |
| Main mass of bedrock | 18 | 31 |
| Overall erosion rate, m$^3$/km$^2$/year | 40–70 | — |

## ARCTIC AREAS

| | Percentage of annual total Somerset Island[j] |
|---|---|
| Surface and regolith | 100 |
| Overall erosion rate m$^3$/km$^2$/year | 2 |

a. Authors.
b. Jennings, 1972a, 1972b.
c. Sweeting, 1966.
d. Aubert, 1967, 1969.
e. Williams, 1968.
f. Includes river channels.
g. Smith, Drew and Atkinson, 1972.
h. Corbel and Muxart, 1970.
j. Smith, 1972a.

erosion in most cases and, as Williams has written (1970, p. 109), they form 'a mainline of transport, not a locus of corrosion'.

Estimates of the distribution of erosion are even scarcer for Tropical than for Temperate areas. However, Smith, Drew and Atkinson (1972) provide a rough breakdown of erosion in the limestone areas of northern Jamaica based upon water samples collected at different sites in the Cockpit Country and adjacent areas of cone karst, but not upon detailed hydrological measurements. In this area an estimated 82% of erosion occurs close to the surface (i.e. within 10 m depth). While larger by 13% than the corresponding figure for the Mendip Hills, this value lies at the upper end of the range found in Temperate areas, and cannot really be regarded as indicating a significantly greater concentration of erosion at the surface in Tropical areas. This conclusion is supported by Corbel and Muxart's breakdown of the average erosion budget in the Caribbean area (based upon water samples from Yucatan, Jamaica, Cuba and Puerto Rico), which shows a distribution almost identical with that of the Mendip Hills. Sixty-nine per cent of erosion occurs close to the surface, although a higher proportion is removed from the soil profile and loose regolith than in the Mendips. Corbel and Muxart also discuss the case of marsh deposits ('*marais et tourbieres*') in dolines and consider that when these are present 90% of erosion occurs close to the surface (Corbel and Muxart, 1970).

The data from Tropical areas indicate that solutional erosion is distributed through the karst landscape and drainage system in a fashion similar to that of Temperature areas. Thus, the hypothesis that differences between Tropical and Temperate karst landforms are due to differences in the distribution of erosional processes cannot be maintained, at least on the basis of present evidence. However, it is worth noting from Table 13.6 that in Arctic areas where permafrost is present, all the erosion takes place at the surface, probably in the top few centimetres of the regolith. Limestone pebbles from the surface rubble permafrost areas often show slight solutional fretting on their upper sides with thin skins of deposited calcite on their lower surfaces. No detailed figures are available for Alpine areas, but even in actively glaciated areas underground drainage has been recorded, e.g. the Castleguard Cave in Alberta which drains subglacial meltwater from the Columbia Icefield (Ford, 1971b). Inevitably, some subsurface erosion must occur in Alpine areas.

## 13.5 Limestone landforms, landscapes and climate

Studies of the limestone solution process are well understood as regards both detail and rate, in comparison with other branches of geomorphology. It is therefore disappointing to review the literature on this topic and to find that the progress made in process studies has had so little effect on our understanding of the morphology of limestone terrains. It is clear that the reason for possible morphoclimatic differences in limestone landscapes does not lie in a consideration of erosion rates alone. Indeed, it must be stressed that geomorphologists are very far from agreement as to the precise nature of morphoclimatic differences between limestone areas and it may be that the reasons for such differences as do exist are so complex that a satisfactory explanation of them will never be found. This philosophy of despair is sounded by Jennings (1967) in summarizing an excellent account of the karst of Australia:

> the number of variables in geomorphology are so manifold and often so little susceptible of measurement and calculation that the question arises whether the whole earth will provide a large enough sample for unravelling all the possible interactions of rock nature and arrangement, land and sea-level movements, climatic and vegetational history, in the present landscape, blurred and blotted as it is today in both form and process by the ever more powerful hand of man.

The history of the development of ideas concerning various evolutionary sequences of landforms in limestone areas is now almost a subject in its own right. Until the last few years this subject was particularly difficult for English-speaking geomorphologists as the initial contributions were written by Slav and German workers, often in journals which were difficult to obtain and in languages difficult for most to fully understand. Happily this situation has been remedied by the publication by Jennings (1971) and Sweeting (1972) of two textbooks devoted to the study of karst. Both writers give detailed reviews of the evolution of ideas on karst morphology. The account which follows is, therefore, a brief review in which particular emphasis is given to the part played by climatic influences, and which draws attention to recent research into limestone morphometry which may prove a fruitful basis for future investigations.

The two workers who were responsible for formulating the basic concepts of limestone geomorphology were Grund and Civijic, both writing at the beginning of this century. The impact of these two writers upon subsequent work is closely comparable to that of Davis and Penck in the formulation of early ideas on the evolution of slope morphology. The similarity even extends to the overstatement, and occasional mis-statement, of what they wrote, due more to the zeal of later disciples than to the views of the originators of the concepts concerned. For Grund (1914), the evolution of a karst landscape began the moment a limestone mass became open to sub-aerial attack. Isolated dolines formed, leaving large areas of the original surface unaffected by karst erosion processes. These dolines grew in size, often coalescing and diminishing the area of original surface. The mature stage of development is reached when the original surface is completely consumed by the continued growth of the enclosed depressions. Once a landscape is established that consists solely of enclosed depressions of various sizes the surface is lowered more or less evenly until further reduction is halted by the centres of the depressions reaching the water table. At this stage the floors of the depressions enlarge until only vestigial ridges and hills of limestone are left overlooking a nearly flat corrosional surface ('*vorfluter*') that corresponds to the water table. The major significance of Grund's Cycle of Erosion in a morphoclimatic setting is that he was undoubtedly influenced by descriptions of Tropical karst, so much so that he terms the third stage of his evolutionary sequence the '*Cockpitlandschaft*' after the Cockpit Country of northern Jamaica, which is entirely formed of closed depressions. Thus, Grund regards the doline karst of Temperate areas as an earlier stage of the mature karst which is typified by cockpit landforms. On the other hand, Civijic (1893) differs from Grund in that he considers that the initial stage was composed of a surface stream network which progressively changes to an underground drainage network through the development of dolines and swallets (stream sinks). His final stage is not developed at the water table but when the influence of underlying impermeable rocks becomes dominant. Thus, both workers recognize a Cycle of Erosion, although they differ as regards the necessity for the initial stage to include a surface drainage network. Other differences between them are related more to the nature of the groundwater hydrology than to the morphology of the resultant landscape. It is also perhaps true to say that Grund acknowledges the possible influence of climate while the studies of Civijic are firmly rooted in examples drawn from the Dinaric Karst of Yugoslavia.

Detailed morphological studies of limestone areas that specifically discuss the part played by climate in the formation of landforms really begin with the work of Lehmann (1936) in a classic study of the Goeneng Sewoe area of south-central Java. Other workers, many of them pupils of Lehmann, studied the karst of other parts of the world and there was a tendency for each worker to suggest a distinct evolutionary sequence for each morphoclimatic area. This work was mainly undertaken by central European scholars and the dominance of climatic thought was in full agreement with the general geomorphological philosophy of that period. This dominance can be seen in the results of the Karst Symposium held in Frankfurt in 1953 (Lehmann, 1954). The corresponding Symposium meeting in Stuttgart in 1963 clearly reflected a change of emphasis towards the study of process, though it is of interest to note that the process studies themselves were dominantly oriented towards the relationship with climate (Lehmann, 1964).

Brief quotations which have been selected to typify a particular school of scientific thought are often misleading, but the dominance of the idea of morphoclimatic control in studies concerned with karst morphology are illustrated by the view of Lehmann (1954), who suggested that each climate gave rise to a 'clima-specific' karst development. Moreover, for each clima-specific region it was possible to outline specific characteristics and landforms. Thus, the 1953 Karst Symposium collectively put forward the following limestone morphogenetic regions: Periglacial and Polar, High Alpine, Cool Oceanic West European type, Mediterranean, Dry Desert, and Humid Tropical. To these could be added further climatically controlled regions. For example, Jennings and Sweeting (1963) distinguish the limestone ranges of the Fitzroy Basin in Western Australia as representing a 'tropical semi-arid karst type' and comment that, 'it is unfortunate that no study of a substantial area of karst in closely comparable climatic conditions is known to the writers for comparison'. They conclude that the Bom de Jesus Lapa in Brazil, described by Tricart and

Da Silva (1960), is similar but comprises only one large hill of limestone. Nevertheless, this limited occurrence was distinguished by Tricart and Da Silva as representing a semi-arid karst type. Jennings and Bik (1962), from studies in New Guinea, attempt to relate differences in morphology to differences in climate over an altitudinal range of 3000 m. They conclude that, 'there seems to be an altitudinal climatic zonation of karst in New Guinea, though at each level the picture is not a simple one, complicated as it is by cross-cutting influences of lithology, structure and evolutionary history'. Virtually all workers in this field have sought to explain the lack of any clear climatic control on landforms by stressing the problems caused by climatic change. Where climate has changed during the evolution of a particular area, the landforms it contains could perhaps be a compound of development under successive ranges of climatic conditions. This problem is generally thought to apply most forcibly to the much-studied areas of Europe in which Pleistocene climatic changes are known to have been particularly marked. Conversely, it is thought to have applied less to the humid Tropical areas.

To summarize the state of the literature and of the knowledge that it ostensibly embodies, one might say that as the number of studies of individual karst areas increases it becomes more and more apparent that simple morphoclimatic classifications of landforms and landscapes are less than helpful in explaining variations in karst morphology. We would agree with Jennings (1971, p. 195) when he states, 'the efforts to distinguish climato-morphogenic systems other than those associated with climatic extremes must be regarded as tentative in terms of present knowledge'.

After an extensive review of various classificatory schemes for limestone regions, Sweeting (1972, p. 256) presents the following groups 'according to the dominant process or climatic regime which has fashioned the landscape'. The divisions are:

(i) True karst or Holokarst;
(ii) Fluviokarst;
(iii) Glaciokarst or Nivalkarst;
(iv) Tropical karst;
(v) Arid and semi-arid karst.

There is doubt as to whether this concise classification is useful except as a basis for the most broad comparisons. As examples of the various groups one might cite the Dinaric Karst of Yugoslavia as a *true karst* area, in which the landscape is the result of the dominance of the single karst process of solution. However, the pre-eminence of the solution process is as likely to be true in Tropical karst areas as for the classic Dinaric area. *Fluviokarst* is formed by the combined action of fluvial and solutional processes and Sweeting gives the Jurassic limestone and Chalk areas of Britain as an example, although she qualifies them as being 'more fluvial than karstic'. *Glaciokarst* shows the dominant influence of glaciation upon limestone areas, and the Ingleborough area of northwest Yorkshire and the Burren of Co. Clare in western Ireland are given as relict examples. Ford (1971b) describes active glaciokarst areas in the Canadian Rockies. *Tropical* karsts and those of *arid* and *semi-arid* areas are extensively reviewed by Sweeting (1972). In our view, tropical karst regions represent the nub of the morphoclimatic debate. It is quite clear that the relative importance of non-karstic processes such as glaciation varies with climate. It is also clear that fluviokarst areas occur in the Tropics just as they do in the Temperate zone, and also that fluvial elements occur in high latitude Arctic karst areas (for examples, see Sweeting (1958, 1972), Versey (1959), Smith (1972a) and Cogley (1972). However, there are also differences in landforms between areas in the Tropics and Temperate zone *which are formed dominantly by the same process*, that of solution. Without doubt, certain karst landforms are restricted in their occurrence to the Tropics, the main examples being cockpit and cone karst, tower karst and other related variants (for definitions and examples see Sweeting (1972) and Jennings (1971)). The converse is not true; most Temperate forms of doline and dry valleys (fossil fluviokarst) are well represented in Tropical regions. Thus, if there is a morphoclimatic control over limestone landforms, it is Tropical landforms that will provide the key to understanding it.

A clue to the problem is provided by the area of north central Jamaica, which has been described by both Sweeting (1958, 1972) and Smith (1969, 1972b). Here the geological structure juxtaposes two different, but very pure, limestone rocks on either side of a major fault line. Both lithologies are subject to the same, or very similar, climatic conditions. The type area of the mature stage of Grund's Cycle occurs in this region. It is the Cockpit Country, an impressive landscape of cone karst developed upon the massive beds of the White Limestone Formation (Eocene). Across the major fault line are the chalks of the Montpelier Beds, which are of the same age as the White Limestone proper, but have a higher primary porosity and fewer major fractures, fissures and joints. The landscape upon the Montpelier Limestone is a fossil fluviokarst with a well integrated dry valley network. Its nearest homologue is probably to be found in the dry valleys of the Chalklands of Salisbury Plain, southern England. The erosion rates and the hardness of runoff waters in the two regions are very similar (Smith, 1970) and cannot be used to explain the difference in landforms. In fact the difference appears to be due to the change in lithology from one limestone type to the other and one may generalize from this and other examples in the Tropics and state that the distinctive Tropical landforms of cockpit and tower karst are only found in certain Tropical karsts and that their occurrence depends upon *a combination of Tropical climate with suitable lithology*. The lithological controls are not fully understood, but those factors which control the form of groundwater circulation appear to be of crucial importance. Numerous studies have shown that massive, jointed limestones with low primary (intergranular) porosity give rise to rapid underground drainage via caverns and conduits with the concomitant development of holokarst landforms such as dolines and closed depressions. In limestones with a greater primary porosity, such as chalks, underground drainage is more diffuse conforming to a greater extent to the normal Darcian model, and dry valleys predominate over dolines in the surface landforms. Examples of the different types of limestone drainage are discussed by Smith, Atkinson and Drew (1975) while Ineson (1962) describes the groundwater hydrology of the English Chalk. Certainly in the tropics, massive limestones seem to give rise to cockpit and cone karst, in which dolines are so well developed that they cover almost the entire land surface, to the exclusion of other forms such as dry valleys, which occur on chalk lithologies.

Several decades of research have now been directed towards the classification of karst landforms and the explanation of possible differences in their morphology. Many authors have attempted to include the climate as an integral aspect of their classification and theories of morphological evolution. In spite of all this effort further descriptive classifications are unlikely to result in any more progress. The salient features of the karst landscape are twofold, namely the dry valley system and the enclosed depression. Indeed, both Grund and Cvijic recognized that these two forms, with all their sub-types and variations in scale, were the elements that typified the morphology of karst areas. This being so, it is perhaps surprising that so few attempts have been made to apply the morphometric methods described by Horton (1945) to limestone regions.

## 13.6 Morphometric analysis of limestone landscapes

The morphometric analysis of fluvial landscapes has undoubtedly proved a fruitful technique for the description and comparison of the shapes and forms of river basins at a variety of scales. It is perhaps true to say that for fluvial landscapes the techniques are now sufficiently well established for the future to lie in investigating the relationships between morphometry, hydrology, erosion and the evolution of the landscape: in short, in the relationship of form to process. For limestone areas, process studies are well advanced but objective morphometric description is in its infancy.

The necessity for some form of objective description of tropical karst landforms has been urged by Balazs (1973a). He outlines the problems by stating that the concepts of *Kegelkarst* and *Turmkarst* are not suitable for delimiting types of Tropical karst regions. The *Kegel* (cone) and *Turm* (tower) often occur within the same karst region and one may develop into the other with the passage of time. Balazs suggest a very simple index, termed the *Morphogenetic Index*, with which to distinguish four different forms which are characteristic of Tropical plateau karstlands. The index is simply the ratio of the diameter of the form to its height. Figure 13.7 shows diagrammatically the four forms involved and indicates the range of morphogenetic index typical of each, together with details of the heights of the features and their numbers per unit area.

La Valle (1967, 1968) analysed the morphometry of dolines on Temperate karst in Kentucky. He related the shape and frequency of the depressions to lithological, structural and hydrological factors of the area with a fair degree

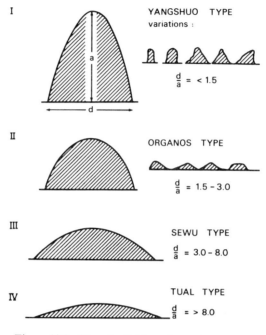

Figure 13.7. The sub-division of morphogenetic relief types, as proposed by Balazes (1973a).

of success. However, both his study of depressions and the morphogenetic index of Balazs represent only the mere beginning of a truly morphometric approach to karst studies.

The only published attempts to modify the morphometric methods of Horton for use in limestone areas are those of Williams (1966, 1969). His work, and that of subsequent authors in the field, can be divided into two parts: first, a description of the *drainage network* and, second, the application of morphometric methods to the *surface topography*. Such a division is to some extent arbitrary, but it does separate the two major elements of limestone terrain, the dry valley and the enclosed depression.

## 13.6.1 *The morphometry of drainage networks*

The methods of analysing the limestone drainage network were described by Williams in 1966. They will be reviewed briefly here. The following measures are used to characterize the drainage net:

(1) The *area of the limestone* $(A_L)$.
(2) The *number of stream swallets* $(S)$ and the *order of the surface stream network feeding the swallets* $(S_0)$, determined by the methods normally used for fluvial morphometry (Strahler, 1957). The majority of the swallet drainage

net will normally be on impermeable (non-limestone) strata, but any part of the stream which flows over limestone prior to its engulfment is included. Dry valleys on the limestone, which are tributary to the main stream are excluded.
(3) The *length of active stream channel on the limestone* ($L_L$).
(4) The *number of karst springs or resurgences* ($\Sigma K$).

From these quantities the following attributes of the limestone drainage network can be calculated.

(1) The *swallet density* $D_S = \Sigma S/A_L$.
(2) The *karst resurgence density* $D_R = \Sigma K/A_L$.
(3) The *stream density of limestone* $D_L = \Sigma L_L/A_L$.
(4) The *swallet/resurgence ratio* $R_{SR} = \Sigma S/\Sigma K$.
(5) The *swallet ratio* $R_{S_o} = \Sigma S_o/\Sigma S_{o+1}$.

Williams (1966) presents the results obtained by applying this analysis to the Ingleborough district of Yorkshire. The same methods have been applied by Hellden (1973) to the Artfjället area of Swedish Lapland. The only difference in method between these two studies is that Hellden's was based upon field mapping whereas Williams worked from 1:25,000 maps. The results of both are compared in Table 13.7. They give a simple measure of some of the salient drainage features of limestone terrain, although it could be argued that the swallets are essentially determined by the form of drainage on adjacent non-limestone strata and are little affected in their order or number by the limestone itself. Measures of the stream density on the limestone itself are clearly important, although the density is usually low and many streams are intermittent. If studies of the surface drainage are not undertaken in the field they are particularly susceptible to the quality of the data available. There is no doubt that if field information at defined values of runoff can be obtained, the value of morphometric data would be greatly enhanced. This has been attempted by Hellden, who recognizes 'permanent', 'intermittent' and 'fossil' sub-categories of the network. Field mapping under flood conditions may enable valuable

Table 13.7. Comparison of morphometric data on limestone drainage in an Arctic and a Temperate area

| | Area of limestone $A_L$ (km$^2$) | Swallet density $D_S$/km$^2$ | Karst resurgence density $D_R$(km/km$^2$) | Stream density on limestone $D_L$(km/km$^2$) | Swallet/Resurgence ratio $R_{SR}$ |
|---|---|---|---|---|---|
| Ingleborough N. W. York., UK (Williams, 1966) | 31·1 | 2·96 | 2·5 | 1·5 | 1.12 |
| Artfjället, Lapland, Sweden (Hellden, 1973) | 2·4 | 8·2 | 6·1 | 1·1 | 1·5 |

comments to be made regarding the degree of development of the underground flow net as well as the surface streams (Hanwell and Newson, 1970). It may also prove possible to categorize the groundwater flow for those areas where detailed groundwater tracing has been undertaken. Baker (1973) has attempted to analyse the groundwater flow network in terms of Horton's Laws of Morphometry in a small area of limestone in New York State. In this area, there was sufficient information available from dye tracing and cave exploration for graphs to be produced relating stream order to stream number, stream length and basin area of both the surface and subsurface parts of the drainage system. The results show a general agreement to those obtained from similar studies of surface catchments. This type of study appears to be very promising in expanding our understanding of karst hydrology and its influence on morphology, although it is only practical to undertake it where dye tracing and cave surveys are available, and where the conduit-dominated form of groundwater flow is important (White and Schmidt, 1966).

### 13.6.2 The morphometry of surface forms

The application of morphometric techniques to surface forms has progressed further than studies concerned with drainage networks. It now seems strongly possible that such studies will not only provide an adequate description of the surface morphology and allow meaningful comparison between areas, but also that they may give an insight into the evolution of the forms within individual regions.

The necessary basic measurements for describing the surface forms are listed below. Details of the technique of analysis are given by Williams (1966, 1969), who first invented many of the measures employed.

(1) The *area of limestone* ($A_L$).
(2) The *number of closed depressions* ($C_d$).
(3) The *number of summits* ($T$).
(4) The *area of closed depressions* ($A_{cd}$).

These quantities can be combined to give the following morphometric measures,

(1) The *closed depression density* $D_{cd} = \Sigma C_d / A_L$.
(2) The *pitting density* $R_p = \Sigma A_L / \Sigma A_{cd}$.
(3) The *summit/depression ratio* $R_{sd} = \Sigma T / \Sigma C_d$.

They are useful in describing the morphometry of specific areas. A selection of results from the work of Williams (1966, 1971), Hellden (1973) and unpublished data of the present authors' is given in Table 13.8.

Williams (1971, 1972a, 1972b) also ordered the channel network that occurred in each depression in Tropical karst areas. The relationships between the depression order and the other morphometric parameters demonstrate that

Table 13.8. Comparison of morphometric data on surface forms in limestone areas in the Arctic, Temperate and Tropical zones

|  | Area of limestone $A_L$ (km$^2$) | Summit/ depression Ratio, $R_{sd} = \Sigma T/\Sigma C_\alpha$ | Closed depression area $A$ (km$^2$) | Closed depression density $D_{cd}$ /km$^2$ | Pitting density $R_p = \dfrac{A_L}{A_{cd}}$ |
|---|---|---|---|---|---|
| Ingleborough N. W. Yorks., UK (Williams, 1966) | 31·1 | — | 0·013 | 20·8 | 2,500 |
| Artfjället, Lapland, Sweden (Hellden, 1973) | 2·4 | — | 0·008 | 57·0 | 297 |
| Mt. Kaijende, New Guinea | 13·2 | 0·40 | 13·2 | 13·05 | 1·0 |
| Darai Hills, New Guinea (Williams, 1972a, 1972b) | 13·8 | 0·85 | 13·8 | 13·5 | 1·0 |

the limestone terrains that Williams studied have a high degree of spatial organization and do not exhibit 'the chaos that has too often been ascribed to karst' (Williams, 1972b, p. 158).

The results in Table 13.8 do not give a full description of the karst areas from which they have been taken. But they do provide a useful approach to the description of karst morphology by objective means. The preliminary ideas of morphometric analysis upon which they are based have been further developed by Williams in later papers (1971, 1972a, 1972b) on studies of the polygonal karst of New Guinea. The number of morphometric parameters has been increased in these studies so that it is now possible to formulate a complete description of the spatial organization of the land surface. Williams suggests the use of the following variables in distinguishing between different sub-areas of polygonal karst.

(1) The *mean product of symmetry of depressions*.
(2) The *mean length to width ratio of depressions*.
(3) The *mean number of residual hills per depression*.
(4) The *mean distance from a stream sink to its nearest neighbour of the same order*.
(5) The *mean area of depressions of a given order*.

With the aid of these parameters and the relationship between them and other measures discussed above, it is possible to distinguish sub-regions within a major area of polygonal karst. In the areas of New Guinea studied, Williams terms these sub-types Pinnacle, Intersecting Linear, and Conic karst types,

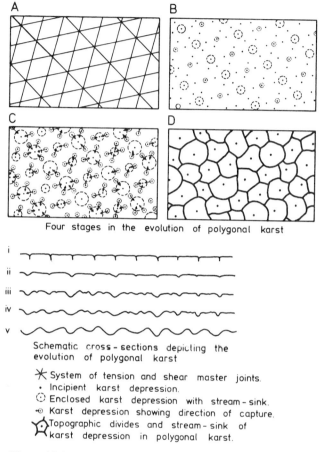

Figure 13.8. Four stages in the evolution of polygonal karst in New Guinea (redrawn after Williams (1972a)).

using a new terminology in order to avoid the many connotations of terms like 'Conekarst' and 'cockpit'.

Of even greater importance than the success of the morphometric method in discriminating fine differences in landforms within a region is the fact that Williams was able to use his results from New Guinea to support a model of the evoltion of the polygonal karst. This model is shown in Figure 13.8 (Williams, 1972a, Figure 18, p. 789). In the initial stage (Figure 13.8A) the limestone mass is intersected by a simple system of joints, which become enlarged by solution. The enlargement gives rise to a progressive increase in secondary permeability and the runoff from the area becomes progressively concentrated towards developing stream sinks. The favoured sinks tend to be located at joint intersections (Figure 13.8B) and the less favoured swallets are 'captured' by expanding stream sinks to give the pattern shown diagrammatically in Figure 13.8C. Eventually, the 'successful' sinks expand to fill the whole limestone surface (Pitting index = 1) to give the spatial pattern shown in Figure 13.8 D. The ideal shape for each depression would be a hexagon, but this shape is normally modified by the effects of the original slope of the region and the structural inhomogeneity of the rock mass. It should be stressed that this model of landscape evolution, which bears a remarkable similarity to the Karst Cycle suggested by Grund (1914), was arrived at by inductive reasoning from the results of morphometric analysis and not by a *priori* deduction.

Williams stresses that karst landscapes of the polygonal type which are frequently found in the Tropics are well ordered and that they originate essentially through the solutional action of local stream systems. Indeed he suggests that the polygonal enclosed depressions are best considered as 'small river basins of a special kind' (Williams, 1972, p. 155). This view draws attention to the dominant role played by hydrological factors in the evolution of karst forms. The progressive development of secondary permeability that underlies the evolutionary sequence shown in Figure 13.8 is, in our opinion, the most important factor of all in the development of karst landforms. That secondary permeability may increase still further once the fourth stage (Figure 13.8D) has been reached may be illustrated by another example from Jamaica. The Cockpit Country has a similar morphometry to the polygonal karst of New Guinea. However, in Jamaica, field observations revealed almost no integrated surface streams or even rivulets on the side slopes of cockpits, even during intense rainfall. The very few surface streams that were observed flowed temporarily on soil overlying the limestone and in only one out of several score of cockpits examined was there any feature which could be described as even an ephemeral stream channel. It is possible that the Jamaican karst has developed even further than the New Guinean examples and that secondary permeability has increased throughout the rock mass to such an extent that surface streams no longer flow. This view is described further by Smith, Drew and Atkinson (1972).

## 13.7 Conclusions

In the first part of this essay, we examined the effect of climate upon the limestone solution process. It appears that the traditional latitudinal divisions of climate do not reflect a corresponding division of processes or erosion rate. The principal climatic parameter affecting erosion rate is the mean annual runoff, while temperature has some influence, depending upon whether or not the mean annual temperature is high enough to support a continuous cover of soil and vegetation. So far, the study of processes has not been of great help in explaining the reasons why some distinctive landforms and landscapes occur only in the Tropics, whereas others are found throughout the Temperate and Tropical zones. Part of the difficulty lies in defining the problem. Hitherto it has been possible to recognize overall differences between, say, dry valley and doline topography found in many Temperate areas, and the cockpit or polygonal topography of some Tropical areas. With only verbal tools of analysis at our disposal, it has proved almost impossible to describe and distinguish different sub-types of karst objectively. In future, geomorphologists should be able to use the morphometric measures developed by Williams, and perhaps others which may be suggested in the future, to make very fine distinctions between different landscape types. Then, and only then, will it be possible to frame precise questions about the influence of climate upon landscape and the interrelation of landforms and erosion processes, erosion rates and hydrology. Until that time, we must content ourselves with a few hypotheses which may be properly tested when sufficient morphometric data have been collected from different parts of the world. Firstly, there is the hypothesis, put forward by Grund and Cvijic, that karst landscapes evolve through a Cycle of Erosion. This can certainly be tested by comparing areas of similar runoff and lithology, but which can be shown on geological evidence to have been subjected to solutional attack for differing periods of time. If the Cycle of Erosion is one in which there is an initial dissection of the surface by active streams which are engulfed as secondary permeability increases, then comparable areas should show a progressive reduction of the drainage density of active streams on the limestone from an initially high value to zero, with a concomitant increase in the dry valley density. Subsequent development of dolines in dry valley floors, which seems to be occurring in at least some temperate karst areas (Ford and Station, 1968) should lead to a progressive decrease in the index of pitting and other measures of closed depression topography. If the closed depressions subsequently expand to obliterate the dry valley network and form a polygonal karst, the pitting index should approach unity in the oldest areas. In this way, it should be possible to use morphometric analysis to establish the existence or otherwise or the Karst Erosion Cycle.

Secondly, there is the question of why polygonal karst landscapes appear to be confined to massive limestones in the tropics. It may be that the Tropical regions have provided a long period of uninterrupted climatic conditions in which the solutional process has been dominant, whereas Temperate regions have experienced violent fluctuations of climate in the Pleistocene. These may have caused the obliteration of partially formed polygonal karsts through the action of glaciation, nival processes, fluvial action and so on. Certainly, some writers (e.g. Gilewska (1971)) have claimed to be able to identify relict fragments of earlier polygonal karst landscapes in parts of Europe, but the interpretation of the evidence for these is very doubtful and open to argument. An alternative explanation is that there are differences in the operation of solution processes between Tropical and Temperate limestones which act to produce polygonal karst on the one hand and Temperate dry valley and doline karst on the other. This hypothesis could best be tested by examining regions of differing morphometry, but with the same geological age, runoff, erosion rate and lithology and comparing detailed studies of the distribution of erosion in each area.

Some evidence has already been examined which suggests that the lithology of the limestone exerts a profound influence upon the landscape, probably through its control upon hydrology and especially the development of secondary permeability and underground drainage. This possibility can again be tested by comparing the morphometric measures of surface topography and drainage network in areas of differing lithology but comparable runoff, erosion rate and geological age.

W. M. Davis (1899) described fluvial landscape as being produced by the interaction of three groups of variables, namely Structure, Process and Stage. We have already seen that quantitative studies of process are not enough to explain the variety of karst landscapes. However, the tools of process studies and morphometric analysis, when used together, may prove powerful enough to unravel the interaction of lithology (or Structure), erosion rate and distribution (Process) and time (Stage) in sculpting the complex variety of landscapes found in limestone regions. Only then will it be possible to determine with certainty whether latitudinal variations of climate have had more than a coincidental effect upon the development of limestone landforms and landscapes.

## Appendix

Erosion rates reported in the literature

| Locality | Erosion rates ($m^3/km^2$/year) | Source |
| --- | --- | --- |
| TEMPERATE ZONE | | |
| British Isles: | | |
| Derbyshire | 115; 136; 42; 128; 73; 98; 24; 121; 111; 108; 193 | Pitty, 1968 |
| | 77 | Dearden, 1963 |
| | 79 | Douglas, 1964 |
| | 60 | Ford, 1966 |
| Mendip Hills | 81 | Atkinson, unpublished |
| N. W. Yorkshire | 83 | Sweeting, 1966 |
| R. Mellte, South Wales | 16 | Groom and Williams, 1965 |
| R. Frome, Dorset | 60 | Casey, 1969 |
| N. W. Scotland | 100; 88; 96 | Smith, unpublished |
| R. Fergus, Eire | 51 | Williams, 1968 |
| Europe: | | |
| Yugoslavia | 109; 66; 93; 66; 55; 58; 66; 48; 90; 28; 14 | Corbel, 1965 |
| Dinaric Karst, Yugoslavia | 43; 43; 63; 55; 64; 19; 56; 59; 64; 37; 42; 44; 9; 72; 126 | Gams, 1969, 1972 |
| Joux Valley, Swiss Jura Mountains | 98 | Aubert, 1967, 1969 |
| Krakow Plateau, Poland | 20 | Corbel, 1965 |
| | 21 | Oleksynowa and coworkers, 1971 |
| | 19; 16; 32; 21; 17; 10; 24 | Pulina, 1971 |

| Locality | Erosion rates (m³/km²/year) | Source |
|---|---|---|
| Tatra Mountains, Poland | 95; 86 | Kotarba, 1971 |
| | 50 | Oleksynowa and coworker, 1971 |
| R. Punkva, Czechoslavakia | 25 | Stelcl and coworkers, 1969 |
| Aggtelek, Hungary | 20 | Balazs, 1973b |
| R. San Antonio, Texas, USA | 4 | Corbel, 1971 |
| E. Siberia, USSR | 32; 6; 1; 4 | Pulina, 1968 |
| Australia: Cookeman Plain, N. S. W. | 24 | Jennings, 1972a, 1972b. |

## TROPICAL ZONE

| Locality | Erosion rates (m³/km²/year) | Source |
|---|---|---|
| Indonesia | 63; 99; 86; 83 | Balazs, 1971 |
| Florida | 35 | Runnels, 1971 |
| | 5; 8 | Corbel, 1959b |
| Jamaica, West Indies | 69; 39; 86; 96 | Smith, 1969, 1970 |
| Puerto Rico | 42 | Corbel, 1971 |
| Yucatan, Mexico | 30; 16 | Corbel, 1959b |

## ARCTIC/ALPINE ZONE

| Locality | Erosion rates (m³/km²/year) | Source |
|---|---|---|
| Alaska | 530 | Corbel, 1959a |
| | 40; 8 | Corbel, 1959b |
| Canada: Somerset Island, | 2 | Smith, 1972a |
| Great Bear Lake | 7 | Corbel, 1959b |
| Spitsbergen | 320; 16 | Corbel, 1960 |
| | 30 | Corbel, 1959b |
| Yugoslavia | 47; 83; 82; 60; 90; 79; 93; 77; 83; 104; 104 | Gams, 1969, 1972 |
| | 129 | Corbel, 1965 |
| Tatra Mountain, Poland | 96 | Corbel, 1965 |
| | 36; 38 | Kotarba, 1971 |

## TROPICAL DESERT ZONE

| Locality | Erosion rate (m³/km²/year) | Source |
|---|---|---|
| In Salah, Sahara | 3 | Corbel, 1971 |

# References

Atkinson, T. C., 1971, The dangers of pollution in limestone aquifers, *Proc. Univ. Bristol Spelaeo. Soc.*, **12**, 281–290.

Atkinson, T. C. and D. P. Drew, 1974, Underground drainage of limestone catchments in the Mendip Hills, in 'Fluvial processes in instrumented watersheds', K. J. Gregory and D. E. Walling (Eds.), *Inst. Brit. Geogr. Spec. Pubn.*, **6**, 87–106.

Atkinson, T. C. and D. I. Smith, 1975, The erosion of limestones, in *Speleology: the Science of Caves*, T. D. Ford (Ed.), David and Charles, Newton Abbot, *in press*.

Aubert, D., 1967, Estimation de la dissolution superficielle dans le Jura, *Bull. Soc. vaud. Sci. nat.*, **69**, 365–376.

Aubert, D., 1969, Phénomènes et formes du karst jurassien, *Ecl. geol. Helv.*, **62**, 325–399.

Baker, V. R., 1973, Geomorphology and hydrology of karst drainage basins and cave channel networks in east central New York, *Water Resources. Res.*, **2**, 549–560.

Balazs, D., 1971, Intensity of the tropical karst development based on caves of Indonesia, *Karst es Barlangkutatas, Hungarian Speleo. Soc.*, **6**, 1–67.

Balazs, D., 1973a, Relief types of tropical karst areas, in *Symposium on Karst morphogenesis, Intnl. Geogr. Union*, Hungary, 16–32.

Balazs, D., 1973b, Comparative morphogenetical study of Karst regions in the tropical and temperate zones, *Trans. Cave Res. Gt. Br.*, **15**, 1–8.

Bauer, F., 1964, Kalkabtragungsmessungen in den Osterreichischen, *Erdkunde*, **18**, 95–112.

Bögli, A., 1960, Kalklösung und Karrenbildung, *Zeit. für Geom.—Suppl.*, **2**, 4–21.

Boussingault, J., and B. Levy, 1852, Sur la composition de l'air confine dans la terre végétale, *C. R. Acad. Sci., Paris*, **35**, 765.

Boynton, D. and O. C. Compton, 1944, Normal seasonal changes of oxygen and carbon dioxide percentages in gas from the larger pores of three orchard soils, *Soil. Sci.*, **57**, 107–117.

Casey, H., 1969, The chemical composition of some southern English Chalk streams and its relation to discharge, *Ass. Riv. Authorities Year Book*, 100–113.

Chulakov, Sh. A., 1959, The problem of the formation of soil structures, *Soil Fertil. Abstr.*, **1960**, 134.

Cogley, J. G., 1972, Processes of solution in an arctic limestone terrain, in 'Polar geomorphology' R. J. Price and D. E. Sugden (Eds.), *Inst. Brit. Geogr. Spec. Pubn.*, **4**, 201–211.

Cooke, H. J., 1971, A study of limestone solution under tropical conditions in northeast Tanzania, *Trans. Cave Res. Gp. Gt. Br.*, **13**, 265–276.

Corbel, J., 1957, Les Karsts du nord-ouest de l'Europe, *Mem. Inst. Etudes Rhodaniennes*, **12**, 1–544.

Corbel, J., 1959a, Erosion en terrain calcaire, *Ann. Geogr.*, **68**, 97–120.

Corbel, J., 1959b, Vitesse de l'érosion, *Zeit. für Geom.*, **3**, 1–28.

Corbel, J., 1960, Nouvelles recherches sur les karsts arctiques Scandinaves, *Zeit. für Geom.—Suppl.*, **2**, 74–80.

Corbel, J., 1965, Karsts de Yougoslavie et notes sur les Karsts tcheques et polonias, *Rev. Geogr. de l'Est.* **5**, 245–294.

Corbel, J., 1971, Les Karsts des régions chaudes (des deserts aux zones tropicales humides), *Studia Geomorphologica Carpatho-Balcanica*, **5**, 49–74.

Corbel, J. and R. Muxart, 1970, Karsts des zones tropicales humides, *Zeit. für Geom.*, **14**, 411–474.

Cvijic, J., 1893, Das Karstphänomen, *Geogr. Ab.*, **5**, 217–329.

Davis, W. M., 1899, The geographical cycle, *Geogr. J.*, **14**, 481–504.

Dearden, J., 1963, Derbyshire limestone—its removal by man and nature *East Midland Geographer*, **3**, 199–205.

Douglas, I., 1964, Intensity and periodicity in denudation processes with special reference to the removal of material in solution by rivers, *Zeit. für Geom.*, **8**, 453–473.

Douglas, I., 1968, Some hydrologic factors in the denudation of limestone terrains, *Zeit. für Geom.*, **12**, 241–255.

Drew, D. P., 1968, Tracing percolation waters in karst areas, *Trans. Cave. Res. Gp. Gt. Br.*, **10**, 107–114.

Drew, D. P., 1970, Limestone solution within the east Mendip area, Somerset, *Trans. Cave. Res. Gp. Gt. Br.*, **12**, 259–270.

Drew, D., M. D. Newson and D. I. Smith, 1968, Mendip Karst Hydrology Research Project, *Occ. Publ. Wessex Cave Club*, Series 2 (2), 1–28.

Ede, D. P., 1972, Comments on 'Seasonal fluctuations in the chemistry of limestone springs' by Shuster and White in *J. Hydrol.*, **16**, 53–56.

Ek, C. M., 1969, L'effet de la loi de Henry sur la dissolution de $CO_2$ dans les eaux naturelles, *Problems of the Karst denudation, Studia Geographica (Brno)*, **5**, 53–56.

Ford, T. D., 1966, The underground drainage systems of the Castleton area, Derbyshire, and their evolution, *Cave Sci.*, **5**, 369–396.

Ford, D. C., 1971a, Characteristics of limestone solution in the southern Rocky Mountains and Selkirk Mountains, Alberta and British Columbia, *Can. J. Earth Sci.*, **8**, 585–609.

Ford, D. C., 1971b, Alpine karst in the Mount Castleguard–Columbia Icefield area, Canadian Rocky Mountains, *Arctic Alpine Res.*, **3**, 239–252.

Ford, D. C., and W. I. Stanton, 1968, The geomorphology of the south central Mendip Hills, *Proc. Geol. Ass. Lond.*, **79**, 401–427.

Gams, I., 1969, Ergebnisse der neueren Forschungen der Korrosien in Slowenien (N. W. Jugoslavien), *Problems of the Karst Denudation, Studia Geographica, (Brno).*, **5**, 9–19.

Gams, I., 1972, Effects of runoff on corrosion intensity in the northwestern Dinaric Karst, *Trans. Cave. Res. Gp. Gt. Br.*, **14**, 78–83.

Gerstenhauer, A., 1960, Der tropische Kegelkarst in Tabasco (Mexico), *Zeit. für Geom. Suppl.*, **2**, 22–48.

Gerstenhauer, A., 1969, Offene Frage der Klimagenetischen Karstgeomorphologie der Einfluss der $CO_2$ Konzentration in der Bodenluft auf die Landformung, *Problems of the karst denudation, Studia Geographica, (Brno)*, **5**, 43–52.

Gerstenhauer, A., 1972, Der Einfluss des $CO_2$-Gehaltes der Bodenluft auf die Kalklösung, *Erdkunde*, **26**, 116–119.

Gilewska, S. S., 1971, The palaeogeographic conditions of karst evolution in Poland, *Studia Geomorphologia Carpatho-Balcanica*, **5**, 1–23.

Groom, G. E., and V. H. Williams, 1965, The solution of limestone in South Wales, *Geogr. J.*, **131**, 37–41.

Grund, A., 1914, Der geographische Zyklus im Karst, *Z. Ges. Erdk. Berl.*, **52**, 621–640.

Hanwell, J. D. and M. D. Newson, 1970, The storms and floods of July 1968 on Mendip, *Occasional Publication of the Wessex Cave Club*, Series I, No. 2.

Harmon, R. S., J. W. Hess, R. W. Jacobsen, E. T. Shuster, C. Haygood and W. B. White, 1972, Chemistry of carbonate denudation in north America, *Trans. Cave Res. Gp. Gt. Br.* **14**, 96–103.

Hellden, U., 1973, Artfjället, Lapland, Sweden: karst hydrology and morphometry. *Lunds Universitets Naturgeografiska Inst.*, **20**, 1–23.

Horton, R. E., 1945, Erosional development of streams and their drainage basins; a hydrophysical approach to quantitative morphology, *Bull. geol. Soc. Am.*, **56**, 275–370.

Ineson, J., 1962, A hydrogeological study of the permeability of the Chalk, *J. Instn. Wat. Engrs.*, **16**, 449–463.

Jacobson, R. L. and D. Langmuir, 1972, An accurate method for calculating saturation levels of groundwaters with respect to calcite and dolomite, *Trans. Cave Res. Gp. Gt. Br.*, **14**, 104–108.

Jakucs, L., 1973, The role of climate in the quantitative and qualitative control of karstic corrosion, *Symposium on Karst Morphogenesis, Intnl. Geogr. Union, European Regional Conf. Hungary*, 122–152.

Jennings, J. N., 1967, Some karst areas of Australia, in *Landform Studies in Australia and New Guinea*, J. N. Jennings and J. A. Mabbutt (Eds.), Canberra, 256–292.
Jennings, J. N., 1971, *Karst*, M. I. T. Press, New York.
Jennings, J. N., 1972a, Observations at the Blue Waterholes, March '65–April' 69 and and limestone solution on the Cooleman Plain, N. S. W., *Helictite*, **10**, 1–46.
Jennings, J. N., 1972b, The Blue Waterholes, Cooleman Plain, N. S. W., and the problem of karst denudation rate determination, *Trans. Cave Res. Gp. Gt. Br.*, **14**, 109–117.
Jennings, J. N. and M. J. Bik, 1962, Karst morphology in Australian New Guinea, *Nature*, **194**, 1036–1038.
Jennings, J. N. and M. M. Sweeting, 1963, The limestone ranges of the Fitzroy Basin, Western Australia, *Bonn. Geogr. Abh.*, **32**, 1–60.
Kotarba, A., 1971, The course and intensity of present day superficial chemical denudation in the Western Tatra Mountains, *Studia Geomorphologica Carpatho-Balcanica*, **5**, 111–127.
Krumbein, W. C. and F. A. Graybill, 1965, *An introduction to statistical models in geology*, McGraw-Hill, New York.
La Valle, P., 1967, Some aspects of linear karst depression development in South Central Kentucky, *Ann. Ass. Am. Geogr.*, **57**, 49–71.
La Valle, P., 1968, Karst depression morphology in South Central Kentucky, *Geog. Annlr*, **50A**, 49–71.
Leather, J. W., 1915, *India Dept. Agric. Mem. Ser.*, **4**, 85 (1915), cited by Russell (1961, p. 366).
Lehmann, H., 1936, Morphologische Studien auf Java, *Geogr. Abh (Stuttgart)*, **9**, 1–114.
Lehmann, H., 1954, Der tropische Kegelkarst der verschiedenen Klimazonen, *Erdkunde*, **8**, 130–139.
Lehmann, H., 1960, Introduction to meeting on karst phenomena, 1959, *Zeit. für Geom. Suppl.*, **2**, 1–3.
Lehmann, H., 1964, States and tasks of research on karst phenomena, *Erdkunde*, **16**, 81–38.
Lehmann, H., 1970, Kegelkarst and Tropengrenze, *Tübinger Geogr. Stud.*, **34**, 107–112.
Matskevitch, V. B., 1957, Carbon dioxide regime in soil air of steppe and semi-desert under tree and herbaceous coenoses, *Soil Fertil. Abstr.*, **1961**, (718).
Maximovich, G. A., 1966, Hydrochemical platform Karst water zones, *Problems of the Speleological Research (Brno)*, **2**, 129–146.
Muir, A., H. G. M. Hardie, R. L. Mitchell and J. Phemister, 1956, The limestone of Scotland: chemical analyses and petrography, *Mem. Geol. Surv. Spec. Repts. Min. Res.*, **37**, 1–150.
Muxart, R., T. Stchouzkoy and J. Franck, 1969, Contribution a l'étude de la dissolution des calcaires par les eaux de ruissellement et les eaux stagnantes, *Problems of the Karst denudation, Studia Geographica (Brno)*, **5**, 21–42.
Newson, M. D., 1971, A model of subterranean limestone erosion in the British Isles based on hydrology, *Trans. Inst. Brit. Geogr.*, **54**, 55–70.
Newson, M. D., 1972, Discussion of 'Seasonal fluctuations in the chemistry of limestone springs' by Shuster and White, *J. Hydrol.*, **16**, 49–51.
Nicholson, F. H. and H. M. Nicholson, 1969, A new method of measuring carbon dioxide for limestone solutional studies, with results from Jamaica and the United Kingdom, *J. Brit. Speleo. Ass.*, **VI**, 43/44, 136–148.
Oleksynowa, K. and B. Oleksynowna, 1971, A tentative comparison of Karst in the Tatra Mountains with Krakow–Czestochowa Plateau, *Studia Geomorphologica Carpatho-Balcanica*, **5**, 93–104.
Paterson, K., 1971, Some considerations concerning percolation waters in the Chalk of North Berkshire, *Trans. Cave Res. Gp. Gt. Br.*, **13**, 277–282.
Paterson, K., 1972, Responses in the chemistry of spring waters in the Oxford region to some climatic variables, *Trans. Cave Res. Gp. Gt. Br.*, **14**, 132–140.

Picknett, R. G., 1964, A study of calcite solutions at 10 °C, *Trans. Cave Res. Gp. Gt. Br.*, **7**, 39–62.

Picknett, R. G., 1972, The pH of calcite solutions with and without magnesium carbonate present, and the implications concerning rejuvenated aggressiveness, *Trans. Cave Res. Gp. Gt. Br.*, **14**, 141–150.

Pitty, A. F., 1968, The scale and significance of solutional loss from a limestone tract of the south Pennines, *Proc. Geol. Ass. Lond.*, **79**, 153–177.

Pulina, M., 1968, The Eastern Siberian Karst, *Geographica Polonica*, **14**, 109–117.

Pulina, M., 1971, Observations on the chemical denudation of some Karst areas of Europe and Asia, *Studia Geomorphologica Carpatho-Balcanica*, **5**, 77–92.

Rocques, H., 1969, A review of present-day problems in the physical chemistry of carbonates in solution, *Trans. Cave Res. Gp. Gt. Br.*, **11**, 139–164.

Runnels, D. D., 1971, Chemical weathering of the Biscayne Aquifer, Dade County, Florida, *Southeastern Geol.*, **13**, 167–174.

Russell, E. W., 1961, *Soil Conditions and Plant Growth*, 9th ed., Longmans, London.

Schwarzenbach, G., 1957, *Compleximetric Titrations*, London.

Sheikh, K. H., 1969, The responses of *Molinia caerulea* and *Erica tetralix* to soil aeration and related factors: gas concentrations in soil air and soil water, *J. Ecol.*, **57**, 727.

Shuster, E. T. and W. B. White, 1971, Seasonal fluctuations in the chemistry of limestone springs: a possible means for characterizing carbonate aquifers, *J. Hydrol.*, **14**, 93–128.

Siegel, S., 1956, *Nonparametric statistics for the behavioral sciences*, McGraw-Hill, New York.

Smith, D. I., 1965, Some aspects of limestone solution in the Bristol region, *Geog. J.*, **131**, 44–49.

Smith, D. I., 1969, Solutional erosion in the area around Maldon and Maroon Town, Jamaica, *J. Brit. Speleo. Ass.*, **VI**, 43/44, 120–133.

Smith, D. I., 1970, The residual hypothesis for the formation of Jamaican bauxite—a consideration of the rate of limestone erosion, *J. Geol. Soc. Jamaica*, **11**, 3–12.

Smith, D. I., 1972a, The solution of limestone in an Arctic environment, *Inst. Brit. Geogr. Spec. Pub.*, **4**, 187–200.

Smith, D. I., 1972b, Where limestone fashions landscape, *Geogr. Mag. Lond.*, **45**, 31–5.

Smith, D. I. and D. G. Mead, 1962, The solution of limestone, *Proc. Univ. Bristol Spelaeo. Soc.* **9**, 188–211.

Smith, D. I., D. P. Drew and T. C. Atkinson, 1972, Hypotheses of karst landform development in Jamaica, *Trans. Cave Res. Gp. Gt. Br.*, **14**, 159–173.

Smith, D. I. and M. D. Newson, 1974, The dynamics of solutional and mechanical erosion in limestone catchments on the Mendip Hills, Somerset, in 'Fluvial processes in instrumented watersheds', K. J. Gregory and D. E. Walling (Eds.), *Inst. Brit. Geogr. Spec. Pub.*, 6, 155–167.

Smith, D. I., T. C. Atkinson and D. P. Drew, 1975, Hydrology of limestone terrain, in *Speleology: the Science of Caves* T. D. Ford (Ed.), David and Charles, Newton Abbot, *in press*.

Stelcl, O., 1972, Solution intensity on various types of rocks in Czechoslovakia, *Trans. Cave Res. Gp. Gt. Br.*, **14**, 174–175.

Stelcl,O., V. Vlcek and J. Pise, 1969, Limestone solution intensity in the Moravian karst, *Problems of the Karst Denudation, Studia Geographica (Brno)*, **5**, 71–86.

Strahler, A. N., 1957, Quantitative analysis of watershed geomorphology, *Trans. Am. Geophys. Un.*, **38**, 913–920.

Sweeting, M. M., 1958, The karstlands of Jamaica, *Geogr. J.*, **124**, 184–199.

Sweeting, M. M., 1964, Some factors in the absolute denudation of limestone terrains, *Erdkunde*, **18**, 92–95.

Sweeting, M. M., 1966, The weathering of limestones, with particular reference to the Carboniferous Limestones of Northern England, in *Essays in Geomorphology*, G. H. Dury (Ed.), Heinemann, London, 177–210.

Sweeting, M. M., 1972, *Karst Landforms*, Macmillan, London.
Sweeting, M. M. and A. Gerstenhauer, 1960, Zur Frage der absoluten Geschwindigkeit der Kalk korrosion in verschiedenen Klimaten, *Zeit. für Geom.—Suppl.*, **2**, 66–73.
Thornthwaite, C. W., 1948, An approach towards a rational classification of climate, *Geogr. Rev.*, **38**, 55–94.
Thornthwaite, C. W., and J. R. Mather, 1957, Instructions and tables for computing the potential exapotranspiration and the water balance, *Drexel Inst. Tech. Pubn. in Climatology*, **10**, 185–311.
Tricart, J. and T. C. Da Silva, 1960, Une example d'evolution karstique en milieu tropical sec: Le morne de Bom Jesus da Lapa (Bahia, Brasil), *Zeit. für Geom.*, **4**, 29–42.
Trudgill, S. and T. C. Atkinson, 1975, The solution of limestone beneath a soil cover, in *Karst Geomorphology and Hydrology: A Process Approach*, T. M. L. Wigley (Ed.), Edward Arnold, London.
Versey, H. R., 1959, The hydrological character of the White Limestone formation of Jamaica, *Trans. 2nd Caribbean Congr., Kingston, Jamaica*, 59–68.
Vine, H., Thompson, H. A., and F. Hardy, 1943, Studies and aeration of cacao Soils in Trinidad—2, Trop. Agric. Trin., **19**, 215–223.
Weyl, P. K., 1958, The solution kinetics of calcite, *J. Geol.*, **66**, 163–176.
White, W. B. and V. A. Schmidt, 1966, Hydrology of a karst area in east central West Virginia, *Water Resources. Res.*, **2**, 549–560.
Williams, P. W., 1966, Morphometric analysis of temperate karst landforms, *Irish Speleol.*, **1**, 23–31.
Williams, P. W., 1968, An evaluation of the rate and distribution of limestone solution and deposition in the River Fergus Basin, W. Ireland, in *Contributions to the study of Karst*, Dept. Geog. Publ. G/5, Australian Nat. Univ. Canberra, 1–40.
Williams, P. W., 1969, The geomorphic effects of ground water, in *Water, Earth and Man*, R. J. Chorley (Ed.), Methuen, London, 269–284.
Williams, P. W., 1970, Limestone morphology in Ireland, in *Irish Geographical Studies*, N. Stephens and R. E. Glasscock (Eds.), (Queens University, Belfast) 105–124.
Williams, P. W., 1971, Illustrating morphometric analysis of karst with examples from New Guinea, *Zeit. für Geom.*, **15**, 40–61.
Williams, P. W., 1972a, Morphometric analysis of polygonal karst in New Guinea, *Bull. geol. Soc. Amer.*, **83**, 761–796.
Williams, P. W., 1972b, The analysis of spatial characteristics of karst terrains, in *Spatial Analysis in Geomorphology*, R. J. Chorley (Ed.), Methuen, London, 135–163.
Yanat'eva, O. K., 1954, Solubility of dolomite in water in the presence of carbon dioxide, *Bull. Acad. Sci. U.R.S.S. (Chem)*, 977–978.
Zonn, S. F. and G. K. Li, 1960, Characteristics of the energy relations of biological processes of tropical forest soils, *Soil Fertil, Abstr.*, (716).

CHAPTER FOURTEEN

# Criteria for the Recognition of Climatically Induced Variations in Granite Landforms

MICHAEL F. THOMAS

## 14.1 Introduction

The study of landforms developed on granitoid rocks has for long been of particular interest in geomorphology. Reasons for this include the wide distribution of granites and similar acid plutonic rocks over the earth's landsurface; the variety and striking character of much granite scenery; and, in particular, the individuality of the rock landforms of granites. A considerable part of this variety and individuality undoubtedly derives from the distinctive structural characteristics of this intrusive rock suite, and especially from the pattern and frequency of joints. Other characteristics reflect the mineralogy and texture of the rock and its behaviour within denudation systems.

The emplacement of granites below one or more kilometres of country rock means that they become subject to stress release and weathering under atmospheric pressures and in the presence of oxygen and water. Thus, although granites are resistant to mechanical abrasion, they are generally subdivided into a hierarchy of structural compartments, varying in size from blocks of less than 1 metre square to monolithic domes having a basal area of perhaps $10^3$ square metres. The formation of separated rock plates (sheeting or exfoliation), and the development of small structural microcracks (microfissures or potential joints) also occurs as a result of stress release. All these weaknesses contribute to the physico-chemical disintegration of the rock.

Considerations such as these have led some writers to the conclusion that the distinctive characteristics of granitoid rocks impose a degree of uniformity upon landforms over a wide range of climatic types (King, 1948, 1957). Variations in form are to be sought, therefore, in terms of structure, texture or mineralogy. Conversely, however, it can be reasoned that, precisely because structural variations can be accounted for, the effects of varying denudation systems dependent upon climate can readily be investigated. In fact, many students of granite geomorphology have developed hypotheses to demonstrate climatic control over regolith type (Bakker, 1960, 1967; Bakker and Levelt, 1964) and the development of rock landforms (Linton, 1955) in granites.

Accordingly, this review aims to consider how far it is possible to substantiate claims for the recognition of climatogenic (Büdel, 1963, 1968) forms and deposits within areas of granitoid rocks. A central part of such a discussion must concern the susceptibility of granites to chemical weathering processes and the particular ways in which climate may influence the rate of weathering and the depth and nature of the weathered mantle. Because many rock landforms comprised of granite are commonly regarded as having been exhumed from such regoliths, a study of the varying balance within the entire denudation system must be attempted. However, it is first necessary to consider in more detail some geological features of the granitoid rocks.

## 14.2 The nature of granite and granitoid rocks

The first problem encountered in all studies of granite relief concerns the comparability of observations obtained from different areas. No single, agreed definition of 'granite' as a rock is yet available, but even the most rigorous definitions encompass a group of rocks of varying composition and texture. Many studies, especially in geomorphology, have been content to accept a very wide range of acid plutonic rocks within the definition. By general assent, these are better described as 'granitoid' and range from quartz diorite and granodiorite, through 'true granites' to quartz monzonites. Granitoid rocks are also often taken to include foliated rocks such as primary gneisses.

True granite has been described as a mixture of quartz, feldspar and mica but even this allows for very wide variation. Genetic classifications do not simplify the problem. Several authors (Eskola, 1932; Marmo, 1971) have classified granites on a tectonic basis. *Synkinematic granites* are regarded as early orogenic rocks which include the primary gneisses and are often dominated by quartz diorites and granodiorites. According to Marmo (1971) these rocks are commonly characterized by microcline and oligoclase, and possess biotite and also hornblende. *Late kinematic granites* are often diapiric in structure and commonly consist of uniform, potash-rich rocks, containing microcline and albite with muscovite and epidote. *Post-kinematic* granites are also generally true granites containing orthoclase and microcline in association with albite. These rocks have been called 'disharmonious' (Walton, 1955) because they were intruded into already cooled country rock. They possibly include rocks not associated with orogenic activity, such as the granites in cratons recognized by Raguin (1966). According to Oen (1965) this class of granites is often characterized by well developed sheeting, resulting from decompression of the upper volumes of granite instrusions. Such rocks become compressed owing to upward movement during emplacement and show important density differences, notably between the lighter granite and the superincumbent country rocks.

Close definitions of granite may be of little value to geomorphological studies, because many landforms associated with granites are found over the whole range of granitoid rocks. Yet contrasts between different areas of granite terrain demand that close attention be given to the composition, texture and structure of individual occurrences of granite. It is generally accepted that differences in mineral composition, such as the nature of the feldspars and of the accessory minerals will commonly affect the reaction of granite to chemical weathering processes. Moreover, the tendency for the rock to disaggregate has often been associated with texture, the coarser textured rocks being more prone to this kind of disintegration. Structural features of granite often interfere with predictions based on composition and texture, however, for jointing frequencies and the contribution of structural micro-cracks (microfissures) to rock porosity (Birot, 1962) both affect the rates of chemical decay and frost weathering.

Amongst the granitoid rocks, true granites comprising orthoclase, microcline and muscovite often prove most resistant to chemical decay, while those containing calcic and potash feldspar together with biotite are more susceptible to weathering. Many authors claim that the plagioclases or biotite decay first, weakening the rock fabric and leading to disaggregation (Ruxton and Berry, 1957). More basic rocks containing olivine and hornblende (such as granodiorite) are noticeably more prone to deep weathering, but many still give rise to the characteristic rock forms of granites. In fact, it is often pointed out that large batholiths are commonly dominated by granodiorites rather than by true granite. Nevertheless, the basic rocks such as diorite and gabbro cannot be included within any definition of granitoid rocks, and should generally be excluded on geomorphological grounds as well.

## 14.3 Some structural features of granitoid rocks

Granite relief commonly reveals the clear imprint of rock structure at varying scales, from the overall morphology of the intrusion, and a strong compartmentation along fracture zones, down to the detailed effects of individual joint sets. Many granitoid masses are associated with disharmonious, ellipsoidal intrusions such as batholiths; others with diapiric structures of smaller dimensions (stocks). These are generally late kinematic or post-kinematic granites, but the occurrence of granitoid rocks that have possibly been granitized in place is also widespread. These rocks are commonly foliated, may be associated with migmatites and have no clear boundaries. Many such rocks give rise to rock domes and occasionally to tors. In fact, it has been claimed (King, 1975) that migmatized rocks form most of the higher and more massive domed inselbergs.

### 14.3.1 Joint patterns

It has long been recognized that characteristic joint patterns may develop in granites. Some of these have been associated with the intrusion of the mass and with its cooling (Cloos, 1923; Balk, 1937), while others may be imprinted during later tectonic events (Twidale, 1964, 1971). Granite masses remaining undivided by other joints often develop curvilinear sheeting fractures which may be closely related to the configuration of the rock landforms. It has been suggested that the deep interiors of many intrusive granites are substantially free from joints. If widespread, this implies that the depth of erosion into a granite mass is likely to affect the landform patterns produced. Several categories of rock fractures may be recognized in granites.

(1) Two or more nearly vertical joint sets commonly subdivide the rock into structural compartments varying in size from perhaps $10^{-1}$ to $10^4$ square metres. These joints influence the early penetration of groundwater and largely control the progress of weathering.

(2) Flat-lying joints, possibly of varying origins (Balk, 1937), subdivide the granite into cuboid blocks. These will vary in size and location both spatially and in depth.

(3) Sheeting (exfoliation) joints which are generally curved in general conformity with the topographic surfaces, have commonly developed on larger rock masses, but may be present on quite small blocks. Many controversies have surrounded this phenomenon, both in terms of the causes of sheeting and also in respect of its influence upon landform. Sheeting appears to be a result of stress relaxation in rocks formerly under compression. The sources of compressive stress may be many, including the superincumbent load of country rock (Gilbert, 1904); the effects of emplacement and of density differences between the granite and the country rock (Oen, 1965); and lateral stresses stored in the rock during later phases of crustal upheaval (Twidale, 1964, 1971). Although the simple concept of unloading is inadequate to account for all instances of sheeting, stress from whatever source is released wherever boundary conditions permit and these may be influenced by the course of denudation, in particular by the nature of the denudation system. Thus sheeting has developed on recently glaciated rock surfaces; in fact it is most impressive in glaciated areas (Matthes, 1930; White, 1945; Oen, 1965). On the other hand, sheeting in warm climates may have developed in response to deep weathering patterns (Thomas, 1965). In this case a structural phenomenon cannot be regarded as independent of climate.

(4) 'Potential joints' were described by Chapman (1958, p. 555) as 'highly discontinuous faces or zones of weaker material... represented by cracked grains, cleared grains, disjointed grain boundaries, tiny faults, and layers of tiny fluid inclusions'. These appear to be similar to the 'microfissures' of Birot (1958) or the 'micro-cracks' described by Bisdom (1967). Chapman's (1958) original argument was that many directions of potential jointing are impressed upon rocks such as granite, but that only those favoured by boundary conditions of stress existing at a particular time would become open joints. Thus major tensional fractures may admit groundwater at an early stage of denudation within the rock mass, while other potential joints open up and lead to further rock disintegration or decay only at a later stage in landform development.

This concept of sequence in the development of joints is closely related to that of a hierarchy of fractures. These ideas cannot be fully explored here, but they influence the interpretation of rock forms in granites. For instance, major fractures may delineate compartments that, through the separation of sheeting and weathering penetration, develop into domes which in turn undergo subdivision by weathering and tensional collapse along newly opened plane joints (Thomas, 1965). Similarly, during weathering penetration into rectilinear joint blocks, spheroidal modification of the blocks takes place. Spheroidal weathering results in enwrapping layers of partially rotted rock passing outwards into structureless sand and clay; but it is not ubiquitous, and its development may depend on the interaction of weathering with stress relaxation on a small scale within individual joint blocks (Oen, 1965). Once exhumed, such corestones may fracture as other potential joints open up under the tensional forces acting on the core boulder.

These and other observations make it difficult to separate the purely structural features of granitoid rocks from those produced by denudation. Structures are not independent of landform, because the morphology of the landsurface creates its own stress fields as well as providing boundary conditions for the release of stored stresses (Gerber and Scheidegger, 1969). Nevertheless, it is important to the enquiry into the effects of climate upon granite landforms that the contribution of structures to such forms are correctly identified and evaluated.

## 14.4 The alteration of granites

Many of the questions surrounding the evolution of granite landforms turn upon the response of granitoid rocks to weathering agents in different environments. Before these can be adequately assessed, however, the effects of autometamorphism or hydrothermal alteration need to be considered.

The formation of kaolin deposits in areas such as southwest England has commonly been attributed to hydrothermal alteration rather than to deep groundwater weathering (Palmer and Neilson, 1962; Exley, 1964; Eden and Green, 1971). Such interpretations have not gone unchallenged (Konta, 1969) and, in reality, the situation is complex. In the first place, hydrothermal alteration and weathering cannot be regarded as mutually exclusive processes: the effects of both may be superimposed upon a single regolith. Second, the assumption that alteration from deep-seated causes will always produce advanced kaolinization (Eden and Green, 1971) conflicts with observations of slight oxidation of minerals such as biotite in certain granites (Eggler and coworkers, 1969). If the rock is weakened during emplacement, the granite will then be susceptible to more rapid penetration by groundwater weathering, and this seems to have been the case in Wyoming, where contiguous granites exhibit contrasting landforms and deposits (Eggler and coworkers, 1969).

Furthermore, criteria for distinguishing one form of alteration from the other are far from universally accepted. Brunsden (1964) has suggested that the occurrence of a marked basal surface of weathering below the regolith cover will indicate penetration of weathering from above. However, two agruments

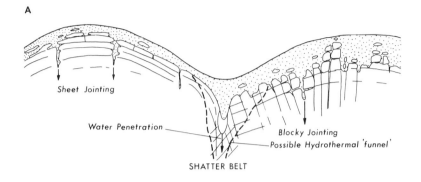

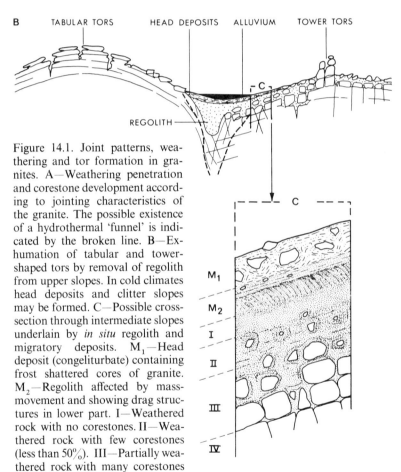

Figure 14.1. Joint patterns, weathering and tor formation in granites. A—Weathering penetration and corestone development according to jointing characteristics of the granite. The possible existence of a hydrothermal 'funnel' is indicated by the broken line. B—Exhumation of tabular and tower-shaped tors by removal of regolith from upper slopes. In cold climates head deposits and clitter slopes may be formed. C—Possible cross-section through intermediate slopes underlain by *in situ* regolith and migratory deposits. $M_1$—Head deposit (congeliturbate) containing frost shattered cores of granite. $M_2$—Regolith affected by mass-movement and showing drag structures in lower part. I—Weathered rock with no corestones. II—Weathered rock with few corestones (less than 50%). III—Partially weathered rock with many corestones (over 50%). IV—Unweathered granite. (Adapted from Ruxton and Berry (1959), Fitzpatrick (1963), Waters (1965) and Bristow (1968)).

can be used to throw doubt on some interpretations of this kind. First, it has been suggested that alteration of the granite by hydrothermal agencies will take place only where pressure conditions permit widespread permeation of the rock mass via open joints. Second, and perhaps more important, Bristow (1968) has argued that the effects of hydrothermal alteration spread laterally beneath the cover of aureole rocks from narrow funnels at depth (Figure 14.1). It must be admitted, however, that such effects will have been localized and widespread decomposition of granites in many landscapes can hardly be attributed to this cause.

Distinctive mineral assemblages have also been claimed as evidence for deep-seated alteration. Thus Exley (1958, 1964) considered the assemblage tourmaline, lithium–mica, flourite, topaz and albite among the kaolins of southwest England to be indicative of a hydrothermal origin. At the same time, Konta (1969) has suggested that such mineral combinations exist in unaltered rocks and, from a comparison of the kaolins of southwest England with those of the Karlovy Vary area of Czechoslovakia, he concluded that 'no method, in particular no laboratory one, has yet been established to clarify the genesis of kaolins'.

It is possible that Palmer and Neilson (1962), Bristow (1968) and Eden and Green (1971) are correct in suggesting that the spatial and morphological characteristics of the highly kaolinized granites of Dartmoor and Bodmin Moor are indicative of hydrothermal alteration, although such arguments do not necessarily oppose the concept of the widespread development of a subaerially produced regolith from which rock landforms may have been exhumed. Nonetheless, the problematic orgins of kaolinized granite impinge on some morphoclimatic interpretations of granite regoliths and landforms.

## 14.5 Regolith characteristics and climate

The depth and character of granite regoliths have become important to many discussions concerning the climatic environment of rock decay and landform development. This is clearly a special case of the general relationship between weathering products and climate and, by broadening the basis of the discussion, the particular problems of hydrothermal alteration can be seen in a much wider context. It has long been known that the clay content of soils on stable sites increases with mean annual temperature (Jenny, 1941), and also with total precipitation (Sherman, 1952). It has also proved possible to define a broad regionalization of weathering products based largely upon the characteristic clay minerals formed in different environments (Pedro, 1968; Tardy and co-workers, 1973).

Such relationships permit comparison of regoliths from different sites in an attempt to reconstruct the climate of formation in each case. However, several factors affect regolith composition, including the position of the sample within the weathering profile (Lumb, 1962; Ruddock, 1967; Flach and coworkers, 1968), the angle and position of slope of the original site (West and Dumbleton, 1970; Tardy and coworkers, 1973) and the age of the regolith or soil (Wambeke,

1962; Haantjens and Bleeker, 1970). Where contemporary soils or a very large number of samples are being considered, these factors can be accounted for, but in many instances samples from relict and probably truncated profiles have to be compared and the absence of detail concerning the character of the original site makes it necessary to view with caution any attempt to draw conclusions based on minor differences between regoliths.

Fortunately, several studies of relict granite regoliths in Europe and elsewhere have indicated major differences in weathering products and it has proved possible to place these within a global context of regolith zones. Thus it is known that humid tropical conditions favour the formation of kaolinitic clays (which may be mainly halloysite) and also of gibbsite, where leaching is intense. In less humid conditions, especially within the seasonally arid tropical climates, montmorillonite replaces much of the kaolinite. According to Sanches Furtado (1968) granite regoliths in the tropics vary with rainfall thus:

annual rainfall 800–1000 mm kaolinite plus montmorillonite
annual rainfall 1000–1200 mm kaolinite dominant
annual rainfall 1200–2000 mm kaolinite plus gibbsite

With an annual precipitation of less than 800 mm, the dominance of 2:1 lattice clays such as montmorillonite will increase, while in very wet tropical climates the proportion of gibbsite may rise on well drained sites. In temperate areas 2:1 clays persist, but montmorillonite is in general replaced by illitic clays (hydrous mica), while the total clay is much reduced.

The use of gibbsite in palaeoclimatic reconstruction has to be approached with caution because it also appears as an initial product of granite decomposition (Loughnan, 1969; Green and Eden, 1970). However, regoliths containing important quantities (up to 30%) of this mineral may generally be regarded as products of prolonged and intensive leaching by groundwater low in dissolved silica (Parham, 1969). Many regoliths classed as kaolinitic are now known to contain important quantities of halloysite and the significance of this is not certain, although Parham (1969) has suggested, from studies of granite weathering in Hong Kong, that feldspars are altered initially to allophane which develops into halloysite. However, comparison of the geologically recent weathering of Hong Kong with ancient regoliths in Minnesota has shown that the latter is dominated by kaolinite. Thus, Parham (1969) makes the suggestion that the halloysite is gradually changed to kaolinite with time. If this is so then it may be possible to assess whether certain regoliths are of recent or ancient origin.

A series of studies undertaken by J. P. Bakker and his colleagues in Amsterdam has established the composition of many granite regoliths in Europe (Bakker, 1960, 1967; Bakker and Levelt, 1964). Bakker (1967) made a clear distinction between widespread sandy regoliths containing 2%–7% clay less than 2 µm, and pockets of clayey material containing 15%–30% clay of mixed illite/kaolinite composition. These were compared (Bakker and Levelt, 1964)

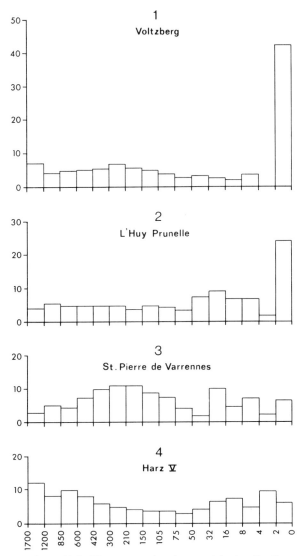

Figure 14.2. Histograms showing particle size distribution in different granite weathering types (after Bakker, (1967). 1—Surinam; humid tropical. 2—Central Morvan; relict clayey type. 3—Creusot Region; relict sandy type. 4—Harz; relict sandy type. Vertical scale—percentage. Horizontal scale—diameter in millimetres.

with regoliths from the humid tropics (Figures 14.2 and 14.3) containing 30%–50% clay, nearly all of which is kaolinite or halloysite, but with some samples containing up to 30% gibbsite.

The clayey regoliths collected from the Massif Central of France were attribu-

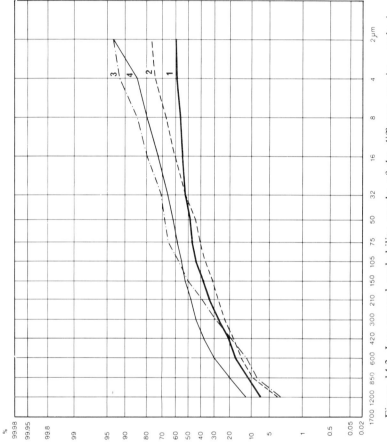

Figure 14.3. Log normal probability graphs of the different granite weathering types shown in Figure 14.2 (after Bakker, 1967).

ted by Bakker (1967) to marginal tropical conditions (*c.* 20°N latitude) prevalent during the Miocene in western and central Europe. The sandy gruss was attributed to more recent (but not contemporary) weathering under moist sub-tropical conditions. Conversely, it has been argued by Piller (1951) that this 'arenization' of the granite could be a result of weathering under present-day cool temperate conditions and this question remains unresolved. In a recent study of the products of granite alteration on Dartmoor, Eden and Green (1971) recognized the similarity of the 'growan' of southwest England to the sandy weathering of other parts of Europe. Compared with tropical regoliths this material has a low clay content and a high feldspar to quartz ratio. These authors compared the composition of the growan with that of the china clay deposits and concluded that the latter were likely to have been hydrothermal in origin. Their figures are summarized in Table 14.1. Eden and Green (1971) appear to concur with Bakker's (1967) view that the sandy regolith cannot be tropical but may have been a result of conditions warmer than those of the contemporary climate in southwest England.

Regoliths from the Separation Point Granite of the South Island of New Zealand display similar characteristics. The area, west of Tasman Bay, has been subject to strong crustal movement and uplift, probably during the Pliocene and into the Pleistocene. Zones of deep arenization (Thomas, 1974a) occur over a wide altitudinal range, and the gruss displays a clay-plus-silt content of 2·0%–10·0%. The material is dominantly feldspar and quartz with small amounts of clay, but is entirely disaggregated. Within these deposits are bands of a red clayey material which has a clay-plus-silt content of 32·0%–62·0% and a clay mineral content approaching 70%. Most of the clay is kaolinite or halloysite (which dominates the 2 μm size fraction). Although a hydrothermal origin for these clays may be involved, their present appearance is modified by weathering which has lead to the liberation of iron oxides (probably geothite) imparting a strong colouration (5YR). This red weathering probably involved a warm climate, having a mean annual temperature of at least 15 °C (Pedro, 1968). The present-day climate has a mean annual temperature of 12 °C, with a rainfall of around 1500 mm, and it seems probable that this environment could be responsible for the comparatively rapid development of the deep sandy gruss in the shattered granite (see Thomas (1974b)). The important distinction between type and degree of alteration in rocks has recently been emphasized by Pedro and coworkers (1975).

Table 14.1. Composition of growan and china clay on Dartmoor (after Eden and Green (1971); reproduced by permission of SSAG (The Swedish Geographical Society))

|  | Growan | | China clay | |
|---|---|---|---|---|
|  | Range | Average | Range | Average |
| clay + silt (percentage) | 13·5 –28·0 | 21·5 | 32·0–72·0 | 47·0 |
| clay/2 μm (percentage) | 2·0 –10·5 | 4·5 | 1·5–30·5 | 11·0 |
| feldspar: quartz | 0·73– 1·26 | 0·97 |  | 0·20 |

Other criteria can be used to determine the environment of weathering, such as the persistence of heavy minerals and the etching, through corrosion, of quartz grains. The micromorphology of quartz grains may offer important clues to many of the problems alluded to in this section, but few results are as yet available (Konta, 1969; Doornkamp and Krinsley, 1971; Krinsley and Doornkamp 1973) and the criteria used have not gone unchallenged (Brown, 1973).

### 14.5.1 The significance of regolith depth

The depth of alteration in granite is particularly important to many accounts of granite landforms, but taken alone it may be of limited significance to discussions of the climate of weathering. Ruxton and Berry (1961) have suggested that within the tropics regolith depths vary with the humidity of climate. They used examples of plains below major hillslopes (Figure 14.4) and advanced the following figures:

| | |
|---|---|
| Humid tropics (Hong Kong) | well over 30 m |
| Wetter savannas (Ingessana Hills) | up to 26 m |
| Drier savannas (Central Sudan) | 6–9 m |
| Arid regions (Red Sea Hills) | 3–6 m |

There are many arguments to support the view that weathering depth will

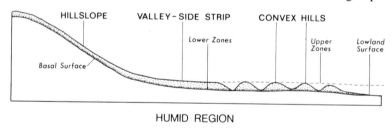

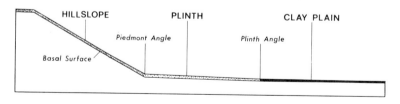

Figure 14.4. Landforms and weathering profiles on granite in arid (Qala en Nahl, Sudan) and humid (Hong Kong) regions (after Ruxton and Berry (1961)).

increase with the humidity and warmth of climate (Strahkov, 1967, Thomas, 1974b), but these figures can be no more than a guide to conditions beneath surfaces of comparable age and geomorphic position. It is virtually impossible to satisfy these restrictions in any global comparison of weathering depths and, for this reason, such tabulations may be very misleading (Leopold, Wolman and Miller, 1964, Table 4.6).

Many granite weathering profiles in high latitudes have been assumed to be relict, and in these cases both the environments of weathering and their duration remain matters of speculation. Many such profiles have also been truncated. Depths of the sandy weathering are extremely variable. Thus Eden and Green (1971) recorded only 2 or 3 m of growan on Dartmoor, while Eggler and coworkers, (1969) found at least 60 m in Wyoming. In Central Europe commonly quoted depths are in the region of 6 m (Bakker, 1967), but very deep pockets locally exceeding 30 m are known. This depth is also found in the Separation Point granite discussed above.

Within the kaolinized troughs or funnels in southwest England alteration at depths exceeding 200 m is known (Bristow, 1968), sometimes below aureole rocks or fresh granite. Instances of this kind certainly appear to be the result of hydrothermal activity and cannot on available evidence be used in palaeoclimatic reconstruction.

In the tropics, depths of alteration again vary widely as part of a pattern of undulation in the weathering front (Thomas, 1966) and also according to the age and degree of dissection of the land surface. Beneath many ancient plains, sometimes protected by duricrusts, depths of weathering may exceed 50 m, although 15–30 m is more common. Even in areas with frequent rock outcrops, troughs and basins of weathering reach depths of 15 m in places. Patterns of decomposition appear similar in both granites and gneisses in the Basement Complex of west Africa (Thomas, 1966) and signs of autometamorphism are generally absent. On the other hand these deposits are often highly kaolinized (Lumb, 1962; Ruddock, 1967) and have a high clay content (Bakker, 1967; and Figures 14.2 and 14.3).

It must be concluded that the depth of weathering is not necessarily or always related to degree of alteration. It is equally clear that different granites disintegrate or decompose at widely varying rates. General principles should also warn against the implication that all weathering penetrates vertically from the landsurface. Rock that is deeply and closely shattered by joints will admit water to some depth before weathering has really begun. Decomposition will therefore proceed outwards from such troughs and basins as well as downwards from the surface. Thus, while depth of decay is often important to arguments about the evolution of rock landforms in granites, it cannot be used alone to indicate the age or environment of weathering.

The only major discontinuity in the distribution of weathered granite is related to the boundary of the Pleistocene ice sheets (Feininger, 1971). Within the glaciated area only isolated pockets of regolith remain. There is, however, sufficient to be of significance to arguments concerning the development of landforms and soils (Fitzpatrick, 1963). In the areas subjected to periglacial conditions during the glacial advances of the Pleistocene Period, most of the regolith mantle appears to have been lost from summits as a result of strong mass-movement (gelifluction). In a comparable manner powerful surface denudation (sheetwash?) in the semi-arid zones has stripped many interfluves of former weathered mantles. In neither of these cases, however, has the cover been subject to wholesale excavation and redistribution in till as it has in the glacial zone.

*14.5.2 The time scale of weathering and the occurrence of relict profiles*

The rates of weathering penetration into granite bedrock remain uncertain, and very few studies present quantitative information. According to Leneuf and Aubert (1960) complete ferrallitization of 1 m of granite under seasonal tropical conditions takes from 22,000–77,000 years, while Trendall (1962) calculated that the duricrusted profiles over Ugandan granites have advanced at the rate of 9 m per million years. Haantjens and Bleeker (1970) found negative evidence in the tectonically active island of New Guinea to corroborate such findings. It may be concluded, therefore, that highly kaolinized residues (not of hydrothermal origin) require a prolonged period of tropical or near-tropical conditions to attain observed depths of 5–15 m. Such a conclusion would invalidate the concept of deep rotting of granites during the comparatively brief interglacial periods, for interglacials of 25,000–60,000 years would achieve only 1–3 m of kaolinized granite.

The situation is less straightforward than this, however. In the first place, rates of weathering will vary according to humidity of the climate and position in the landscape (Tardy and coworkers, 1973). They will also be influenced by the mineralogy and fabric of the rock. Furthermore, as the weathering front advances, the rate of decay is likely to become retarded, because of the accumulation of solutes in discrete groundwater compartments, the increase of confining pressures in the rocks, and the reduction in amount of fresh groundwater reaching deep levels.

Also fundamental is the question of the time taken to produce 'arenized' granite in which little chemical alteration is evident. According to Haantjens and Bleeker (1970) tropical conditions achieve 'skeletal' weathering (comparable with arenization) in 5000 years; 'immature' weathering in 5000–20,000 years and 'mature' weathering (ferrallitization) in periods exceeding 20,000 years. Depth data are not presented in relation to these figures from New Guinea, but widespread, immature profiles of about 3 m depth were recorded.

Even if we allow for a slower rate of decay under temperate conditions, it appears possible that 2–5 m of sandy gruss could be produced in granites during periods of 25,000 years or less. Contrasts between individual granites make such predictions difficult, however (Eggler and coworkers, 1969), and the time taken to produce deep arenization must remain uncertain. That it is a relatively rapid process is apparent from the deep profiles found in the recently uplifted areas of New Zealand (Thomas, 1974a) and in the wide altitudinal range of weathering along the mountain front of the Sierra Nevada (Wahrhaftig, 1965).

Compared with known rates of surface denudation, the penetration of weathering into granites is slow and, in many areas of contemporary dissection, the profiles are clearly relict. In fact, all profiles of more than a metre or so must be relict in the sense that the period of weathering commenced long before the contemporary climate was established. The possibility of deep profiles having resulted from prolonged tropical or subtropical weathering during the Tertiary Era in Europe is strong, and arguments of a similar kind have been advanced to account for deep profiles in Australia (Ollier, 1965).

Nevertheless, the wider view suggests that the weathering of granites should be considered continuous through geological time (Nikiforoff, 1959; Thomas, 1974a) and, although the rate of decay will have varied with changes of climate, oscillations in the rate of surface denudation may have been much greater. Evidence for development and truncation of individual profiles indicates the presence in denudation systems of little known thresholds which control the sequence of weathering and exhumation of rock landforms in granite areas. Similarly, striking spatial discontinuities in the occurrence of regolith mantles suggests that conditions of stability have fluctuated across the landscape with local as well as regional changes in denudation systems. The alternative, that all rock forms are to be regarded as products of differential erosion (see Hack (1960)), is not satisfied by the field evidence concerning the occurrence of rock landforms.

## 14.6 The landforms of granitoid rocks

Because granites are prone to deep decay, the landforms associated with these rocks are commonly developed over waste mantles of varying thickness. These mantles exhibit a characteristic zonation over jointed rock (Ruxton and Berry, 1957; and Figures 14.1 and 14.5) and the composition and erosional mobility of the material varies according to depth (Ruxton and Berry, 1961; Lumb, 1962; Ruddock, 1967). The rock landforms of granites commonly appear to have been exhumed from such a regolith mantle and slopes developed over granitoid rocks can be described in terms of the degree of development or removal of such weathered profiles.

Such descriptions are implicit or explicit in many studies, and the categories of slope material listed below are built up from work by Birot (1958), Demek (1964) and Young (1972):

(1) Smoothly convex or concave slopes developed over a continuous regolith cover, generally many metres thick and containing few cores of fresh rock near the surface (Zones I and II of Ruxton and Berry (1957)); see also Figure 14.5.
(2) Slopes characterized by detached corestones within a residual regolith (Zones II and III).
(3) Rocky slopes dominated by tors (with rock foundations) and frequent,

426

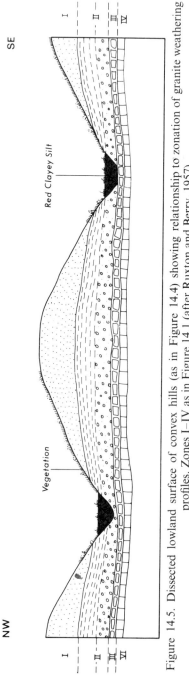

Figure 14.5. Dissected lowland surface of convex hills (as in Figure 14.4) showing relationship to zonation of granite weathering profiles. Zones I–IV as in Figure 14.1 (after Ruxton and Berry, 1957).

large corestones (Zones III/IV), or by other rock outcrops exhibiting a rectilinear jointing pattern.
(4) Smooth rock surfaces, commonly domical (multi-convex) in form and associated with sheeting.
(5) Frost riven angular cliffs and rock towers. This category may be similar to and therefore confused with (3) above (Figure 14.6).

Such forms as these may be regarded as hillslope facets (Beckett and Webster, 1965) or as unit landforms (Lueder, 1959), and as individuals can be considered meso-scale forms. It is necessary, however, to consider forms which become apparent at different scales of enquiry (see also Jahn (1962)). These include microforms that diversify individual rock surfaces, and the macroforms or landform systems which embrace the combination of slope facets within an organized spatial pattern. At each level of resolution clues may be sought concerning the effects of climate.

The setting for landform systems or patterns is generally the intrusive body in the case of granites, but diversification within the boundaries of these is often apparent. Such landform systems contain particular combinations of facets and individual landforms and these may frequently be of greater palaeoclimatic significance than the existence of individual forms alone (Thomas, 1974a).

The detailed examination of such systems of forms commonly reveals distributional patterns among the individual facets, together with an indication of any dominance within a given system by particular forms or materials. In the first place, however, some account of the mesoforms (facets and landforms) associated with granites is required.

### 14.6.1 Granite regoliths and the emergence of rock forms

Large areas of granite terrain exhibit no rock exposures at all, but must be considered in terms of slope forms on sedentary or transported mantles. Although such forms may be related to specific characteristics of granites and their weathering products, such relationships need to be carefully demonstrated. The study of slope characteristics on unconsolidated materials, and of the possible relationships between these and climate, must lie outside the scope of this paper, but certain comments will be offered in recognition of the functional unity of hillslopes on which rock and regolith covered slopes may alternate.

Three aspects of this situation may be considered: first, the occurrence, location and proportionate extent of individual convex, constant or concave slopes; second, the characteristics of complex hillslopes involving slope facets developed on different materials; and, third, the composite patterns of such facets displayed by distinctive landform systems. In addition the dynamics of the slope forms are important. Much simplified, this can be viewed in terms of stability and profile development, instability and profile truncation, and dynamic equilibrium within which a delicate balance between regolith formation and removal is maintained over a recognizable area. Such considerations are important to all geomorphological enquiry, but in the present study they relate to questions surrounding the emergence of rock forms or the maintenance of regolith covered slopes, and also to possible genetic relationships which may exist between the rock landforms of granites and contiguous slopes on mantle deposits.

The development of convexity or concavity on individual slopes is not to be sought in a single cause (Young, 1972). Convexity, for instance, appears likely to develop as a result of drainage incision into a formerly planate landscape, but may also reflect the operation of particular slope processes such as chemical weathering and creep. In addition, convex rock faces possessing parallel sheeting fractures are prominent in granite landscapes. There may therefore be more than one causal mechanism leading to an equilibrium form which is domical or cupola-shaped (White, 1945). Concavity, although often related to surface water flow, also probably reflects the weakness of channel cutting in an area. These considerations must be borne in mind when evaluating some of the following observations.

Convexity in granite landscapes has been attributed to incision into deeply weathered landsurfaces of low relief and instances of this kind have been mainly identified from humid tropical and subtropical areas (Ruxton and Berry, 1957, 1961; Hurault, 1967; Young, 1972). Hurault (1967) described such landscapes as 'alveolate' and demonstrated a strong structural control over the cupola-shaped compartments. Ruxton and Berry (1961) sought to constrast such landscapes with those of semi-arid areas. They suggested that piedmont slopes in humid areas would become deeply weathered and might be dissected into a lower storey landscape of convex compartments, while piedmonts of semi-arid areas developed only shallow weathering profiles and remained as low angled and convex plinths or pediment slopes (Figure 14.4). Further sapects of this observation are considered below.

It is difficult to relate the emergence of rock forms in granites to slope conditions. Particularly in the seasonal tropics, rock pavements may break the surface of low angled slopes, while under forest, both in tropical and temperate climates (Bakker, 1967; Thomas, 1974a) slopes exceeding 20° may retain a regolith cover. However, the abrupt transition from weathered to rocky slopes is a common feature of many granite landscapes and the relationships of rock landforms to these contiguous slopes are central to our understanding of granite geomorphology. Such patterns possibly result from present or past climatic conditions through the establishment and disruption of stability or equilibrium within denudation systems.

Under stable conditions deep, zoned profiles develop in jointed granites (Figure 14.1). These may be truncated by surface denudation, if this is accelerated by stream incision, ecologic change or both. According to Ruxton and Berry (1957), dissection into a deeply weathered piedmont will create weathered compartments across different zones of the regolith (Figure 14.5), stream channels commonly flowing over solid rock. If the compartments are rock cored, then thinning of the regolith on valley sides will expose progressively lower zones of the profile over time. In this way different levels in the former profile may become exposed on hillslopes. It is this development that partially explains the list of granite slope forms given above (Section 14.6). Each slope may be regarded in terms of its status within the contemporary system and of the existing balance between weathering penetration and regolith removal.

Thus shallow regoliths containing many corestones can be regarded in terms of a sequence of stripping towards the exposure of solid rock forms, or as existing within a dynamic system in which corestone formation is matched closely by the tendency for detached cores to be ejected on to the surface and removed by gravitational transfer to lower slopes or into stream channels. These alternatives also have significance for any account of tors and other apparently exhumed rock surfaces.

Before carrying this discussion further it is useful to define the main categories of rock landforms in granites (see Figure 14.6). These are possibly only three and distinctions between them may in certain instances be arbitrary.

(1) *Tors (boulder inselbergs)* comprising groups of contiguous corestones having a rock base. These forms generally exhibit clearly the pattern of joints in the rock mass. Morphologically, they fall into three broad groups: *tower-like* forms defined by a blocky, rectilinear jointing; *tabular* forms induced by strong lateral fractures such as sheeting joints that although not necessarily horizontal run parallel with the ground surface (Ruxton and Berry, 1959; and Figure 14.1); and *hemispherical* forms in which a domical profile is evident among the constituent boulders (Thomas, 1965).

(2) *Domes (domed inselbergs or 'bornhardts')* and larger spheroids that rise as monolithic forms to heights varying form a few metres to 500 m or more, Many possible subdivisions of this large group of ladforms have been attempted, often using criteria such as size, or position rather than morphology. Morphologically, domes may be very complex, but they vary from *rock-pavements* (sometimes called 'ruwares') that may scarcely break the profile of the encircling landsurface to low, elongate *whalebacks* and cupola-like *bornhardts*. Many of these forms are found in foliated rocks such as gneisses and migmatites which are closely related to some (synkinematic) granites. In these cases asymmetry is often striking. Most domical forms are associated with sheeting and tabular tors may surmount many domes (Thomas, 1965). Similar half-domes appear on escarpments around the margins of some granite instrusions.

(3) *Stacks (rock towers or 'castle koppies')* are angular, often castellated forms which exhibit few if any signs of spheroidal modification of constituent blocks. No general agreement concerning a descriptive term for these features exists, partly because they are often confused with or referred to as tors. Tropical examples have been called *koppies* (King, 1948; Thomas, 1965). Linton (1955) described isolated examples as *stacks*, and those interrupting hillslopes as *buttresses*, and Jahn (1962) used the term *rock tower*. Since tor and tower are similar

Figure 14.6. Jointing patterns are usually clearly revealed in granite landforms, particularly where chemical weathering has been active. The photographs illustrate specific aspects of this joint control. (a) Rectilinear joint patterns commonly define tower-like *tors* composed of spheroidally-weathered boulders of widely varying size. Here, the rock cores are of similar size, reflecting the regularity of jointing, but contiguous cores are more commonly of different sizes. (b) In (b) the main tor is of abnormal size, dwarfing the human figures in the right foreground. The height of this feature exceeds 20 metres. Both tor groups (a) and (b) are from the Jurassic Younger Granite Suite on the Jos Plateau, Nigeria. (c) Radial stress release often affects large rock cores, giving rise to sheeting joints (exfoliation). These may appear on spheroidal cores but are characteristic of domed inselbergs (bornhardts). The small dome, around 20 metres high, illustrated here is from the Cambrian Older Granite Suite on the Jos Plateau of Nigeria.

It should be noted that the large core in (b) is higher than the small dome in (c), illustrating the difficulty of distinguishing tors from domes in terms of size. However, the girth of the dome is greater, and the occurrence of an incurved under side is unlikely. Nevertheless perception of such morphological differences depends on the depth of exposure.

words with common origins, stack may be regarded as the preferable term, although there is a slight risk of confusion with sea stacks.

The difficulties surrounding the separate recognition of these landforms will be evident. These are mainly questions of size and morphology. Domes and tors may appear quite separate, except when small domes and large spheroids are being compared. In some cases the incurved underside of small domes may be concealed by the mantle, so that distinction between spheroidal and domical forms can be illusory (Thomas, 1974a, 1974b). Similarly, distinctions between the angularity of rock stacks and the spheroidal form of tors are often difficult to define and this has contributed to the debate over the origins of tors in southwest England (Linton, 1955; Palmer and Neilson, 1962).

*14.6.2 The climatic significance of rock landforms in granites*

If the various rock landforms are regarded exclusively as exposures of some former weathering front, then the argument surrounding their appearance as surface forms must centre upon the conditions required for alternate regolith development and removal. However, while this remains important to the present theme, other and different possibilities must be considered.

It must be envisaged that some forms may be structural in the sense that they develop within a wide range of denudation systems and are largely independent of them (King, 1957, 1975). Others may be the end products of convergent development from different previous states of the landforms via quite separate sequences of events within contrasting denudation systems (White, 1945; Cunningham, 1969).

These questions are focused by the problems surrounding the development of granite domes which occur over a wide range of climatic environments (White, 1945, 1972; King, 1957). Many domes in tropical regions are associate with deep regoliths (Ollier, 1960; Thomas, 1966) and some have been revealed in artificial excavations (Boye and Fritsch, 1973). Others have been associated with slope retreat in unweathered rock (King, 1948, 1966) and most of those in glaciated regions appear to have no connection with former regolith covers. Possible interactions between glaciation and sheeting may be important in these areas (Oen, 1965), while relationships between deep weathering and sheeting can be suggested for domes that have developed in tropical or subtropical environments (Thomas, 1965). Few geomorphologists have suggested that sheeting alone is responsible for the domed profile and, while no single hypothesis satisfies all examples of the form, it is unlikely that the development of domes can be regarded as independent of the denudation systems that have operated within the areas they occupy.

*14.6.3 Tors and associated landforms*

The hypothesis for tor exhumation from a deep regolith has been argued many times (e.g. Falconer (1912), Handley (1952), Linton (1955), Demek (1964) and Thomas (1965)). It has come to be known as the 'two-stage hypothesis' of tor formation, and has been related to a specific sequence of climatic change in high latitudes. According to Linton (1955) deep rotting of granites took place throughout Europe during the warm and periodically humid Tertiary Era. As the cooler Pliocene Period gave way to the increasing cold of the Pleistocene, rates of weathering would have been retarded until, with the onset of periglacial conditions in many upland areas, strong mass-movement (gelifluction) removed most of this regolith cover to expose tors on summits and flanks of many granite hills (see Figure 14.1).

This account was seen to beg a number of questions. In the first place surviving regoliths in southwest England have been considered primarily a result of hydrothermal alteration, while the absence of spheroidal rounding and of closely associated deep weathering around the tors appeared to render unnecessary the argument adduced by Linton (see Palmer and Neilson, 1962). Even the recent discussion by Eden and Green (1971) may seem to leave these questions unresolved. However, if a broader view is taken, it is readily shown that deep weathering in granites is widespread as a relic of Tertiary climates (see above) and, further, that surviving shallow weathering profiles are associated with the Dartmoor tors. After all, the former deep weathering need have been no deeper than the heights of existing tors.

The exhumation hypothesis for tor development postulates that granite weathering proceeds more rapidly beneath a mantle of moisture-retaining debris than on exposed rock surfaces, so that, once exposed, the constituent boulders weather only slowly and persist in the landscape. To expose the tors in the first place an acceleration of surface erosion is required to remove the regolith mantle. Such a development is generally considered *a priori* to depend upon a reduction in vegetation cover and an increase in the rate of either mass-movement or surface erosion.

If such an argument is accepted then core formation will take place during warm and moist periods of long duration and exposure of the tors during comparatively short phases of periglacial or semi-arid climate. Evidence for cold climatic conditions in high latitude uplands is widespread and includes the occurrence of coarse, frost-shattered block fields (felsenmeer or clitter) around the tors and the occurrence of head deposits beneath intermediate and lower slopes (Brunsden, 1964; Waters, 1962, 1965). However, the prominence of these features has been used to argue that the entire tor and blockfield complex is a result of periglacial conditions (Palmer and Neilson, 1962).

Evidence for greater aridity is present in many tropical and subtropical environments. Tor fields on the Jos Plateau of Nigeria (11°N) are associated with thick wash deposits on lower slopes and fans below the major escarpments (Thomas, 1974a). The former importance of aeolian processes is marked by lunette forms to the lee of tor girt basins on the Monaro Plateau in New South Wales.

Possible limitations upon the application of the two-stage hypothesis should

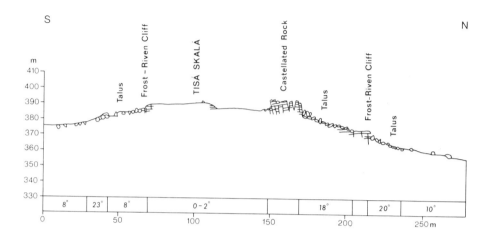

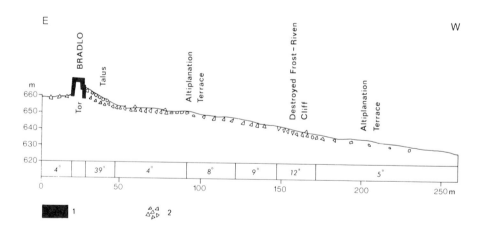

Figure 14.7. Profiles illusrating possible periglacial forms in granites from the Bohemain Highland (Czechoslovakia), (after Demek (1964, 1969)). 1—granite outcrop; rock stack. 2—blockstream; slopes.

be considered. If the tors are to be related to the differential weathering of the smaller joint blocks and the survival and later emergence of larger ones, then for a given rate of surface denudation there should be a threshold size of block which will emerge during a continuous and possibly constant lowering of the landsurface. Thus under given conditions rock forms could become exposed *without* major ecologic change. In fact the continual formation, decay and re-formation of tors during systematic slope retreat may be envisaged (Thomas, 1965, 1974a). This possibility should be excluded wherever a wide range of block sizes occurs together, and where they are distributed widely over a gently undulating landsurface. Such situations are common in many granite landform systems (see below).

Differential weathering and erosion may clearly give rise to rock towers resembling tors without a prior phase of deep rotting of the granite. Such a concept can arise from the general hypothesis of slope retreat and pediplanation as advanced by King (1953, 1958) among others. These 'summit tors' would therefore be the last remnants of former hillslopes that have undergone slope retreat. King considered this hypothesis to have wide application, but it was derived from studies of seasonal and semi-arid, subtropical climates. Detailed studies have not always been able to demonstrate the efficacy of this process in granite terrain, however.

It has also been seriously suggested that cryoplanation (or altiplanation) is the periglacial equivalent of pediplanation (Demek, 1969) and that the retreat of frost-riven cliffs may isolate residual rock towers (stacks), in some respects resembling tors, standing above gently sloping cryoplanation terraces (Figure 14.7). Such features have been described as 'palaeoarctic tors' as opposed to 'palaeotropical tors' already considered.

It should be emphasized that neither hypothesis is necessarily confined to accounts of granitoid landforms. Tor exhumation is possible in the case of feldspathic grits (Linton, 1964) for instance, while accounts of cryoplanation (Demek, 1969) make it clear that the process is a general one and that the rock stacks have no special association with granites. In the case of cryergic processes, the most important consideration is probably the jointing of the rock and this would favour the survival of more massive blocks of granite or other rock. A recent study of tor formation in southern Victoria Land, Antarctica (Derbyshire, 1972), aptly illustrates these considerations. The tor-like forms in this polar environment occur both as 'woolsacks' and as angular, frost-shattered rock stacks. The local rock is a quartz dolerite and variations in tor morphology cannot be ascribed to petrographic differences. Signs of incipient kaolinitic alteration were found in samples taken from joint faces within the woolsack tors and Derbyshire (1972) considered that the chemical weathering may have continued slowly since the last glaciation. The simultaneous operation of cryergic processes would have been more effective at sites of snow accumulation, where frost-shattered forms predominate. This argument was advanced to account for the occurrence of woolsack forms on exposed ridges and more angular stacks along cols and on sheltered sites. Conclusions from this study were that pre-weathering may not be required for the production of rounded tor blocks, that the periglacial climate does not exclude chemical alteration nor does it prevent the formation and survival of rounded forms, and that variety in tor morphology may result from local site conditions as well as from different regional climates or climatic oscillations.

It must be concluded, therefore, that not all tors formed by differential chemical weathering result from a formal two-stage process involving major ecologic change and, further, that not all rock stacks on high latitude uplands are tors in this sense. The six possibilities envisaged here are set out below.

(1) Tors arise by exhumation during periglacial conditions following upon a a prolonged period of deep weathering under a warm moist climate.
(2) Tors arise by exhumation during semi-arid or arid conditions following upon a prolonged period of deep weathering under a warm moist climate.
(3) Tors arise by differential denudation during simultaneous chemical weathering and surface erosion without ecologic change.
(4) Rock stacks (towers, kopjes, inselbergs) arise during slope retreat within variably weathered and irregularly jointed rock, especially in seasonal or semi-arid climates.
(5) Rock stacks arise during the retreat of frost-riven cliffs under periglacial conditions.
(6) Rock exposures including tors arise as a result of random events leading to removal of regolith.

The last possibility was envisaged by Wahrhaftig (1965) and also by Hurault (1967) in respect of the larger granite domes. True randomness among such events has not been demonstrated; if it could be then tors might arise irrespective of structural controls and in response to local disturbances of the denudation system.

It is clear from all these possibilities, many of which can be applied to the larger granite domes, that before climatic control over the emergence of rock landforms in granites can be demonstrated other criteria are required in addition to the rock form itself. Five important criteria can be listed:

(1) The morphology and weathering stage (Melton, 1965; Ollier, 1969) of the boulders or other rock surfaces.
(2) The character (or absence) of microforms on exposed rock surfaces.
(3) The nature of the landform systems (macroforms) within which the individual mesoforms are found.
(4) The nature and distribution of any associated regolith mantle.
(5) The nature and distribution of any associated deposits resulting from mass-movement or sedimentation.

It will be evident that several, indeed perhaps all, of the above characteristics need to be taken into account in order to substantiate a particular case. Furthermore, if a two-stage hypothesis is being advanced, then evidence of two contrasting environments is required, with an indication of a rapid change in the state of the denudation system. In these cases (which are the majority) the final position of the enquiry may remain ambiguous because rock forms are erosional and generally lack spatial continuity with the depositional features which commonly form the basis of any chronology or sequence of landform development.

*14.6.4 Morphology of rock surfaces*

Morphological criteria for climatic reconstruction are notoriously uncertain, yet the widespread occurrence of spheroidal forms which closely resemble buried corestones has commonly been taken as an indication of exhumation. The lack of rounding on the tors of southwest England was one circumstance leading to doubts about their subsurface origins (Palmer and Neilson, 1962).

Two considerations become important here. First, it is evident that spheroidal weathering is not ubiquitous although it is very common in granite regoliths. The phenomenon is not well understood (Ollier, 1971), but the presence of compressive stress within the joint blocks may be as important as the progress of chemical weathering. Thus an absence of well-rounded 'woolsack' forms does not necessarily imply the lack of a former regolith mantle. Second, the exhumation hypothesis for tor development implies that the forms have been subject first to groundwater weathering in the absence of abrasion and frost and later to sub-aerial weathering and erosion under varying conditions of frost, humidity, or both. Linton (1955) saw this as leading to the partial destruction of spherodial forms following their exhumation. On the other hand, Jahn (1962), has found evidence for the exhumation of 'prismatic' blocks which, under atmospheric weathering, became rounded into 'woolsack' forms.

This last observation is uncommon, however, and evidence can be adduced to demonstrate the splitting of exposed corestones along plane joints and the development of distinctive surface microforms discussed below. There is also some evidence to show that many exhumed corestones are already partially rotted. They can be broken easily with a hammer and the interior of the rock may be stained with iron oxide released by weathering of biotites. Case hardening, due to diffusion of salts towards the surface of the boulder, where evaporation and crystallization occur is also common (White, 1944; Ollier, 1965). This process probably occurs during subsurface spheroidal weathering (Augistithis and Otterman, 1966; Ollier, 1971). Hydrothermal alteration along joint planes has also been shown to increase resistance to subaerial weathering (Marchand (1974)).

*14.6.5 Microforms on rock surfaces*

Common microforms on granite domes and boulders can mostly be described as weathering hollows, but they also include *rillen* and the effects of scaling. Hollows on granite surfaces include shallow pans and deep, cylindrical pits on gentle slopes, armchair-shaped hollows on steeper slopes, and cavernous weathering (*tafoni*) beneath overhangs or at joint intersections on rock walls.

It is clear that these forms have a wide climatic range (Bakker, 1960), but they are certainly not a result of frost action and it would appear that few are inherited from the form of the weathering front. Thus Bakker and Levelt (1964) comment on relic hollows (*kociolki*) beneath a newly-formed mantle in the tropical rainforest climate of Surinam.

It seems likely that many such hollows, especially those formed on sub-horizontal surfaces, develop from irregularities on the rock surface resulting from the effects of subsurface weathering on weaker parts of the rock fabric. On exposed surfaces such irregularities retain water in contact with the rock and permit the growth of algae and lichens throughout a wide climatic range. A more rapid rate of weathering within the hollow results and it becomes deepened. On some granite surfaces these features interconnect to produce a highly irregular micro-relief. This phenomenon can be illustrated from the cool, cloudy and humid climates of southwest England, where the summit boulders of Rough Tor on Bodmin Moor are highly modified in this way. In such areas as this, low evaporation rates combined with high rainfall keep the rock surfaces moist and frequently renew the water within weathering hollows.

On the other hand, cavernous weathering is commonly associated with arid or maritime environments where salt weathering is active (Wellman and Wilson, 1965). Selby (1972), for instance, attributed spectacular cavernous weathering, producing 'cyclopean masonry' on granite stacks in Antarctica to this cause. However, the cavernous weathering produced in this polar desert environment is quite unlike that developed on humid temperate and tropical tors. Both forms of weathering strongly suggested that sub-aerial environments lead to the diversification of rock surfaces and may destroy forms exposed by exhumation or even perhaps by frost shattering. But the apparent subaerial rounding of Antarctic tors recorded by Derbyshire (1972) should warn against over-generalization on this question.

The occurrence of *rillen* (pseudo-lapiés) on the steep flanks of tors and boulders has been taken to indicate former warm climatic conditions (Demek, 1964), but the reason for this deduction is not clear. Incipient channels may be formed under a shallow regolith cover, as groundwater is induced to flow across the sharply defined basal surface of weathering, and these features may be exploited by surface flow after exhumation. On the other hand, runoff may quite rapidly produce *rillen* on boulders exhumed in a slightly altered state. In New Zealand channels of this kind have been observed on boulders being exhumed in steeply dissected country. Such features had to be considered contemporary and therefore formed in a climate with a mean annual temperature of 12 °C, but with a fairly high rainfall exceeding 1500 mm.

The climatic significance of many of these forms is elusive, although some can be related to prevailing climatic conditions. It must be remembered, however, that it is the microclimate of the rock surface which is important, and that the weathering stage of the exposed rock surface may influence the rate of sub-aerial modification.

*14.6.6 Landform systems in granites*

The combination of individual mesoforms into distinctive landform systems (macroforms) may reveal more about the environments of landform development than the study of individual forms alone (Harpum, 1963; Hurault, 1967; Thomas, 1974a). No complete typology can be offered but the following patterns may be widely observed:

(1) *Multi-convex (dome form or 'alveolate') systems* with weathered compartments appear particularly in warm, humid climates (Ruxton and Berry, 1961; Hurault, 1967). They mark the incision of drainage into a deep regolith under conditions that favour continued weathering of the granite. Tors and other rock forms may outcrop at structurally determined locations or may be absent altogether.

(2) *Multi-concave (basin-form) systems* probably reflect the operation of efficient slope retreat within base-levelled landscapes. Some basins of this kind have rock floors, others remain deeply weathered (Thomas, 1974a). Basin forms are undoubtedly also related to the structural compartmentation of the granite (Waters, 1957; Thorp, 1967) and probably to former basins of weathering (Thomas, 1966) or hydrothermal alteration (Palmer and Neilson, 1962; Bristow, 1968). Most basins are only semi-enclosed and appear to have developed by fluvial action. Excavation of enclosed basins may have taken place as a result of glaciation (Feininger, 1971) or aeolian action.

These landscapes often result from periods of strong surface erosion and display frequent rock exposures, including corestones, tors, rock pavements and domes, according to the jointing characteristics of the rocks. It is not without significance to their interpretation that a wide range of block sizes may have been exhumed and that tors, for instance, can occur at all levels in the landscape from interfluves to basin floors. Thus while the individual tor may not require a two-stage development, tor landscapes of this type probably do.

(3) *Planate surfaces containing residual granite domes* ('inselberg landscapes') are common in the tropics and are mentioned here because many domes may be much older than the landscapes within which they are found. It has been suggested that some are exhumed from stratigraphic covers (Savigear, 1960; Jeje, 1972), while others have probably undergone several periods of development (Thomas, 1965). These landscapes have also been considered characteristic of the tropics (Büdel, 1957; Stoddart, 1969), but they are also associated with the shield areas which have remained comparatively stable in recent geological time. White (1972) has recently argued that the absence of bornhardts from the northern shields is a reflection of the erosive power of the Pleistocene ice sheets, but this comment remains speculative.

This account does not attempt to be exhaustive, and these are illustrations only of the range of granite landscapes. The morphology and morphometry of such systems, and the recognition of different generations of landforms within them, may be important, however, to discussions of origins and climatic controls over landform development.

### 14.6.7 Regolith distribution and characteristics

These questions have been discussed above and in some detail. If the granite weathering can be attributed to a limited range of climate in any given case, this certainly assists arguments based upon climatic control over landform development; but two points intervene. The exhumation of rock forms only requires a difference in weathering stage (Ollier, 1969) between the corestones and the regolith. This difference may be very slight and a prior phase of weathering cannot be assigned to a specific climatic regime unless a diagnostic weathering product survives. With sedentary deposits of this kind, there is likely to be a close juxtaposition of rock and regolith, but in the case of the Dartmoor tors for instance, the absence of a surviving regolith immediately around many of the tors has led to ambiguity in the argument. It must also be recognized that weathering of footslopes around the tors could be subsequent to their formation.

### 14.6.8 The nature of associated deposits

As in all cases of landform development, the climatic significance of associated deposits is often critical to accounts of erosional forms. But the spatial separation of rock forms from these deposits leads to ambiguity. For example the blockfields (clitter slopes) and head deposits on Dartmoor are clear indications of periglacial conditions, but this has not resolved the question of tor formation. On this evidence alone they could have arisen prior to, or during, the cold conditions.

Less ambiguity surrounds the significance of some other transported materials in warm climates. Thus the accumulation of slope-wash deposits or of fans below stripped zones (Thomas, 1974b) must be regarded as strong evidence for the emergence of rock forms during stripping. If mineralogical affinities between sediments and regoliths can be established then the link becomes very strong indeed. Climatically accelerated stripping in warm climates is generally related to decreased atmospheric humidity and a reduction in the vegetation cover.

## 14.7 Conclusions

It must be concluded that few granite landforms can be unambiguously associated with closely defined climatic environments. It would therefore be mistaken to attempt a climatic classification of granite landforms. On the other hand, some features can be attributed to climates lacking in strong frost action, or can be ascribed principally to the effects of groundwater weathering. Thus subsurface development of rock cores may clearly take place over a wide range of climates and only careful examination of relict regoliths can establish the limits more closely. Similarly, conditions under which stripping can occur vary widely and close attention to the deposits resulting from this phase is necessary to any conclusive argument. This may also be corroborated by patterns of slope forms and rock exposures within the landform systems.

The difficulties alluded to in this account result from three major considerations: first, the structural and mineralogical characteristics of granites tend to give rise to similar features wherever granites are found; second, few geomorphological processes can be closely circumscribed in climatic terms: and, third, different processes are capable of producing similar forms. Taken together these conclusions pose formidable problems in geomorphology. The only method by which pitfalls in analysis can be avoided is to recognize the different components of specific landscapes and to search for different generations of forms and deposits that reveal a succession of contrasting environments of denudation. Commonly this will require attention to features varying in scale from microforms to the macro-scale patterns of entire landform systems which can be represented in morphometric terms. It is also essential to consider erosional forms, sedimentary regoliths and transported deposits as components of a single but complex picture.

When all these factors are accounted for, the latitude within a given argument may be very limited, but the problem of missing evidence can never be overcome. Nevertheless it would be perverse to argue for the subsurface origin of angular frost shattered rock stacks, associated with cryoplanation terraces and clitter slopes (blockfields) unless a surviving regolith can be demonstrated (Demek, 1964, 1969). Similarly arguments against a subsurface origin for many tors and domes in the tropics and subtropics are misplaced in the face of available evidence (Thomas, 1966, 1974b; Feininger, 1971; Boye and Fritsch, 1973).

Finally, the question of recognizing climatically controlled landforms in granitoid rocks is but part of the wider problem surrounding climatic or climatogenic (Büdel, 1963) geomorphology. Most landform systems are genetically composite and an oscillatory model of climatic change is required to account for the different generations of forms and deposits. Erosional forms produced at an early stage in the development of a landscape will pass through all succeeding stages and will be modified by each. Only occasionally, however, is a separate, diagnostic imprint available to mark the passage of each climatic phase.

## References

Augustithis, S. S. and J. Otterman, 1966, On diffusion rings and spheroidal weathering, *Chemical Geology*, **1**, 201–209.

Bakker, J. P., 1960, some observations in connection with recent Dutch investigations about granite weathering in different climates and climatic changes, *Zeit. für Geom. Supplement*, **1**, 69–92.

Bakker, J. P., 1967, Weathering of granites in different climates, in *L'Évolution des versants*, P. Macar (Ed.), Congrés et Colloques de L'Université de Liège, **40**, 51–68.

Bakker, J. P. and Th. W. M. Levelt, 1964, An enquiry into the problems of a polyclimatic development of peneplains and pediments (etchplains) in Europe during the Senonian and Tertiary Period, *Publications Services du Carte Géologique de Luxembourg*, **14**, 27–75.

Balk, R., 1937, *The structural behaviour of the igneous rocks*, Ann Arbor.
Beckett, P. H. T. and R. Webster, 1965, *A classification system for Terrain*, Military Engineering Experimental Establishment, Christchurch, Rep. 872.
Birot, P., 1958, Les dômes crystallines, *Centre National de la Recherche Scientifique (C.N.R.S.), Mémoires et Documents*, **6**, 8–34.
Birot, P., 1962, *Contribution a l'étude de la désagregation des roches*, Centre Documentation Universitaire, Paris.
Bisdom, E. B. A., 1967, The role of microcrack systems in the spheroidal weathering of an intrusive granite in Galicia (N. W. Spain), *Geologie en Mijnbouw*, **46**, 333–340.
Boye, M. and P. Fritsch, 1973, Dégagement artificiel d'un dôme crystallin au Sud-Cameroun, *Travaux et Documents de Géographie Tropical (Bordeaux)*, **8**, 31–62.
Bristow, C. M., 1968, Kaolin deposits of the United Kingdom of Great Britain and Northern Ireland, *Congrès Géol. Int., Prague 1968*, **15**, 275–288.
Brown, Joan E., 1973, Depositional histories of sand grains from surface textures, Nature, **242** (5397), 396–398.
Brunsden, D., 1964, The origin of decomposed granite on Dartmoor, in Dartmoor Essays, I. G. Simmons (Ed.), *Devon Association Advancement of Science Literature and the Arts*, 97–116.
Büdel, J., 1957, Die 'doppelten einebnungsflächen' in den feuchten Tropen, *Zeit. für Geom. N. F.*, **1**, 201–288.
Büdel, J., 1963, Klima-genetische Geomorphologie, *Geographische Rundschau*, **15**, 269–285. Reproduced in English as Climatogenetic Geomorphology, in *Climatic Geomorphology*, E. Derbyshire (Ed.), Macmillan, London, 202–227, (1972).
Chapman, C. A., 1958, The control of jointing by topography, *J. Geol.*, **66**, 552–558.
Cloos, H., 1923, Das Batholiten Problem, *Fortschritteder Geologie und Palaeontologie*, **1**, 1–80, Borntraeger, Berlin.
Cunningham, F. F., 1969, The Crow Tors, Laramie Mountains, Wyoming, U. S.A., *Zeit. Geom. N. F.*, **13**, 56–74.
Cunningham, F. F., 1971, The Silent City of Rocks, a bornhardt landscape in the Cotterel Range, South Idaho, U.S.A., *Zeit. Geom. N. F.*, **15**, (4), 404–429.
Demek, J., 1964, Slope development in Granite areas of the Bohemian massif, Czechoslovakia, *Zeit. Geom. N. F.*, Supplement, **5**, 82–106.
Demek, J., 1969, Crypolanation Terraces, their geographical distribution, genesis and development, *Rozpravy Československé Akademie Věd, Rada Matematických a Prirodnich Věd*, Ročnik 79–Sesit, 4, Prague.
Derbyshire, E., 1972, Tors, rock weathering and climate in southern Victoria Land, Antarctica, in Polar Geomorphology, R. J., Price and D. E. Sugden (Eds.), *Inst. Brit. Geog. Spec. Pub. No. 4*, 93–105.
Doornkamp, J. C. and D. Krinsley, 1971, Electron microscopy applied to quartz grains from a tropical environment, *Sedimentology*, **17**, 89–101.
Eden, M. J., and C. P. Green, 1971, Some aspects of granite weathering and tor formation on Dartmoor, England, *Geog. Annlr.*, **53A**, 92–99.
Eggler, D. H., E. E. Larson and W. C. Bradley, 1969, Granites, grusses and the Sherman erosion surface, southern Laramie Range, Colorado–Wyoming, *Am. J. Sci.*, **267**, 510–522.
Eskola, P., 1932, On the origin of granite magmas, *Mineralogische und Petrographische Mitteilungen*, **42**, 455–481.
Exley, C. S., 1958, Magmatic differentiation and alteration in the St. Austell granite, *Quart. J. Geol. Soc. Lond.*, **114**, 197–230.
Exley, C. S., 1964, Some factors bearing on the natural synthesis of clays, *Clay Min. Bull.*, **5**, 411–426.
Falconer, J., 1912, The origin of kopjes and inselbergs, *British Association for the Advancement of Science, Transactions*, Section C, 476.

Feininger, T., 1971, Chemical weathering and glacial erosion of crystalline rocks and the origin of till, in *Geological Survey Research 1971, U.S. Geol. Surv., Prof. Paper 750-C*, C65–C81.

Fitzpatrick, E. A., 1963, Deeply weathered rock in Scotland, its occurrence, age and contribution to the soils, *J. Soil Sci.*, **14**, 33–43.

Flach, K. W., J. G. Cady and W. D. Nettleton, 1968, Pedogenic alteration of highly weathered parent materials, *Ninth International Congress of Soil Science, Transactions*, **4**, 343–351.

Gerber, E., and A. E. Scheidegger, 1969, Stress induced weathering of rock masses, *Ecologae geologicae Helvetiae*, **62**, 401–414.

Gilbert, G. K., 1904, Domes and domed structures of the high Sierra, *Bull. geol. Soc. Am.*, **15**, 29–36.

Green, C. P. and M. J. Eden, 1970, Gibbsite in the Dartmoor granite, *Geoderma*, **6**, 315–317.

Haantjens, H. A. and P. Bleeker, 1970, Tropical weathering in the Territory of Papua and New Guinea, *Austral. J. Soil Res.*, **8**, 157–177.

Hack, J. T., 1960, Interpretation of erosional topography in humid temperate regions, *Am. J. Sci.*, **258-A**, 80–97.

Handley, J. R. F., 1952, The geomorphology of the Nzega area of Tanganyika with special reference to the formation of granite tors, *Congrès géologique International, Compte Rendus 21e*, Algiers, 201–210.

Harpum, J. R., 1963, Evolution of granite scenery in Tanganyika, *Geological Survey of Tanganyika Records*, **10**, 39–46.

Hurault, J., 1967, L'érosion régressive dans les régions tropicales humides et la genèse des inselbergs granitiques, *Études Photo-interpretation 3*, Institut Géographique National, Paris.

Jahn, A., 1962, Geneza Skałek Granitowych (Origins of granite tors), *Czasopismo Geograficzne*, **23**, 19–44.

Jeje, L. K., 1972, Landform development at the boundary of sedimentary and crystalline rocks in south-western Nigeria, *J. Trop. Geog.*, **34**, 25–33.

Jenny, H., 1941, *Factors in soil formation*, McGraw-Hill, New York.

King, L. C., 1948, A theory of bornhardts, *Geog. J.*, **112**, 83–87.

King, L. C., 1953, Canons of landscape evolution, *Bull. geol. Soc. Am.*, **64**, 721–752.

King, L. C., 1957, The uniformitarian nature of hillslopes, *Trans., Edinburgh Geol. Soc.*, **17**, 81–102.

King, L. C., 1958, The problem of tors, *Geogr. J.*, **124**, 289–291.

King, L. C., 1975, Bornhardt landforms and what they teach, *Zeit. für Geom.*, N. F., **19**, 299–318.

Konta, J., 1969, Comparison of the proofs of hydrothermal and supergene kaolinisation in two areas of Europe, *Proc. Intnl Clay Conf. 1969*, 281–290.

Krinsley D. and J. C. Doornkamp, 1973, *Atlas of quartz sand surface textures*, Cambridge University Press.

Leneuf, N. and G. Aubert, 1960, Essai d'évaluation de la vitesse de ferrallitisation, *Proceedings 7th Intnl Congr. Soil Sci.*, 225–228.

Leopold, L. B., M. G. Wolman and J. P. Miller, 1964, *Fluvial processes, in Geomorphology*, Freeman, San Francisco.

Linton, D. L., 1955, The problem of tors, *Geog. J.*, **121**, 470–487.

Linton, D. L., 1964, The origin of the Pennine tors: an essay in geomorphological analysis, *Zeit. für Geom.*, Sonderheft, 5–23.

Loughnan, F. C., 1969, *The chemical weathering of the silicate minerals*, Elsevier, New York.

Lueder, D. R., 1959, *Aerial Photographic Interpretation*, McGraw-Hill, New York.

Lumb, P., 1962, The properties of decomposed granite, *Geotechnique*, **12**, 226–243.

Lumb, P., 1965, The residual soils of Hong Kong, *Geotechnique*, **15**(2), 180–194.

Marchand, D. E., 1974, Chemical weathering, soil development and geochemical fractionation in a part of the White Mountains, Mono and Inyo Counties, California, *U.S. Geol. Surv., Prof. Paper 352-J*.

Marmo, V., 1971, *Granite Petrology* Elsevier, Amsterdam.
Matthes, F. E., 1930, The geologic history of the Yosemite Valley, *U.S. Geol. Surv., Prof. Paper 160*.
Melton, M. A., 1965, Debris covered hillslopes of the Southern Arizona desert—consideration of their stability and sediment contribution. *J. Geol.*, **73**, 715–729.
Nikiforoff, C. C., 1959, Reappraisal of the soil, *Science*, **129**, 186–196.
Oen, I. S., 1965, Sheeting and exfoliation in the granites of Sermersôq, South Greenland, *Meddelelser om Grønland*, **179**, 1–40.
Ollier, C. D., 1960, The inselbergs of Uganda, *Zeit. für Geom. N. F.*, **4**, 43–52.
Ollier, C. D., 1965, Some features of granite weathering in Australia, *Zeitschrift für Geomorphologie N. F.*, **9**, 285–304.
Ollier, C. D., 1969, *Weathering*, Oliver and Boyd, Edinburgh.
Ollier, C. D., 1971, Causes of spheroidal weathering, *Earth Science Reviews*, **7**, 127–141.
Palmer, J. and R. A. Neilson, 1962, Origin of granite tors on Dartmoor, Devonshire, *Proc. Yorkshire Geol. Soc.*, **33**, 315–340.
Parham, W. E., 1969, Halloysite-rich tropical weathering products of Hong Kong, *Proc. Intnl. Clay Conf. 1969*, 403–416.
Pedro, G., 1968, Distribution des principaux types d'altération chimique à la surface du globe, *Rev. Géog. Phys. et Géol. Dynam.*, **10**, 457–470.
Pedro, G., A. Delmas, F. K. Seddoh, 1975, Sur la nécessité et l'importance d'une distinction fondamentale entre type et degré d'altération. Application on problème de la définition de la ferralitisation, *C. R. Acad. des. Sciences*, Paris, t. **280**, Serie D, 825–828.
Piller, H., 1951, Über Verwitterungsbildungen des Brockengranits nördlich St. Andreasberg, *Heidelberger Beiträge Z. Mineralogie und Petrographie*, **11**, 498–522.
Raguin, E., 1966, Sur la classification des granites et l'importance des granites de cartons, *C. R. Academie des Sciences*, **262**, 333–336.
Ruddock, E. C., 1967, Residual soils of the Kumasi district in Ghana, *Geotechnique*, **17**, 359–377.
Ruxton, B. P., 1958, Weathering and sub-surface erosion in granite at the Piedmont Angle, Balos, Sudan, *Geol. Mag.*, **95**, 353–377.
Ruxton, B. P. and L. Berry, 1957, Weathering of granite and associated erosional features in Hong Kong, *Bull. geol. Soc. Am.*, **68**, 1263–1292.
Ruxton, B. P. and L. Berry, 1959, The basal rock surface on weathered granitic rocks, *Proc. Geol. Ass.*, **70**, 285–290.
Ruxton, B. P., and L. Berry, 1961, Weathering profiles and geomorphic position on granite in two tropical regions, *Rev. Géom. dynam.*, **12**, 16–31.
Sanches Furtado, A. F. S., 1968, Alteration des granites dans les régions interpropicales sous differents climats, *9th International Congress of Soil Science*, Adelaide, *Transactions*, **IV**, 403–409.
Savigear, R. A. G., 1960, Slopes and hills in West Africa, *Zeit. für Geom. Supplement 1*, Morphologie des versants, S156–171.
Selby, M. J., 1972, Antarctic Tors, *Zeit. für Geom., Supplement*, **13**, 73–86.
Sherman, G. D., 1952, The genesis and morphology of the alumina rich laterite clays, *Clay and Laterite Genesis*, American Institute of Mining and Metallurgy, 154–161.
Stoddart, D. R., 1969, Climatic geomorphology: review and re-assessment, *Progress in Geography*, **1**, 160–222.
Strahkov, N. M., 1967, *Principles of lithogenesis*, Vol. 1, Oliver and Boyd, Edinburgh.
Tardy, Y., G. Bocquier, H. Paquet and G. Millot, 1973, Formation of clay from granite and its distribution in relation to climate and topography, *Geoderma*, **10**, 271–284.
Thomas, M. F., 1965, Some aspects of the geomorphology of domes and tors in Nigeria, *Zeit. für Geom.*, **9**, 63–81.
Thomas, M. F., 1966, Some geomorphological implications of deep weathering patterns in crystalline rocks in Nigeria, *Trans. Inst. Brit. Geog.*, **40**, 173–193.

Thomas, M. F., 1968, 'Tor', Encyclopedia of Geomorphology, R. W. Fairbridge (Ed.), Reinhold, New York, 1157–1159.
Thomas, M. F., 1974a, Granite landforms; a review of some recurrent problems of interpretation, in Progress in Geomorphology—Essays in Honour of David L. Linton, E. H. Brown and R. S. Waters, *Inst. Brit. Geog. Spec. Pub.*, **7**, 13–37.
Thomas, M. F., 1974b, Tropical geomorphology, Macmillan, London.
Thorp, M. B., 1967, Closed basins in Younger Granite Massifs, northern Nigeria, *Zeit. für Geom. N. F.*, **11**, 459–480.
Trendall, A. F., 1962, The formation of 'Apparent peneplains' by a process of combined lateritisation and surface wash, *Zeit. für Geom. N. F.*, **6**, 183–197.
Twidale, C. R., 1964, A contribution to the general theory of domed inselbergs, *Trans. Inst. Brit. Geog.*, **34**, 91–113.
Twidale, C. R., 1971, *Structural landforms*, M. I. T. Press, Cambridge, Mass.
Wahrhaftig, C., 1965, Stepped topography of the southern Sierra Nevada, California, *Bull. geol. Soc. Am.*, **76**, 1165–1190.
Walton, M., 1955, The emplacement of 'granite'. *Am. J. Sci.*, **253**, 1–18.
Wambeke, A. R. van, 1962, Criteria for classifying tropical soils by age, *J. Soil Sci.*, **13**, (1), 124–132.
Waters, R. S., 1957, Differential weathering and erosion on oldlands, *Geog. J.*, **123**, 501–509.
Waters, R. S., 1962, Altiplanation terraces and slope development in Vestspitzbergen and southwest England, *Biul. Periglacialny*, **11**, 89–107.
Waters, R. S., 1965, The geomorphological significance of Pleistocene frost action in southwest England, in *Essays in Geography for Austin Miller*, J. B. Whittow and P. D. Wood (Eds.), University of Reading, 39–57.
Weliman, A. W., and A. I. Wilson, 1965, Salt weathering, a neglected geological erosive agent in coastal arid environments, *Nature*, **205**, 1097–1098.
West, G. and Dumbleton, J. J., 1970, The minerology of tropical weathering illustrated by some West Malaysian soils, *Quart. J. Eng. Geol.*, **3**, 25–40.
White, W. A., 1944, Geomorphic effects of indurated veneers on granite in the southeastern States, *J. Geol.*, **52**, 333–341.
White, W. A., 1945, The origin of granite domes in the south east piedmont, *J. Geol.*, **53**, 276–282.
White, W. A., 1972, Deep erosion by continental ice-sheets, *Bull. geol. Soc. Am.*, **83**, 1037–1056.
Young, A., 1972, *Slopes*, Oliver and Boyd, Edinburgh.

CHAPTER FIFTEEN

# The Climatic Factor in Cirque Variation

EDWARD DERBYSHIRE AND IAN S. EVANS

## 15.1 Introduction

The recognition of cirques as one of the characteristic landforms of glaciated mountains is almost as old as the glacial theory itself (Agassiz, 1840; Close, 1870; Gastaldi, 1873; Helland, 1877). Their association with the snowline, one of the earth's two primary climato-hydrological boundaries according to Penck (1910), gave cirques a special place in climatic geomorphology. Given that striated and polished rock surfaces are an integral part of their definition, these distinctive concavities constitute one of the few landforms with a claim to be considered diagnostic of a climatically-determined suite of processes.

The postulate that climate, acting through earth-surface processes, produces distinctive landforms, is perhaps most readily accepted for the forms of glacial erosion. Yet even these landforms, of which the cirque is one of the more rudimentary, are not the product of one but of several processes. Indeed, there is a substantial body of opinion which regards the cirque as a product as much of periglacial processes (notably frost shattering) as of glacial processes (Tricart, 1963). This point of view may be of particular value in assessing the climatic factor in cirque genesis by drawing attention to those regions characterized, either currently or in the Pleistocene, by climatically-marginal glaciation.

'Cirque' is a generic term adopted from the French language to describe a hollow that is open on the downslope side and bounded upslope by a steep slope of arcuate form usually referred to as the headwall. The term cirque and its equivalent in other language is popularly used to refer to such hollows over a wide range of sizes and in a variety of situations including fluvial valley heads, arcuate sea cliffs and major scars resulting from rock-fall, landsliding and associated processes (Gregory, 1913; Freeman, 1925; Hinds, 1925).

The specific term 'nivation cirque' is already well established in the scientific literature (Russell, 1933; Watson, 1966; Derbyshire, 1968; Davies, 1969), so that, on grounds of consistency, cirques produced predominantly by glacial action should be qualified by the epithet 'glacial'. They are distinguished from large nivation cirques by the distinctive evidence of glacier flow in the form of ice-polished and striated rock surfaces, *roche moutonnée* forms and, in some cases, true rock basins. Although sliding of snowpacks can produce striation, it rarely leads to rock polishing, and *moutonnée* forms and basins in solid rock are unknown. Both glacial and nivation cirques will be considered in their climatic context in this chapter.

## 15.2 Climate and glacier distribution

A glacier is generated wherever snow accumulation exceeds snow ablation (melting, sublimation and calving) over a period of years. Low temperatures encourage the preservation of snow, but there is no necessary relation between glacier generation and any 0 °C isotherm. If a local equilibrium line is defined as an altitude below which ablation exceeds accumulation, a glacier may be defined as a system permitting transport of ice (derived from snow) under the influence of gravity, from altitudes of surplus accumulation across an equilibrium line to lower altitudes of surplus ablation. It is therefore dependent on an overall decrease in glacier balance (accumulation minus ablation) with decreasing altitude. There may be local inversions in this relationship, but the highest point on any glacier flow line must have a positive balance.

The distinction between a glacier and a perennial snowpatch is at first confusing, especially since not only perennial but also seasonal snowpatches may move downslope under gravity and erode bedrock (Costin and coworkers, 1964). But a perennial snowpatch lacks a distinction between (upper) accumulation and (lower) ablation zones: its surface has built up until there is a balance between accumulation and ablation throughout, usually because a sheltered concavity has been filled. Also the thickness of a snowpatch is inadequate for firnification to run its course and generate glacier ice (Sharp, 1960), or it may ablate completely in some years so that although persistent, it never contains really old snow.

Glacier distribution reflects the balance between accumulation and ablation. The accumulation of snow results from direct snowfall, from snow drifting off more exposed areas upwind, and from avalanching of snow off steep slopes above. Ice accumulates by sublimation of water vapour, or freezing of rain or of meltwater. Ablation of snow and ice results from melting, sublimation, or calving (flotation into water bodies, within which ablation is completed).

The heat sources for ablation may be radiation, molecular conduction from adjacent air (or water), convection (turbulent eddy transfer) and condensation. Molecular conduction is unimportant, while condensation usually supplies less than 30% of the heat, with an average of 8% in the 32 measurements on glaciers tabulated by Paterson (1969). In 24 of these cases, radiation supplies over half the heat for ablation; in tropical or continental climates, such as Central Asia and the Rocky Mountains, it supplies even more. Convection is of course more important in windy, warm climates, for example in maritime areas where heavy snowfall lowers the snowline. The cloudiness of such areas reduces the importance of radiation, and Ahlmann (1948) suggested that convection accounts for some 60% of glacier ablation in Norway and Iceland.

Although a multiplicity of factors is involved, the distribution of glaciers accords with a number of simple rules. Glaciers occur:

(i) at *higher latitudes*, because low temperature and low radiation encourage preservation of snow;

(ii) at *higher altitudes*, because of lower temperature and increased snowfall. Although at very high altitudes the reduced moisture content of thin air leads to reduced snowfall, e.g. on Mount McKinley (Alaska), temperatures are then so low that glacier balance remains positive. Likewise, increasing solar radiation is unable, even above levels of maximum cloudiness, to produce upper ice-free zones above zones of glacier generation;

(iii) in more *maritime* climates, because summer ablation is reduced by cloudiness and, at higher altitudes, snowfall is heavier because the air has a higher moisture content;

(iv) on *poleward aspects* because reduced temperature and solar radiation inhibit ablation. This effect is stronger where radiation is more important, as in continental climates and at high altitudes (Figure 15.1);

(v) on *leeward aspects* because their shelter permits wind-drifted snow to accumulate and reduces ablation by hindering turbulent transfer of heat from

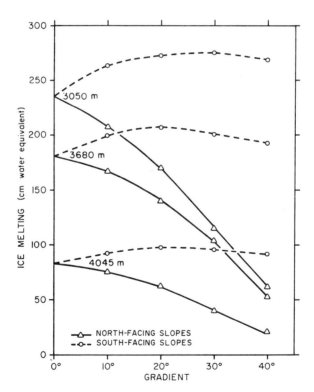

Figure 15.1. Observed ice melting due to solar radiation, as a function of gradient, altitude and aspect on Elbrus, at 43°N in the Caucasus. Based on data from Lyubomirova (1964).

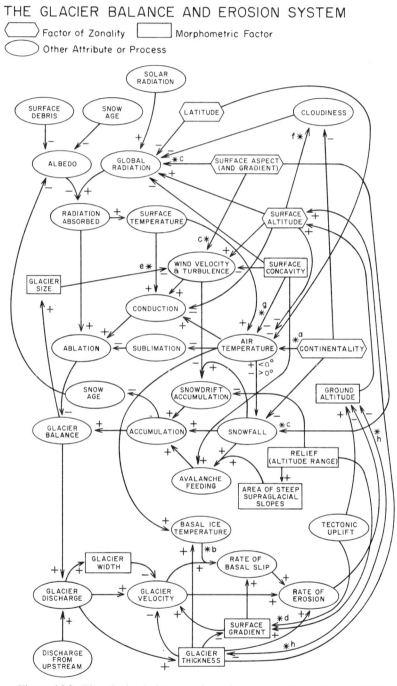

Figure 15.2. The glacier balance and erosion system, from Evans (1974). Notes. *a—increases variability of air temperature. *b—glacier velocity increases the rate of basal slip *unless* basal ice temperature is below freezing. *c—circular relation. *d—tectonic uplift may increase or decrease surface gradient. *e—a more marked glacier wind reduces turbulence. *f—parabolic relation. *g—thermal inversions.

the air. Snow drifting involves essentially winter and spring winds, while summer winds are more relevant in ablation;

(vi) on *eastward aspects* because ablation is less effective during the morning, when east slopes are exposed to the sun;

(vii) in topographic *concavities*, which provide increased shelter and shade.

Exceptions to one of these rules are usually explicable in terms of interaction with one of the others. The favouring of leeward slope aspects, for example, may work either with or against tendencies to poleward or to eastward aspects, depending upon the regionally dominant wind directions during accumulation or ablation. The eastward tendency is strengthened by morning cloudiness in maritime climates, reducing ablation from solar radiation, but it may be completely reversed by dominant afternoon (convectional) cloudiness.

The effects of poleward and eastward aspect increase with surface gradient, but effects involving snow drifting are greater where summit and windward slopes are relatively gentle, as well as where snow falls at lower temperatures and winds are stronger and more consistent. The effect of poleward aspect must decline as the equator or the poles are approached. Figure 15.2 is an attempt to portray some of these interactions, including those with the glacial erosion system; see also Hoinkes (1964) and Meier (1965).

Among these seven effects, we may distinguish the four last, which operate

Table 15.1. Mean and concentration of glacier aspects, summarized from Table 8·2 in Evans (1974). The proportionate strength of the resultant vector measures the degree of asymmetry. Each glacier is given unit weight. The Okoa Bay aspects are for cirques with glaciers, within a large rectangular area. The error margins for vector means are 95% confidence limits based on assumptions of independence (unlikely) and linear normality (valid except for low vector strengths). All distributions are significantly non-uniform at the 99% confidence level.

| Latitude | Area | Number of glaciers | Vector mean degrees (east from north) | Percentage strength of resultant vector |
|---|---|---|---|---|
| 68°N | Okoa Bay, Arctic Canada | 48 | 011 ± 15 | 68 |
| 51–52°N | Waputik Ranges, Alberta | 108 | 001 ± 17 | 21 |
| 51°N | Bendor Range, B. C. | 105 | 012 ± 6 | 87 |
| 48–49°N | N. Cascades, USA | 740 | 023 ± 5 | 44 |
| 33°S | Rio Aconcagua area, Chile | 53 | 190 ± 22 | 37 |
| 60–70°N | Scandinavia | 2412 | 048 ± 3 | 45 |
| 64–68°N | Urals, USSR | 66 | 079 ± 9 | 81 |
| 43½°N | N. W. Caucasus, USSR | 164 | 009 ± 8 | 66 |
| 42°N | Tashkent area, USSR | 374 | 022 ± 6 | 57 |
| 42°N | E. Tien Shan (parts), USSR | 231 | 005 ± 12 | 26 |
| 43°N | Alma Ata area, USSR | 194 | 009 ± 8 | 65 |
| 49°N | S. W. Altai, USSR | 133 | 011 ± 7 | 80 |
| 66–70°N | Khrebet Orulgan (Far East USSR) | 74 | 024 ± 10 | 78 |
| 53–59°N | Kamchatka | 401 | 333 ± 10 | 19 |

at a local scale, from the regional effects of latitude and maritime climate: the altitude effect is important both locally and regionally. Hence the type of glaciation (type of local glacier distribution) is influenced by aspect and topographic effects, involving altitude, relief, gradient and concavity. In addition to the common morphological classification of glaciation as cirque, valley or ice sheet, we may distinguish symmetrical from azimuthally asymmetrical glaciation. In different regions, asymmetrical glaciation involves varying combinations of poleward, leeward and eastward tendencies.

Table 15.1 shows the direction and degree of asymmetry in present-day glaciation. Each azimuthal distribution includes all glaciers in a particular region or mountain block, and most are unimodal, i.e. there is one aspect of

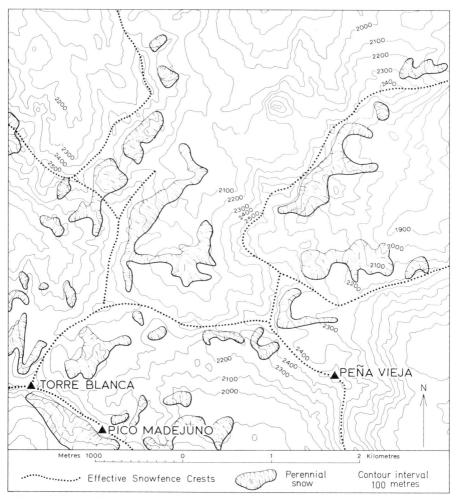

Figure 15.3. Distribution of perennial snow in part of the Picos de Europa, northern Spain.

maximum frequency and frequency declines on both sides of this to a minimum on the opposite side of the compass.

Although data from many regions are lacking, these analyses suggest that strength of asymmetry has little relation to latitude or continentality. Analyses for ranges within these regions suggest that asymmetry is stronger in lightly glaciated mountains, i.e. where the snowline is only just low enough to support glaciers. As glaciation becomes less marginal, glaciers form on all aspects and as they expand, their overall gradients decline and hence the importance of aspect is reduced.

The type of asymmetry is usually slightly east of poleward, reflecting the importance of shade. The eastward tendency, where weak, may be due to morning–afternoon differences, but where strong (as in the Urals and parts of Scandinavia) it involves snow drifting from westerly winds. North–south ridges form 'snowfences' and snow is trapped in sheltered hollows downwind.

The example of the Picos de Europa in northern Spain (Figure 15.3) serves to illustrate the wide variation in perennial snowpatch distribution due to local factors within a single macroclimatic type. Solution, nivation, rock-fall and supranival wash of fines are all active around these snowpatches so that the cirque forms have remained under attack since the end of the last glaciation. Enormous perennial snowpatches are of two main forms in this region. The first are elongated along the contour in the lee of high ridges, while the second are linear patches which run downslope starting on cirque headwalls and extending to the local main valley floor. Examples of both types can be seen west and east of Pico Madejuno respectively.

As snowfall and drifting are dominated by winds from somewhat north of west in this area, the orientation of elongate snowpatches shows a southerly component with ridge alignments between 010° and 125°, but a northerly component with ridge alignments of 130° to 190°. The lower limits of the latter group are some 200–250 m below those of the southerly-facing snowpatches. Hence snowpatch aspect is controlled largely by differential accumulation. For a snowpatch to become a glacier, however, it must survive the differential ablation of summer, after which southerly aspects are much less favoured.

Because ablation is of greater importance during periods of net glacier wastage, differential ablation becomes a stronger control of glacier aspect. Hence snow drifting diminishes in importance and the inhibition of ablation by shade gives the reduced glaciers an increasingly poleward aspect. McGregor (1967) noted the changing aspect of retreat moraines, rock glaciers and protalus ramparts in New Zealand, as did Derbyshire (1973) in Tasmania. Figure 15.4 and 15.5 show a similar sequence in protalus ramparts in Northern Spain: the younger (upper) ramparts have a more northward orientation than the older ramparts and the cirques as a whole. An alternative explanation may be that cloudiness has diminished.

Local complications such as these make it difficult to define 'the snowline' of a region (Hoshiai and Kobayashi, 1957) and this concept is tending to be replaced by two others which may be defined more precisely. The equilibrium

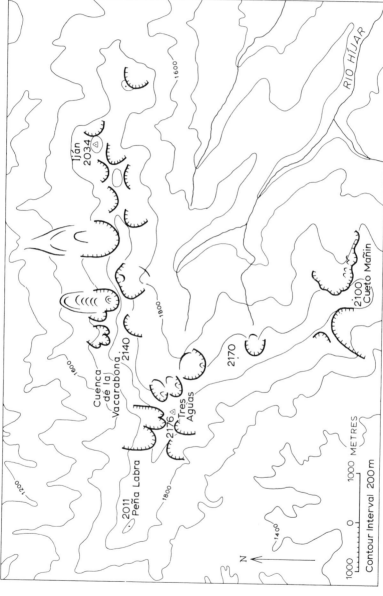

Figure 15.4. Cirque headwalls (pecked) and glacial and nivation moraines (solid arcs) in the Sierra de Peña Labra, Santander province, northern Spain.

Figure 15.5. Sketch of three nivation cirques immediately east of Tres Aguas, Sierra de Peña Labra (Figure 15.4). Note the eccentricity of the nivation moraines with respect to the cirque axes. From a photograph.

line altitude on a glacier separates the accumulation zone from the ablation zone. Because of cold glacier winds and high albedo, the equilibrium line is lower than the snowline on the rock slopes around the glacier. It varies with aspect, gradient and topographic position, but it is usual to quote a single, averaged altitude for each glacier system, especially in mass balance studies.

Secondly, the glaciation limit or level is the altitude which a summit or crest must attain before it can support a glacier on its slopes. Operationally, it is the average of the altitudes of the lowest summit supporting a glacier and the highest lacking a glacier: ideally, these should be located in close proximity. Østrem (1966, 1972) has mapped the glaciation limit for British Columbia and (1964) for Scandinavia. In both cases, it rises rapidly with distance from the coast, indicating that a more continental climate is less favourable to glaciation. This relationship applies generally, and is true also of the snowline.

## 15.3 Climate and cirque aspect

It is generally assumed that the distribution of glacial cirques relates closely to the distribution of former local glaciers. This statement must immediately be accompanied by three provisos. First, the erosion of a cirque may take a long time, and most cirques have been occupied and reoccupied over a succession of glacial periods. Climate varied both within and between these periods, so the distribution of cirques is the composite product of the long-term succession of climates. The time period is tens or hundreds of millenia, whereas (except for ice sheets) that for glacier: climate relations is centuries.

Second, many cirques formed at the sources of valley glaciers, while others were formed by cirque glaciers which later extended beyond the thresholds, to form valley glaciers. It may prove impossible to distinguish these situations, although it can sometimes be established that a cirque glacier never extended beyond its cirque. Accordingly, cirques may represent a glaciologically mixed population.

Third, every erosional form develops through the interplay of force (process) and resistance (structure and form): the latter may prevent development of a cirque, despite the presence of a glacier. Examples in which lithology and structure have prevented cirque formation have not been demonstrated, but cirques must develop more rapidly on well-jointed rocks. It is easier to find examples where slopes were too gentle for cirque development, even though local glaciers were formed, e.g. the head of Measand Beck on High Raise near Hawes Water (Cumbria); see also Grove (1961). The northern part of Garibaldi Park (British Columbia) contains many cirques unoccupied by glaciers, while numerous glaciers occupy sites either too convex or too gentle to be defined as glacial cirques.

Nevertheless, the presence of a well-developed, glaciated cirque may be taken as an indication of the former presence of a glacier, and the distribution of such cirques provides a rough approximation to the distribution of former local glaciers. Ideally, cirque distributions should be related simultaneously to altitude and aspect as in Goldthwait (1970, Figure 8) or, more dubiously,

Table 15.2. Mean and Concentration of cirque aspects, summarized from Table 8·4 in Evans (1974)

| Latitude | Area | Number of cirques | Vector mean degrees (east from north) | Strength of resultant vector, percentage | Source of data |
|---|---|---|---|---|---|
| 51°N | Bridge River, B. C. | 538 | 009 ± 5 | 64 | Evans, 1974 |
| 50°N | N. and N. E. Garibaldi, B. C. | 193 | 006 ± 11 | 59 | Evans, 1974 |
| 47°N | Snoqualmie Pass, Washington, USA | 189 | 033 ± >12 | 33 | S. Porter, personal communication |
| 65°N | Yukon–Tanana upland, Alaska | 1087 | 001 ± > 5 | 25 | Péwé and coworkers, 1967 |
| 65°N | Yukon–Tanana upland (fresh cirques only) | 387 | 017 ± 5 | 68 | Pewe and coworkers, 1967 |
| 68–69°N | Home Bay, Arctic Canada | 79 | 036 ± 15 | 51 | Andrews and coworkers, 1970 |
| 68°N | Okoa Bay, Arctic Canada | 159 | 021 ± 11 | 46 | |
| 55–58°N | Scotland | 347 | 047 ± 7 | 54 | Sissons, 1967 |
| 55–58°N | Scotland | 876 | 044 ± > 6 | 24 | Sale, 1970 |
| 54½°N | Cumbria, England | 104 | 062 ± >16 | 28 | Sale, 1970 |
| 53°N | N. Wales | 118 | 059 ± >15 | 33 | Sale, 1970 |
| 53°N | Snowdonia, N. Wales | 81 | 048 | 57 | Unwin, 1973 |
| 48°N | Vosges and Schwarzwald, Central Europe | 179 | 064 ± 7 | 68 | Zienert, 1967 |
| 50°N | En. Terek Ra., Altai, USSR | 97 | 028 ± 7 | 80 | Ivanovskiy, 1965 |
| 50°N | Katun Range, Altai, USSR | 390 | 027 ± >10 | 33 | Ivanovskiy, 1965 |
| 62°N | Suntar-Khayata, Far East USSR | 233 | 353 ± 8 | 60 | Map in Kornilov (1964) |
| 5–9°S | Papua–New Guinea | 182 | 183 ± >13 | 24 | Maps in Löffler (1972) |
| 42°S | Three regions of Tasmania, Australia | 265 | 109 ± 7 | 64 | Derbyshire, personal communication |
| 32°S | Rio Aconcagua area, Chile | 112 | 168 ± >14 | 37 | Map in Wojciechowski and Wilgat (1972) |
| 52°S | Falkland Is., S. Atlantic | 62 | 077 ± 12 | 68 | Clapperton, 1971 |
| 54°S | South Georgia, S. Atlantic | 163 | 000 (insignificant) | 12 | |

Zienert (1967 Figure 4 and 5)) but we shall take the simpler course of treating aspect first, followed by altitude.

Continuing the approach to glacier aspect taken in the previous section, we find that most frequency distributions of cirque aspect are also unimodal (Table 15.2). Again, these distributions should be for complete mountain ranges or blocks, so that opposing aspects are equally likely and, if glaciation is symmetrical, a vector strength of zero may be expected. There is the additional problem that glacial cirques may be defined differently by different authors (Evans and Cox, 1974), whereas the World Glacier Inventory provides an approach to uniformity for glacier definition.

Cirque aspects from southwest British Columbia, Alaska, Chile and the Soviet Union show strong poleward tendencies, as do most of the distributions of glacier aspect. Interestingly, the same tendency is found quite close to the equator, in Papua–New Guinea, although cirque aspect is highly dispersed as might be expected from the high sun angles and multiple directions of snow bearing winds. On the other hand, asymmetry from differential radiation remains strong around the Arctic Circle, in Canada and Alaska: it is necessary to move further north before radiation on slopes of different aspects approaches uniformity.

A contrasting type of asymmetrical glaciation is found in Tasmania and the Falkland Islands. The very strong and consistent westerlies to which these islands were exposed gave an essentially leeward aspect to the glaciers which eroded their cirques. The cirques of the West Coast Range of Tasmania show quite a wide spread of leeside aspects with respect to the snowfence (Figure 15.17), a characteristic which may reflect their occurrence above broad ice streams and coalescent piedmont ice sheets (Derbyshire, 1972) and so well above the snowline. The Arthur Range, in the southwest of the island, shows an almost symmetrical pattern with a cluster of cirque aspects on both leeward and windward slopes. This derives from severe valley glaciation in a marked maritime regime. The most diffuse pattern is provided by the nivation cirque aspects: this is probably a result of micro-relief rather than mean ridge alignment (cf. Garcia-Sainz (1962) and below).

In Central Europe north of the Alps, the wind effect was almost as great and produced east-northeast facing cirques. Limited evidence from the Alps themselves (Evans, 1974) suggests a weak northward tendency in the generally more symmetrical glaciation; presumably the higher relief increases the effect of shade and reduces the potential for snow drifting by wind. In Britain, the leeward and poleward tendencies are balanced equally. Rather patchy information from Scandinavia gives resultants which are northward in some areas, southeast in others. The Washington Cascades show a leeward component superimposed on a stronger northward tendency.

In general, the wind effect increases with wind strength and consistency, and with decreasing relief. Fairly high concentrations of cirque aspect may occur with either leeward or poleward types of glaciation. Comparisons of neighbouring ranges such as Terek and Katun, or Falklands and South Georgia, suggest that weaker concentrations of aspect relate to heavier glaciation rather than to climatic types. Only in equatorial, and perhaps high polar, climates do we expect an absence of strongly asymmetrical glaciation. Duplicate studies in Britain suggest that operator variance may affect vector strength more than vector mean aspect.

When glaciers and cirques (or, as in Alaska, younger and older cirques) in the same region are compared, the concentration of aspect is almost always less for the (older) cirques, formed during the heavier glaciation. Evans (1974) has formalized this as a 'Law of decreasing asymmetry of glaciation, as the snowline falls'. This emphasizes the association of asymmetry with marginal glaciation, which is most sensitive to local variations in glacier balance. Figure 15.6 gives a further example of this from the Sierra Nevada of California: all aspects are represented among the 85 cirques here, whereas all but one of the 30 residual glaciers face between north-northwest and northeast.

The type and the vertical and horizontal distribution of glacial cirques can sometimes be used to test palaeo-synoptic models. Consideration of the mean aspect and form of discrete cirques in northwest-central Tasmania led Derbyshire (1968) to suggest that they were generated in windier conditions than prevail at the present time and that wind-drifting of snow was of primary importance, the shade factor being secondary. He noted that the dominant snow-bearing winds in present winters come from south and southwest and that the strongest winds, associated with mild temperatures, come from northwest. It was suggested that, given lower mean air temperatures during the

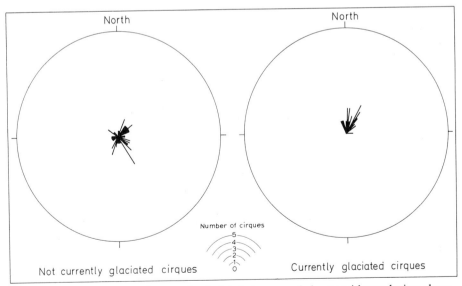

Figure 15.6. Orientation of currently glaciated cirques and cirques without glaciers along the crest of the Sierra Nevada between Mt. Sill and Glacier Divide (Mt. Goddard Quadrangle, California, 1 : 62,500).

Pleistocene, northwesterly winds would have been more important than at present in carrying and re-distributing snowfall. The resultant mean snow bearing wind direction (westerly) is used to build a mean winter circulation pattern for the glacial maxima (Derbyshire, 1971) characterized by an enlarged high over the Australian continent and a more simplified sub-polar depression zone yielding a more vigorous westerly flow (higher zonal index). This accords well with the type, distribution and aspects of the Tasmanian cirques.

## 15.4 Climate and Cirque Altitude

While site and regional factors may combine to produce considerable local variation in cirque climates, detailed data are so sparse that assessment of cirque variation in relation to macroclimate is out of the question at present except in the generalized form of approximations of present and past snowlines (Unwin, 1973).

Most previous work in this field has used cirque floor levels to reconstruct hypothetical former snowlines. Flint (1957, pp. 100 and 309) used the floors of small, independent cirques as an approximation to the 'orographic snowline'. Cirques were acknowledged to be the composite product of several glaciations so that the cirque floor surface indicates a composite snowline. Porter (1964) used only northeast-facing cirques, thus eliminating the effect of aspect on orographic snowline.

More recent attempts at snowline reconstruction involve the fitting of polynomial trend surfaces to cirque floors. This technique is an extension of that used by Linton (1959) and by Flint and Fidalgo (1964), who plotted cirque floor altitudes as projected on to a line of steepest variation with altitude (that is, of greatest dip in the two-dimensional trend). They subjectively estimated gradients of 3·25 m/km for Scotland and 9–15 m/km for the Argentine Andes around 40°S, respectively.

Linear trend surfaces improve the objectivity of this procedure and ensure the derivation of a true (maximum) rather than an apparent dip. Peterson and Robinson used all 379 definable cirque floors in Tasmania to approximate 'the gradient of a composite glacial snowline' (1969, p. 75); the linear trend surface, with a dip of 7·5 m/km to the southwest (236°) accounts for 50·5% of variance in altitude. The quadratic surface, which is inclined more steeply in the southwest (some 13·3 m/km) accounts for 56%, which seems to be significantly more. This is a minimum estimate for the effects of regional climatic factors, which produce a trend more complex than a polynomial of any reasonable number of terms, because each separate mountain group has its own curvilinear effect on snowline. The remaining variation in cirque floor altitudes would be explained largely by lithology, aspect and local situation, as well as by temporal oscillations in snowline.

In three areas of east-central Baffin Land, Andrews and coworkers (1970) found the trends of cirque lip altitudes to be linear. Seaward trend surface dips are roughly 17 m/km for an alpine area (Okoa Bay) and roughly 16 m/km for a lower area (Nudlung), both over distances of some 70 km, while for a larger, relatively low area some 140 km long (Ekalugad) the dip is only 6 m/km. Linear dips may be less for larger areas because of the greater averaging involved. This proposal is supported by the proportions of variance accounted for; 75% and 59% for Okoa Bay (for cirques with ice patches and cirques with glaciers respectively; both have the same trend and dip), 57% for Nudlung and and only 33% for Ekalugad (for all cirques, in both cases). For empty cirques in Okoa Bay the percentage fit falls to 29, but for empty cirques facing between southeast and southwest it is 54. Cirques with glaciers average about 200 m higher than those without, throughout the Okoa Bay area.

Robinson and coworkers (1971) have updated Linton's work by fitting a cubic trend surface (accounting for 62% of variance in altitude) to Scottish Highland cirque floors. In the Northern Highlands this surface rises to the northeast at about 6·5 m/km, but relatively little variation in altitude is shown for the Grampian Highlands.

In Snowdonia (Gwynedd, North Wales), Unwin (1973) similarly fitted polynomial surfaces to the lip (sill) altitudes of all 84 cirques north of the Glaslyn and Lledr valleys. The steep dip of the linear surface (13·3 m/km) may relate to the small extent of the area studied (29 km long, down-dip). This surface accounts for 42% of variance in altitude, while the quadratic accounts for 49% and is probably a significant improvement.

Unwin's map of residuals (1973, p. 91) suggests some autocorrelation, which must distort the significance levels assigned to various trend surfaces. This is only one of several difficulties in the application of this technique (Norcliffe, 1969). However, trend surfaces may be accepted as improvements in the objective description of regional trends.

Superimposed on these trends is a tendency for cirques to be higher in the centres of individual massifs. This tendency is topographic and need not be related to any snowline. Hence the use of *all* cirques, instead of the lowest cirques or those with a certain aspect, produces high estimates for former snowlines. These estimates must be treated with caution. The only upper limit to cirque floor distribution appears to be maximum mountain altitude minus minimum cirque depth. No evidence to the contrary has been found in the Bridge River District, the Alps of Dauphiné, Britain or Tasmania. Cirques are formed at all altitudes above the lowest snowline.

The implication of this distribution is that cirque floors cover a range in altitude from the lowest snowline upward. Unless isolated cirques which never contributed to vally glaciers are carefully selected, cirque floor altitude trend surfaces will tell us no more about former snowlines than they do about regional topography. In the Bridge River District, the floors of cirques at the heads of former valley glaciers, such as those of Tommy Creek and Bobb Creek (Figure 15·7), are generally higher than those of isolated cirques. Presumably they stood higher also in relation to former snowlines.

The distribution of cirque floors may also help explain the difficulties encountered, for example by Unwin (1970) and by Sale (1970), in searching for

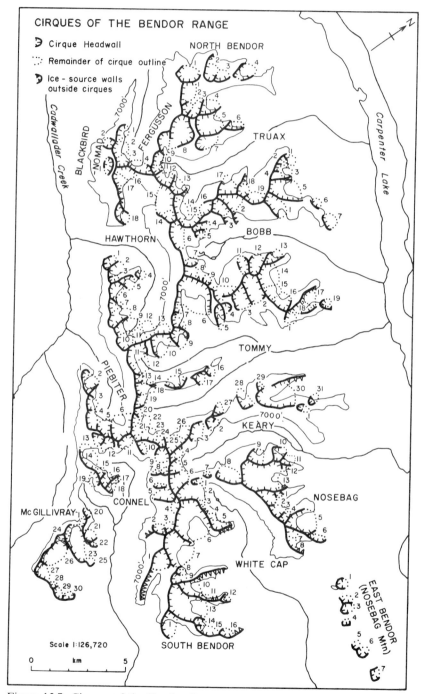

Figure 15.7. Cirques of the Bendor Range, British Columbia, in relation to the 2134 m (7000 feet) contour. From Evans (1974).

regularities in the variation of cirque floor altitude with aspect. Such analyses probably include both isolated cirques facing northwest or southeast, and cirques facing northeast which nourished valley glaciers. By contrast, the altitude of cirque moraines of similar age varies systematically with aspect (Seddon, 1957).

Unless isolated cirques can be reliably distinguished, then cirque floor altitudes are of much less climatic significance than is generally believed. The same considerations apply to cirque summits or crests. No distinctive morphological features of isolated cirques have been discovered; their identification therefore depends upon palaeogeographical reconstructions of glacier distribution.

Because of these limitations to the use of cirque floors, it seems preferable to study the altitudinal relations of cirques in terms of the glaciation limit (see above). The palaeoglaciation limit related to maximum local glaciation is the height above which mountains nourished cirques (and, by implication, glaciers) and below which they did not. Like the glaciation limit, the palaeoglaciation limit can be established precisely only at a few points, where high mountains or ridge crests without cirques are juxtaposed with low mountains with cirques.

There are about a dozen localities in the Bridge River District where both glaciation limit and palaeoglaciation limit have been established. The difference in altitude varies narrowly around 400 m. Although the glaciation limit lies above the snowline, its lowering reflects that of the snowline. Therefore, it may be inferred that the snowline at the time of formation of the lowest cirques in the Bridge River District was 400 m below the present snowline. Along the northeast side of the Bendor Range, there is a series of sites which would have been converted to cirques had the snowline fallen low prior to inundation by ice from the west. On such grounds, it is believed that the palaeoglaciation limit permits more accurate measurement of snowline lowering than other methods.

## 15.5 Climates within cirques

At the broad scale, cirque distribution reflects macroclimate, but at the local scale, the existence of cirques modifies climate drastically. Once cirques form, the shelter and shade which they provide helps preserve cirque glaciers and, in further glaciations, they are the first places where glaciers will form. Yet we have little information about cirque microclimates.

Derbyshire and Blackmore (1974) have reported a short-term experiment in a north-northwesterly facing glacial cirque in County Kerry, southwest Ireland. The cirque, with its floor 500 m above sea level, is markedly concave and the lower part of the steep, 300 m high headwall has a minimum continuous shade period of seven months each year (Figure 15.8). During this period the radiation balance remains close to zero, with negative values in clear sky conditions (with

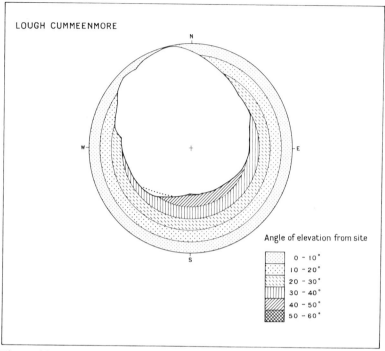

Figure 15.8. Horizon diagram for Cummeenmore cirque, Macgillicuddy's Reeks, Co. Kerry, Ireland. The sun track for 21 April and 21 August is shown by the dotted line. After Derbyshire and Blackmore (1974).

ground temperatures lower than at 2 m) and positive values under cloudy conditions due to back-radiation during daytime.

During the experimental period (late summer), a moderate diurnal temperature range was noted on sunny days but this could not be explained as being due to site heating. As these days displayed a daytime peak in windspeed within the cirque (but not in the free-air record), the explanation of the site heating appears to be that the topographic factors generate a valley wind system. While the free-air flow throughout the experimental period came from between east and south, wind directions within the cirque were from north and west. This topographically-induced up-valley wind was enhanced by large-scale eddies, often channelled by headwall gullies. Thus, while the wind strength of gusts in the cirque frequently approximated the mean wind speed of the free air, a comparison of the overall wind record reveals a 50%–70% reduction in wind velocities within the cirque.

Microclimatic data of this type should be considered in terms of a period of snow and firn accumulation at the foot of a backwall. While this cannot be done directly, conditions approximating this situation may be found at the present time in southern Norway where two sites have been investigated (Derbyshire, unpublished). The first site is a perennial snowpatch within a well developed nivation cirque facing north-northeast and the second, 1000 m away from the first and at the same altitude (1380 m), lies on an exposed ridge adjacent to the snout of Bläisen glacier. Air drains katabatically to the latter from the HardangerJökulen ice-cap, but the nivation cirque is protected from this by a high rock ridge.

While the experimental period was brief (15 August–8 September, 1972) a comparison of the instrumental results from the two sites indicates something of the part played by the nivation cirque and its contained snow and firn. For example, the daily rise in temperature was considerably suppressed, being 2·5 °C lower in the cirque than on the exposed site. Night-time katabatic drift of cool air from the glacier rendered the ridge site 1·5 °C cooler than the sheltered snowpatch. This disparity was destroyed by synoptic changes, however: the deeper air flow provided by a regional wind from the south or southwest minimized these temperature differences.

Owing to the lower temperature of the radiating surface over the snowpatch at night and owing to its higher albedo in the daytime, the radiation balance curve in the nivation hollow showed a lower amplitude than at the open site (see also Eriksson (1958)). An example of an up-valley wind, such as that described above, was recorded in the nivation hollow where it persisted twice as long (6 hours) as in the exposed site. Unique to the nivation site was the rapid fall in temperature and incoming radiation in mid-afternoon each day as the backwall shadow crossed the site. Moreover, the recorded velocities of winds from southerly quarters were somewhat less than half those at the exposed site.

## 15.6 Cirque development

Of fundamental importance to the nature and effects of the glaciers and snowpatches generated in an area is the process of firnification, the diagenesis of freshly fallen snow to produce firn and glacier ice. The process is very slow in high polar environments where only solid precipitation accumulates and no meltwater is produced: firn and ice form by recrystallization under compression. Old firn of this type in an Antarctic cirque glacier was found to have a density of only 0·07 (Dort and coworkers, 1969). Such glaciers have temperatures below pressure-melting point throughout.

In less severe sub-polar and temperate mountains, varying proportions of the annual precipitation may fall in the liquid state. In such conditions percolation of liquid precipitation, with or without meltwater, results in firnification due to regelation as well as to recrystallization. Depending on the temperature regime, the relative importance of these processes varies as does the amount of meltwater lost during the ablation season. Glaciers in these conditions may be wholly or in part at the pressure melting point for part or all of the year.

These climatic considerations, which form the basis of all modern classifications of glaciers (e.g. Ahlmann (1935), Court (1957), Avsyuk (1955) and Shumskii (1964)), are of great significance to processes of cirque development.

They lie at the root of the important generalization that maritime glaciers, with their large throughputs of mass and heat, display high rates of movement and erosion (Ahlmann's (1948) 'dynamic glaciers'), while continental glaciers, with restricted throughputs, are characterized by low movement rates and limited erosion. While the bulk of current knowledge of cirque glacier processes is drawn from the study of glaciers of temperate type, the majority of cirques currently occupied by glaciers occur in polar and sub-polar glaciological environments, a significant proportion being in continental climates characterized by low to very low total precipitation, low annual mean temperatures and low totals of evaporation and solar radiation.

Lewis (1939) and others have explained the characteristic rounding of gully-heads and other depressions in regions with perennial snowpatches in terms of the concentration of the nivation processes around and beneath snowpatch margins. As the margins of a snowpatch actively undercut the slopes, so the form progressively approximates a circular outline and the oversteepened headwall may ultimately extend in an arc of 240°, further enhancing its efficiency as a snowfence. As this process continues the snowpack will tend to grow in volume. With increased densities and thicknesses, the snowpatch becomes capable of debris transport: rock is transported over the ice-crusted surface to form ridges known as protalus ramparts, if they consist predominantly of coarse debris (Bryan, 1934), or nivation moraines, if they include fine debris derived from washing over the snow surface (Russell (1933) and Behre (1933)), or from beneath the margins by solifluction (Lewis, 1939; Botch, 1946). Basal erosion also begins as the thickening snowpack moves as a whole under gravity (Bowman, 1916; Lewis, 1925; Mathews and Mackay, 1963; Costin and, coworkers, 1964; Mackay and Mathews, 1969).

At the extreme northeastern end of the glaciated mountains of southeastern Australia, it has been shown by Costin and coworkers (1973) that in an annual snowpatch in the small cirque on Mount Twynam (Kosciusko massif), which attains thicknesses varying from 6 m to over 30 m and surface gradients on the lower slopes of 10° and over, downslope movement is such as to produce locally very high shear stresses (5·2–11·4 bars). Active planing and scoring of granite bedrock is thus occurring today in a high density, relatively thin snowpatch with low surface gradients.

The steep, often rugged headwall and the smooth concave floor and rock step or threshold, considered characteristic of well developed glacial cirques, led early workers to think in terms of a shift in process from nivation and block removal on the cirque headwall to glacial abrasion on the floor. Direct observation by Johnson (1904) of headwall conditions at the base of a deep bergschrund in a temperate glacier, revealed a shattered rock face with considerable dislodgement of rock which appeared to become incorporated in the glacier below. Johnson ascribed such shatter and incorporation of rock to alternating freeze and thaw, the relative consistency of the base of such action producing the frequently occurring break in slope between cirque headwall and headwall foot zone commonly referred to as the 'schrund line'.

This hypothesis became influential in the work of Gilbert (1904), Penck (1905), Hobbs (1910), de Martonne (1926) and Wright (1937). However, Chamberlin and Chamberlin (1911) questioned the likelihood of oscillation of air temperature deep in a bergschrund, a viewpoint fortified instrumentally some decades later (McCabe, 1939), while the absence of bergschrunds in cirque glaciers in some high altitude tropical mountains and in the Antarctic was pointed out by Bowman (1916) and Priestley (1923) respectively. Haynes (1968) showed that many supposed 'schrund lines' are in fact controlled by rock structure. To overcome the difficulties encountered by the bergschrund hypothesis, Lewis (1938) envisaged meltwater and summer rainwater running down the headwall where it became frozen in rock crevices and so accelerated joint-block removal. However, the experimental results of Battle (Battle and Lewis, 1951; Battle 1960) threw doubt on this explanation except for the wider and shallower parts of bergschrunds. Moreover, the conditions in such sites could well be reproduced, and show greater severity, on the sub-aerial parts of the headwall, especially (by analogy with snowpatch erosion) near the margins of the glacier. Galibert (1960, p. 161) suggested the importance of frost in 'pseudo-rimayes' at the bases of rock walls.

Following a unique field experiment, Fisher (1963) described conditions in firn and ice well above the equilibrium line at 4000 m on the Breithorn in the Pennine Alps. Here, some 30 m of cold ice overlies isothermal ice, the product of frictional heat of the moving glacier, ice thickness and geothermal heat. Fisher suggested that this freeze–thaw boundary is a potentially greater source of rock shatter than the shallow bergschrunds found in temperate glaciers, although the transition from cold to isothermal ice is typically the site of a bergschrund. Thus much of a headwall will be affected by very low frequency freeze–thaw as the cold–isothermal ice boundary oscillates in response to ice thickness and average air temperature. This suggests development of cirques within a limited band of altitudes.

This infrequent but profound freezing can operate on joints produced or opened by pressure release (Lewis, 1954) which may be important in the perpetuation of the arcuate cirque headwall by a form of parallel retreat (Battey, 1960). The action of pressure release also aids entrainment of joint-blocks beneath thick isothermal ice on cirque floors.

Glacier movement occurs as internal plastic deformation and as sliding of the whole mass over the glacier bed. The relative proportion of these two modes of flow in any glacier is the product of the prevailing climate. In polar regions, for example, glaciers may be below the pressure melting point throughout so that they are frozen to their beds and move entirely by plastic deformation and shearing. In milder climates, or in very thick polar glaciers with ice at the pressure melting point at depth, a proportion of flow will be by basal sliding. This may account for as much as 90% of the total flow in some steep temperate glaciers (McCall, 1960). Basal slip is important as it provides the conditions necessary for the smoothing and striation of the rock bed. It is also fundamental to the hypothesis that rotational slip is a major component of cirque glacier movement (Gibson and Dyson, 1939; Garcia-Sainz 1941; Lewis 1949). Given greatest deepening beneath the thickest ice, this hypothesis has been used in explanation of the rock basin (Clark and Lewis, 1951), another common though not universal characteristic of glacial cirques.

In developing a general model for the shear stress ($\tau$) beneath a rigidly rotating cirque glacier, Weertman (1971) has shown that $\tau$ varies with the ice thickness ($h$) and the sine of the surface slope of the ice ($\alpha$). According to the equation

$$\tau = F\rho g h \sin \alpha$$

(where $g$ is the acceleration due to gravity and $F$ is a conversion factor), variation in the ice density ($\rho$) will be of minor influence and the highest basal shear stresses will be found in the thicker, steeper glaciers. In the case of discrete cirques the basal shear stress due to rotation will decline in glaciers with low surface gradients and in steep, thin glaciers, the rotational movement being replaced at very high gradients by mass sliding of thin ice and firn.

The relative importance of headwall retreat and floor deepening in glacial cirques has been a recurrent theme for over seventy years and one might expect variations in relation to glaciology and climate. De Martonne (1910) proposed that cirque development was principally by wall retreat and that although floor deepening by glacial erosion did take place, it was distinctly subordinate. This view was further developed by Hobbs (1910, 1921), whose cycle of glaciation was based chiefly on cirque wall retreat consuming the supposedly preglacial upland surface. Lewis (1938) and Cotton (1942) provided examples of cirque development through headwall retreat. On the other hand, Battle's observations (1960; Thompson and Bonnlander, 1957) suggested that headwall retreat by bergschrund sapping was less active than Johnson had maintained.

The degree of asymmetry present, for example, in the Bridge River District of British Columbia (Evans, 1969, 1974), can be understood only in terms of a very considerable retreat of cirque walls. Deepening alone could not have produced the northward bias in valley direction (Figure 15.7): a shift in drainage divides seems likely also because of their present position southwest of the median line and highest peaks of each range. This is also true of cirques cut into dolerite and horizontal sedimentary rocks in central Tasmania where, in one or two cases, discrete cirques have cut back through a ridge to form a col (Derbyshire, 1963), while others were about to break through a ridge when glaciation last ceased. Mont Olympus (Figure 15.9) provides an example of each of these types.

White (1970) suggested that retreat had been overemphasized and that cirque deepening is more important: in the Uinta Mountains, few cirques cut right through divides. This may be because the pre-glacial Uinta divides were very broad (but see p. 486).

Although we cannot isolate the influence of pre-glacial relief, intersection of cirques appears more common in regions of heavy, symmetrical glaciation. Asymmetry itself discourages cirque intersection in the Bridge River District, because cirques rarely attack a ridge from diametrically opposed directions.

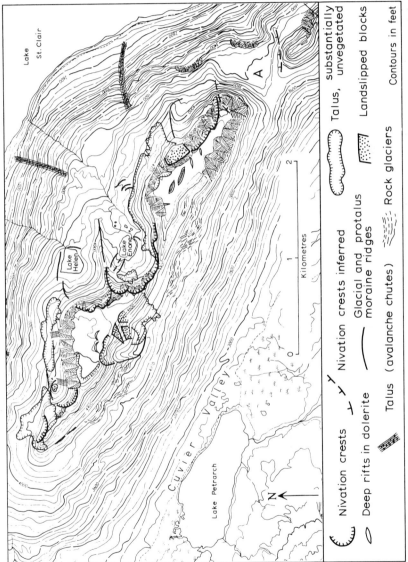

Figure 15.9. Glacial and cryonival forms of the Mount Olympus ridge, west-central Tasmania. The col at A is considered the product of cirque breaching. Modified after Derbyshire (1973).

Figure 15.10. The central Cadwallader Range around 122°43′W, 50°40′N, British Columbia. In the south, cirques have coalesced and may have destroyed a former headwall at*. Scale 1/50,000, contour interval 30·48 m (100 feet). Reproduced by permission of Department of Energy, Mines and Resources, Ottawa.

Nevertheless, divides between Keary Creek, East Tommy Creek and West Tommy Creek (Figure 15.7), are cut by deep cols which reflect cirque wall retreat. Cirque development between Connel Creek and McGillivray Creek seems to have destroyed several ridges, leaving poorly defined cirques. In the more heavily glaciated Cadwallader Range, there are many more examples where cirque walls seem to have been removed, replacing arêtes with smoother divides (Figure 15.10).

However, if it is accepted that cirque deepening has kept pace with headwall retreat then it becomes much easier to account for the form of divides, especially the unbroken ones discussed by White. This would prevent destruction of sharp-crested divides.

White also suggested that avalanche gullies are a frequent feature of cirque headwalls above a 'schrund line'. They occur here in the Dogtooth Range of the Canadian Rockies and in Sweden (photographs in Østrem and coworkers, (1973)). However, in both the Bridge River District and in Dauphiné (France) quite the reverse is true: cirque headwalls are irregularly fractured rockfall faces, and avalanche gullies are well developed only on slopes which were not major ice sources. Evidently, cirque headwalls have been undercut and probably eroded more rapidly than other steep slopes. This undercutting may relate more to sub-glacial abrasion and plucking, especially with rotational flow, than to bergschrund freeze–thaw.

There is little to be said in favour of the 'cycle' of mountain glaciation of Hobbs (1911). 'Fretted' uplands of coalescing cirques are found especially in alpine areas of high relief such as the British Columbia Coast Mountains and most of the Alps. Scalloped, 'grooved uplands' with 'biscuit-board topography' are found in less dissected areas, often in parts of less unstable mountain belts such as New England, Labrador and Scandinavia. The main control of this contrast seems to be tectonic environment and degree of pre-glacial dissection. Climatic influences on pre-glacial relief may also have been important, as in the contrast between east and west Scotland (Linton, 1959).

The main objection to the 'cycle' is the improbability that areas where cirques are now coalescent ever resembled areas where cirques are cut into summit plateaux. Though pre-glacial divides must have been broader than their narrow modern successors, there is no evidence for the former existence of extensive plateaux; indeed, the general fineness of dissection in these areas makes it highly improbable that such plateaux existed when the present cirques began to form. Worse still, Hobbs used the 'puny glaciers of arid regions' and 'high latitude glaciers of excessive precipitation' as end members of his supposed time sequence rather than as end members of a climatic sequence.

In dissected mountain areas, light glaciation (that is marginal glaciation with the snowline little below the summits), produces not a 'grooved upland' but an area of high asymmetric slopes with cirques facing near to a single direction. Resultant vector strength, which measures the degree of asymmetry of the cirque aspect distribution (see above), can be used as a measure of how marginal glaciation was during those phases which were morphogenetically most significant. However, vector strength is not proposed as an index of 'youth, maturity and old age'. Highly marginal conditions might persist for a long time and would produce results different from those of a brief but more extensive glaciation. In practice, ridge altitude is the important control; areas are less asymmetrically glaciated as their ridges lie higher above the snowline or glaciation limit. Coalescence of cirques is more common in higher ranges (Blache, 1952, Figure 34).

Fretted uplands may be glaciated either symmetrically or asymmetrically. Grooved upland plateaux are unlikely to be glaciated symmetrically, since a snowline sufficiently low is likely to permit development of ice-caps. Until the ice-caps form, wind drifting across the plateaux may strengthen the effect of shade in producing asymmetrical glaciation. A relatively much lower snowline is required before ice-caps overwhelm the relief of fretted uplands (Davis and Mathews, 1944), which may therefore suffer relatively symmetrical glaciation with glaciers on all sides.

Hence Hobbs's cycle of mountain glaciation is unrealistic in the way it substitutes space for time and in its overemphasis on cirque wall retreat. A third deficiency is the suggestion that as cirques converge, those parts which are closest to another cirque receive less snow because the upland surface which provided the snow has been reduced in extent. This gives a cirque 'irregularities of outline' dependent primarily upon the initial positions and individual nourishments of its near-lying neighbours (Hobbs, 1911, p. 31).

There are two objections to this suggestion. First, it probably overemphasizes the role of wind drifting in snow accumulation in areas of such dissection. Second, even if accumulation were to be affected adversely, it is unlikely that the central part of the glacier, in the maximum concavity, would disappear. A search has failed to establish a single example of two glaciers separated by an unoccupied zone, with more upland surface above the glaciers than above the unoccupied zone. Buttresses which might have been generated by partition of a cirque in this way are equally elusive.

Hobbs presumably reasoned that large cirques are older than small, and since they are more complicated, cirques become more complex with age. This is perhaps akin to using two simple correlations, when a partial correlation is required. Again, the substitution of space for time is a dubious practice and many would argue that complex cirques have often developed from the coalescence of several small cirques, rather than the growth of a single cirque (Linton, 1963, p. 20). The process of cirque development seems to involve simplification of cirque contours, with projections being attacked more rapidly, until smooth cirque walls are produced. Similarly, glacial erosion of the cirque floor reduces irregularities due to pre-glacial erosion; the only small-scale irregularities to be enlarged are those due to structure.

Grove (1954) discussed situations in which glaciers occupying parts of cirques, which had been cut by larger glaciers, may emphasize and enlarge irregularities in the cirque walls. 'Minor weaknesses in the rock of a headwall may cause ravine formation, leading gradually to irregularities in the plan form of the névé. Once the disposal of the névé is sufficiently complicated to affect the flow pattern in the ice, diversification of the flow pattern must emphasise the original inequalities in the surface of the rock. So a spiral of activity is begun, which may eventually produce a compound basin such as that east of Leirhöe' (Grove, 1954, pp. 226–227).

It can be seen that the circumstances in which cirque complexity may be increased are quite limited. They do not form a final phase of a period of glaciation with a constant snowline, as Hobbs implied. Rather, they relate to a phase of more restricted glaciation, with a snowline higher than that when most cirques in the area were formed. In the Bridge River District there are a number of situations where two glaciers may be attacking a cirque wall individually, but there are no examples where the process of cirque subdivision has progressed far.

It is always difficult to decide to what extent irregularities are inherited, having never been eliminated by the former larger glaciers, and to what extent they have developed recently. At the head of Peridotite Creek (Bridge River District), where glaciers cover convexities on the cirque headwall and a deep concavity contains only a snowpatch, it is clear that headward erosion around present-day glaciers has been minimal. In any case, Hobbs's cycle of mountain glaciation is more misleading than instructive. It must be replaced by a more empirical understanding of spatial variation in landform parameters, based on multivariate studies. These require agreement on procedures for cirque definition and delimitation and on the way morphometric variables are to be measured (Evans, 1974). In the next section, it will be shown that such studies are in their infancy.

## 15.7 Climate and cirque form

The degree to which longitudinal profiles of cirques approximate the arc of a circle has been used in support of the hypothesis of rotational sliding (Lewis, 1960), although it might be objected that cause and effect may have been reversed in this case. Objections to the simple morphological model provided by the semicircle have been raised by Haynes (1968) who found that a series of exponential curves with one inflection ('$k$ curves') provides a better description of the long-profiles of many Scottish cirques.

Profiles of cirques in the middle latitudes of both hemispheres, taken from areas of tabular-unfolded, severely folded and massive rocks (Figure 15.11), suggest that a double concavity is common, even in quite small glacial cirques. While local structural features are always likely to be a factor in the generation of multi-concave glacial forms (Fuller, 1928), the persistence of a double concavity in many structural situations and over a wide range of cirque sizes suggests that it represents a change in process type from nivation at the upper concavity to truly glacial action at the lower.

A similar result might be expected at the contact of cold and isothermal ice

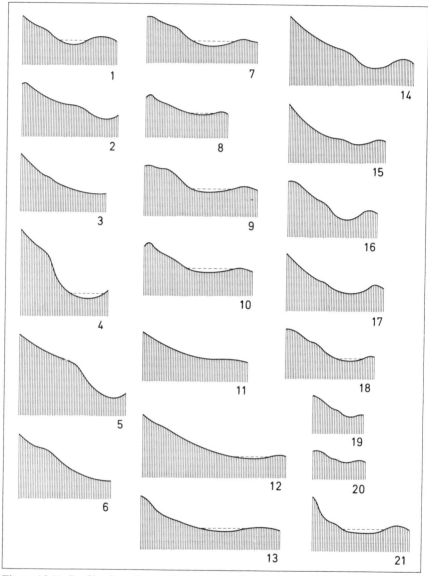

Figure 15.11. Profile along the median axes of selected cirques from Tasmania, Spain and Britain. Key to numbers: 1–6, Mount Murchison (West Coast Range); 7–10, Arthur Range (southwest Tasmania); 11–13, Mount Owen (West Coast Range); 14–17, Hartz Mountains (southern Tasmania); 18–20, Sierra de Urbión (Loñgrono province, north-central Spain); 21, Blea Tarn (English Lake District). Vertical and horizontal scales are equal.

discussed above. As ice thins in polar regions, so a glacier structure produced by rotational slip within the cirque basin may become relict and take on the character of an uncomfortable surface as low density firn overlies it and the whole glacier becomes frozen to its bed (Dort and coworkers, 1969), so bringing the corrasion process to a halt while active nivation continues about the margins. Such glaciers tend to produce shallow cirque basins (high length:depth ratios) as suspected by Priestley (1923). While rotational movement appears to be common in snowpatches, especially those found within glacial and nivation cirques, such snowbanks, while capable of abrasion, appear incapable of overdeepening. Thus the concavities in nivation cirques are markedly less developed than those of glacial cirques.

While it was long ago inferred (Griffith Taylor, 1922; Wright, 1914) that nivation cirques may constitute the first stage in the development of glacial cirques, the relationship still awaits proof. A nivation cirque is a possible but not a necessary stage in the development of a glacial cirque (Davies, 1969; Evans, 1969; Ives, 1973). Transformation of a fluvial or avalanche-gully produces a half-funnel occupied by a niche glacier (Groom, 1959). It should not be inferred that present-day nivation cirques will develop into glacial cirques: many lack adequate relief, or are situated where they will be overwhelmed by ice from upstream (Evans, 1974).

The distinguishing cartographic and photogrammetric characteristic of cirques is the sharp break of slope around the headwall, referred to here as the nivation crest. This characteristic, detected on maps by the sharp inflections in the contour patterns from convex outside to concave or straight within the cirque, is paramount as some cirques may have no exposed floor or footslope and only a poorly developed threshold, i.e. they may be funnel-shaped valley heads.

Cirques have often been described in terms of linear characteristics such as their length, height, breadth and depth (Manley, 1959; Demangeot, 1965; Andrews and Dugdale, 1971). However, there are some real problems of

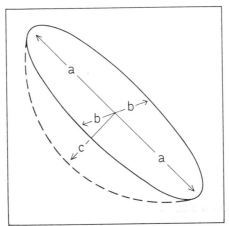

Figure 15.12. Diagram of an idealized cirque form, showing the orthogonals measured for the calculation of the index of flatness $\left(\dfrac{a}{2c}\right)$ and the gradient (a–b plane).

definition, such as determination of cirque 'height'. Moreover, while cirque relative relief and mean altitude may be important in direct measures of some climatic factors (notably the snowline and the snowfence), the aptness of the horizontal plane as the datum for cirque form description can be questioned. As a purely morphometric description is required to form a basis of comparison of numbers of cirques, a series of ratios would appear more appropriate than any absolute measures, as they eliminate the scale factor.

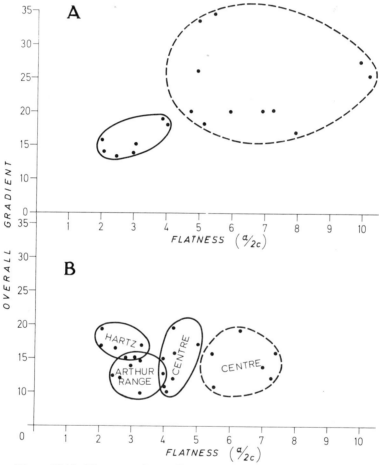

Figure 15.13. Flatness and overall gradient values for some Tasmanian glacial cirques (solid line) and nivation cirques. A—Glacial and nival cirques on Mount Murchison and Mount Owen, western Tasmania. Nival cirques include those developed by firn at heads of valley glaciers. B—Glacial and nivation cirques in the northwest-centre of Tasmania (Mount Olympus, King William Range, Mount Ronald Cross, Mount Field, Raglan Range) and some glacial cirques of the Arthur Range and Hartz Mountains in southwest Tasmania. Note the generally low gradients (cf. p. 485).

Sugden (1969) has already used arcs of circles to show that some Cairngorm glacial cirques exhibit a double concavity in plan, and it is suggested that shape analysis might be taken a step further by referring measurements to a plane fitted to the cirque's nivation crest (Figure 15.12). This surface provides a measure of overall gradient and a datum for the measurement of concavity. Using a ratio between length of the a-axis (median line along the maximum gradient of the a–b surface) and maximum divergence of the cirque floor from the a–b surface (essentially the c-axis of the figure measured normal to the a–b plane: Figure 15.12), concavity may be expressed as an index of flatness in the form $a/2c$. Using this simple formula, well-developed glacial concavities have low numbers while poorly developed concavities, including nivation cirques and certain glacial cirques, have higher numbers. When plotted in combination with overall gradient (gradient of the a–b plane), nivation cirques and glacial cirques emerge as distinct populations (Figure 15.13A). It will be seen from Figure 15.13 that most of the discrete glacial cirques sampled have indices of 4·2 or less.

The relative importance of factors such as components of the heat balance and the nature of the wind regime may have implications for both the form and aspect of cirques. This possibility has been examined for southeastern Australia, a region marked by steep climatic gradients both now and in the Pleistocene (Derbyshire, 1968, 1972). Peterson (1968) has shown that the cirques of the west and south of Tasmania were occupied by vigorously active ice which produced many rock basins and gave rise to overflow of ice beyond the cirque thresholds. In contrast, the cirques of the northeast and east of the island (east of the present 60 inch (152 cm) isohyet) show only minimal glacial modification, less well developed basins and a protalus component in some of their moraines (Figure 15.14). This morphological contrast is interpreted in terms of a Pleistocene glaciological gradient from extreme maritime glacial conditions in the southwest to more continental conditions in the north and east-centre with much lower snowfall totals, lower mean cloudiness, firnification more by compaction than recrystallization and a generally low energy throughput (Derbyshire, 1973). A corollary of this is that western Tasmania should have been glaciated earlier and its glaciers should have persisted longer than those in the east in any single glacial episode. There is some radiocarbon dating to support this contention (Peterson, 1968).

When the flatness index of cirques (see above) is plotted for selected mountain areas from southwest to northeast across Tasmania, the cirques of the north and east display values of 4·0 or more, while those of the south and west generally give values between 2·0 and 4·2 (Figure 15·15). This distinction is not the result of either lithology or orientation of the mountains selected, as three of the mountain areas (Hartz, Mt. Field and the mountains of the northwest-centre) are in horizontally-disposed dolerite, usually above horizontal sediments, while the glaciological snowfence is disposed north–south in two of the ranges (King William Range and the Hartz Mountains), northeast–southwest in three (Mt. Murchison, Mt. Field East and Mt. Owen) and northwest–southeast in the remainder (Arthur Range, Mt. Olympus, Mt. Field West and

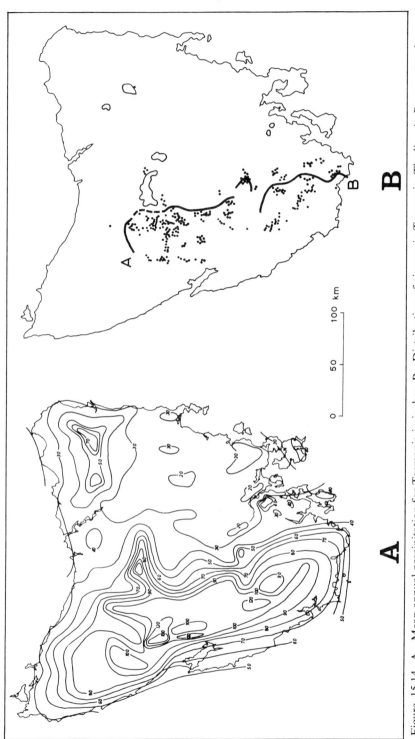

Figure 15.14. A—Mean annual precipitation for Tasmania in inches. B—Distribution of cirques in Tasmania. The line A–B separates cirques of the west and south of the island from those of the east. Glaciers of the overdeepened cirques to the east did not, in general, extend much beyond their thresholds. Redrawn from Peterson (1968).

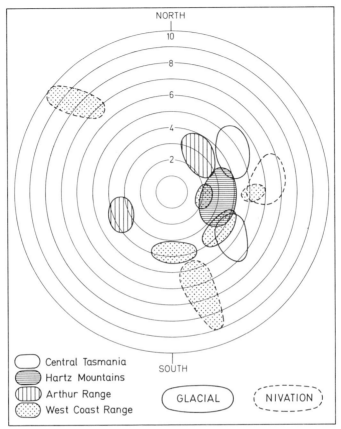

Figure 15.15. Orientation and flatness index of glacial and nivation cirques from selected Tasmanian massifs.

the Raglan Range). The similarity between the flatness index for cirques developed in folded Pre-Cambrian metamorphic rocks of the Arthur Range and that for cirques entirely in dolerite in the Hartz Mountains, is striking, especially in view of the contrast in the degree of glaciation of the two ranges.

The increase in flatness values of the Tasmanian cirques from southwest to northeast, is clearly related to Pleistocene climatic (glaciological) gradients. The decrease in the proportion of overdeepened cirque basins from southwest to northeast is a reflection of the tendency towards thinner ice and lower basal shear stresses due to rotation predicted by Weertman's (1971) model (see above). Figure 15.16 shows the aspect of selected groups of Tasmanian cirques with respect to mean snowfence orientation. Of those of glacial origin, the shallow cirques of continental type occurring in central Tasmania show the simplest relationship to the snowfence, while all other show varying degrees of divergence.

The distinction between maritime-type cirque glaciers with their large annual

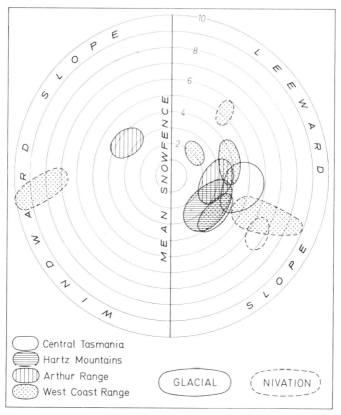

Figure 15.16. Cirque data shown in Figure 15.15 rotated to show relationship of cirque orientation and flatness to mean snowfence crest. Windward accumulation of cirque glacier ice occurred only in the extremely maritime Arthur Range.

balances and subtropical cirque glaciers has been recognized from a study of the glaciated mountains of Iberia. Garcia-Sainz (1949) drew this distinction between the Pyrenees and the glaciated sierras to the south (Sierra del Moncaya, Sierra de Gredos, Sierra Nevada; cf. Obermaier and Carandell (1916)). He explained the distribution of the former ice masses in terms of the interplay between ridge alignment and former mean depression tracks, deducing a trajectory of 76° west of north for the western Pyrenees and 55° west of north for the Sierra Nevada. Elements typical of the subtropical glaciers are relatively shallow, elongate basins bounded by superposed, semicircular moraines (*superposée en terrasse*) and glacier ice occurring at altitudes much lower than the climatic snowline owing to severe drifting across suitably disposed mountain crests. This contrasts with the glaciers of the western Pyrenees which were typically equidimensional with deeper, more symmetrical basins.

In extremely marginal climatic situations such as the Montes Universales (Idubeda), which were characterized by perennial snowpatches in the Pleisto-

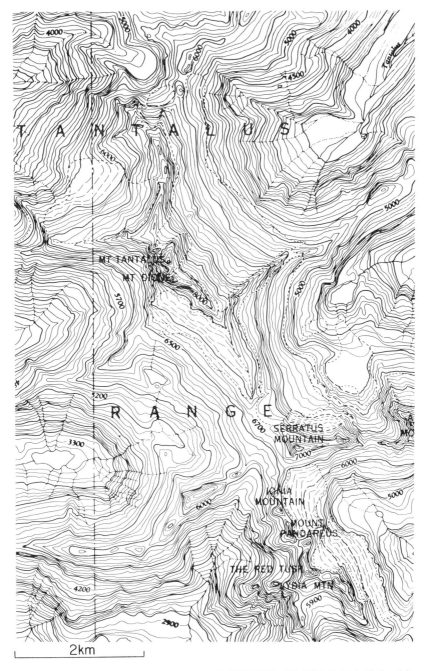

Figure 15.17. The Tantalus Range around 123°19′W, 40°49′N, British Columbia. Note the long 'apron' slopes, gently concave in plan, separating headwalls from trough-head basins.

cene rather than cirque glaciers, the relief factor completely outweighed the wind-direction factor (Garcia-Sainz, 1962).

The shallowing of cirque basins along climatic gradients, towards the east-southeast, has been noted in the Picos de Europa (Garcia-Sainz, 1949) where all occur in a single rock type (limestone).

In British Columbia, Evans (1974, p. 116) has contrasted the conventional though coalescent cirques on the lee side of the Coast Mountains with the complex glaciated valley-heads of the coastal side, e.g. the Tantalus Range (Figure 15.17). The latter have low but long unbroken headwalls, often only 50–100 m high, rising above long, ice-scoured 'aprons' which slope at $26°-31°$ over vertical distances of up to 700 m. These 'aprons' are gently concave in plan and often focus upon a trough-head 150–450 m high around a rock-basin lake about 1 km long, elongated downstream. Sometimes the 'apron' is interrupted by one or two intermediate floors, for example south of Mt. Tantalus.

The strong mid-profile convexity makes the term 'cirque' inappropriate for these features, often 2 km and sometime 4 km across; rather, they are cirque/trough-head combinations, related to a heavier and more active glaciation than that on the lee side of the Coast Mountains. Although there are differences in position relative to the Cordilleran Ice Sheet, differences in glacier balance and basal sliding provide the more likely explanations of this contrast.

A similar contrast exists between the cirque/trough-head combinations of Moskenesöy, the westernmost Lofoten Island (Ahlmann, 1919) and the cirques of Swedish massifs such as Sarek. The combination of strong glacial erosion in the centre, producing the rock basin, with rapid headwall retreat producing the 'apron', may be an extreme development of the 'double spheroid' cirque form discussed above.

Statements on cirque dimensions and shapes have hitherto been based on either small or subjectively selected samples. Drawing upon unpublished data from J. Peterson (Tasmania) and I. S. Evans (British Columbia), Figure 15.18 shows that in each region, cirque areas follow a logarithmic-normal frequency distribution. Most cirques have areas near the minimum, but there is an upper 'tail' of much larger cirques. On a logarithmic scale of area, then, the number of cirques declines symmetrically on either side of the median.

The two structural provinces in Tasmania are indistinguishable and have a median cirque area of $0.46$ km$^2$ compared with $0.4$ for the Bendor Range. In each case 5% of cirques have areas below $0.12$ km$^2$. In Tasmania, another 5% exceed $2.1$ km$^2$, compared with $1.6$ km$^2$ in the Bendor Range, and the largest cirques are around 4 km$^2$. The smoothness of the frequency distribution implies that, despite the great range in size, we are dealing with a single type of landform, for which steady growth in area is possible.

A more variable distribution of cirque areas comes from unpublished data by J. Gordon for the Kintail–Affric–Cannich area of Scotland. These cirques have a median area of $0.3$ km$^2$. A small break in slope at 1 km$^2$ on the cumulated graph probably reflects inclusion of 'cirque complexes' and 'cirque troughs'.

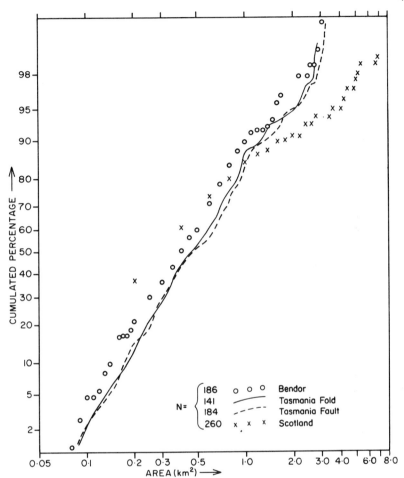

Figure 15.18. Cumulated frequency distributions of cirque area in the Bendor Range, B. C. (data from I.S. Evans), in Tasmania (data from J. Peterson, divided into two structural provinces), and in the Kintail–Affric–Cannich area of Scotland (data from J. Gordon). This plot has a logarithmic horizontal scale and a probability vertical scale, so that a logarithmic-normal distribution plots as a straight line (Bliss, 1969). The middle parts of the probability scale are based on larger numbers of occurrences, and the appearance of the lower tail (small areas) is sensitive to the resolution and accuracy of area measurement. The Tasmania and Bendor plots are similar and logarithmic-normal. The flatter plot for Scotland contains more small cirques and more large cirques than the others.

This supports the suggestion that these latter form a population of forms distinct from cirques.

Frequency distributions of cirque length, breadth and amplitude are almost normal, with slight positive skews. Gordon's Scottish cirques (etc.) have a median length of 590 m compared with 740 m for Bendor cirques and 1053 m for 165 'well developed' cirques in the Okoa Bay area of Baffin Island (Andrews and Dugdale, 1971). Lengths range from 280 m to 2250 m in the Bendor Range, slightly more widely in Okoa Bay, and much more so in Scotland.

Sugden (1969) gave only cirque breadth, which varies between 309 and 1120 m for 31 cirques in the Cairngorm Mountains (on the more continental side of Scotland). This compares with a range of 130 m to 3180 m (median 560 m) for Gordon's 255 cirques, cirque complexes and cirque troughs. Since cirques are fairly compact forms, their breadths are comparable to their lengths. Length/breadth ratios cluster tightly around one, with medians of 1·08 (Gordon) or 1·3 (Okoa Bay; range 0·26 to 5·9). This lack of variability means that, in both studies, length/breadth ratios have low correlations with other morphometric variables.

Rather more information is available for variations in cirque amplitude, the range in altitude between highest and lowest points in a cirque. Median values (ranges in parentheses) are Bendor 421 m (152–1006), Kintail–Affric Cannich 274 m (92–915), Okoa Bay 257 m (40–600), Tasmania fold 262 m (60–640), Tasmania fault 226 m (50–670), Crystalline Vosges (France) 250 m (65–555), Sandstone Vosges 180 m (137–234), Crystalline Schwarzwald 215 m (111–351), and Sandstone Schwarzwald 135 m (80–245). Data for the four latter areas, containing 58, 8, 22 and 91 cirques respectively, come from Zienert (1967), who excluded marginal forms and cirques with very poor thresholds. In New Hampshire, Goldthwait's (1970) headwall height data give a median of 224 m (90–366) for 14 cirques, which may be almost doubled for comparison with the amplitude data.

In its effect on glacial erosion, the most important cirque morphometric parameter is gradient, which is usually measured along a centre line. For the Tasmanian cirques, measures of length were not available, but area and amplitude were. Hence a generalized gradient in degrees was calculated as arctan (amplitude/area$^{0.5}$). The results in Figure 15.19 show that most Bendor Range cirques are steeper than the median Tasmanian cirques, which slope at 20·2° in the fold province, 16·5° in the fault. The Bendor Range median is 33·5° and only one cirque is steeper than 46 . The difference is due essentially to the greater amplitudes of Bendor cirques. If not due simply to different approaches to cirque delimitation, this difference probably relates to the greater relief of the Bendor Range which is some 1600 m within circles, 10 km in diameter, centred on major summits, compared with 1100 m in Tasmania.

Because of cirque compactness, these generalized gradients are well correlated with, but some 12% greater than, conventional gradients calculated as arctan (amplitude/length). Median results for the latter are 25° in the Kintail–Affric–Cannich area (range 8°–50°), and 13° in Okoa Bay (range 2°–58°). Hence (despite low regional relief) the Scottish cirques compare with those of the Bendor Range, but Okoa Bay cirques are more like those of Tasmania.

Floor gradients for the British Columbian, Scottish and New Hampshire cirques all have medians around 9·5°. Gordon includes some with gradients

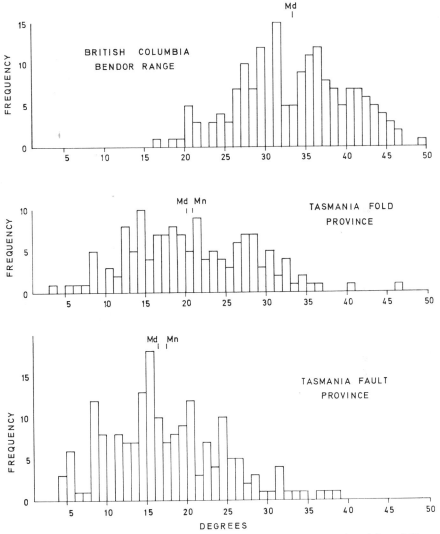

Figure 15.19. Frequency distributions of generalized gradient (length/(area$^{0.5}$)) of cirques in the Bendor Range (data from I. S. Evans) and in Tasmania (data from J. Peterson). Although cirques in the fold province of Tasmania are only slightly steeper than those in the fault province, those in the Bendor Range (B. C.) are much steeper and have a more symmetrical distribution.

up to 35°, but floor gradients above 20° are rare. Median cirque headwall gradients are 66° (range 44°–78°) in the Bendor Range, and 45° (range 18°–73°) for the Scottish examples. Finally, profile closures (differences between maximum headwall and minimum floor gradients) have medians of 56° (range 24°–78°) for the 66 northern Bendor cirques, and 33° (8°–70°) in Kintail–Affric–Cannich; plan closures (range in headwall azimuth of longest unbroken

Figure 15.20. North face of Mount Truax (2880 m), the 590 m high headwall of a large cirque containing a diminished glacier. The southwest slopes of this mountain are gentle and debris-covered. The minor summit in the middle of the photograph, viewed from across two glacial troughs, is also asymmetric. These forms are cut in metasedimentary rocks of the Bendor Range, at 50° 50′N in British Columbia (2 August, 1966).

contour, measured over 100 m increments by Evans; Gordon used the contour at the middle of the range in altitude) are 132° (25°–285°) and 108° (26°–247°) respectively. This suggests that the Bendor cirques are better developed both in plan and profile; the Scottish (and also some Cumbrian) cirques are more like the 43 cirques of the Tyax–Gun block, a lower neighbour of the Bendor Range.

If we make a dangerous equation of time with space, viewing larger cirques as more developed forms, it seems unlikely that cirques reach an equilibrium form as suggested by Linton (1963, p. 18). Cirque development involves positive feedback from form to process, whereas equilibrium is attained more readily through negative feedback. Gordon's morphometric correlations suggest that as cirque size increases, length increases slightly faster than breadth and amplitude. Demangeot (1965) also found, in the Apennines, that cirque breadth (ranging 100–800 m) increases faster than cirque amplitude (100–300 m). This refutes White's (1970) suggestion that cirques develop downward rather than headward. Concavity increases, especially in plan, and overall gradient

Figure 15.21. A cirque on the northeast slope of Big Dog Mountain (c. 2880 m) at 51°04′N in the Shulaps Range, British Columbia. The serpentinized peridotite of this range is easily broken up by freeze–thaw, providing abundant talus which a small glacier has taken further to form a fat, almost complete morainic arc. The shallow cirque in which it lies is thus being deepened, although it forms part of a larger cirque whose floor is seen on the left. The view is eastward, towards Yalakom Mountain, on 18 July, 1966.

decreases as cirque size increases. The decline in floor gradient is more marked than the increase in headwall gradient.

In conclusion, glacial cirque length and amplitude have approximately a tenfold range (less in small areas). We should be wary of classifying as 'glacial' cirques less than 200 m long (and wide), or with amplitudes below 100 m. More commonly, these figures are around 700 m and 250 m, cirque area is around 0·4 km$^2$, and generalized gradient is about 20°–30°. Gradients above 45° or much below 12° should be viewed with some suspicion because they make rotational flow unlikely in any glacier that fills the cirque.

Variations between regions are small in relation to variability within each region, and may be explained by operator variance, topography (especially vertical relief) and geology, rather than by climate. Attempts to relate cirque morphometry to aspect (and hence microclimate) have generally been unsuccessful. Regional trends in morphometric parameters are usually weak, perhaps because most study areas cover only a small variation in regional climate. It is therefore necessary for individual researchers to compare separate areas, different in climate but as alike as possible in relief and geology. Because of variations in cirque definition (Evans and Cox, 1974), statistical comparisons of

Figure 15.22. An active glacier filling and overflowing a complex of steep cirques at the head of Keary Creek, at 50° 43′N in the Bendor Range, British Columbia. Note the four bergschrunds, the firn covering part of the headwall above (on 31 July, 1965), and the dirt lines from rockfalls which crossed bergschrunds before they fully opened. Debris buried in the accumulation zone joins material eroded from the glacier bed and in glaciers of this gradient it is usually transported to the glacier snout.

the results of different researchers are not justifiable at present. The figures given here serve merely as approximate descriptions of the nature of cirques as currently defined.

## 15.8 Conclusion

It can be seen that despite the large number of regional studies of cirque form and distribution, it is not easy to make generalizations about cirque variation in relation to climate. Different studies have taken different measurements and have been based perhaps on different definitions of cirques. We now require, on the one hand, detailed studies of cirque development processes and of glacier and snow balance inside and outside cirques and, on the other hand, broad-scale studies of large samples of cirques in various climates, for which identical operational definitions are applied so that comparable measurements are obtained. There remains the question of designing observations so as to control for variations in geology and topography.

Present knowledge of cirque distribution, although incomplete, is much further advanced than knowledge of variability in cirque form. The sensitivity

Figure 15.23. Transition from relict glacial cirque (Laguna Helada, foreground) to relict nivation cirques with well developed arcuate protalus ramparts (background) with reduction in altitude of the effective snowfence (from 2080 m to 1960 m) and its height above the leeside bench (from 100 m to 50 m). Sierra de Urbián, Burgos province, north-central Spain.

of cirques to climate, via glacier balance, is such that, in a particular position, there is only a certain range of aspects which can be occupied and there is a limiting altitude below which cirques cannot develop. Cirques have been especially useful in reconstructing former snowlines. It is much more difficult, because of the interplay of factors, to reconstruct former wind directions from cirque aspects: the automatic equation of palaeowind direction with the opposite of cirque vector mean aspect, found in some early papers, was premature.

Cirque development appears to require steep ice surface gradients; it slows down considerably in the interiors of major ice sheets. Hence development has gone further where active glaciers persisted through glacial maxima, as along mountainous coasts. The development of isolated cirques, for example in the western USA or in the Schwarzwald, has often proceeded as far as that of cirques glaciated for longer periods, but buried under ice sheets for much of that time, for example in the Bridge River District or in Scotland. Knowledge of rates of cirque development is very tenuous and it is difficult to verify suggestions of polar versus temperate differences (Andrews, 1972). Apart from the maritime–continental distinction, we cannot substantiate climatic variations in cirque form; but this may be because so little work has yet been done. Variations in cirque form related to geology (Haynes, 1968) and to topography (Chevalier, 1955) are better known because they can be detected within a single regional study. A great deal of palaeoclimatic information remains locked away in cirques: we are still working on the key.

## Acknowledgements

The authors wish to thank James A. Peterson and John Gordon for allowing them to use their unpublished measurements of cirques in Tasmania and Scotland, respectively.

## References

Agassiz, L., 1840, *Études sur les glaciers*, Neuchatel.
Ahlmann, H. W., 1919, Geomorphological studies in Norway; *Geog. Annlr*, **1**, 1–148 and 193–252.
Ahlmann, H. W., 1935, Contribution to the physics of glaciers, *Geog. J.*, **86**, 97–113.
Ahlmann, H. W., 1948, *Glaciological research on the North Atlantic coasts*, London, R. Geog. Soc. (R. G. S. Res. Series No. 1).
Andrews, J. T., 1972, Glacier power, mass balances, velocities and erosion potential, *Zeits. für Geomorph.*, Supplementband, **13**, 1–17.
Andrews, J. T. and R. E. Dugdale, 1971, Quaternary history of northern Cumberland Peninsula, Baffin Island, N. W. T., Part V: Factors affecting corrie glacierization in Okoa Bay, *Quaternary Research*, **1**, 532–551.
Andrews J. T., R. G. Barry and L. Drapier, 1970, An inventory of the present and past glacierization of Home Bay and Okoa Bay, east Baffin Island, N. W. T., Canada and some climatic and paleoclimatic considerations; *J. Glaciol.* **9**, (57), 337–362.
Avsyuk, G. A., 1955, Classification of glaciers, translated into English from *Izvestiya Akademii Nauk S.S.S.R., Seriya Geogr.*, **1**, 14–31, in *J. Glaciol*, **3** (21), 7 (1957).
Battey, M. H., 1960, Geological factors in the development of Veslgjuv-botn and Vesl-Skaut-botn, in W. V. Lewis (Ed.), *Norwegian cirque glaciers.*, R. Geogr. Soc. Res. Ser., **4.**, pp. 5–10.
Battle, W. R. B., 1960, Temperature observations in bergschrunds and their relationship to frost shattering, in W. V. Lewis (Ed.), *Norwegian cirque glaciers;* R. Geog. Soc. Res. Ser., **4**, pp. 83–95.
Battle, W. R. B. and W. V. Lewis, 1951, Temperature observations in bergschrunds and their relationship to cirque erosion, *J. Geol.*, **59**, 537–545.
Behre, C. H., 1933, Talus behaviour above timber in the Rocky Mountains, *J. Geol.*, **41**, 622–635.
Blache, J., 1952, La sculpture glaciaire, *Rev. Geogr. alpine*, **40**, 31–123.
Bliss, C. I., 1969, *Statistics for biologists*, Wiley, New York.
Botch, S. G., 1946, Les névés et l'érosion par la neige dans la partie nord de l'Oural, *Bull. Soc. geogr. USSR*, **78**, 207–234 (French translation from Russian original by C.E.D.P., Paris).
Bowman, I., 1916, The Andes of southern Peru, *Am. Geogr. Soc. Spec. Pub.* **2**, 295 pp.
Bryan, K., 1934, Geomorphic processes at high altitudes, *Geog. Rev.*, **24**, 655–656.
Chamberlin, T. C. and R. T. Chamberlin, 1911, Certain phases of glacial erosion, *J. Geol.*, **19**, 193–216.
Chevalier, M., 1955, Le glaciaire des Pyrénées du Couserans :1: les cirques, *Rev. geogr. des Pyrenees et du Sud-ouest*, **25**, 97–124.
Clapperton, C. M., 1971, Evidence of cirque glaciation in the Falkland Islands; *J. Glaciol.*, **10** (58), 121–125.
Clark, J. M. and W. V. Lewis, 1951, Rotational movement in cirque and valley glaciers, *J. Geol.*, **59**, 546–566.

Close, M., 1870, On some corries and their rock-basins in Kerry, *J. roy. geol. Soc. Ireland*, **2**, 236–248.

Costin, A. B., J. N. Jennings, H. P. Black and B. G. Thom, 1964, Snow action on Mt. Twynam, Snowy Mountains, Australia, *J. Glaciol.*, **5**, 219–228.

Costin, A. B., J. N. Jennings, B. C. Bautovich and D. J. Wimbush, 1973, Forces developed by snowpatch action, Mt. Twynam, Snowy Mountains, Australia, *Arctic and Alpine Res.*, **5**, 121–126.

Cotton, C. A., 1942, *Climatic accidents in landscape making*, Whitcombe and Tombs, Christchurch, N. Z., 354 pp.

Court, A., 1957, The classification of glaciers, *J. Glaciol.*, **3**, 2–3.

Davies, J. J., 1969, *Landforms of cold climates*, Australian National University Press, Canberra, 200 pp.

Davis, N. F. G., and W. H. Mathews, 1944, Four phases of glaciation with illustrations from southwestern British Columbia, *J. Geol.*, **52**, 403–413.

Demangeot, J., 1965, *Geomorphologie des Abruzzes Adriatiques;* Centre de Recherches et Documentation Cartographiques et Geographiques; Memoires et Documents, numero hors serie, 400 pp., C.N.R.S. (Paris).

Derbyshire, E., 1963, Glaciation of the Lake St. Clair district, west-central Tasmania, *Austral. Geogr.*, **9**, 97–110.

Derbyshire, E., 1968, Glacial map of northwest-central Tasmania, *Explan. Rpt., Geol. Surv. Tasm., Geol. Record No.* **6**, 46 pp.

Derbyshire, E., 1971, A synoptic approach to the atmospheric circulation of the last glacial maximum in south-eastern Australia, *Palaeogeogr., Palaeoclim., Palaeoecol.*, **10**, 103–124.

Derbyshire, E., 1972, Pleistocene glaciation of Tasmania: review and speculations, *Austral. geogr. Studies*, **10**, 79–94.

Derbyshire, E., 1973, Periglacial phenomena in Tasmania, *Biul. Peryglacjalny*, **22**, 131–148.

Derbyshire, E. and R. S. Blackmore, 1974, A portable weather station for expedition use, *Weather*, **29**, 167–178.

Dort, W., E. F. Roots and E. Derbyshire, 1969, Firn–ice relationships, Sandy Glacier, Southern Victoria Land, Antarctica, *Geogr, Annlr*, **51A**, 104–111.

Eriksson, B. E., 1958, Glaciological investigations in Jotunheim and Sarek in the years 1955 to 1957, *Geographica* (Uppsala Univ. Geogr. Inst.), **34**, 165 pp.

Evans, I. S., 1969, The geomorphology and morphometry of glacial and nival areas, in R. J. Chorley (Ed.) *Water, Earth and Man*, Methuen, London, pp. 369–380.

Evans, I. S., 1974, *The geomorphometry and asymmetry of glaciated mountains*, unpublished doctoral dissertation, University of Cambridge, 527 pp.

Evans, I. S. and N. Cox, 1974, Geomorphometry and the operational definition of cirques, *Area*, **6**, 150–153.

Fisher, J. E., 1963, Two tunnels in cold ice at 4,000 m on the Breithorn, *J. Glaciol.*, **4** (35), 513–520.

Flint, R. F., 1957, *Glacial and Pleistocene Geology*, Wiley, New York, 553 pp.

Flint, R. F. and F. Fidalgo, 1964, Glacial geology of the East flank of the Argentine Andes between latitude 39° 10 S and latitude 41° 20 S, *Bull. Geol. Soc. Am.*, **75**, 335–352.

Freeman, O. W., 1925, The origin of Swimming Woman Canyon, Big Snowy Mountains, Montana, an example of a pseudo-cirque formed by landslide slipping, *J. Geol.*, **33**, 75–79.

Fuller, M. B., 1928, Multiple level cirques, *Illinois State Academy of Science, Transactions*, **20**, 297–303.

Galibert, G., 1960, L'évolution actuelle des 'faces nord' de la haute montagne alpine dans le massif de Zermatt, *Rev. geogr. des Pyrenees et du Sud-Ouest*, **31**, 133–163.

Garcia-Sainz, L., 1941, Las fasas epiglacieres del Pirineos espanol, *Estudios Geograficos*, **2** (2), 209–268.

Garcia-Sainz, L., 1944, Las diferencias del glaciarismo iberico-cuatenario y sus causas, *Scientia*, **74**, 6–16.
Garcia-Sainz, L., 1949, L'origine des glaciers ibériques quaternaires et la trajectoire cyclonale de l'Atlantique, *16th Congr. Intnl. Géogr. (Lisbon)*, C. R. II, 722–730.
Garcia-Sainz, L., 1962, Frostbodenformen in Idubeda-Gebirge (Spanien), *Zeit. f. Geomorph.*, **6**, 33–50.
Gastaldi, B., 1873, On the effects of glacier-erosion in alpine valleys, *Quart. J. Geol. Soc. London*, **29**, 396–399, discussion 399–401.
Gibson, G. R. and J. L. Dyson, 1939, Grinnell Glacier, Glacier National Park, Montana, *Bull. Geol. Soc. Am.*, **50**, 681–696.
Gilbert, G. K., 1904, Systematic asymmetry of crest lines in the High Sierra of California, *J. Geol.*, **12**, 579–588.
Goldthwait, R. P., 1970, Mountain glaciers of the Presidential Range in New Hampshire, *Arctic and Alpine Research*, **2** (2), 85–102.
Gregory, J. W., 1913, *Nature and origin of fjords*, London.
Groom, G. E., 1959, Niche glaciers in Bünsow Land, Vestspitzbergen, *J. Glaciol.*, **3**, 369–376.
Grove, J. M., 1954, *A study of aspects of the physiography of certain glaciers in Norway*, unpublished Ph.D. dissertation 2832, University of Cambridge, 251 pp.
Grove, J. M., 1961, Some notes on slab and niche glaciers, and the characteristics of proto-cirque hollows, *Int. Ass. Sc. Hydrol. Publication 54* (Helsinki General Assembly), 281–287.
Haynes, V. M., 1968, The influence of glacial erosion and rock structure on corries in Scotland, *Geogr. Annlr.*, **50A**, 221–234.
Helland, A., 1877, On the ice-fjords of North Greenland and on the formation of fjords, lakes and cirques in Norway and Greenland, *Quart. J. Geol. Soc. London*, **33**, 142–176.
Hinds, N. E. A., 1925, Amphitheatre valley heads, *J. Geol.*, **33**, 816–818.
Hobbs, W. H., 1910, Cycle of mountain glaciation, *Geogr. J.*, **36**, 146–163, and 268–284.
Hobbs, W. H., 1911, *Characteristics of existing glaciers*, Macmillan, New York, 361 pp.
Hobbs, W. H., 1921, Studies in the cycle of glaciation, *J. Geol.*, **29**, 370–386.
Hoinkes, H. C., 1964, Glacial meteorology, in H. Odishaw (Ed.) *Research in Geophysics*, 2, M. I. T. Press, Cambridge, Mass., 391–424.
Hoshiai, M. and K., Kobayashi, 1957, A theoretical discussion on the so-called 'snowline' with reference to the temperature reduction during the last glaciation in Japan, *Jap. J. Geol. Geogr.*, **28**, 61–75.
Ivanovskiy, L. N., 1965, Rasprostraneniye, morfologiya proiskhozhdeniye Karov Altaya, *Sibirskiy Geograficheskiy Sbornik*, **4**, 152–198.
Ives, J., 1973, Arctic and alpine geomorphology—a review of current outlook and notable gaps in knowledge, in *Research in Polar and Alpine Geomorphology* (3rd. Guelph Symposium on Geomorphology), B. D. Fahey and R. D. Thompson (Eds.), 1–10.
Johnson, W. D., 1904, Maturity in alpine glacial erosion, *J. Geol.*, **12**, 571–578.
Kornilov, B. A., 1964, Investigations along the way from Oymyakon Plateau to the shore of the Okhotsk Sea, in *Freezing of the earth's surface and glaciation in the Suntar-Khyata Range (East Yakutia)* (in Russian), N. A. Grave and coworkers, 106–115.
Lewis, A. N., 1925, Notes on a geological reconnaissance of the La Perouse Range, *Pap. Roy. Soc. Tasmania*, **(1924)**, 9–44.
Lewis, W. V., 1938, A meltwater hypothesis of cirque formation, *Geol. Mag.*, **75**, 249–265.
Lewis, W. V., 1939, Snow patch erosion in Iceland, *Geogr. J.*, **94**, 153–161.
Lewis, W. V., 1949, Glacial movement by rotational slipping, *Geogr. Annlr.*, **31**, 146–158.
Lewis, W. V., 1954, Pressure release and glacial erosion, *J. Glaciol.*, **2**, (16) 417–422.
Lewis, W. V., 1960, Norwegian cirque glaciers, *R. Geogr. Soc. Res. Ser.*, **4**, 104 pp.
Linton, D. L., 1959, Morphological contrasts of eastern and western Scotland, in R. Miller and J. W. Watson (Eds.), *Geographical essays in memory of Alan G. Ogilvie*, Nelson, London, 16–45.
Linton, D. L., 1963, The forms of glacial erosion, *Trans. Inst. Brit. Geogr.*, **33**, 1–28.

Loffler, E., 1972, Pleistocene glaciation in Papua and New Guinea, *Zeit. f. Geomorph. Supplementband*, **13**, 32–58.
Lyubomirova, K. S., 1964, Vliyaniye ekspozitsii, i krutizny sklonov na tayaniye lednikov Elbrusa za schet pryamoy solnechnoy radiatsii, *Materialy Glyatsiol. Issled; Khronika. Obsuzhdeniya*, **10**, 204–208.
McCabe, L. H., 1939, Nivation and corrie erosion in West Spitsbergen, *Geogr. J.*, **94**, 447–465.
McCall, J. G., 1960, The flow characteristics of a cirque glacier and their effect on glacial structure and cirque formation, in W. V. Lewis (Ed.), *Norwegian Cirque Glaciers*, R. Geogr. Soc. Res. Ser. 4, 39–62.
McGregor, V. R., 1967, Holocene moraines and rock glaciers in the central Ben Ohau Range, south Canterbury, New Zealand, *J. Glaciol.*, **6** (47), 737–748.
Mackay, J. R. and W. H. Mathews, 1969, Observations on pressure exerted by creeping snow, Mount Seymour, British Columbia, Canada, in *Physics of Snow and Ice*, H. Oura, (Ed.), Institute of Low Temperature Science, Hokkaido University, Sapporo 1185–1197.
Manley, G., 1959, The late-glacial climate of North-West England, *Lpool. Manchr. geol. J.*, **2**, 188–215.
Martonne, E. de, 1910, Sur la genèse des formes glaciaires alpines, *C. R. Acad. Sci. Paris*, **150**, 243–246.
Martonne, E. de, 1926, *Traite de geographie physique*, 4th ed., Paris.
Mathews, W. H. and J. R. Mackay, 1963, Snowcreep studies, Mount Seymour, B. C.: Preliminary field investigations, *Geogr. Bull.*, **21**, 58–75.
Meier, M. F. 1965, Glaciers and climate, in *The Quaternary of the United States*, H. E. Wright and D. G. Frey (Eds.), Princeton University Press, New Jersey, 795–805.
Norcliffe, G. B., 1969, On the use and limitations of trend surface models, *Can. Geogr.*, **13**, 338–348.
Obermaier, H. and J. Carandell, 1916, Contribucion al estudio del glaciarismo cuatenario de la Sierra de Gredos, *Trab. del Museo Nacional de Ciencias Naturales. Ser. Geol.*, **14**, 54 pp.
Østrem, G., 1964, Ice-cored moraines in Scandinavia, *Geogr. Annlr.*, **46**, 282–337.
Østrem, G., 1966, The height of the glaciation limit in southern British Columbia and Alberta, *Geogr. Annlr.*, **48A**, 126–138.
Østrem, G., 1972, Height of the glaciation level in northern British Columbia and southeastern Alaska, *Geogr. Annlr.*, **54A**, 76–84.
Østrem, G., N. Haakensen and O. Melander, 1973, Atlas over Breer i Nord-Skandinavia, *Norges Vassdrags og Elektrisitetsvesen, Hydrologisk avdeling, Meddelelse*, **22**, 315 pp.
Paterson, W. S. B., 1969, *The physics of glaciers*, Pergamon, Oxford, 250 pp.
Penck, A., 1905, Glacial features in the surface of the Alps, *J. Geol.*, **13**, 1–19.
Penck, A., 1910, Versuch einer Klimaklassification auf physiographischer Grundlage, *Preussen Akademie der Wissenschaft Sitz. der physikalisch-mathematischen*, **12** (1910), 236–246.
Peterson, J. A., 1968, Cirque morphology and Pleistocene ice formation conditions in southeastern Australia, *Austral. geogr. Studies*, **6**, 67–83.
Peterson, J. A., and G. Robinson, 1969, Trend surface mapping of cirque floor levels, *Nature*, **222** (5188), 75–76.
Pewe, T. L., L. Burbank and L. R. Mayo, 1967, Multiple glaciation of the Yukon–Tanana upland, Alaska, *United States Geol. Surv. Misc. Geol. Investigations Map*. I–507, scale 1:500,000.
Porter, S. C., 1964, Composite Pleistocene snowline of Olympic Mountains and Cascade Range, Washington, *Bull. Geol. Soc. Am.*, **75**, 477–482.
Priestley, R. E., 1923, Physiography (Robertson Bay and Terra Nova Regions), *Brit. (Terra Nova) Antarctic Exp. (1910–1913)*, London.
Robinson, G., J. A. Peterson and P. M. Anderson, 1971, Trend surface analysis of corrie altitudes in Scotland, *Scott. Geogr. Mag.*, **87**, 142–146.

Russell, R. J., 1933, Alpine landforms of western United States, *Bull. Geol. Soc. Am.*, **44**, 927–949.
Sale, C., 1970, *Cirque distribution in Great Britain: a statistical analysis of variations in elevation, aspect and density*, unpublished M.Sc. dissertation, Department of Geography, University College, London, 122 pp.
Seddon, B., 1957, The late-glacial cwm glaciers in Wales, *J. Glaciol.*, **3**, 94–99.
Sharp, R. P., 1960, Pleistocene glaciation of the Trinity Alps of Northern California, *Am. J. Sci.*, **258**, 305–340.
Shumskii, P. A., 1964, *Principles of Structural glaciology*, (trans. D. Kraus), Dover, New York.
Sissons, J. B., 1967, *The evolution of Scotland's scenery*, Oliver and Boyd, Edinburgh, 259 pp.
Sugden, D. E., 1969, The age and form of corries in the Cairngorms, *Scott. Geogr. Mag.*, **85**, 34–46.
Taylor, T. Griffith, 1922, Physiography of the McMurdo Sound and Granite Harbour Region, *Brit. (Terra Nova) Antarctic Exp. (1910–1913)*. London.
Thompson, H. R., and B. H. Bonnlander, 1957, Temperature measurements at a cirque bergschrund in Baffin Island: some results of W. R. B. Battle's work in 1953, *J. Glaciol.*, **2**, 762–769.
Tricart, J. 1963, *Géomorphologie des regions froides* (transl. by E. Watson as *Geomorphology of Cold Environments*, 1969, Macmillan, London), Presses Universitaires de France, 289 pp.
Unwin, D. J., 1970, *Some aspects of the glacial geomorphology of Snowdonia, North Wales*, unpublished M. Phil. thesis, University of London, 294 pp.
Unwin, D. J., 1973, The distribution and orientation of corries in northern Snowdonia, Wales, *Trans. Inst. Brit. Geogr.*, **58**, 85–97.
Watson, E., 1966, Two nivation cirques near Aberystwyth, Wales, *Biul. Peryglacjalny.*, **15**, 79–101.
Weertman, J., 1971, Shear stress at the base of a rigidly rotating cirque glacier, *J. Glaciol.*, **10** (58), 31–37.
White, W. A., 1970, Erosion of cirques, *J. Geol.*, **78**, 123–126.
Wojciechowski, K. H. and T. Wilgat, 1972, Mapa geologiczno-hydrograficzna dorzecza gornej Rio Aconcagua, *Przeglad Geograficzny*, **14** (4), 635–648.
Wright, W. B., 1914, *The Quaternary Ice Age*, Macmillan, London, 464 pp.
Wright, W. B., 1937, *The Quaternary Ice Age*, 2nd ed., Macmillan, London, 478 pp.
Zienert, A., 1967, Vogesen und Schwarzwald-Kare, *Eiszeitalter und Gegenwart*, **18**, 51–75.

# Index

Ablation zone, 456
Accelerated erosion, 339
Accelerating creep, 115
Accumulation zone, 456
Acidification, 73, 76, 93
Activation energy, 44
Actual evapotranspiration, 255–256
Adelbert Range, New Guinea, 155
Adrar, Mauretania, 356, 357
Aerial photographs, 290, 323
*Agathis borneensis* Warb, 275
Aggressiveness, 86
Air mass, 15, 212, 213, 221, 231, 317
Alabama, 298
Alaska, 294, 295, 298, 303, 458
Alaska earthquake, 118
Alberta, 383
Aldabra Atoll, Indian Ocean, 85
Allegheny Plateau, 301, 355
Allophane, 418
Alpine humus soil, 142
Alps, 458, 471
Alps of Dauphine, 461
Altiplanation, 435
Altitudinal climatic zonation of karst, 393
Alveolate landscapes, 428
Amazon, 354
Analysis of variance, 174, 354, 355
Andesite basins, 179
Angle of internal friction, 114, 120, 122, 124
Angular cliffs, 427
Angular values of slopes, classification of, 183
Anhydrite, 49, 65, 66
Anion, 30
Anisotropic crystal structure, 34
Annual runoff, 374, 386, 387, 388, 401
Antarctic tors, 438
Ants, 276
Antwerp, 207
Apennines, 224, 237, 239, 486
Appalachians of Virginia, 188

Applied chemical thermodynamics, 40–41
Aqueous solubility, 36
Aqueous solutions, 48, 54
Arable layer, 216, 228
Arable soils, 215
Aravalli Mountains, 351
Areal geomorphology, 4
Areal mapping, 6
Arenization, 421, 425
Arêtes, 471
Argentine Andes, 460
Arid and semi-arid karst, 393
Arid climate, 45, 47, 211, 212
Arizona, 176, 177, 179
Arrhenius equation, 44
Artesian pressures, 117, 121
Artesian slope angles, 121, 130
Artfjället area, Swedish Lapland, 397
Arthur range, Tasmania, 458
Artificial slopes, 266
Aruba (W. Indies), 351
Aspect, 173, 351, 449, 451, 452, 456, 458, 459, 460, 461, 463, 477, 487, 489
Asymmetrical glaciation, 458, 459, 472
Asymmetrical valleys, 158–162, 180–199
 causes of, 184–190
 forms of, 184–190
 indices, 183
 statistical definition of, 184
Atherton Tableland, North Queensland, 360, 361
Atlantic Coastal Plain, United States of America, 189
Atmosphere, 28
Atomic arrangement, 31
Atomic nucleus, 27
Atomic species, 32
Atomic structure, 32
Attenuating creep, 115
Aufbereitung concept, 163
*Austroicetes cruciata*, 276
Auto-asymmetry, 187
Autometamorphism, 415, 424

495

Available relief, 257
Avalanche, 117, 124, 125, 229, 230, 231, 239, 240
Avalanche gullies, 471
'Avalanche years', 229–230
Avalanching of snow, 448
Average angle of valley-side slope, 176
Average maximum slope angles, 179, 182, 184
Azimuthally asymmetrical glaciation, 452
Azonal landforms, 358

Backwall, 465
Bacteria, 350
Bacterial products, 63
Baffin Island, 484
Balki, 289
Banjaran Benom, 337
Banjaran Gunung Bintang, 337
Banjaran Tahan, 319
Banjaran Timor, 331, 336, 337
Banjaran Titiwangsa, 319, 326, 331, 337
Bankfull stage, 203
Banks Island, Canada, 158, 159, 184
Barbados, 308
Basal attack, 195
Basal concavity, 181, 264
Basal erosion, 185, 196, 197, 466
Basal lowering, 262
Basal sapping, 185
Basal sliding, 467, 482
Basal surface of weathering, 415, 438
Basalt basins, 179, 225
Basalt landforms, 360–361
Base flow, 269, 290, 303, 323
Base-level, 237, 262
Basement Complex of west Africa, 424
Basic network, 311
Basin area, 294, 295, 306, 309
Basin form, 171, 174, 176, 337, 395
Basin outlet, 171
Basin perimeter, 171
Basin relief, 272
Bassari Mountains, Senegal, 359
Batholiths, 413
Beauraing, Belgium, 359
Bed load, 227
Bendor Range, British Columbia, 462, 482, 484, 485, 486, 488
Bergschrund, 466, 467, 488
Bergschrund hypothesis, 467
Bergschrund sapping, 468
Berkshire Downs, 381
Bifurcation ratio, 326

Bingham deformation, 107
Bingham solid, 115
Biogenic carbon dioxide, 373
Biological activity, 347
  seasonal, 276
Biological productivity, 80
Biosphere, 28, 30
Biscuit-board topography, 471
Black earth, 142
Block-falls, 172, 357
Blockfields, 360, 433, 440, 441
Blockstreams, 351
Blue-green algae, 88
Bodmin Moor, 417, 438
Bog soils, 142, 165
Bogoé, West Africa, 359
Bom de Jesus Lapa, Brazil, 392
Bornhardts, 145, 429, 439
'Bottle slides', 219
Boulder controlled slopes, 158
Boulder inselbergs, 429
Boutonniere, 354
Bowl-slides, 224
Breached dome, 354
Breithorn, 467
Bridge River District, British Columbia, 461, 463, 468, 473, 489
Brittany, 359
Brown Mountain, southern New South Wales, 360
Brown podzolic soil, 142
Bulk density, 117, 371, 372
Buried soils, 269
Burren, Co. Clare, Ireland, 394
Burrowing mammals, 276

Cadwallader Range, British Columbia, 470, 471
Cairngorms, 477, 484
Calabber Valley, 359
Calcicoles, 72
Calcifuge, 72, 73
Calcite, 47
Calcium, 28
Caliche, 47
California, 295, 298, 459
Canada, 458
Canadian Rockies, 394
Canberra, Australia, 278
Caneleira, 108, 117
Caneleira failure, Brazil, 104
Capillary slope angles, 121
Carbon dioxide, 367, 368, 369, 370, 374, 377, 378, 380, 381, 382, 383

Carbonate dissolution, 86
Carbonate staining, 86
Carbonate systems, 81
Carbonation, 56
Carbonic acid, 368
Carboniferous Limestone, 53
Caribou Hills, N.W.T., Canada, 195
Case hardening, 437
Castle kopjes, 436
Castle koppies, 429
Castleguard Cave, Alberta, 390
Catalonia, 352
Catalysis, 44
Catchment studies, 15
Catena, 137–144, 147, 148, 151–153, 159, 162, 163, 165, 166, 349
Cation clays, 47
Cation exchange, 62
Catombal Range, New South Wales, 352
Caucasian republics, USSR, 219
Caucasus, 240
Cavernous weathering, 351, 438
Caves, 383, 394, 398
Central Europe, 458
Central Otago, 308
Central Tasmania, 468
Chalk valleys, southern England, 188
Chalkland soils, 147
Champlain Clays, 108
Channel-forming discharge, 299
Channel geometry, 291, 299, 300
Channel network, 171
Channel pattern, 291, 299, 300
Characteristic form, 254
Characteristic slope forms, 259–260, 262, 263, 266
Chelates, 348
Chelation, 62, 276
Chemical bonding, 27, 31
Chemical denudation, 269
Chemical elements, 27
Chemical energy, 28
Chemical environments, 25
Chemical equilibrium, 61, 65, 74, 95
Chemical kinetics, 42, 44
Chemical lowering, 257, 262
Chemical products, 30
Chemical reactions, 25, 31
Chemical removal rate, 256, 261, 262
Chemical stability, 25
Chemical systems, 31, 54
Chemical theory, 27
Chemical thermodynamics, 25, 35, 37, 40
Chemical weathering, 28, 45, 52, 85, 162, 227, 232, 239, 240, 345, 347, 351, 352, 359, 361, 412, 413, 435, 436, 437
Chemoautotrophic organisms, 62
Cherozem, 63
Chezy uniform flow equation, 271
Chile, 458
Chiltern Hills, 151, 308
China clay, 421
Chinle badlands, Arizona, 177
Chi-square test, 375, 383
Chlorite, 80
Chocolate soil, 142
Cirque altitude, 460–463
Cirque amplitude, 484, 486, 487
Cirque aprons, 482
Cirque area, 483, 487
Cirque aspect, 456–460, 471
Cirque breaching, 469
Cirque climates, 460
Cirque compactness, 484
Cirque complexes, 482, 484
Cirque definition, 473, 487
Cirque development, 465–473, 489
Cirque distributions, 456
Cirque floor altitudes, 460, 461, 463
Cirque floor deepening, 468, 471
Cirque floor gradients, 484, 485
Cirque floor surface, 460
Cirque form, 488, 489,
Cirque glacier, 17, 187, 190, 456, 467, 468, 482
Cirque gradient, 484, 486, 487
Cirque height, 476
Cirque length/breadth ratios, 484
Cirque microclimates, 463
Cirque moraines, 463
Cirque morphometry, 484, 487
Cirque relative relief, 476
Cirque/trough-head combinations, 482
Cirque troughs, 482, 484
Cirque variation, 447
Cirques, 240
  climates within, 463–465
  distribution of, 478
Clay-humus complex, 61
Clay minerals, 28, 45–47, 63, 80, 140, 347, 348, 417, 421
Clay mobility, 140
Clay slopes, 122
Clay soils, cracking of, 347
Climate, 1–10, 14, 17, 25, 30, 60–63, 67, 70, 73, 77, 79, 81, 89–95, 124, 128–132, 137, 145, 162, 165, 166, 171, 173, 174, 180, 182, 199, 247, 248, 251, 266, 269,

272, 300, 301, 307, 310, 317, 318, 337, 340, 347–356, 370, 376, 381–383, 392, 393, 395, 401–403, 414, 417, 425, 427, 436, 447, 456, 487, 488, 489
and sediment yield, 257–259
arid, 47
classification of, 2, 3, 15
continental, 15
humid, 28
mediterranean, 15
seasonal, 15
semi-arid, 5
surface layers, 172
synoptic, 17
wet-dry, 15
Climate-diagnostic landforms, 17
Climate-process-form relationship, 90
Climate-process systems, 13, 232, 317
Climatic asymmetry, 192, 195, 197, 213
Climatic change, 91–93, 130, 145, 239, 283, 307, 311, 319, 354, 360, 393, 433, 441
Climatic data, 10
Climatic gradients, 477, 479, 482
Climatic region, 6
Climatic seasonality, 275
Climatic sequence, 471
Climatic snowline, 480
Climatic zones, 197, 348
Climatically-determined processes, 353
Climatically-dominated system, 73
Climatically-marginal glaciation, 447, 453, 459, 471
Climato-genetic zone, 292
Climato-geomorphological zones, 290
Climato-morphological zones, 297
Climosequence, 64
Clitter, 433
Clitter slopes, 440, 441
Closed depressions, 367, 392, 398, 402
Closed depression density, 398
Closed system, 82, 84
Cloudiness, 448, 449, 451, 453, 477
Coalescence of cirques, 472, 482
Coast Mountains, British Columbia, 482
Coastal Peru, 351
Cockpit and cone karst, 394, 395
Cockpit Country, Jamaica, 390, 392, 394, 401
Cockpit topography, 388
*Cockpitlandschaft*, 392
Cohesion, 112
Cohesive strength, 120
Col, 468, 469, 471
Cold-isothermal ice boundary, 467

Colloidal systems, 79
Colorado, 176, 177, 179
Colorado Plateau, 103
Columbia Icefield, 390
Compleximetric method, 370
Composite snowline, 460
Computer simulation, 249, 253–254
Concavo-convex slopes, 154
Concentration of aspect, 459
Conceptual models, 270
Condroz, Belgium, 188
Conduits, 383, 394, 398
Cone karst, 367, 388, 390, 394
Congruent dissolution, 73, 74
Congruent solution, 88
Conic karst, 399
Connecticut, 300
Consolidation, 114
Constant slope, 181
Continental climates, 228, 232
Continental denudation equation, 30
Continental glaciers, 466, 477
Continuity-of-mass transport equation, 270
Continuous rainfalls, 221–228, 240
Convection, 448
Convergence, concept of, 350
Convergent landforms, 350–352, 432
Convexo-concave profile, 254
Convexo-concave slope, 238, 263, 265
Coober Pedy, Australia, 152
Cordilleran Ice Sheet, 482
Core boulders, 351
Corestones, 319, 352, 426, 427, 429, 437, 439, 440
Coriolis force, 180, 186, 190, 191
Cotswolds, 381
County Kerry, Ireland, 463, 464
Covalent bonds, 32, 34
Cracking of clay soils, 347
Crana Basin, 359
Creep, 101, 128, 130, 132, 137, 138, 141, 142, 147, 154, 158, 162, 163, 164, 166, 197, 248, 254, 261, 262, 289, 428
Cryoplanation, 435
Cryoplanation terraces, 435, 441
Crystal structures, 33
Crystaline Schwarzwald, 484
Crystalline solids, 31, 36
Crystalline Vosges, 484
Cuba, 390
Cuestas, 362
Cycle diagram, 28
Cycle of erosion, 391, 392, 402
Cycle of glaciation, 468, 471

Cyclic moisture changes, 101
Cyclone tracks, 225
Cyclones, 204, 210, 212, 274, 275, 279
Cyclopean masonry, 438

Dacite basins, 179
Dambo, 141, 289
Darcian model, 395
Darling River, New South Wales, 363
Dartmoor, 301, 356, 417, 421, 424, 433, 440
Dauphiné, France, 471
Davis, W. M., 114, 153, 199, 289, 347, 391, 403
Debris avalanches, 219, 220
Debris-creep, 125, 126, 132
Debris flows, 205, 218, 219, 224, 228, 237, 240
Debris slope, 152
Decelerating creep, 116
Deconsolidation, 112, 113, 118
*Dellen*, 307, 310
Dendrochronology, 219
Denudation index, 219, 235
Denudation rate, 26, 205, 206, 215, 216, 225, 227, 234, 235, 237, 241, 282, 355, 426, 435
Denudation system, 8, 317–340, 411, 412, 441, 426, 428, 432, 436, 437
Denudational efficiency, 235
Depth-creep, 103, 117, 132
Derbyshire, 129, 130
Dessication, 114
Detention, 215
Devon, England, 7
Devon Island, Canada, 383
Diagenesis, 114
Differential dissolution, 83
Differential solution, 88
Diffusion rates, 69, 82
Dinaric Karst, Yugoslavia, 392, 393
Dindings, 337
Disaggregation, 25
Discordant extreme processes, 240
Discordant structures, 353
Discrete cirques, 459, 468, 477
Dissolution, 35
Dissolution potential, 88
Dissolved load, 28, 30, 48, 70, 278
Dissolved solids, 30, 349
Distribution of erosion, 388–390
Distribution of glacial cirques, 459
Distribution of precipitation, 180
Dnieper valley, 216

Dogtooth Range, Canadian Rockies, 471
Dolines, 92, 390–395, 401, 402
Domed inselbergs, 350, 352, 429
Domes, 429, 432, 439, 441
Dorrigo, 361
Double concavity, 473, 477
Double spheroid cirque form, 482
Drainage basins, 15, 180
Drainage density, 5, 176–179, 182, 289–307, 311, 317–319, 323–334, 336, 337, 354–356, 402
Drainage density index, 176, 179
Drainage networks, 7, 110, 189, 289–311, 396
Drainage network changes, 307–310
Drainage network expansion, 304–310
Drainage system, 29
Drainage texture, 318, 319, 323–329, 334–340, 356
Drifting of snow, 459
Driftless Area of Wisconsin, 188
Dromilly Clear Spring, Jamaica, 381
Dry valley, 290, 307, 309, 310, 394, 395, 397, 401, 402
Dry valley density, 308, 309, 402
Dumaresq Creek, Armidale, New South Wales, 278
Duricrust, 166, 357, 424, 425
Dust-storm, 234
Dynamic equilibrium, 61, 127, 203, 206, 207, 213, 227, 239, 427, 428
Dynamic metastable equilibrium, 207
Dynamic model, 77–79
Dynamic process models, 81
Dynamics of ecosystems, 275, 277

Earth-avalanche, 104, 105
Earth-falls, 103, 104, 110
Earth-flow, 106, 108, 110, 116
Earth rubble, 124, 128
Earth rubble slopes, 122
Earth-slide, 104, 126
Earth-slumps, 110
Earthquake, Alaska, 118
  New Guinea, 118
East Africa, 351
East-central Baffin Land, 460
Eastern Carpathians, 229
Ebor, 361
Eden Shale belt, Kentucky, 192, 195
Effective climate, 13
Effective normal stress, 117
Effective precipitation, 274
Effective rainfall, 306

Effective solute transport, 252
Eh, 85, 348, 349
Eh-pH diagrams, 348
Eigenvalues, 327
Eigenvectors, 327
Elba, 351
Electron, 27
Electron donation and acceptance, 35
Electron transfer, 56
Electronic systems, 17
Electrostatic attraction, 55
Electrostatic bonds, 35
Electrostatic charges, 32
Elm, Switzerland, 104
Empirical model, 259
Enclosed depressions, 391, 395
Energy balance, 6, 17
Energy budget approach, 298
Energy changes, 31
Energy models, 59, 60
Energy system approach, 340
England, southwest, 417
Entropy, 41
Equatorial regions, 212
Equifinality, 17
Equilibrium, 35, 39, 44, 46, 51, 61, 66, 70, 77, 79, 82, 85, 90, 91, 153, 154, 164, 198, 204, 219, 240, 241, 252, 308, 339, 368, 370, 377, 380, 428, 453
Equilibrium concentration, 41
Equilibrium curve, 368
Equilibrium form, 428, 486
Equilibrium line, 448, 467
Equilibrium modelling, 60, 62
Equilibrium process, 61
Equilibrium product, 61
Equilibrium profiles, 126
Equilibrium reactions, 62
Equilibrium slopes, 164, 221
Equilibrium soils, 254, 260, 261
Equilibrium solubility, 368
Erosion budget, 388, 390
Erosion catena, 140
Erosion intensity, 270, 347
Erosion mechanics, 270
Erosion potential, 82, 85–88
Erosion rate, 269, 271, 272, 306, 372, 373–388, 390, 394, 401, 402
Erosivity of rain, 13
Étagement, 3
Etching of quartz grain, 422
Etchplain, 145
Evaporation, 318, 322, 331, 349, 466
Evaporation rate, 258, 260, 333

Evaporite, 49
Evapotranspiration, 5, 248, 251, 252, 254, 266, 306, 374, 388
Exaration-slopes, 240
Exfoliation, 112, 411, 414
Exhumation of rock, 224, 239
Exhumation hypothesis of tor formation, 437
Exmoor, England, 104, 123, 124, 128, 129, 154, 298
Expanded network, 311
Experimental catchments, 294
Extreme meteorological events, 7, 203, 206, 213–232

Factor modelling, 62
Falkland Islands, 458
Falls, 103, 124, 125
Fault scarp, 188
Faulting, 188
Feedback, 10, 50, 62, 70, 307, 311, 486
Feldspar to quartz ratio, 421
Felsenmeer, 433
Felsite basins, 179
Ferrallitization, 425
Ferricrete, 143–146, 155, 156, 166
Firnification, 465, 477
Fissures, 112–114, 118, 120, 348, 352, 382, 383, 394
Fissuring, 382, 383, 394
Flatness, 476
Flatness index, 477, 479
Flood-oriented model, 279
Flood stage, 227
Flood wave, 278
Flow duration curve, 371
Flow-duration/sediment method, 283
Flow-through time, 73–81
Fluidization, 117
Fluvial regime, 10
Fluviokarst, 393, 394
Fluviothermal erosion, 158
Flysch Carpathians, 215, 220, 224
Folkestone Warren, England, 105
Form-change sequence, 18
Fossil asymmetry, 191
Fossil valleys, 240
Fouta Djalon, Guinea, 356, 357
Fracture zones, 413
Frank, Alberta, 104
Fraser Experimental Forest, Colorado, 279
Free energy, 40, 41
Free face, 149, 151, 181, 182
Free meanders, 185

501

Free-thaw, 92, 101, 104, 129, 196, 300, 360, 466, 467, 471
French Alps, 231
Frequency of extreme events, 239
Fretted uplands, 471, 472
Frictional slope angle, 120
Frictional strength, 114, 120
Frost action, 92, 165, 190, 359, 438, 440
Frost-creep, 132
Frost heave, 108
Frost-riven cliffs, 435, 436
Frost shattering, 352, 438, 447
Frost weathering, 413
Frozen ground, 228, 231, 308
'Functional' method, 18
Furre, Norway, 112

Gangplank, Wyoming, 195
Garibaldi Park, British Columbia, 456
Gelifluction, 108, 127, 132, 425, 433
General circulation of the atmosphere, 272
General model, 81, 468
General systems theory, 6, 8
Geobotanical zonality, 296
Geochemical cycles, 26
Geochemical types of rock weathering, 347
Geographical geomorphology, 10
Geological geomorphology, 10
Geological structure, 171
Geometry of basin form, 171
Geomorphology areal, 4
    process, 4
Geosystems, 6
Geothite, 421
Gibbsite, 41, 47, 350, 418, 419
Gippsland, Australia, 138, 151
Gisborne Plains, New Zealand, 269
Glacial abrasion, 466, 471
Glacial asymmetry, 190, 191, 453
Glacial cirque, 224, 448, 456, 458, 463, 468, 475, 477, 479, 487
Glacial drainage channels, 290
Glacial drift, 93, 164
Glacial erosion, 112, 180, 451, 472, 482
Glacial processes, 355
Glacial troughs, 190, 237, 240
Glaciated rock, 414
Glaciated valleys, 190
Glaciation, 190, 452
Glaciation limit, 456, 463
Glacier aspect, 453, 458
Glacier balance, 448
    and erosion system, 450
Glacier-calimate relations, 456

Glacier distribution, 452
Glacier ice, 171
Glacier system, 456
Glaciokarst, 393, 394
Glassy solids, 31
Goeneng Sewoe, Java, 392
Goethite, 47, 55
Gondwanaland, 352
Gradient, 475
Grain diminution, 25
Grampian Highlands, 461
Granite, alteration of, 415–417
Granite basins, 177–179
Granite-gneiss basins, 177
Granite landforms, 426–440
Granite rock surfaces, 437–438
Granite weathering profiles, 424
Granular disintegration, 92, 351, 352, 358
Gravity stress, 116
Grey-brown podzolic soil, 142
Grey-brown soil, 158
Grey podzolic soil, 142
Gritstone tors, 358
Grooved upland, 471
Gros Ventre, Wyoming, 106
*Grosswetterlagen*, 17
Ground ice, 129
Ground-thaw, 204
Groundwater flow network, 398
Groundwater weathering, 415, 440
Growan, 421, 424
Gruss, 421, 425
G-scale, 8
Guil valley, France, 231, 275
Guyra-Armidale, 361
Gypsum, 49, 65, 66

Halloysite, 81, 418, 419, 421
Hardanger Jökulen ice-cap, southern Norway, 465
Hardness of water, 370, 394
Hardware models, 6, 82
Hawaii, 360
Hawes Water, Cumbria, 456
Headwall, 447, 464, 466, 471, 475, 481, 482
Headwall gradients, 485, 487
Headwall retreat, 468, 471, 472, 482
Heat balance, 477
Heave, 107
Heavy minerals, 422
Herbert River, 360
Heyliedsgraben, Eichsfield, central Germany, 218
Hillslope development, 266

Hillslope hydrology, 247, 248
*Hohlblockbildung*, 351
*Hohlkehlen*, 358
Holokarst, 393
Homogeneous samples, 182
Hong Kong, 130, 218, 351, 418
Horizon digram, 464
Humid climates, 28, 44, 212
Humidity, 221
Humus, 63
Humus acid, 63
Humus type, 64
Hunua ranges, New Zealand, 154
Hydration, 54, 56, 347, 351
Hydraulics, principles of, 270
Hydrograph, 12, 79, 272, 278, 279
Hydrological budgeting model, 249–251
Hydrological flow components, 253
Hydrological map of Poland, 291
Hydrological modelling, 248
Hydrology, 10, 17, 90, 94
Hydrolysis, 62, 85, 347, 348
Hydrometeorological approach, 278–281
Hydromorphic catena, 140
Hydrosphere, 28, 36, 49
Hydrothermal alteration, 415, 417, 433, 439
Hydrous mica, 418

Ice caps, 190
Ice sheets, 361
Iceland, 351, 360, 361
Illite, 47, 51, 80, 81, 418
Illuviation, 164
Incongruent dissolution, 74
Incongruent solution, 73
Index of flatness, 475, 477
Index of pitting, 402
Infilling, rates of, 269
Infiltration, 116, 117, 130, 204, 205, 213, 214, 215, 221, 232, 298, 355
  rock types, 5
Infiltration capacity, 176, 218, 241
Infiltration rate, 67, 205, 215, 271
Ingleborough district, Yorkshire, 393, 397
Insects, 275, 276
Inselberg, 145, 219, 436
Insolation, 175, 180, 186, 187, 190, 198, 351
Instrumental record, 10
Instrumentation, 8, 15, 17, 69, 173
Interflow, 278
Internal plastic deformation, 467
Intersecting Linear karst, 399
Ionic bonds, 32, 33

Ionic concentrations, 354, 355
Ionic radius, 32, 33
Ionic radius ratio, 34
Iowa, 300
Iran, 91
Iregolith depth, 422–425
Iron, 348
Ischia, 352
Isothermal ice, 467

Jabatanarah Pemetaan Negara, Malaysia, 323
Jackfield slide, England, 104
Jamaica, 380, 381
Johor, 336, 339
Joint sets, 414
Julian Alps, northwest Yugoslavia, 383

$k$ curves, 473
$K$ cycle, 156
Kajang, 336
Kaolinite, 41, 47, 48, 66, 80, 81, 162, 349, 418, 419, 421
Kaolinization, 415
Kaolinized funnels, 424
Kärkevagge massif, Scandinavia, 224, 239
Karlovy Vary, Czechoslovakia, 417
Karren, 92
Karst denudation, 270
Karst hydrology, 398
Karst landforms, 269, 385, 395
Karst landscape, 391
Karst morphology, 391–393, 399
Karst resurgence density, 397
Karst topography, 345
Karst towers, 367
Katum, 458
*Kegelkarst*, 395
Kelantan, 326, 339
Kentucky, 192, 395
Killybegs, 359
Kimberely district, northwestern Australia, 358
Kinetic factors, 25
Kinetic stability, 25, 56
Kinetic theory, 37
Kinetics, 37, 42
Kinshasa, 207
Kinta Valley, 331, 333, 334, 337
Kintail-Affric-Canich, Scotland, 482, 484, 485
Kline Canyon, Verdugo Hills, California, 182
Kociolki, 438

Korea, 352
Kosciusko massif, 466
Kountkouzout, Niger, 271
Krasnozem, 142
Kuala Lumpar, 326, 336, 339, 359
Kuala Pilah, 336
Kuala Trengganu, 336

Labile systems, 91
Lability, 90, 91, 93
Laboratory experiments, 67, 76
Labrador, 471
Lahars, 231
Landform assemblages, 270
Landform classification, 10
Landform systems, 427, 429, 441
　climate-diagnostic, 17
　in granite, 439–440
　limestone, 26
Landslides, 5, 7, 101, 118, 154, 155, 166, 205, 206, 218, 219, 224, 227, 228, 240, 252, 254, 281
Langbein–Schumm Rule, 281
Laramine Mountains, Wyoming, 123, 124, 129, 195
Late kinematic granites, 412
Latent heat of vaporization, 35
Laterite, 52, 81
Lateritic soils, 218
Lava flows, 187, 361
Law of contributing areas, 294
Law of decreasing asymmetry, of glaciation, 459
Law of soil zonality, 167
Leaching, 41, 47, 48, 51, 52, 55, 56, 60, 62, 65, 70, 73, 76, 79, 81, 93, 140, 162, 418
Leaching columns, 67, 76
Leaching systems, 61–82
Leaching water flow rate, 70
Le Cantal, France, 361
Leda Clays, 108, 116, 117, 120
Limestone, 392
Limestone basins, 177–179
Limestone drainage networks, 396–398
Limestone hydrology, 388
Limestone landforms, 26, 390–401
Limestone morphometry, 391
Limestone pavement, 92, 367
Limiting slope angles, 221
Line, 456
Linear model for solution, 252
Liquefaction of sands, 219
Liquid limit, 109, 218

Liquidity Index, 108
Lithological systems, 61–81
Lithologically dominated system, 73
Lithology, 8, 53, 60, 61, 73, 81, 90, 93–95, 138, 176, 177, 179, 180, 182, 187–189, 204, 216, 224, 248, 272, 282, 300, 345, 347, 348, 352–356, 363, 381, 382, 393, 394, 402, 403, 456, 460, 477
Lithosols, 142, 153, 155, 165, 166
Local erosional environment, 185
Localized asymmetry, 185–188, 192
Locusts, 276
Loess, 127, 147–149, 187, 188, 215, 216, 228, 230, 235, 240, 273, 350
Lofoten Island, 482
Logarithmic-normal frequency distribution, 482
London Clay, 117, 120
Longitudinal profiles, 473
Lowering rates, 253, 259
Lubrication, air, 117
Lunette, 433
Lynmouth flood, 306

Macroclimate and asymmetry, 197–199
Madang earthquake, 155
Magnesium, 28
Magnitude and frequency, 6–8, 13, 18, 236, 237, 361
Main range, Malaysia, 352
Malaysian Borneo, 353
Malaysian Meteorological Service, 326
Mali, 358
Manitoba, 192
Manna Volcanics, Papua, 361
Manuherikia valley, South Island, New Zealand, 308
Marigot, 289
Maritime glaciers, 466, 477, 479
Mass balance, 456
Mass balance equation, 248, 253, 254
Mass movement, 13
Mass movement mechanism, 101, 110–118
Massif Central, France 350, 419
Mathematical models, 7, 69, 163, 247, 261–266
Mauretania, 358
Maximum slope angle, 176, 178, 179, 182, 188
Maximum stability, 39
Mean maximum slope angles, 177, 178
Mechanical lowering, 257, 262
Mechanical removal rate, 256
Mechanical weathering, 166, 345, 369

504

Medicine Hat, Alberta, 174, 175
Mediterranean basin, 307
Mediterranean climate, 210, 224
Mediterranean region, 237
Melaka, 336
Melting, 35
Mendip Hills, 381, 390
Mentakab, 334
Mesa Verde, Colorado, 122, 123, 129
Metal complexing, 52
Metastable dynamic equilibrium, 239
Metastable states, 39
Meteorological climatic events, 8
Meteorological events, 17, 18, 172, 205, 304, 305
Meteorological services, 10
Meteorological stations, 12
Microclimate, 17, 65, 158, 160, 162, 165, 176, 187, 190, 196, 199, 351, 438, 487
Microclimatic zones, 198
Micro-erosion meter, 89
Microfissures, 411, 413, 414
Micro-organisms, 63, 350
Microsaturation, 83
Military Ridge, Wisconsin, 195
Millstone Grit Edges, Derbyshire, 358
Mineral species, 47
Mineral stability, 46, 346
Minimal prairie soil, 156, 158
Minnesota, 418
Model, 63, 65, 77, 79, 165
 of chemical weathering, 240
 of hillslope profiles, 254
 of landscape development, 155
 of landscape evolution, 152
 of polygonal karst, 400
 of slope profile development, 124–128
Model building, 6
Moisture balance, 6
Monaro Plateau, New South Wales, 433
Monsoon regions, 213, 221
Montes Universales Idubeda, Spain, 480
Montmorillonite, 47, 66, 67, 80, 81, 162, 348, 349, 418
Moraines, 453
Moraines superposée en terrasse, 480
Morphoclimatic regions, 241
Morphoclimatic systems, 337
Morphoclimatic zones, 298, 300–307, 311
Morphogenetic index, 395
Morphogenetic model, 241
Morphogenetic potential, 213
Morphogenetic regions, 337–340, 392
Morphogenetic system, 204, 357

Morphological effectiveness of rainfall, 207
Morphological systems, 203
Morphometric analysis, 395–401
Morphometric description, 476
Morphometry, 4, 110, 355, 396–401, 439
Mount Crawford, South Australia, 161
Mount McKinley, Alaska, 449
Mount Olympus, Tasmania, 468, 469
Mount Rainier, Wellington, 231
Mount Warning, Richmond River valley, New South Wales, 361
Mountain environments, 275
Mountain regions, 213
Movements of sea level, 171
Muar Valley, 331
Muckish-Errigal, 359
Mudflow, 101, 106, 107, 110, 117, 129, 172, 205, 218–221, 227, 228, 231, 239, 240, 360
Mud lobes, 108
Mud-slides, 106, 108, 129
Multi-concave systems, 439
Multi-convex systems, 439
Multiple regressions, 277
Multivariate approach, 90, 340
Multivariate models, 281
Muscovite, 41, 81, 413

Namurian clayshales, Derbyshire, 123
Namurian shales, 53
Natal Hills, 151
Natural hazards, 281
Navation crest, 477
Nebraska, 298
Needle ice, 17
Neotectonics, 363
Network analysis, 308
Névé, 473
New Caledonia, 279
New England, New South Wales, 361, 471
New Guinea, 118, 353, 361, 393, 401, 425
New Hampshire, 484
New Mexico, 112, 176, 177, 179
New Zealand, 307, 453
Newtonian fluid, 115
Niche glacier, 475
Nivalkarst, 393
Nivation, 158, 453, 466, 473, 475
Nivation cirques, 447, 448, 455, 458, 465, 475, 477, 479
Nivation crest, 475
Nivation moraines, 454, 455, 466
Niveofluvial processes, 310
Non-Newtonian fluid, 115

North central Jamaica, 394
North Downs, 310
Northeast Queensland, 278, 352, 354
Northern Ireland, 359, 361
Northwest-central Tasmania, 459
Northwest Territories, 192
Northwest Yugoslavia, 381
Nowra, New South Wales, 152, 158
Numerical methods, 7

Oceanic air-masses, 204
Oceanic climates, 228, 232
Ohio, 319
Onion-skin weathering, 352
Ontario, 129, 130
Open system, 83, 84, 85
Organic acids, 52, 62, 63
Organic matter, 63, 64
  decomposition, 62
  production, 60
Organic processes, 63
Organic productivity, 52
Orientation, 171, 173
Orientation of slopes, 185, 186, 189, 191, 192, 198
Oriented surfaces, 190
Orographic snowline, 460
Overall gradient, 476, 477
Over-consolidation, 113
Overland flow, 14, 74, 75, 151, 215, 248–257, 262, 266, 278, 290, 306, 318, 363
Oxidation, 347
Oxidation-reduction, 55, 56

Pahang valley, 326
Palaeoartic tors, 435
Palaeoglaciation limit, 463
Palaeosols, 156, 284
Palaeo-synoptic models, 459
Palaeotropical tors, 435
Pamir Mountains, 231, 234
Papua-New Guinea, 458
Parallel retreat, 467
Parallel retreat of slopes, 153, 155, 166
Paris Basin, 357
Partial pressure, 40, 47, 73, 74, 368, 377
Partial source area concept, 270
Pays de Bray, 354
Peak sediment yield, 258, 259
Peat, 348
Pediment, 150, 166, 204, 216, 428
Pediplain, 145
Pediplanation, 240, 435
Pembina Basin, S. Manitoba, 195
Penang, 351

Peneplain, 145
Peninsular Malaysia, 353, 355, 359
Pennine Alps, 467
Pennines, England, 123, 124, 128, 129, 188
Pennsylvania, 298
Perak, 331, 337
Percolation rates, 79
Percolation water, 370, 377
Perennial snowpatch, 480
Perennially frozen ground, 13
Periglacial asymmetry, 197
Periglacial climate, 436
Periglacial landforms, 351
Periglacial morphogenesis, 238, 241, 310
Periglacial processes, 197, 355, 447
Periodic soils, 156–158
Perlis-Kedah lowlands, 331, 339
Permafrost, 192, 196, 210, 310, 369, 373, 383, 390
Permeability, 93
  of regolith, 232
  of soil and regolith, 221
  of substratum, 228
Perth Amboy, New Jersey, 121, 291
Peru, 352
Peshan desert, Central Asia, 351
Petrology, 345
pH, 40, 41, 64, 72, 73, 76, 80, 84, 85, 88, 93, 140, 347, 348, 349, 382
Phase change, 35
Photosynthesis, 28, 30, 49
Physical products, 30
Physical weathering, 28, 52, 232
Picos de Europa, Northern Spain, 453, 482
Piedmont slopes, 428
Pinnacle karst, 399
Pinnate drainage pattern, 189
Pipe networks, 290
Pitting density, 398
Planate surfaces, 439
Planezes, 361
Plateau des Coirons, Massif Central, France, 360
Plinthite, 143, 145, 146
Pluvial erosion, 13
Plynlimon, central Wales, 17
Podzol, 74, 138, 165
Podzol soils, 63
Podzolic weathering, 62
Polar desert soils, 142, 153, 158
Polar solvents, 36, 37
Poleo basin, New Mexico, 177
Polygenetic landforms, 350
Polygenetic landscape, 93

Polygonal karst, 402
Polygonal karst of New Guinea, 399
Ponors, 92
Pore-fluid chemistry, 114
Pore pressure, 104, 116–118, 121, 129, 130
Porewater, 48, 49
Porewater pressures, 224
Porosity, 5, 60
Port Moresby, Papua, 158
Portugal, 359
Post-kinematic granites, 412
Potassium, 28
Potassium-argon dating, 269
Potential energy, 39, 44
Potential evaporation, 13, 248, 257, 259
Potential evapotranspiration, 253, 254, 372
Prairie mounds, 174–176
Prairie soil, 142
Precipitation, 172, 186, 190, 355
  amount, 5
  distribution of, 186
  frequency, 13
  intensity, 5
  type, 275
  variation, 195, 198, 206, 279
Precipitation duration/amount: intensity ratio, 228
Precipitation-effectiveness (P-E) index, 176–179, 298, 300, 307, 311
Precipitation/evaporation balance, 6
Precipitation intensity, 210, 241
Precipitation minus evapotranspiration, 311
Prediction, 18
Prediction equations, 281
Preferential orientation, 116
Pressure melting point, 465, 467
Pressure release, 467
Pressure ridges, 361
Pre-weathering, 436
Principal components analysis, 327–329
Principal components solution, 327, 328
Principle of equifinality, 350
Process dynamics, 95
Process geomorphology, 4
Process rates, 17
Process-response models, 6
Profile closure, 485
Profile of cirques, 473
Profile of equilibrium, 357
Profile thickness, 153
Protalus ramparts, 453, 466
Pseudo-lapies, 438
Puerto Rico, 390

Pulau Pinang, 326, 336, 339
Pyrenees, 480

Quantification, 164
Quartz, 28, 41, 42, 46, 51, 88, 89, 348, 421
Quartzite basins, 178
Quasi-equilibrium, 126, 300
Queenston Shale, Ontario, 123

Radiation, 192, 195, 298, 448, 449
Radiation balance, 463, 465
Radiation index of dryness, 298
Raindrop impact, 271, 275
Raindrop size, 270
Rainfall, morphological effectiveness of, 207
  simulated, 67
Rainfall data, 12
Rainfall detachment, 271
Rainfall duration, 214
Rainfall effectiveness, 271
Rainfall energy, 271
Rainfall erosion index, 271
Rainfall event, 13, 76, 207, 220, 226, 228, 248
Rainfall frequency, 250, 329
Rainfall intensity, 5, 87, 88, 206, 213–226, 254, 257, 264, 271, 272, 298–301, 306, 308, 311, 319, 355, 356
Rainfall magnitude, 270
Rainfall rate, 259
Rainfall-runoff relationship, 271
Rainfall simulators, 270
Rainfall type, 87
Rainsplash, 14, 132, 172, 252, 254
Rainwash, 196, 197
Rajasthan, 219
Rapid temperature changes, 212, 228–233
Rates of lowering, 254
Razorback ridges, 319
Reaction rate, 42, 43
Recurrence interval, 5, 280, 299
Red-brown earth, 156
Red earth, 156
Red podzolic soil, 142, 156, 158
Redox reactions, 55
Regelation, 465
Regosols, 158
Regression analysis, 279
Regression equations, 334
Relative relief, 176–179
Relative solubility, 41
Relaxation time, 7
Relict granite regoliths, 418

507

Relict landforms, 355, 362
Relict profiles, 426
Relief, 182
Relief ratio, 209
Remoulding, 103, 104, 113, 116
Rendzina, 74, 151
Rendzina soil, 148
Residence time, 42
Residence time of water, 70
Residual strength, 114
Resource management, 340
Return period, 216, 220, 321, 331, 334
Rhine Rift Fault, 357
*Rhizocarpon* lichen, 224
Rhodesia, 319
Rhyolite basins, 179
Richter slope, 152, 153, 165
Rill action, 196, 197, 199, 216, 271
Rill initiation, 172
Rillen, 437, 438
Rill-wash, 228
Rio Ameca, Mexico, 349
Rising stage, 279, 283
River basin, 172, 173, 176–180, 205, 270, 272, 289, 306, 311, 319, 323, 326, 395, 401
River channels, 13, 14
River meanders, 188
River profiles, 357
Roche moutonnee, 447
Rock-avalanche, 104, 112
Rock basins, 447, 468, 477, 482
Rock cement, 345, 348, 352, 356
Rock-fall, 103, 104, 125, 128, 240, 453, 471
Rock glaciers, 453
Rock pavements, 428, 429, 439
Rock-slide, 112, 117, 124, 125, 127
Rock slopes, 122
Rock stacks, 441
Rock step, 466
Rock surface erosion systems, 61
Rock system, 85
Rock towers, 427, 429, 436
Rotational flow, 487
Rotational slip, 467, 473, 474, 487
Roumanian Carpathians, 231
Roumanian Sub-Carpathians, 237, 239
Roztocze of Goraj, Poland, 307
Rubble slopes, 122
Run-off, 5, 26
Runoff coefficient, 374
Runoff detachment, 271
Runoff intensity, 215
Runoff rate, 371, 388

Runoff ratio, 321, 333
Ruwares, 429

Sahara, 351
St. Lawrence Lowlands, 106
Salisbury Plain, southern England, 394
Salt weathering, 153, 165, 166, 351, 438
Salts, 28, 30
Sampling, 17
Sandstone basins, 177–179
Sandstone landforms, 356–359
Sandstone Schwarzwald, 484
Sandstone Vosges, 484
Saprolite, 151
Sarek, Sweden, 482
Saturation overland flow, 266
Savanna, 5
Scale, 10, 17, 18
    geographical, 8, 15
    time, 8, 15
Scaling, 437
Scandinavia, 458, 471
Schist basins, 177
Schist landforms, 359, 360
Schrund line, 471
Schwarzwald, 489
Scotland, 460, 471, 489
Scottish Highland, 461
Scree, 231
Scree-slopes, 240
Seasonal-climates, 276, 277, 319, 352, 435
Seasonal effects, 248
Seasonal incidence, 13
Seasonal precipitation, 298
Seasonality, 274–278, 294
    of precipitation, 274
    of storm events, 275
    of sunshine, 333
Second Malaysia Plan, 339
Sedihydrogram, 15, 277
Sediment budget, 265
Sediment concentration, 272, 279
Sediment discharge, 252
Sediment hydrographs, 278
Sediment load, 271
Sediment production, 276
Sediment removal, rate of, 247
Sediment transport, 253, 254, 270, 272
Sediment transport model, 252, 253
Sediment transport rates, 249
Sediment yield, 13–15, 254, 272–283, 298, 319, 337
Sediment yield rate, 259
Sedimentation rate, 269

Seepage pressure, 218, 226
Selangor, 331
Self-regulation, principle of, 207
Semi-arid Australia, 295
Semi-arid regions, 219, 225
Semi-frictional slope angle, 120
Sensors, 17
Separation Point Granite, South Island, New Zealand, 421
Serendipity, 265
Serozem, 63
Settling, 107
Shale basins, 177, 179
Shallow slide stability analysis, 119
Shear slide, 7
Shear strain, 108
Shear strength, 116, 122, 128, 130, 154
Shear stress, 116, 117, 120, 126, 271, 466, 468, 479
Shear surfaces, 117
Shearing, 467
Sheet erosion, 186
Sheeting, 411–415, 427–429, 432
Sheeting joints, 414
Sheet-slides, 218, 224
Sheetwash, 138, 142, 152, 154
Shenandoah, 354
Shock, 117, 118, 166
Shulaps Range, British Columbia, 487
Sierra de Gredos, 480
Sierra de Peña Labra, 454, 455
Sierra del Moncaya, 480
Sierra Nevada, California, 426, 459, 480
Silcrete, 166
Silty-sand slopes, 122
Silty-sandy, 124
Simulated rainfall, 67
Simulation experiments, 82–85
Simulation model, 265, 306
Siwalik Hills, 237
Sjels, 219, 228, 231, 237, 240
Slab-falls, 103, 104, 112, 127
Slides, 103–106
Slope analysis, 237
Slope angle, 172, 177, 178
Slope budget, 164
Slope decline, 125, 126, 128, 153, 154, 196
Slope degradation, 237
Slope development, 163, 241, 247
Slope dimensions, 187
Slope disequilibrium, 235
Slope elements, 149, 150, 152
Slope equilibrium, 203, 205, 237
Slope evolution, 137, 138, 149, 150, 153, 203, 204, 240, 241
Slope evolution/development, 154
Slope form 173–180, 187, 189, 199, 204, 263, 356, 427
Slope hydrology, 129
Slope infiltration model, 205
Slope length, 257, 270
Slope maps, 183
Slope microclimate, 196
Slope processes, 172, 173, 176, 195, 198, 206, 213, 228, 248, 266, 269
Slope profile, 181, 182, 185, 186, 196, 237, 248, 251, 252, 254, 257, 264
Slope profile development, 110, 114, 125, 126, 266
Slope replacement, 124
Slope retreat, 124, 128, 161, 196, 227, 237, 239, 432, 435, 436, 439
Slope stability, 115, 116, 120, 121, 156, 221, 240
Slope system, 206
Slopes, 4, 5, 7, 14, 17
Slopewash, 141, 145
Slopewash rate, 221
Slumgullion, Colorado, 110
Slump density, 224
Slump-earthflows, 106, 110
Slump-slide, 105, 106
Slumps, 103, 105, 106, 110, 117, 125, 138, 172, 205, 218, 221, 224, 226
Slush-avalanches, 125, 228
Slush-flood, 230
Slush flows, 229
Smectite, 80, 81
Snake River district, USA, 360
Snow and ice, melting of, 204
Snow bearing wind, 460
Snow-debris avalanches, 231
Snow density, 229
Snow drifting, 187, 451, 453, 458
Snowdonia, 451
Snowfence, 453, 458, 466, 476, 477, 479
Snowfence crest, 480
Snowline, 447, 448, 453, 456, 458, 460, 461, 463, 471, 476, 489
Snowmelt, 206, 207, 210, 219, 228–232, 235, 241, 275, 303, 310
Snowpatch, 448, 453, 464–466, 475
Snowpatch aspect, 453
Snowpatch erosion, 467
Snowy Mountains, Australia, 275, 352
Sodium, 28
Soil aggregates, 229
Soil air, 377, 378, 380, 381

509

Soil climate, 172
Soil compaction, 215
Soil-creep, 5, 7, 101, 103, 105, 108, 127, 172, 252, 306
Soil depth, 155, 172
Soil development, 60, 65, 252
Soil drainage, 226
Soil erodibility, 270, 275
Soil erosion, 14, 252, 279, 318, 339, 340
Soil-flows, 228, 229, 233
Soil formation, 52, 56, 137, 138, 154, 163, 233, 308
Soil horizons, 254, 271
Soil mechanics, 5, 67
Soil microclimate, 172, 173
Soil minerals, 47
Soil moisture, 163, 187, 195, 215, 216, 306
Soil particle detachment, 270
Soil perfusion apparatus, 67
Soil profile, 29, 30, 42, 45, 47, 52, 65, 70, 140, 141, 150, 164, 165, 253, 377, 390
Soil/rock systems, 73
Soil series, 149
Soil sterilization, 68
Soil storage capacity, 264
Soil structure, 205
Soil system, 49, 60
Soil texture, 172, 383
Soil thickness, 252, 254
Soil transmissibility, 67
Soil type, 64, 65, 94, 152, 158, 301, 306, 321, 367, 382, 383
Soil wash, 132
Soil water, 5, 45, 47–49, 64, 70, 256
Soil water storage, 266
Soil water storage capacity, 248, 249, 251, 254
Soils, 7
   zonal, 2
   zonal concept of, 137
Solar energy, 28
Solar radiation, 171, 182, 186, 276, 277, 317, 349, 449, 451, 466
Solifluction, 5, 108, 109, 117, 125, 128, 151, 153, 158, 165, 197, 231, 240, 308, 466
Solifluction mantles, 240
Solifluction slopes, 204
Solubility, 60, 66
   relative, 41
Solute concentration, 74
Solute denudation, 269
Solute load, 354, 355
Solution, 5, 26, 27, 65, 85, 252, 254, 264, 266, 269, 369, 453

Solution chemistry, 371
Solution concentration, 40
Solution kinetics, 82
Solution load, 371
Solution lowering rates, 265
Solution potential, 86
Solution rates, 248
Solution velocity, 65, 66, 69, 77, 78, 82, 88
Solutional disintegration, 84
Solutional erosion rates, 367, 371
Solvated ions, 36
Somerset Island, Canada, 383
South Georgia, 17, 458
Southeast Devon, 304, 310
Southeastern Australia, 466
Southeastern New South Wales, 354
Southeastern United States, 307
Southern Rocky Mountains, 219
Southern Victoria Land, Antartica, 435
Southwest British Columbia, 458
Southwest England, 279, 295, 301, 424, 432, 438
Southwest North America, 351
Soviet Union, 458
Soxhlet extractor, 67, 68
Spatial scale, 173, 282, 283, 294, 346, 352–355, 362, 427
Specific runoff, 224
Spheroidal weathering, 415, 437
Spitsbergen, 358
Splash erosion, 214, 218, 271
Spring sapping, 146
Sri Lanka, 300, 353
Stable electronic configuration, 32
Stability, 44
   concept of, 37
   maximum, 39
Stability charts, 117, 118
Stability continuum, 37, 38
Stacks, 429, 432, 435, 436
Staining techniques, 81
Statistical analysis, 327, 328
Statistical thermodynamics, 41
Steady creep, 115, 116
Steady state, 42
Steady state slope profile, 262
Stochastic methods, 6
Stocks, 413
Stony rises, 360, 361
Storage capacity, 256
Storm autocorrelation, 248
Storm hydrograph, 248
Strahler maximum angle, 181, 191
Strain-softening, 113, 115, 116

Stratified random samples, 192, 323
Stratified sampling, 326
Stratigraphy, 345
Stream density, 397
Stream discharge, 272
Stream frequency, 300
Stream length, 397, 398
Stream spacing, 318
Stress field, 415
Stress release, 411
Stress-strain curve, 115, 116
Striation, 447, 467
Structural compartments, 411
Structural controls, 176, 187, 188
Structural factors, 353, 354, 357, 361, 395, 415, 428
Structural influences, 179, 180, 191
Structural provinces in Tasmania, 482
Structural sjels, 219, 226, 227
Structure, 176, 180, 188–190, 199, 206, 317, 345, 347, 348, 352, 353, 356, 357, 363, 393, 394, 411, 412, 456, 467, 472
Subsoil erosion, 93
Substitution, 347
Subsurface erosion, 269
Subsurface flow, 220, 221, 239, 240, 248, 250, 251, 254–256, 318, 363
Subsurface runoff, 252
Subtropical cirque glaciers, 480
Suffosion, 219, 220
Suffosion channels, 205
Suffosion holes, 221
Suffosion valleys, 307
Sugar Belt of Natal, 138
Summit/depression ratio, 398
Summit tors, 435
Sungai Gombak, west Malaysia, 278
Supranival wash, 453
Surface aggregation ratio, 275
Surface lowering, 249, 388
Surface runoff, 204–206, 213–216, 220, 221, 228, 232, 239, 258, 274, 275, 290, 306, 308, 310, 318, 319, 323, 355, 438
Surface system, 82–89
Surface wash, 5, 141, 248
Surficial-creep, 103
Suspended load, 28, 30, 203, 205, 207
Suspended sediment discharge, 279
Suspended solids, 30
Swallet density, 397
Swallet ratio, 397
Swallet/resurgence ratio, 397
Swallet water, 370
Swallets, 392, 396, 400

Swelling, 113
Symmetrical glaciation, 458, 468, 472
Synkinematic granites, 412, 419

*Tafoni*, 351, 438
Talus-cones, 224, 228
Talus-creep, 103
Talus slopes, 122
Taluvium, 122, 128
Tantalus Range, British Columbia, 481, 482
Tanzania, 218
Tasmania, 453, 458, 460, 461, 484, 485
Tatra Mountains, 224
Tectonic escarpments, 237
Temperate glaciers, 467
Temporal scale, 173, 352–354, 361–363, 425, 426
Tenerife, Canary Islands, 358
Tenmile Range, Rocky Mountains, 224
Tension cracks, 112
Terek, 458
Termites, 132, 141, 270, 276
Terraces, 10
Terracette, 101, 126, 138, 147, 221
Texas, 192
Thames Estuary, 108
Theoretical methods, 69
Theoretical model, 150
Thermal energy, 31
Thermodynamic analysis, 37
Thermodynamic principles, 346
Thermodynamic stability, 25, 56
Thermodynamics, 37
Threshold, 456, 466, 484
Threshold slope angle, 119–122
Threshold stress, 119
Threshold values, 204
Throughflow, 77, 78, 79, 142, 166, 218, 278, 290
Throughflow hydrology, 70
Tianshan Mountains, 231
Tibesti, 358
Till, 130, 149, 174, 221, 224, 425
Tillite, 358
Topograph, 89
Toposequence, 137
Tor exhumation, 432, 435, 436
Tor formation, 416, 440
Tors, 319, 358, 413, 426, 429, 432–441
Total channel length, 294
Total dissolved solids, 26
Total pressure, 40, 47
Total stream length, 295
Tower karst, 319, 394

511

Trans-Baikal Mountains, 224
Transport capacity, 271
Trend surfaces, 460, 461
Trengganu Plateau, 319, 326, 336, 337
Trinidad, 377
Tropical karst, 393, 394
Tropical weathering, 51
Truncated profiles, 418
Tsheri valley, Caucasus, 219
Tundra gleys, 142
Tundra soils, 63, 158
Turbulent eddy transfer, 448
Turbulent sjels, 219, 226, 227
*Turmkarst*, 395
Two-stage hypothesis of tor formation, 433
'Typhoon-eye' downpours, 212

Uinta Mountains, 468
Ultimate landforms, concept of, 350–352
Ulvadal, western Norway, 220, 221, 239
Unaka Mountains, 356
Unconcentrated wash, 252
Underground drainage, 369, 390, 392, 394
Underground streams, 370
Undrained loading, 117
Undrained shear, 117
Uniclinal shifting, 188, 189, 195
Uniformitarianism, 203
Unit hydrograph, 278
Unloading, 112, 114, 126, 127
Uruguay, 351
Utah, 176, 178

Vaiont Dam, Italy, 112
Valley asymmetry, 173, 180
Valley density, 239, 300, 303, 308–310
Valley glaciers, 185
Valley-side slopes, 171, 172, 180, 195, 351
  average angle of, 176
  maximum angle of, 176
Valley-wide asymmetry, 185, 188
*Vallons en berceau*, 307, 308
Varimax rotation, 328, 329
Vegetation, 2–8, 14, 64, 67, 72, 79, 80, 93, 94, 104, 138, 147, 155, 158, 172, 176, 182, 196, 216, 225, 234, 240, 248, 250–258, 266, 273, 276, 281–283, 290–292, 296–307, 317, 337, 355, 356, 367, 370, 374, 383, 401, 433, 440
Vegetation recycling, 80
Vegetation-soil-rock system, 70
Vegetational Life Zones, 177, 178
Verdugo Hills, California, 122–124, 129
Vermiculite, 80, 81, 162

Vertical zonation, 213, 233
Victoria Land, Antarctica, 153
Viscoplastic flow, 107
Viscous flow, 107
Void ratio, 112
Volcanic cones, 361
Volume expansion, 114
*Vorfluter*, 391
Vosges, France, 356
Vosges, 357, 358
Vosges Horst, 357

Warping, 189
Wash slope, 181
Washes, 289
Washington Cascades, 458
Water, chemical analysis of, 26
  residence time of, 70
Water balance, 65
Water blow-outs, 219
Water bursts, 220
Water deficit, 271
Water flow rate, 67, 68, 73, 77
Water hardness, 373–388
Water-holding capacity of soils, 218
Water resources, 340
Water tracing, 70
Water yield, 318, 337
Weathering, 5, 6, 13, 25–30, 41–56, 61–67, 79, 80, 90, 94, 101, 104, 112, 113, 116, 118, 126, 128, 130, 138, 145, 153, 154, 164, 173, 176, 205, 240, 270, 347, 348, 350, 351, 356, 357, 360–362, 377, 411, 414, 415, 421, 438, 440
  basal surface, 415
  geochemical types, 346
  tropical, 47, 48
Weathering dynamics, 77
Weathering environment, 346, 347, 350
Weathering equations, 49, 50
Weathering front, 144, 424, 425, 432, 438
Weathering-induced swelling, 113
Weathering intensity, 347
Weathering potential, 81, 89
Weathering processes, 56
Weathering product, 345–347, 440
Weathering profile, 61, 80, 81, 349, 417, 428, 433
Weathering rate, 239, 345, 348, 412, 425, 433, 438
Weathering reactions, 47–56, 65
Weathering regions, 13
Weathering solutions, 49
Weathering systems, 59–63, 67, 79

West Coast Range, Tasmania, 458
West Malaysia, 319–321
Western Carpathians, 221
Western Tasmania, 477
Western Victoria, Australia, 360
Westpitsbergen, 351
Whalebacks, 429
White Mountains, California, 219
Wind, 171, 172, 186, 192, 195, 198, 206, 212, 213, 234, 235, 241, 270, 275, 357, 358, 464, 465, 477, 489
Wind action, 13
Wind consistency, 458
Wind direction, 451
Wind drifting, 472
Wind effect, 458
Wind erosion, 276
Wind strength, 458
Windward Islands, 295, 292
Wisconsin, 192
Woolsack tors, 435
Woolsacks, 435

World Glacier Inventory, 458
Wrightwood, California, 107
Wyoming, 192, 424

X-ray diffraction, 80, 86
X-ray fluorescent spectrometry, 86

Yarrangobilly catchment, New South Wales, 281
Yellow podzolic soil, 142, 158
Yorkshire, 394
Yorkshire Wolds, 308
Yucatan, 390

Zonal concept, 1
Zonal concept of soils, 137
Zonal index, 460
Zonal soils, 2
Zonality, 1
Zone of highest precipitation, 213
Zoned profiles in weather granite, 428, 429

# BRITISH GEOMORPHOLOGICAL RESEARCH GROUP

The British Geomorphological Research Group was founded in 1961 to encourage research in geomorphology, to undertake large-scale projects of research or compilation in which the cooperation of many geomorphologists is involved, and to hold field meetings and symposia.

It is open to all involved in geomorphological research and has an international membership of some 500. Details of membership of the Group may be obtained from the editor of this volume.

## PUBLICATIONS
The Group has sponsored the following publications:

*A bibliography of British Geomorphology,*
    K. M. Clayton (Ed.), (George Philip), 1964

*Slopes: Form and Process,*
    Denys Brunsden(Ed.), (Institute of British Geographers, Special Publication No. 3), 1971

*Polar Geomorphology,*
    R. J. Price and D. E. Sugden (Eds.), (Institute of British Geographers, Special Publication No. 4), 1972

*Spatial Analysis in Geomorphology,*
    R. J. Chorley (Ed.), (Methuen) 1972

*Current Research in Geomorphology with Register of Research Equipment,*
    Peter Beaumont and Edward Derbyshire (Eds.), (Geo Abstracts Ltd.), 1973

*Fluvial processes in instrumented watersheds,*
    K. J. Gregory and D. E. Walling (Eds.), (Institute of British Geographers, Special Publication No. 6), 1974

*Progress in Geomorphology,*
    E. H. Brown and R. S. Waters (Eds.), (Institute of British Geographers, Special Publication No. 7), 1974

*The Unquiet Landscape,*
    Denys Brunsden and John Doornkamp (Eds.), (David and Charles—Geographical Magazine), 1974

*Nearshore sediment dynamics and sedimentation,*
    J. R. Hails and A.. P. Carr (Eds.), (Wiley) 1975

## TECHNICAL BULLETINS
A continuing series, include the following:
1. Field methods of water hardness determination Ian Douglas, 1969
2. Techniques for the tracing of subterranean drainage David P. Drew and David I. Smith, 1969
3. The determination of the infiltration capacity of field soils using the cylinder infiltrometer
   Rodney C. Hills, 1970
4. The use of the Woodhead sea bed drifter
   Ada Phillips, 1970
5. A method for the direct measurement of erosion on rock surfaces
   C. High and F. K. Hanna, 1970
6. Techniques of till fabric analysis
   J. T. Andrews, 1970
7. Field method for hillslope description
   Luna B. Leopold and Thomas Dunne, 1971

8. The measurement of soil frost-heave in the field
   Peter A. James, 1971
9. A system for the field measurement of soil water movement
   Brian J. Knapp, 1973
10. An instrument system for shore process studies
    Robert M. Kirk, 1973
11. Slope profile survey
    Anthony Young, 1974
12. Electrochemical and fluormetric tracer techniques for streamflow measurement
    Michael Church, 1974
13. The measurement of soil moisture
    L. F. Curtis and S. Trudgill, 1974
14. Drainage basin morphometry
    V. Gardiner, 1974.

The Group also publishes quarterly, in association with John Wiley and Sons Ltd., *Earth Surface Processes*—A journal of Geomorphology.